Transport and Climate Change

Maxim A. Dulebenets · Szymon Wiśniewski ·
Marta Borowska-Stefańska
Editors

Transport and Climate Change

New Mitigation and Adaptation Strategies

Springer

Editors
Maxim A. Dulebenets
Department of Civil and Environmental Engineering
FAMU-FSU College of Engineering
Tallahassee, FL, USA

Szymon Wiśniewski
Faculty of Geographical Sciences
University of Lodz
Łódź, Poland

Marta Borowska-Stefańska
Faculty of Geographical Sciences
University of Lodz
Łódź, Poland

ISBN 978-3-032-00562-5 ISBN 978-3-032-00563-2 (eBook)
https://doi.org/10.1007/978-3-032-00563-2

This Springer imprint is published by the registered company Springer Nature Switzerland AG
The registered company address is: Gewerbestrasse 11, 6330 Cham, Switzerland

Preface

Urban areas are substantially impacted by greenhouse gas emissions produced by the transportation sector. Global warming and rapid urbanization are all contributing factors that heighten the likelihood of compound extreme weather events. As climate change events become more frequent and intense, communities around the world become increasingly vulnerable due to their everyday dependence on transportation systems. The aim of this book is to overview the existing status of the transportation sector impacts toward accelerated climate change and the main solutions that have been proposed over the years to address the problem (e.g., emission reduction targets, electrification of the transportation sector, elimination of vehicles with traditional fossil fuel engines, renewable energy sources, and introduction of emission control measures). Some of these solutions are quite difficult to implement due to the associated costs. Certain approaches may seem rather ambitious and need more thorough assessment. Therefore, innovative mitigation and adaptation strategies are presented in this book to reduce the future effects of climate change resulting from the transportation sector and promote adaptive transport systems to minimize human as well as economic losses caused by climate change. The new strategies are inspired by intelligent supply chain management solutions, transportation network operational alternatives, and sustainability-driven interdisciplinary approaches. More emphasis is given to low-risk and high-payoff solutions.

The long-term reliability and functioning of transportation systems will increasingly need to consider and plan for climate change and extreme weather events. The proposed research requires input from a variety of disciplines, including inter alia geography, engineering, and climatology. The results of this research will be particularly useful for decision-makers responsible for transport management as well as for stakeholders responsible for protecting the population against the effects of extreme phenomena. On the one hand, it is important to mitigate the impacts of the transportation sector on the environment and accelerated climate change. On the other hand, it is also essential to adapt transport infrastructure to these changes. Overall, this book will provide essential guidance on how to make transportation systems more resilient to the impacts of climate change. It will also highlight the importance of sustainable transport planning and the need for collaboration between different disciplines to address this pressing issue from various angles.

Tallahassee, FL, USA
June 2025

Maxim A. Dulebenets
Lead Editor

Book Organization

This book has a total of 19 chapters covering diverse issues related to transport and climate change, and the chapters are structured as follows. Chapter "Moving Toward Sustainability: A Methodological Review of the Pollution-Routing Problem in the Optimization Literature" focuses on the Pollution-Routing Problem (PRP), which is an extension of the Vehicle Routing Problem with Time Windows that aims to minimize both routing costs and environmental pollutants while meeting customer demands. The PRP decision problem is explained it detail with the key modeling assumptions and supporting mathematical model, which can serve as a point of reference for the future PRP studies. The existing PRP literature is then critically reviewed with a particular emphasis on problem features and classifications, objective functions considered, and solution methodologies adopted. Limitations of the existing PRP studies are pointed out, and constructive recommendations for the future PRP research are proposed. The study concludes with a discussion of how innovative, scalable, and sustainable routing solutions are vital for shaping a greener future for global transportation systems.

The next two chapters explore the role of electromobility in climate change mitigation and adaptation. Chapter "Adoption of Electric Taxis in Hong Kong: Challenges and Opportunities" discusses challenges and opportunities of adopting electric taxis in Hong Kong. The chapter starts with a thorough overview of the electric vehicle progress in Hong Kong, including passenger cars, buses, minibuses, e-trucks, and e-public light buses. After that, the focus shifts toward the development of electric taxis, the role of government in this development, existing challenges, and opportunities. Furthermore, the study conducts in-street interviews with Hong Kong taxi drivers to explore the licensed taxi owner intentions to buy electric vehicles for operations and to investigate the driving behavior of Hong Kong taxi drivers. Interesting insights gained from the in-street interviews are presented. The study findings will be useful to the relevant stakeholders and will facilitate the development of electric taxis across Hong Kong.

Chapter "Electromobility of Urban Public Transport in Poland: Implementation Issues in the Context of Climate Change Challenges" overviews the evolution of electromobility of urban public transport in Poland. The study provides a detailed analysis of the key electric vehicle manufacturers in Poland, some of which produce electric vehicles for other countries as well. After that, the first-generation electromobility developments (e.g., urban rail carriages, trams, and trolleybuses) in Poland are described, followed by the introduction of the means of the second-generation electromobility (e.g., hydrogen buses, electric buses, liquefied natural gas buses, and hybrid buses). A comprehensive spatial analysis is conducted to determine electromobility developments across various cities in Poland. Toward the end, this chapter discusses the future of electromobility in Poland and some important areas that require more attention in order to ensure a smooth implementation of electromobility in the following years.

Chapter "Analyzing the Impact of the Co-evolution Policy of Land Use Allocation and Autonomous Vehicle Adoption on Carbon Dioxide Emission Reduction" evaluates the impacts of co-evolution policy of land use allocation and autonomous vehicle adoption on reduction in carbon dioxide emissions. Transportation and land use systems have substantial environmental impacts. Emerging transportation technologies, such as autonomous vehicles, offer an opportunity to mitigate transportation-related environmental impacts that affect land use patterns. The study proposes a bi-level optimization model that directly captures the interconnection and hierarchy between land use allocation and traffic assignment decisions. The upper-level model aims to minimize the emission of the whole system (land use together with transportation), whereas the lower-level model focuses on mixed traffic assignment. The results from the conducted sensitivity analyses show that optimized land use allocation and appropriate autonomous vehicle adoption rates could be effective in decreasing the carbon dioxide footprint.

Chapter "Smart Parking as a Tool for Climate Change Mitigation and Preparedness" assesses the potential of smart parking systems in managing the environmental impacts of the transportation sector and enhancing climate change preparedness. The chapter starts with a review of the relevant literature focusing on smart cities and the role of smart parking in these smart cities. A conceptual model of a smart parking system is presented, and its components are thoroughly discussed. Another part of the study specifically describes how smart parking systems could be used in crisis management, including disruptions caused by climate change. The relevant applications include information sensing, direction guidance, parking reservation for emergency services, and post-crisis data collection for emergency services.

Chapter "From Traffic Jams to Climate Solutions: The Micromobility Revolution in Urban Areas Based on Experience from Slovakia" explores the role of micromobility in mitigating the impacts of climate change and lessons learned in Slovakia. Different micromobility options, including bicycles, shared bicycles, and scooters, can serve as a promising solution for curbing greenhouse gas emissions and promoting sustainable urban mobility. The chapter employs a data-driven approach to understand the environmental advantages of micromobility, assess the adoption rates, and determine the success of micromobility programs in the City of Trnava (Slovakia). Social-demographic characteristics of micromobility users are evaluated with a particular emphasis on age, gender, education, employment, access to passenger vehicles, and cycling frequency. The cycling utilization is investigated by time of day and season. Overall, the study underlines that micromobility can provide more space-efficient and healthier transport options, contributing to a more sustainable and livable urban environment.

The next two chapters concentrate on greening last-mile deliveries. Chapter "Greening Last-Mile Deliveries and Urban Transport: Evaluation of the Potential of E-Cargo Bikes for Sustainable Mobility" evaluates the potential of e-cargo bikes for sustainable mobility. A systematic literature review is performed, aiming to investigate the deployment of e-cargo bikes in different countries. The collected studies are classified in different groups, including the following: (1) logistics and mobility; (2) transportation and health; (3) transportation and life cycle assessment; (4) travel behavior; and (5) relevant survey studies. The study finds that e-cargo bikes when powered by renewable energy sources can improve environmental sustainability even further. However, more attention should be dedicated to consumer perception and community engagement to ensure that e-cargo bikes will have adequate adoption rates.

Chapter "The Impact of Decarbonization Instruments on Shaping Innovative and Zero-Emission Last-Mile Delivery Strategies" continues the discussion regarding the decarbonization of last-mile delivery strategies. In-depth interviews are conducted with the representatives of entities carrying out last-mile deliveries. Various decarbonization methods and tools are evaluated in terms of their effectiveness, including legal and policy measures, transport organization and last-mile delivery management, logistics in last-mile deliveries, infrastructure and modern technologies, new business and financial models, and social and behavioral changes. All the decarbonization methods are assessed taking into account economic, environmental, and social considerations. Transport organization and management of last-mile deliveries are found to be the most effective decarbonization tool based on the stakeholder input.

Chapter "Redesigning a Closed-Loop Supply Chain Network in the Tire Industry" presents a new optimization model for redesigning a closed-loop supply chain network in the tire industry. Most of the existing studies deal with traditional closed-loop supply chain network models. However, in the network redesign concept, the problem is more complex as it involves both existing facilities that are already open and candidate facilities that may extend this network. Key differences between network design and redesign problems include the necessity to evaluate the closure of existing facilities, where the associated costs differ significantly from those incurred when opening new facilities. The study analyzes the network redesign process using the developed optimization model and assesses the effects of recycling, which are expected to facilitate sustainable and environment-cognizant logistics practices.

Chapter "Alternative Forms of Transport and Possibilities of Sustainable Transport in the European Union" discusses sustainable transport in the European Union. The chapter begins with a detailed overview of related policies and regulations introduced in the European Union to minimize the environmental impacts from transport and eventually reach net-zero emissions. After that, a detailed review of biofuel options is provided, highlighting their advantages and limitations. Then, a methodology is proposed to rank the European Union countries based on their sustainable transport potential. The following indicators are directly considered: number of electric vehicles per one thousand people, number of chargers per one thousand people, and share of renewable energy resources in the energy mix. The countries are classified as developing markets, fast-growing markets, high-potential markets, and emerging markets. The proposed methodology can be used to monitor the development of sustainability goals and progress made toward these goals for each member state year by year.

Chapter "Low Emission Zones as a Tool for Climate Change Mitigation and Adaptation" provides a thorough review of low emission zones as a tool for climate change mitigation and adaptation. A historical implementation of low emission zones in different countries is discussed. Following the examples of low emission zone implementations, the study evaluates diverse effects of low emission zones, including the effects on the natural environment, urban transport system elements, local economy and urban development, and health of city residents. Moreover, the impacts of low emission zones on local communities are carefully explained, such as environmental and social justice, mobility and social inclusion, and public acceptance. The study highlights that policies related to low emission zones should be dynamic and adaptable to ensure that the new emission reduction targets are successfully met.

Chapter "Strategic Transport Management as a Tool for Climate Change Mitigation" presents a framework for sustainable transport management for climate change mitigation. The framework includes three major elements, including prevention (avoiding impacts), correction (minimizing impacts), and compensation (compensating for impacts). Preventive mitigation is considered as a preferred line of action against climate change. Preventive mitigation could be applied at the design, construction, and operational stages of transport projects. Corrective measures for transport projects are generally limited. Compensatory measures aim to balance or remedy the environmental impacts of transport projects. The chapter provides a detailed description of preventive, corrective, and compensatory measures and discusses the main implementation challenges that should be further addressed by the relevant stakeholders to ensure sustainability of transport projects.

Chapter "Sectorial Climate Policy Targets in Germany: Recent Developments and Opportunities" focuses on the challenges surrounding sectorial climate policy targets in Germany. The chapter underlines the difficulties in meeting climate policy targets in the transportation sector, as the allowable emission limits have been surpassed. The lessons learned from other sectors are discussed as well, including the areas for wind energy, quantitative targets for the development of new housing, and the 30-hectare land-saving target. The study concludes that it is rather difficult to state whether disaggregating targets is an effective approach to achieve them. It is a broad question that depends on many factors, including political risks.

Chapter "Harnessing Digital Twins in Transportation: Adaptation and Mitigation Strategies for Sustainability and Resilience" explores the potential digital twin technology in transportation and how it can be used for improving sustainability and resilience. Digital twins are dynamic virtual replicas of physical systems that incorporate real-time data and advanced analytics to enhance transportation network performance, adaptability, and efficiency. The study is mainly based on the review of the recent and relevant literature published over the past years. Real-world applications are presented to demonstrate how digital twins can enhance energy consumption and streamline logistics operations. This chapter contributes to the current understanding of digital twin technology by critically examining its dual benefits and highlighting its role in adaptation and mitigation strategies to improve transportation resilience against climate change and extreme weather conditions.

Chapter "A Comparative Analysis of Transport Operations, Practices, and Environmental Sustainability in the Rhine-Alpine and the Delhi-Mumbai Industrial Corridors" performs a comparative analysis of transport operations, practices, and environmental sustainability in the Rhine-Alpine and the Delhi-Mumbai Industrial Corridors. The Rhine-Alpine Corridor is viewed as a mature European model, whereas the Delhi-Mumbai Industrial Corridor is an emerging model. The study adopts a mixed-method approach to evaluate emission reduction strategies, renewable energy integration, and climate resilience measures across these two corridors, revealing distinct approaches shaped by their developmental contexts. A set of constructive recommendations are provided for both corridors, including prioritizing rail transport electrification, implementing corridor-wide environmental monitoring systems, and establishing robust climate adaptation frameworks.

The last four chapters shift their attention to maritime transportation and climate change, decarbonization of maritime transportation, and environmental performance of maritime shipping services. Chapter "Enhancing Maritime Supply Chain Resilience to Climate Change in the Greater Bay Area and Southeast Asia Region" discusses the maritime supply chain resilience to climate change in the Greater Bay Area and Southeast Asia region. A systematic review is conducted for the studies dealing with maritime supply chain resilience. The systematic review identifies the most influential scholars, countries, and institutions working on this topic. Furthermore, a series of in-depth, semi-structured interviews are performed with industrial practitioners, policymakers, logistics associations, researchers, and environmental representatives to investigate the key challenges of the maritime supply chain in the Greater Bay Area and Southeast Asia region, recognize port infrastructure disruptions because of climate change, and point out how to develop resilience-building capacities. The valuable insights obtained during the conducted interviews are expected to help developing a theoretical framework and proposing possible ways to mitigate the negative impacts of climate change in emerging areas and coastal regions.

Chapter "A Customized Hyperheuristic Algorithm for the Green Berth Allocation and Scheduling Problem with Carbon Dioxide Emission Considerations" presents an optimization model for green berth allocation and scheduling at marine container terminals, which directly captures carbon dioxide emissions produced during berth allocation and scheduling operations. Due to the computational complexity of the model developed, a Customized Hyperheuristic Algorithm is introduced as a solution methodology. This advanced algorithm employs a dynamic selection mechanism that adaptively chooses heuristic operators based on their ongoing performance, enhancing the traditional framework of typical heuristic and metaheuristic algorithms. Extensive experiments are conducted to evaluate the developed algorithm. Moreover, a set of sensitivity analyses are performed for the unit penalty on carbon dioxide emissions and the unit penalty on delayed time in departures of vessels.

Chapter "Decarbonization Transition Strategy of Container Vessels with Vessel Size, Route, and Policy Considerations" describes the evolution of environmental regulations in the maritime industry and specifically concentrates on decarbonization transition strategies of container vessels. A variety of important technical and operational factors are directly considered in the analysis, including retrofit costs, vessel size, route features, new shipbuilding prices, efficiencies in new ship design, maintenance, and downtime costs. The study shows how engine and fuel type decisions can be impacted by the implementation of the Carbon Intensity Indicator and penalties related to the violation of different potential emissions reduction policies. It is highlighted that tailored approaches must be proposed for vessels to ensure a successful decarbonization process without negatively impacting the economic performance of shipping services.

Chapter "Green Port Transformation: A Case Study for the Port of Xiamen" focuses on green port initiatives at Xiamen Port (China). The green port initiatives are assessed with a particular emphasis on alternative fuels, onshore power supply, and policy improvement. The importance of Xiamen Port in the cruise shipping network is evaluated based on the conducted analysis. Moreover, the economic and social benefits brought by the green initiatives at Xiamen Port are reviewed in detail. Finally, a set of constructive recommendations are proposed to facilitate the green transformation of Xiamen Port. These recommendations mainly focus on monitoring of ship emissions, green shipping infrastructure development, improvement of green shipping policies, financial incentives, and collaborative initiatives.

All the chapters complement each other and comprehensively evaluate the challenges imposed by climate change on the transportation sector from various angles. The readers of this book will have the opportunity to familiarize themselves with the recent developments and a broad array of strategies for sustainable and environment-oriented passenger and freight transportation without substantially compromising the economic and social performance.

Contents

Editors and Contributors

About the Editors

Maxim A. Dulebenets is an Associate Professor and Graduate Program Director in the Department of Civil and Environmental Engineering at Florida A&M University-Florida State University (FAMU-FSU) College of Engineering. Dr. Dulebenets holds B.Sc. and M.Sc. degrees in Railway Construction from the Moscow State University of Railroad Engineering (Moscow, Russia) and M.Sc. and Ph.D. degrees from the University of Memphis (Memphis, TN—USA) in Civil Engineering with concentration in Transportation. His research interests include, but are not limited to, operations research, optimization, simulation modeling, NP-hard problems, mathematical programming, metaheuristics, hybrid algorithms, evolutionary computation, transportation engineering, freight transportation, intermodal freight facilities, railroads, liner shipping scheduling, and GPS data processing. Dr. Dulebenets has been involved in a variety of research projects with the overall value ~$19.5 million (~$2.9 million as a PI), sponsored by the United States Department of Defense, United States Department of Transportation, National Science Foundation, Center for Accessibility and Safety for an Aging Population, Florida Department of Transportation, Tennessee Department of Transportation, Center for Freight and Infrastructure Research and Education, Federal Highway Administration, Canadian National Railway, Federal Express, Intermodal Freight Transportation Institute, North Caucasus Railroad Authority, and Moscow Railroad Authority.

He serves as a referee for 250+ international journals. According to the 2024 Scopus citation report, he is ranked as #14 in the world out of 28,891 scholars in Logistics and Transportation based on the composite c-score c(ns) and last year performance and #148 based on the lifetime performance. His work has been published in the leading international journals, including the International Journal of Production Economics, Information Sciences, Reliability Engineering and System Safety, and IEEE Transactions on Intelligent Transportation Systems. Dr. Dulebenets is actively involved in activities of more than 10 Standing Committees and Subcommittees of the Transportation Research Board of the National Academies of Sciences, Engineering, and Medicine. He is an Affiliate Member of the INFORMS Optimization Society and a Senior Member of the IEEE. Dr. Dulebenets is a licensed Professional Engineer registered in the State of Michigan (License No. 6201063266).

Szymon Wiśniewski is an Associate Professor at the University of Łódź, serving in the Institute of Built Environmental and Spatial Policy within the Faculty of Geographical Sciences. He holds the position of Vice Dean for Development and Promotion of the Discipline of Socio-Economic Geography and Spatial Management. Additionally, he directs the Research Centre for European Spatial Policy and Local Development. His research focuses on transport infrastructure planning and development, transport geography, urban logistics, transport accessibility, spatial mobility, intelligent transport systems, and traffic engineering. His studies have addressed various issues, including the role of low-emission transport zones in shaping urban residents' mobility behavior, the sensitivity of transport systems to urban floods, and changes in travel behavior in Łódź following the implementation of Sunday trade restrictions. As Vice Dean, he oversees the research activities of academic staff, ensuring alignment with institutional goals and disciplinary standards. He monitors research progress, assesses publication and project activities, and documents the societal and economic impact of scientific endeavors. He also supervises the preparation of reports demonstrating the societal relevance of research, assisting researchers in compiling the necessary evidence. Szymon Wiśniewski has authored and co-authored over 130 publications, including four academic monographs. He has participated in numerous research projects, serving as the principal investigator in grants funded by the National Science Centre (NCN) and the National Agency for Academic Exchange (NAWA). His academic journey includes a bachelor's degree in Spatial Management (2009) and Logistics (2011), a master's degree in Spatial Economy (2011), a Ph.D. in Geographical Sciences (2014), and a habilitation in Social Sciences (2022). He has also completed postgraduate studies in Traffic Engineering and Transport Planning (2019) and National Security (2023). As an educator, Szymon Wiśniewski delivers courses aligned with his research interests, equipping students with knowledge to address contemporary challenges in transport planning and spatial development.

Marta Borowska-Stefańska is an Associate Professor and Deputy Director of the Institute of the Built Environment and Spatial Policy at the University of Lodz. Her research interests encompass topics related to flood hazard areas, flood risk management, evacuation, spatial development, daily mobility, transport geography, and GIS. Her academic journey includes a bachelor's degree in Geography (2008) and Spatial Management (2009), a master's degree in Spatial Management (2010), and a Ph.D. in Geographical Sciences (2014). She has also completed postgraduate studies in National Security (2023). Marta Borowska-Stefańska has authored and co-authored over 100 publications, including three academic monographs. She has participated in numerous research projects, serving as the principal investigator for grants funded by the National Science Centre (NCN) and the National Agency for Academic Exchange (NAWA). She actively collaborates with an international network of researchers from countries such as the USA, Germany, Austria, Slovakia, Romania, and Sweden. Her commitment to advancing scientific dialogue is evident in her active participation in over 30 international conferences and her role as a reviewer for prominent journals, including *Transportation Research Part D*, *Safety Science*, and *Environmental Hazards: Human and Policy Dimensions*. Since 2010, she has been teaching at the Faculty of Geographical Sciences of the University of Lodz. Her lectures primarily focus on topics such as Spatial Information Systems and Transport Geography. She has supervised 18 bachelor's theses, 27 engineering projects, and 14 master's theses.

Contributors

Payam Afkhami Department of Civil and Environmental Engineering, Florida A&M University-Florida State University (FAMU-FSU) College of Engineering, Tallahassee, FL, USA

Ahmet Onur Agca Engineering Management, College of Engineering and Physical Sciences, Aston University, Birmingham, UK

Lasse Berntzen USN School of Business, Department of Business, History and Social Sciences, University of South-Eastern Norway, Borre, Norway;
USN School of Business, Department of Business, Marketing and Law, University of South-Eastern Norway, Borre, Norway

Marta Borowska-Stefańska Faculty of Geographical Sciences, University of Lodz, Łódź, Poland

Guoqiang Chen Navigation College, Jimei University, Xiamen, China

Qiong Chen Navigation College, Jimei University, Xiamen, China

Lijuan Deng Navigation College, Jimei University, Xiamen, China

Christian Diller Urban Geography and Spatial Planning, Institute for Geography, Justus-Liebig-University Gießen, Gießen, Germany

Maxim A. Dulebenets Department of Civil and Environmental Engineering, Florida A&M University-Florida State University (FAMU-FSU) College of Engineering, Tallahassee, FL, USA

Amir M. Fathollahi-Fard School of Business Administration, Laurentian University, Sudbury, ON, Canada;
Department of Engineering Science, Faculty of Innovation Engineering, Macau University of Science and Technology, Taipa, Macau, China

Łukasz Fiedeń Wydział Geografii i Geologii, Instytut Geografii i Gospodarki Przestrzennej, Uniwersytet Jagielloński w Krakowie, Kraków, Poland

Jiaqi Gao Navigation College, Jimei University, Xiamen, China

Anne Xaviera Go Department of Systems Innovation, Graduate School of Engineering, The University of Tokyo, Hongo, Tokyo, Japan

Robert Guzik Wydział Geografii i Geologii, Instytut Geografii i Gospodarki Przestrzennej, Uniwersytet Jagielloński w Krakowie, Kraków, Poland

Krzysztof Gwosdz Wydział Geografii i Geologii, Instytut Geografii i Gospodarki Przestrzennej, Uniwersytet Jagielloński w Krakowie, Kraków, Poland

Reem Hadeed Engineering Management, College of Engineering and Physical Sciences, Aston University, Birmingham, UK

Gábor Horváth Department of Meteorology, Faculty of Science and Technology, Institute of Earth Sciences, University of Debrecen, Debrecen, Hungary

Marcel Horňák Department of Economic and Social Geography, Demography and Territorial Development, Comenius University Bratislava, Bratislava, Slovakia

Marius Rohde Johannessen USN School of Business, Department of Business, History and Social Sciences, University of South-Eastern Norway, Borre, Norway

Maneerat Kanrak Faculty of Interdisciplinary Studies, Khon Kaen University, Nong Khai, Thailand

Bharadwaj Kasula Supply Chain Management Expert, Aston University, Birmingham, UK

Tomoya Kawasaki Department of Systems Innovation, Graduate School of Engineering, The University of Tokyo, Hongo, Tokyo, Japan

Razieh Khayamim Department of Civil and Environmental Engineering, Florida A&M University-Florida State University (FAMU-FSU) College of Engineering, Tallahassee, FL, USA

Arkadiusz Kołoś Wydział Geografii i Geologii, Instytut Geografii i Gospodarki Przestrzennej, Uniwersytet Jagielloński w Krakowie, Kraków, Poland

Yui-Yip Lau Division of Business and Hospitality Management, College of Professional and Continuing Education, The Hong Kong Polytechnic University, Kowloon, Hong Kong

Bokang Li Department of Civil and Environmental Engineering, Florida A&M University-Florida State University (FAMU-FSU) College of Engineering, Tallahassee, FL, USA

Wentao Liao Navigation College, Jimei University, Xiamen, China

Magdalena Magiera Faculty of Geographical Sciences, University of Lodz, Łódź, Poland

Marin Marinov SFHEA, Senior Lecturer, EM Research Lead, Aston University, Birmingham, UK

Csongor Máthé Department of Geography in Hungarian, Faculty of Geography, Babeş-Bolyai University, Cluj-Napoca, Romania

Miroslava Mikusova Department of Road and Urban Transport, Faculty of Operation and Economics of Transport and Communications, University of Zilina, Zilina, Slovakia

Rei Miratsu Nippon Kaiji Kyokai Administration CenterAnnex, Research Institute, Nipon Kaiji Kyokai (ClassNK), Tokyo, Japan

Ryuji Miyake Nippon Kaiji Kyokai Administration CenterAnnex, Research Institute, Nipon Kaiji Kyokai (ClassNK), Tokyo, Japan

Urszula Motowidlak Faculty of Economics and Sociology, Department of Logistics and Innovation, University of Lodz, Łódź, Poland

Roland Németh Department of Economic and Social Geography, Demography and Territorial Development, Comenius University Bratislava, Bratislava, Slovakia

Adam R. Parol Wydział Geografii i Geologii, Instytut Geografii i Gospodarki Przestrzennej, Uniwersytet Jagielloński w Krakowie, Kraków, Poland; Szkoła Doktorska Nauk Społecznych, Uniwersytet Jagielloński w Krakowie, Kraków, Poland

János Pénzes Department of Social Geography and Regional Development, Faculty of Science and Technology, Institute of Earth Sciences, University of Debrecen, Debrecen, Hungary

Alireza Sahebgharani Department of Transportation Engineering, Isfahan University of Technology, Isfahan, Iran

Reza Shahin Department of Mathematics and Industrial Engineering, Polytechnique Montréal, Montréal, QC, Canada

Fatemeh Shekoohi Department of Civil and Environmental Engineering, Florida A&M University-Florida State University (FAMU-FSU) College of Engineering, Tallahassee, FL, USA

Jarle Løwe Sørensen USN School of Business, Department of Business, History and Social Sciences, University of South-Eastern Norway, Borre, Norway

Jakub Taczanowski Wydział Geografii i Geologii, Instytut Geografii i Gospodarki Przestrzennej, Uniwersytet Jagielloński w Krakowie, Kraków, Poland

Guilhermina Torrao Engineering Management, College of Engineering and Physical Sciences, Aston University, Birmingham, UK

Tamás Tóth Department of Meteorology, Faculty of Science and Technology, Institute of Earth Sciences, University of Debrecen, Debrecen, Hungary

Szymon Wiśniewski Faculty of Geographical Sciences, Institute of the Built Environment and Spatial Policy, University of Lodz, Łódź, Poland;

Research Center for European Spatial Policy and Local Development, University of Lodz, Łódź, Poland

Yang Andrew Wu Senior Lecturer, Division of Science, Engineering and Health Studies, School of Professional Education and Executive Development, The Hong Kong Polytechnic University, Kowloon, Hong Kong

Máté Zakar Department of Meteorology, Faculty of Science and Technology, Institute of Earth Sciences, University of Debrecen, Debrecen, Hungary

Pengfei Zhang Navigation College, Jimei University, Xiamen, China

Abbreviations

ACO	Ant Colony Optimization
AFV	Alternative Fuel Vehicle
AHP	Analytic Hierarchy Process
AIS	Automatic Identification System
ALNS	Adaptive Large Neighborhood Search
AV	Autonomous Vehicle
B&P	Branch-and-Price
B2B	Business-to-Business
B2C	Business-to-Consumer
BASP	Berth Allocation and Scheduling Problem
BC	Black Carbon
CB	Cargo Bike
CCS	Carbon Capture and Storage
CDM	Clean Development Mechanism
CDV	Circular Development Value
CG	Column Generation
CHA	Customized Hyperheuristic Algorithm
CII	Carbon Intensity Indicator
CLSC	Closed-Loop Supply Chain
CO_2	Carbon Dioxide
CO	Carbon Monoxide
CRT	Common Reporting Table
CT	Carbon Tax
CVRP	Capacitated Vehicle Routing Problem
CW	Clarke and Wright Algorithm
DMIC	Delhi-Mumbai Industrial Corridor
DPF	Diesel Particulate Filters
DSS	Decision Support System
EA	Evolutionary Algorithm
EAFO	European Alternative Fuels Observatory
ECA	Emission Control Area

ECB	Electric Cargo Bike
EC	European Commission
EEDI	Energy Efficiency Design Index
EEXI	Energy Efficiency Existing Ship Index
EIA	Environmental Impact Assessment
ELT	End-of-Life Tires
EMS	Environmental Management Systems
ERTMS	European Rail Traffic Management System
ETS	Emissions Trading System
EU	European Union
EV	Electric Vehicle
FA	Flow Averaging
FIS	Fuzzy Inference System
FLP	Facility Location Problem
GA	Genetic Algorithm
GBASP	Green Berth Allocation and Scheduling Problem
GDP	Gross Domestic Product
GDPR	General Data Protection Regulation
GHG	Greenhouse Gas
GIS	Geographic Information System
GNSS	Global Navigation Satellite System
GPA	Gradient Projection Algorithm
GZM	The Upper Silesian and Zagłębie Metropolitan Area
HA	Hybrid Algorithm
HC	Hydrocarbon
HDV	Heavy-Duty Vehicle
HEV	Hybrid Electric Vehicle
HFO	Heavy Fuel Oil
HLH	High-Level Heuristic
HLS	Hybrid Local Search
HOV	High-Occupancy Vehicles
HVO	Hydrotreated Vegetable Oil
ICE	Internal Combustion Engine
IEA	International Energy Agency
ILUC	Indirect Land Use Change
IMO	International Maritime Organization
IoT	Internet of Things
IPCC	Intergovernmental Panel on Climate Change
JI	Joint Implementation
KA	Keshtel Algorithm
LEZ	Low Emission Zone
LLH	Low-Level Heuristic
LNG	Liquefied Natural Gas
LPG	Liquefied Petroleum Gas
LP	Linear Programming

LSA	Life Cycle Assessment
LS	Local Search
LSZ	Low-Speed Zone
LTZ	Limited Traffic Zone
LULUCF	Land Use, Land-Use Change and Forestry
MACC	Marginal Abatement Cost Curve
MBI	Market-Based Instrument
MCT	Marine Container Terminal
MDO	Marine Diesel Oil
MFEA	Multi-factorial Evolutionary Algorithm
MGA	Modified Genetic Algorithm
MILP	Mixed-Integer Linear Programming
MINLP	Mixed-Integer Nonlinear Programming
MIP	Mixed-Integer Programming
MOICA	Multi-objective Imperialist Competitive Algorithm
MSA	Method of Successive Averages
NFC	Near Field Communication
NFU	Non-forest Use
NGO	Non-governmental Organization
NO_2	Nitrogen Dioxide
NO_x	Nitrogen Oxides
NSGA-II	Non-dominated Sorting Genetic Algorithm II
NS	Neighborhood Search
NT	Nature-based Target
NZE	Net Zero Emissions
OPS	Onshore Power Supply
PAPI	Paper and Pen Personal Interview
PES	Payment for Ecosystem Services
PHEV	Plug-In Hybrid Electric Vehicle
PM_{10}	Particulate Matter 10
$PM_{2.5}$	Particulate Matter 2.5
PM	Particulate Matter
PPP	Polluter Pays Principle
PRP	Pollution-Routing Problem
PSO	Particle Swarm Optimization
PUDO	Pick-Up and Drop-Off
PVC	Photovoltaic
PV	Private Vehicle
RAC	Rhine-Alpine Corridor
RDA	Red Deer Algorithm
REDDT	Reducing Emissions from Deforestation and Forest Degradation
RED	Renewable Energy Directive
RFID	Radio Frequency Identification
RMP	Restricted Master Problem
RO	Robust Optimization

RSM	Response Surface Method
RT	Renewable Technology
RUC	Road Usage Charge
SA	Simulated Annealing
SAV	Shared Autonomous Vehicle
SDG	Sustainable Development Goal
SEV	Smart Electric Vehicle
SFS	Stochastic Fractal Search
SOA	Speed Optimization Algorithm
SOC	Soil Organic Carbon
SPLEZ	Superilla Zona de Bajas Emisiones (Superilla Low Emission Zone in Italian)
SSA	Sample Average Approximation
SUE	Stochastic User Equilibrium
TFL	Transport-Freight-Logistics
TOD	Transit Oriented Development
TOPSIS	Technique for Order Preference by Similarity to Ideal Solution
TS	Tabu Search
TtW	Tank-to-Wake
TW	Time Window
UAM	Urban Air Mobility
ULEZ	Ultra-Low Emission Zone
UNFCCC	United Nations Framework Convention on Climate Change
UN	United Nations
UPT	Urban Public Transport
V2G	Vehicle-to-Grid Technology
VLSFO	Very Low Sulfur Fuel Oil
VMT	Vehicle Miles Traveled
VNS	Variable Neighborhood Search
VOC	Volatile Organic Compounds
VRPTW	Vehicle Routing Problem with Time Windows
VRP	Vehicle Routing Problem
VR	Virtual Reality
WHO	World Health Organization
WOA	Whale Optimization Algorithm
WtW	Well-to-Wake
WWO	Water Wave Optimization
ZEZ	Zero-Emission Zone
ZFE	Zone à Faible Émissions (Low Emission Zone in French)
ZPA	Zone de Protection de l'Air (Air Protection Zone in French)

Moving Toward Sustainability: A Methodological Review of the Pollution-Routing Problem in the Optimization Literature

Reza Shahin

Abstract Recent advancements in the Vehicle Routing Problem have increasingly focused on incorporating environmental concerns in the transportation planning process. Traditionally, transportation optimization efforts centered on minimizing costs, time, or distance. However, as sustainability and environmental cost control have gained importance, logistics companies and retailers are progressively adopting environmentally sustainable practices. This paradigm shift has resulted in the development of the Pollution-Routing Problem (PRP), which seeks to align economic and environmental objectives in transportation. Despite the substantial body of research on PRP, a comprehensive systematic review of its methodological developments has been lacking. This review chapter identifies current gaps by mapping the progression of the PRP literature since its introduction in 2011, classifying studies based on the methodologies utilized over time. Furthermore, this chapter highlights several promising research directions, offering a framework for future exploration in this domain.

Keywords Vehicle routing · Pollution-routing problem · Green vehicle routing problem · Literature review · Sustainability · Optimization · Environmental sustainability · Sustainable transportation

1 Introduction

The pressing issues of global warming and climate change have become a primary focus for researchers worldwide (Ouertani et al., 2022). Consequently, immediate global actions are essential to mitigate the risk of future environmental crises. A major contributor to climate change and global warming is the recent rise in greenhouse gas (GHG) emissions (Lo & Shih, 2021). In response to escalating environmental

R. Shahin (✉)
Department of Mathematics and Industrial Engineering, Polytechnique Montréal, Montréal, QC, Canada
e-mail: Rezaa.shahin.1992@gmail.com

M. A. Dulebenets et al. (eds.), *Transport and Climate Change*,
https://doi.org/10.1007/978-3-032-00563-2_1

concerns, businesses across various sectors have adopted numerous strategies aimed at reducing their GHG emissions (Frölicher, 2016; Hsiang et al., 2017). The transportation industry, in particular, is a significant contributor to carbon dioxide (CO_2) emissions, responsible for 79% of all human-induced GHG emissions (Demir, 2018). CO_2 emissions are directly tied to vehicle fuel consumption, which is affected by factors such as vehicle speed, size, and payload (Demir et al., 2011, 2014a).

Throughout the earlier periods of the COVID-19 pandemic, there was a notable reduction in air pollution, varying between 9 and 64%, primarily due to decreased traffic volumes (Kanniah et al., 2020). This decrease highlights the strong link between atmospheric pollution and human activities, underlining the critical need to reduce GHG emissions and adopt more sustainable solutions (Dündar et al., 2022). In the context of vehicle routing, it is essential to consider how factors like vehicle size, speed, and payload affect fuel consumption (Bektaş & Laporte, 2011). Therefore, the optimization of transportation routes requires an objective function that captures both cost efficiency and GHG emission reductions. Improving such decisions is identified as one of the top ten strategies to reduce CO_2 emissions of heavy-duty trucks by 23% by 2030 (Allen et al., 2017). With oil prices at approximately 450 USD per tonne, even a 1% reduction in worldwide fuel consumption could save around \$1.2 billion and decrease CO_2 emissions by 10.5 million tonnes (Laporte, 2016). Even small decrements in CO_2 emissions can make substantial contributions toward promoting sustainable transportation.

The Vehicle Routing Problem (VRP) is a well-known optimization problem of significant economic importance, mainly in last-mile logistics services (Cherkesly et al., 2015). Due to its practical significance, VRP has attracted considerable attention in academic research. The primary goal of VRP is to determine a sound routing plan that ensures each customer is served exactly once while satisfying all given constraint sets. An interesting extension of VRP is the VRP with Time Windows (VRPTW), in which services must be completed within specified time windows set by customers. The Pollution-Routing Problem (PRP), formally defined by Bektaş and Laporte (2011), extends VRPTW by seeking to minimize both routing costs and environmental pollutants while meeting customer demands (Boru İpek, 2022). This comprehensive objective integrates emission levels and operational cost components (Franceschetti et al., 2013; Kramer et al., 2015a, 2015b; Qiu et al., 2020).

Key factors affecting these objectives include vehicle-specific parameters such as payload, speed, size, and route selection, each influencing both economic and environmental outcomes. Constraints like flow restrictions, time windows, and capacity limits further shape feasible routing decisions, ensuring customer demands are met while adhering to operational boundaries. The ultimate goal of the PRP is to achieve an optimized vehicle route that balances cost efficiency with environmental sustainability, resulting in a practical and impactful solution to reduce emissions in transportation logistics.

To situate our research within the existing body of knowledge, we present a detailed comparison of recent and relevant literature reviews in Table 1. This comparison not only highlights the progress made in understanding the pertinent issues but also underscores the specific objective of our study, which is to deepen the understanding of the PRP. As illustrated in Table 1, numerous studies have been thoroughly reviewed by various authors in recent years. However, despite the significant attention PRP has garnered within the scientific community, a comprehensive literature review addressing this specific topic is conspicuously absent. The table emphasizes the surveys conducted on related subjects, revealing a clear gap in the existing reviews, as no dedicated literature review focused on PRP has yet been conducted.

The contributions of this study are organized into four key areas, underscoring its significance and novelty. First, this study traces the evolution of PRP research since its inception in 2011, providing a comprehensive historical perspective on how the problem has been conceptualized and addressed over time. This allows researchers and practitioners to understand the progress made and the foundational principles that shaped the field. Second, it identifies recent advancements in the PRP studies that have improved its real-world applications, focusing on innovative models, techniques, and technologies that bridge the gap between theory and practice. By highlighting these advancements, the study demonstrates how PRP research has adapted to evolving industrial and environmental challenges. Third, the present study evaluates the most popular methods developed to solve the PRP. Finally, this study highlights the key research gaps in the PRP domain and proposes potential directions for the future investigations. By addressing underexplored areas, such as the integration of emerging technologies like electric and autonomous vehicles, this study paves the way for innovative approaches to tackling pressing environmental and logistical issues. Together, these contributions provide a well-rounded and forward-looking perspective, advancing the state of knowledge in the PRP domain.

In the following sections of the chapter, the following aspects will be discussed in detail. First, we elaborate on how the literature search was conducted during this study in Sect. 2. Then, we present a base optimization model for the problem of interest in Sect. 3. We provide the results from the conducted literature survey in

Table 1 Focus of different literature review papers

References	Problem	# of reviewed paper	Focus
Eguia et al. (2013)	EVRP	78	Reviewed employed methods to address electric VRP (EVRP)
Lin et al. (2014)	GVRP	284	New classification of green VRP (GVRP)
Park and Chae (2014)	GVRP	42	Reviewed solution approaches
Asghari et al. (2021)	GVRP	313	GVRP variants taxonomy with a particular emphasis on the technological attributes of vehicular options
Moghdani et al. (2021)	GVRP	309	Review based on methodology and uncertainty
Current study	PRP	75	Reviewed all the PRP papers based on methodology

Sect. 4. The directions for the future research are introduced in Sect. 5. The major concluding remarks for this research are presented in Sect. 6.

2 Literature Search Strategy

To conduct a thorough review of the literature on the PRP, we applied a systematic search strategy across multiple databases, including Google Scholar, Scopus, and Web of Science. The primary search term utilized was "pollution-routing problem". Initially, we identified studies that directly addressed the PRP, followed by a detailed analysis of their references to uncover the additional relevant research through a recursive reference mining technique. This iterative method allowed us to progressively expand our collection of papers. At the initial stage, we identified 120 papers using the keywords "Pollution-Routing Problem" and "PRP" across the databases mentioned earlier. Subsequently, we reviewed the abstracts of these papers to assess their relevance more thoroughly. This process narrowed the selection to 80 papers. Approximately 40 papers were excluded because they were either not in English, inaccessible, or did not address the PRP directly within their scope. Additionally, 6 duplicate papers were identified and removed, resulting in a final collection of 75 papers.

By adhering to this structured and rigorous process, we ultimately compiled a database containing a total of 75 papers. Figure 1 illustrates the annual publication trends of PRP-related articles, reflecting a cumulative increase academic interest in this area, although some fluctuations in the number of publications per year are noticeable. Taking into account the growing concerns over global warming, climate change, and pollution, the PRP and its variants are expected to receive more and more attention in the following years.

3 Representative Mathematical Model

The PRP presents a unique set of challenges compared to traditional VRP models by accounting for environmental characteristics, including fuel consumption and GHG emissions, along with the costs varying based on the adopted vehicle speed. These complexities reflect the PRP's focus on optimizing both operational efficiency and environmental sustainability dimensions. For example, fuel consumption of vehicles and emission levels are not only influenced by the overall distance they travel but also by their speed and load, which requires the introduction of speed-dependent variables directly in the mathematical formulation. This distinction makes the PRP particularly relevant nowadays for various transportation facilities that put emphasis on eco-friendly logistics.

This section of the chapter introduces an optimization model to emulate the PRP. This base model was developed by Bektaş and Laporte (2011) for the first time in

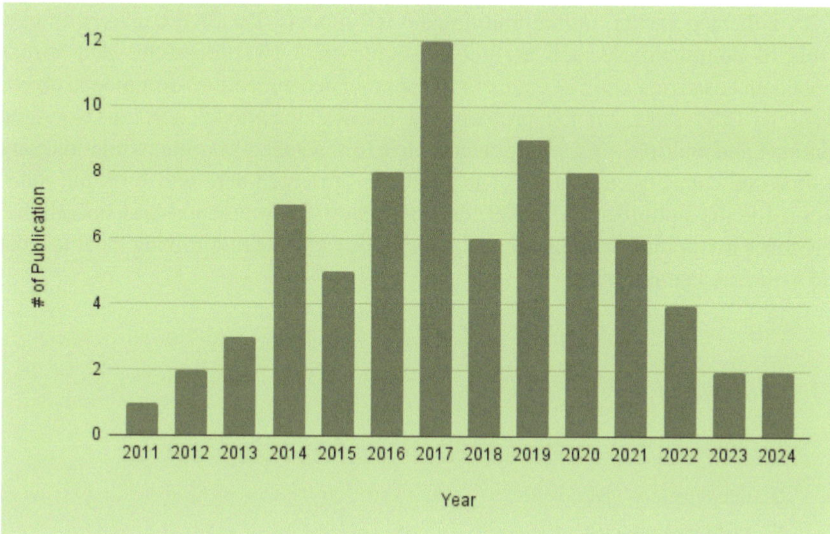

Fig. 1 Number of PRP publications per year

the literature. Below, we define the necessary notations to accurately model the PRP. The PRP is formulated using a graph, in which the depot and customer locations are represented by nodes. Moreover, arcs connect pairs of nodes. Figure 2 presents an example illustrating the construction of the graph for the problem under consideration. In this representation, Node 0 denotes the actual depot, whereas all other nodes correspond to the locations of customers. The arcs connecting the nodes are identified through optimization to determine suitable routes of vehicles.

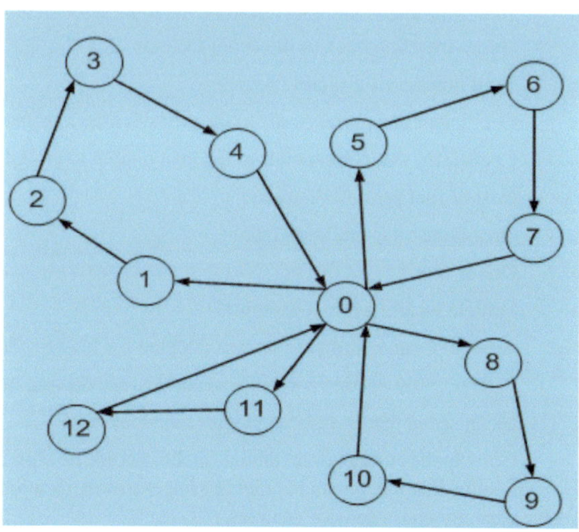

Fig. 2 Distribution of nodes in a graph

As indicated earlier, unlike traditional VRP models, the PRP comes with new levels of complexities, such as fuel consumption, GHG emissions, and speed-dependent costs, reflecting its concentration associated with the environmental objectives. To fully represent the problem's dynamics, the model accounts for time windows and multiple trips, where each vehicle follows specific routes while adhering to time and capacity constraints. The notations presented here will form the foundation for the optimization model, encompassing environmental and operational objectives to optimize the planning of vehicle routes while minimizing the total cost and associated emissions.

3.1 Notations

R Set that represents speed levels.

$\mathcal{N}$ Set that represents customers.

$\mathcal{N}_0$ Set that represents customers along with the actual depot.

$\mathcal{A}$ Set that represents arcs in the transportation network.

v_{ij} Vehicle speed on arc (i, j).

y_j Service start time at node j.

d_{ij} Distance between consecutive nodes i and j in the transportation network.

q_i Demand at customer location i.

t_i Time necessary for service of customer location i.

Q Vehicle capacity.

a_i Start time window at customer location i.

b_i End time window at customer location i.

s_j Total time spent while traveling on the route with node j as the last visited node before traveling back to the depot location: $s_j = (y_j + t_j + d_{j0}/v_{j0})x_{j0}$.

m Total number of available vehicles.

α_{ij} Coefficient that corresponds to the cost of emissions per unit load and distance.

β Coefficient that corresponds to the cost related to emissions and speed.

c_f Cost of fuel per unit distance.

e Greenhouse gas emission factor.

p Driver cost.

w Vehicle weight when it is empty.

M_{ij} A large positive number that was adopted for linearization purposes.

L A large positive constant that was adopted for driving time estimation purposes.

$\overline{v_r}$ Mean speed that is associated with speed level r.

f_{ij} Corresponds to the total amount of commodity flow on arc (i, j). This component assures that the demand is met at each customer location, taking into account the capacity constraints of vehicles.

z_{ij}^r Binary variable that shows if a vehicle travels along arc (i, j) with speed level r. This variable connects the choice of speed levels with emission and fuel consumption costs. For instance, when choosing a higher speed level r, fuel costs will increase as a result of higher aerodynamic drag, but the total travel time can be decreased.

x_{ij} Binary variable that is set to 1 when a given vehicle moves directly from node i to node j and is set to 0 otherwise. This variable is used to assure route connectivity and feasibility across the entire transportation network.

3.2 *Assumptions of the Model*

1. Speed levels r are predefined and discrete, representing practical operating speeds on various arcs.
2. The emission factor e is assumed constant across arcs, which simplifies computation while providing a reasonable approximation.
3. Vehicles have a maximum capacity Q, and any overflow is strictly prohibited.
4. All arcs in the graph are traversable, ensuring no infeasible routes are considered.

3.3 *Model Formulation*

After having introduced the notations, the actual optimization model can be presented as follows:

$$\text{Minimize} \sum_{(i,j)\in\mathcal{A}} (c_f + e)\alpha_{ij}d_{ij}wdx_{ij} + \sum_{(i,j)\in\mathcal{A}} (c_f + e)\alpha_{ij}f_{ij}d_{ij} + \sum_{(i,j)\in\mathcal{A}} (c_f + e)d_{ij}\beta\left(\sum_{r\in\mathcal{R}} \overline{v}_r^2 z_{ij}^r\right) + \sum_{j\in\mathcal{N}_0} ps_j \quad (1)$$

Subject to:

$$\sum_{j\in\mathcal{N}} x_{0j} = m \quad (2)$$

$$\sum_{j\in\mathcal{N}} x_{ij} = 1 \quad \forall i \in \mathcal{N}_0 \quad (3)$$

$$\sum_{i\in\mathcal{N}} x_{ij} = 1 \quad \forall j \in \mathcal{N}_0 \quad (4)$$

$$\sum_{j\in\mathcal{N}} f_{ji} - \sum_{j\in\mathcal{N}} f_{ij} = q_i \quad \forall i \in \mathcal{N}_0 \tag{5}$$

$$q_j x_{ij} \le f_{ij} \le (Q - q_i)x_{ij} \quad \forall (i,j) \in \mathcal{A} \tag{6}$$

$$y_i - y_j + t_i + \sum_{r\in R} (d_{ij}/\overline{v_r}) z_{ij}^r \le M_{ij}\big(1 - x_{ij}\big) \quad \forall i \in \mathcal{N}, j \in \mathcal{N}_0, i \neq j \tag{7}$$

$$a_i \le y_i \le b_i \quad \forall i \in \mathcal{N}_0 \tag{8}$$

$$y_j + t_j - s_j + \sum_{r\in R} (d_{j0}/\overline{v_r}) z_{j0}^r \le L\big(1 - x_{j0}\big) \quad \forall j \in \mathcal{N}_0 \tag{9}$$

$$\sum_{r\in\mathcal{R}} z_{ij}^r = x_{ij} \quad \forall (i,j) \in \mathcal{A} \tag{10}$$

$$x_{ij} = 0 \quad \forall (i,j) \notin \mathcal{A} \tag{11}$$

$$f_{ij} \ge 0 \quad \forall (i,j) \in \mathcal{A} \tag{12}$$

$$z_{ij}^r \in \{0,1\} \quad \forall (i,j) \in \mathcal{A}, r \in \mathcal{R} \tag{13}$$

The objective of the model (1) minimizes the overall cost, which is comprised of four key elements. In particular, the first element incorporates the emission and fuel costs associated with the empty vehicle weight over all arcs considered in the transportation network; the second element represents the fuel and emission costs for the load transported on arcs; the third element corresponds to the speed-induced emission costs, where different speed levels are considered; and finally, the last element accounts for the driver's wage costs based on the time spent on the route.

Constraints (2) assure that precisely m vehicles leave the depot, and every vehicle is expected to start its route from the node corresponding to the depot. Constraints (3) ensure that every customer location i is visited exactly one time by a vehicle. It ensures that there are no missed customers in the routing plan. Constraints (4) guarantee that every customer location j is left exactly once, aiming to prevent vehicles from getting stuck or revisiting the same customer. Constraints (5) ensure that the total demand q_i of every customer location is fulfilled. The total amount of flow from i to j is related to the customer's demand. Constraints (6) ensure that the total load carried by the vehicle when leaving customer location i does not exceed the total capacity of the vehicle denoted as Q. It also ensures that the load is only considered when the assigned vehicle traverses arc (i, j). Constraints (7) calculate the start service times at consecutive customer locations considering the service times and distance traveled. The decision variable x_{ij} is used to determine if arc (i, j) is incorporated in the planned route or not. If $x_{ij} = 1$, then the left-hand side of this constraint set calculates the time of departure from customer location j, including the overall travel time from customer location $i$to location j. The term $M_{ij}(1 - x_{ij})$, where M_{ij} is a sufficiently large positive constant, deactivates this constraint set when $x_{ij} = 0$, effectively ensuring that the compliance with time-windows is enforced only when arc (i, j) is part of the route. The large constant M_{ij} should be set carefully to be larger than the maximum possible difference in time to assure the numerical stability while maintaining the overall correctness of the optimization model.

Constraints (8) enforce the restrictions on time windows for all customer locations, making sure that the actual service at customer location i begins within the allocated time period. Constraints (9) ensure that the every vehicle will travel back to the depot after visiting all the assigned customer locations. The decision variable x_{j0} shows if the return arc $(j, 0)$ is used or not. If $x_{j0} = 1$, then the left-hand side estimates the overall time necessary for the vehicle to travel back to the depot node, including the actual time spent on the route s_j along with the travel time computed as d_{j0}/v_r. The large positive constant L assures that this condition is applicable if and only if arc $(j, 0)$ is used. This mechanism guarantees that the planned vehicle route ends at the depot node and respects the overall time constraints. Constraints (10) link the speed variable z_{ij}^r to the arc variable x_{ij}, making sure that a given vehicle will travel using exactly one speed on arc (i, j). Constraints (11) eliminate any arcs that are not included in the network graph corresponding to the decision problem, ensuring that vehicles do not traverse routes that do not exist. Constraints (12) ensure that the flow among the considered nodes remains non-negative, implying that vehicles cannot deliver loads that are negative. Constraints (13) ensure that the speed variable is binary, i.e., it is either 1 if the vehicle uses speed level r on arc (i, j), or 0 otherwise.

This mathematical formulation integrates operational and environmental considerations, providing a comprehensive approach to solving the PRP. By combining load-dependent costs, speed-dependent emissions, and practical constraints, the model addresses modern logistics challenges. These features enhance its applicability to real-world scenarios, where companies seek to minimize both costs and environmental impacts. In practice, this model could be applied to a logistics company aiming to minimize emissions while meeting delivery deadlines. For example, a fleet delivering groceries in an urban area might use the model to determine optimal routes, balancing fuel efficiency, driver costs, and timely deliveries. The inclusion of speed-dependent emissions allows the company to select routes and speeds that align with sustainability goals, demonstrating the model's versatility and relevance.

4 Literature Review

Various methodologies have been employed to tackle the PRP. In this section, we classify the collected literature based on the techniques applied to solve these problems. The studies are classified into three main categories: (1) Mixed-Integer Linear Programming (MILP) formulations combined with Robust Optimization (RO) techniques, as described in Sect. 4.1; (2) Exact optimization algorithms, which are discussed in Sect. 4.2; and (3) Heuristic and metaheuristic algorithms, covered in

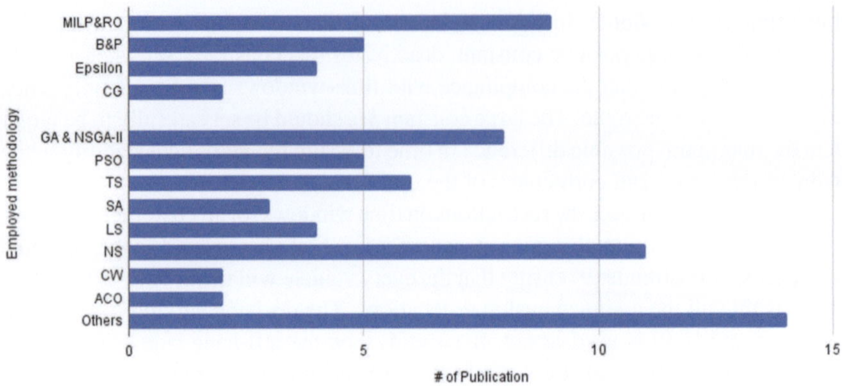

Fig. 3 Distribution of employed methodologies to address the PRP variants. *Notes* Mixed-Integer Linear Programming (MILP); Branch-and-Price (B&P); Column Generation (CG); Genetic Algorithm (GA); Non-Dominated Sorting Genetic Algorithm II (NSGA-II); Particle Swarm Optimization (PSO); Tabu Search (TS); Simulated Annealing (SA); Local Search (LS); Neighborhood Search (NS); Clarke and Wright (CW); Ant Colony Optimization (ACO)

Sect. 4.3. Figure 3 provides an extensive review of the optimization methods that have been adopted in the PRP literature, organizing the studies according to their methodological approaches. Each method is treated in its own section if it has been applied in at least two studies; otherwise, it is grouped under the "others" category. Notably, the "others" category is the most commonly used one. Neighborhood Search (NS) is viewed as the second most commonly applied method, accounting for 14.6% of the total distribution. MILP combined with RO techniques, alongside Genetic Algorithms and the Non-dominated Sorting Genetic Algorithm (NSGA-II), represent 12% and 10.6% of the total methodologies used, respectively. Collectively, these three approaches constitute approximately 37% of all methods applied across the studies. However, certain methodologies have either been minimally explored or entirely neglected, which will be discussed further in the section related to the directions for the future research.

Emerging trends are evident in the rising popularity of metaheuristic methods, such as GA, NSGA-II, and NS, particularly during the later years (2019–2023), reflecting their increasing application in solving optimization challenges. Conversely, methods like CG appear as niche or rarely used techniques, with minimal representation across the timeline. Additionally, the consistent presence of the "others" category, representing hybrid or unclassified methods, suggests a trend toward innovation and combining traditional methodologies to address evolving research needs.

4.1 MILP Formulations and RO

MILP is a powerful mathematical framework widely utilized in Operations Research to address complex optimization problems, particularly in domains such as transportation, scheduling, and logistics. MILP formulations involve linear equations that define both the objective and constraints of a problem, using a mix of continuous and integer decision variables. This hybrid structure enables MILP models to represent real-world scenarios where some decisions are binary or discrete (e.g., route selection or vehicle assignment), while others are continuous (e.g., fuel consumption or travel distance). MILP's primary strength lies in its ability to produce optimal or near-optimal solutions for large-scale, multifaceted problems. As a result, it is often favored for addressing problems like the PRP, where decisions regarding operations significantly affect costs, fuel consumption, and environmental emissions.

In contrast, RO is an advanced technique developed to manage uncertainties within optimization models, ensuring that solutions remain feasible and effective despite variability in conditions. Unlike traditional deterministic models, which assume static inputs, RO explicitly accounts for uncertainties in parameters such as demand, travel time, or vehicle speed. By optimizing for the worst-case scenario within a specified uncertainty set, RO produces solutions that are more resilient to changes, thus enhancing reliability and performance. For transportation and routing problems, RO ensures that decisions remain feasible and efficient even when subject to fluctuating conditions like traffic disruptions or changes in demand, ultimately yielding more sustainable and cost-effective outcomes over time. In many Operations Research problems, including the PRP, MILP models are often used as a foundation, with RO techniques integrated to handle uncertainties.

Bektaş and Laporte (2011) were among the first to introduce four distinct MILP formulations for the PRP, with the objectives of minimizing the following components: (1) driver wages and fuel consumption, (2) load of vehicles, (3) distance to travel, and (4) energy consumption. These MILP models were evaluated both with and without Time Window (TW) constraints. The results showed that incorporating TW constraints reduced energy consumption and related costs by 2% in the third and fourth models. The first MILP formulation was the most effective at reducing fuel consumption and emissions, while the third and fourth models achieved a substantial decrease in energy consumption, by 20% and 30%, respectively. Kopfer et al. (2014) developed an MILP model focused on minimizing fuel consumption and travel distance, demonstrating that employing a heterogeneous vehicle fleet could significantly reduce fuel consumption. Similarly, Salimifard and Raeesi (2014) formulated an MILP model aimed at minimizing CO_2 emissions along with the associated costs for a fleet of bi-fuel vehicles, resulting in a 19% cost reduction and a 27% decrease in CO_2 emissions.

Shamsi et al. (2014) introduced the Pollution-Inventory Routing Problem with Perishable Products (PRIRP), integrating GHG emissions with transportation and inventory costs. Their model led to a 61% reduction in CO_2 emissions and a 23% decrease in empty vehicle trips by prioritizing larger shipments. Ge et al. (2020)

studied the PRP in a multi-route delivery system, using simulations to show that smart route planning can significantly improve urban delivery effectiveness by decreasing fuel consumption, optimizing routing, generating time savings, and reducing the associated costs. Their research demonstrated that the shortest or fastest routes are not always the most fuel-efficient.

Tajik et al. (2014) were pioneers in introducing uncertainties into the PRP, specifically focusing on the speeds of vehicles as the uncertain parameters in the problem. Their model aimed to reduce fuel consumption and driver wages, with the RO technique producing more reliable and superior results compared to fixed-speed approaches. Eshtehadi et al. (2017) expanded on this by incorporating uncertainties related to demand and travel time, finding that minimizing uncertainty produced similar outcomes to deterministic models, though with higher costs for more robust solutions. Their research emphasized that RO methods perform effectively under variable conditions, with their MILP model successfully reducing GHG emissions and costs. Shuang et al. (2019) developed an MILP model to emulate reverse logistics operations, encompassing re-manufacturing and simultaneous delivery and pick-up, with a focus on minimizing CO_2 emissions while optimizing storage, production, and distribution processes. Wu et al. (2021) proposed a model where vehicle maximum speed varied based to a specific probability distribution function. Their findings highlighted that traffic fluctuations and travel distance influence optimal speed patterns, demonstrating that fixed-speed models in uncertain traffic conditions can result in higher fuel consumption and longer travel times.

MILP is well-suited for PRP due to its ability to accurately model the trade-offs between fuel consumption, emissions, and operational costs while adhering to practical constraints, such as vehicle capacities and time windows. Its structured framework ensures optimal solutions, which are particularly critical in minimizing environmental and economic impacts. However, MILP alone assumes deterministic inputs and struggles to address real-world uncertainties like fluctuating vehicle speeds or varying demands. This is where RO complements MILP, offering a robust approach to account for these uncertainties. RO ensures that solutions remain feasible and effective under worst-case scenarios, providing reliability where MILP might otherwise fail. Compared to heuristic or metaheuristic approaches, which are faster but risk suboptimal solutions, MILP and RO provide both precision and resilience. For example, Tajik et al. (2014) demonstrated that RO outperforms fixed-speed deterministic models by producing more reliable results under variable conditions. Similarly, Eshtehadi et al. (2017) highlighted how incorporating uncertainties in travel time and demand through RO leads to robust solutions, albeit at a slightly higher cost.

4.2 *Exact Methods*

We now turn our focus to articles that address the PRP variants using exact methodologies. The primary methods in this group include Branch-and-Price, the ε-constraint method, and column generation. Other less frequently utilized methods are consolidated into a subsection labeled “others”.

While exact methods, such as Branch-and-Price, the ε-constraint method, and column generation, offer optimal solutions and strong theoretical guarantees, they often encounter significant challenges when applied to large-scale problems. One primary issue is the exponential growth of the solution space, which can lead to prohibitively long computational times and memory usage. For example, in Branch-and-Price, the size of the branch-and-bound tree increases rapidly with the problem's complexity, making it computationally expensive to explore all branches. Similarly, column generation relies on solving subproblems iteratively to generate new columns, but in large-scale instances, the subproblem itself can become highly complex, causing convergence to slow significantly. The ε-constraint method faces difficulties as well, particularly in multi-objective optimization, where generating and evaluating the Pareto frontier for numerous objectives in large-scale scenarios can be computationally intensive. These challenges often necessitate tailored problem decomposition techniques, efficient heuristics for initializing solutions, or parallel computing to improve tractability. Acknowledging these limitations is essential when evaluating the applicability of exact methods for solving large-scale instances of real-world problems. In the following sections, we review papers that employed these methods to address the PRP problem.

4.2.1 Branch-and-Price (B&P)

The Branch-and-Price (B&P) method is a decomposition-based approach designed to solve large-scale MILP models, particularly those with complex combinatorial structures. This technique combines two key components: branch-and-bound and column generation (Dell'Amico et al., 2006). In B&P, the problem is divided into a master problem and several subproblems, where column generation is used to iteratively introduce new variables (columns) to the master problem, improving the objective function. These columns represent feasible solutions to the subproblems. The branch-and-bound process is then employed to explore different partitions of the solution space, ensuring the identification of the optimal integer solution (Dell'Amico et al., 2006). This method is especially effective for problems with a large number of variables or constraints, such as vehicle routing or crew scheduling, where traditional MILP solvers may struggle with computational efficiency.

The first application of the B&P algorithm to solve the PRP was introduced by Dabia et al. (2017). Their objective function, designed for a set-partitioning problem, focused on minimizing route-related costs, such as travel time and fuel consumption. To enhance the algorithm's efficiency, they integrated two heuristics: k-node and truncated labeling. The algorithm was tested on Solomon's benchmark instances (Solomon, 1987), demonstrating its effectiveness by solving 82% of instances that have 25 customer locations, 32% instances that have 50 customer locations, and 7% instances that have 100 customer locations.

Lai et al. (2021) further refined the B&P algorithm by integrating it with the Speed Optimization Problem (PRP-SO), where CO_2 emissions and fuel costs were influenced by factors like weight of cargos, speed of vehicles, and road inclination. They presented two solution methods. A Tabu Search (TS) algorithm combined with a fixed-sequence Speed Optimization algorithm was highly effective for larger problem instances. Another method utilized the B&P algorithm that incorporated speed optimization directly into the problem-solving process. While the B&P algorithm performed well in many instances, the TS approach was able to quickly find near-optimal solutions for all cases within one minute. Their study showed that accounting for road gradients and cargo weight could result in fuel savings of up to 53%. Wang et al. (2022) also employed the B&P solution method to investigate the environmental benefits of transitioning from the system with one depot to the system with multiple depots, the ultimate goal of which was to decrease the overall CO_2 emissions. Their computational results revealed that the multi-depot system could cut CO_2 emissions by up to 37.6%, with speed adjustments further reducing emissions by 7.1%.

Gao et al. (2023) introduced an exact Branch-and-Price-and-Cut solution method for the time-dependent PRP. In their approach, the master problem was formulated as a route selection problem using a set-partitioning formulation, while the pricing problem involved solving a time-dependent elementary shortest path decision problem with constraints on resources, where vehicle speeds and customer start times had to be optimized. Similarly, Lai et al. (2024) focused on the PRP that incorporated time windows and speed optimization (PRP-SO), where fuel costs and CO_2 emissions were affected by vehicle speed, payload, and road gradients. They employed a TS-based method along with a fixed-sequence SO solution method, as well as a customized B&P algorithm that integrated speed optimization into the pricing problem. Their findings demonstrated that taking into account payloads and road gradients could lead to fuel savings of up to 53%.

4.2.2 ϵ-Constraint Method

The epsilon-constraint (ε-constraint) method is a widely used approach for solving multi-objective optimization problems by transforming them into a sequence of single-objective problems. In this method, one objective is optimized, while the remaining objectives are treated in constraints, each restricted by an upper limit (ε). By systematically varying the values of ε, a set of Pareto-optimal solutions is generated, allowing decision-makers to evaluate trade-offs between competing objectives (Govindan et al., 2023). This approach is particularly valuable in transportation and logistics, where multiple objectives—such as minimizing costs and emissions or balancing fuel consumption with delivery time—must be addressed concurrently (Govindan et al., 2023). The ε-constraint method provides a mechanism for exploring various scenarios, aiding stakeholders in selecting solutions that align with their priorities.

The study conducted by Soysal et al. (2014) proposed a model with two main objectives: decreasing GHG emissions and minimizing logistics costs. The formulated optimization model was solved using the ε-constraint method. The results revealed several key insights. First, increased export capacity imposed greater demands on the logistics system. Second, insufficient infrastructure negatively impacted fuel efficiency, resulting in higher costs and emissions. Finally, eco-friendly tax incentives were shown to improve environmental and economic performance, underlining the critical role of policy in advancing sustainable logistics operations.

Extending upon this work, Bravo et al. (2019) applied the ε-constraint-based solution method to small-size instances of the problem, including an additional component in the objective to improve service quality for customers. Similarly, Mohsenizadeh et al. (2020) presented an optimization model aimed at reducing CO_2 emissions while optimizing costs. Their application of the ε-constraint method produced feasible and effective solutions, showing that a substantial reduction in emissions could be accomplished even without imposing an excessive economic burden. Xiao et al. (2020) build upon the study conducted by Bektaş and Laporte (2011) by considering diverse components, such as vehicle speed, vehicle load, travel time, fuel consumption rate, along with the relevant scheduling attributes (arrival time, departure time, and waiting time) as continuous variables, taking into account the continuous nature of vehicle speed. Their model also employed the ε-constraint-based solution approach to determine the values of these variables, with the objective to minimize driver wages and fuel consumption costs.

4.2.3 Column Generation (CG)

The column generation technique is a highly effective optimization approach for solving large-scale linear programming (LP) problems that involve an overwhelming number of variables (or columns). It is especially useful in cases where it is computationally infeasible to handle all variables at once (Desaulniers et al., 2006). Instead of solving the full problem with the complete set of variables, column generation starts by addressing a restricted version of the problem, known as the Restricted Master Problem (RMP), which includes only a limited subset of the variables. Once an initial solution is found from the RMP, this solution is evaluated to see if it can be improved by introducing new variables (Desaulniers et al., 2006). This step is done through solving a subproblem, also known as the pricing problem, which seeks to identify variables that, when added to the RMP, can improve the objective function.

In minimization problems, the subproblem searches for variables with a negative reduced cost. If such a variable is identified, it is added to the RMP, and the process is repeated. If no variables with negative reduced costs are found, the current solution is deemed optimal for the entire problem. This method is particularly advantageous for problems that involve a large number of constraints or variables, such as vehicle routing, crew scheduling, and cutting stock problems, where traditional optimization techniques struggle due to the size and complexity of the problem. By iteratively focusing on a smaller subset of variables, column generation significantly reduces computational demands while still achieving optimal or near-optimal solutions.

In their study, Qian and Eglese (2016) proposed schedules and supply routes aimed at improving the total profit and decreasing the total fuel consumption. The study combined the column generation technique with a Tabu Search (TS) method, which allowed them to reorganize the solutions and generate routes with negative costs. This combined approach was mainly efficient in urban distribution contexts, leading to a 3% reduction in GHG-based emissions. Later, Behnke et al. (2021) concentrated on minimizing the overall CO_2 emissions, acknowledging that emissions generated during the transportation process between two points could fluctuate depending on the vehicle's weight. This insight allowed for the identification of multiple low-emission paths. The authors developed a column generation method to address this challenging decision problem using a backward labeling technique. When applied to scenarios involving 100 customers, their method demonstrated exceptional efficiency, yielding near-optimal solutions with an average deviation of only 0.1%, while substantially outperforming exact methods in terms of computational efficiency.

4.2.4 Others

Three studies explored alternative approaches to tackle the problem, which are detailed as follows. Behnke and Kirschstein (2017) examined the effects of employing a heterogeneous vehicle fleet to reduce pollution and enhance travel time effectiveness. This investigation analyzed variations in carbon emissions across different routes, concluding that emissions might be potentially decreased by 2% to 4%. The findings underscored the importance of vehicle-specific routing solution methods, highlighting that smaller vehicles performed more efficiently on shorter routes, while large-size vehicles are recommended to avoid areas experiencing traffic congestion to improve travel time and decrease the total amount of emissions produced.

Zhou and Lee (2017) addressed a nonlinear integer optimization model aimed to minimize the overall GHG emissions by incorporating different elements, such as gravity, three-dimensional customer locations, speed of vehicles, operational hours, and loading capacity. They employed the Lagrangian relaxation-based method to solve the problem. Computational experiments demonstrated that the method was effective, efficient, and robust, showing its potential for solving complex emission-related problems in logistics.

Lastly, Sung and Nielsen (2020) focused on optimizing vehicle speeds for time-sensitive deliveries by determining optimal speeds between nodes while also considering routing decisions. Given the shortcomings of CPLEX in handling medium-sized problem sets, the authors presented an exact optimization approach inspired by the shortest route principle. The outcomes from this study underlined the computational effectiveness of the presented solution method, demonstrating its capability to decrease the overall fuel consumption via the strategic planning of vehicle routes along with speed optimization.

4.3 Heuristic and Metaheuristic Approaches

Given the shortcomings of exact optimization algorithms in solving large-scale PRP problem instances, metaheuristic algorithms have become the preferred approach for handling such instances. This section presents a comprehensive review of the PRP literature that employs metaheuristic algorithms, focusing on their application and effectiveness in addressing the complexities of large-scale PRP scenarios.

4.3.1 Genetic Algorithm (GA) and NSGA-II

A Genetic Algorithm (GA) is a metaheuristic optimization technique inspired by the processes of natural selection and genetics. It is particularly effective for solving complex problems where traditional optimization methods may struggle, such as those involving non-linear, multi-modal, or combinatorial challenges. GA operates on a population of potential solutions, known as individuals or chromosomes, which evolve over successive generations through genetic operations such as selection, crossover (recombination), and mutation (Baker & Ayechew, 2003). The process starts with the initiation of the first population of solutions, either randomly or using heuristic techniques. Each solution is evaluated using a fitness function that measures its quality in relation to the problem's objective. The fittest individuals are then selected for reproduction, passing their traits to the next generation. Crossover combines pairs of parental chromosomes to generate offspring chromosomes, whereas mutation creates random alterations to maintain diversity within the population and prevent the algorithm from being trapped in local optima. The population "evolves" over generations, with better solutions emerging, and inferior ones are discarded. This process stops when a predefined termination criterion is met, such as reaching a target number of generations or achieving an acceptable solution quality (Baker & Ayechew, 2003). GA is especially suited for problems with large search spaces where conventional optimization methods may be inefficient or prone to local optima. It has been applied successfully in various fields, including scheduling, transportation, machine learning, and engineering design.

The NSGA-II is a widely used evolutionary algorithm designed to address multi-objective optimization problems. It builds on the basic concepts of GAs by incorporating techniques that efficiently handle multiple conflicting objectives. NSGA-II maintains a diverse population of solutions and refines them through selection, crossover, and mutation (Srivastava et al., 2021). One of its key features is its fast non-dominated sorting method, which ranks solutions based on Pareto dominance. It also employs a crowding distance mechanism to preserve diversity within the population, ensuring a broad spread of solutions along the Pareto front. By balancing convergence and diversity, NSGA-II generates a set of Pareto-optimal solutions, providing decision-makers with a range of trade-offs between competing objectives (Srivastava et al., 2021). Its robustness and efficiency make it one of the leading algorithms for multi-objective optimization.

In their study, Zhu and Hu (2019) focused on the minimization of the overall costs, which included fuel consumption, by developing a Hybrid Algorithm (HA) that combined GA, Response Surface Method (RSM), and Particle Swarm Optimization (PSO). This hybrid approach effectively reduced fuel consumption. Similarly, Meng et al. (2019) proposed a two-phase solution strategy that began with fuzzy hierarchical clustering applied for grouping customer locations, followed by route optimization within each customer group using the proposed MILP formulation along with the GA approach. Their method successfully reduced both emissions and fuel consumption. Furthermore, Lo and Shih (2021) developed two hybridized algorithms: one combining the k-means and sweep vehicle assignment methods, and the other integrating GA with TS for route optimization. After testing these methods on 30 cases, they found that 29 provided efficient results.

Abad et al. (2018) explored the application of the PRP to cross-docking, demonstrating its potential for reducing inventory costs and ensuring timely delivery of products. For large-scale problems, they applied metaheuristic algorithms such as NSGA-II and Multi-Objective PSO. Rauniyar et al. (2019) enhanced the NSGA-II algorithm to minimize travel distance and fuel consumption, but faced difficulties in optimizing multiple routes simultaneously. To address this, they combined a Multi-Factorial Evolutionary Algorithm (MFEA) with NSGA, enabling the simultaneous planning for multiple vehicle routes. Tirkolaee et al. (2020) integrated the PRP with cross-docking operations to decrease the overall routing and pollution costs while enhancing reliability of the supply chain. For large-scale problems, the authors presented two metaheuristic algorithms—NSGA-II and Multi-Objective Simulated Annealing Algorithm (MOSA)—and used the ε-constraint-based approach for small-size problem instances.

Later in 2022, Boru İpek (2022) utilized the Technique for Order Preference by Similarity to Ideal Solution (TOPSIS) and the Analytic Hierarchy Process (AHP) to identify optimal vehicle routes by considering environmental impacts and operational costs. The Simio software was employed for simulation optimization, while NSGA-II was used for route optimization. AHP prioritized objectives, and TOPSIS selected the most promising solution from the Pareto frontier generated by the NSGA-II method. Finally, Mazinani et al. (2023) introduced a mathematical model for optimizing vehicle location and routing in the banking sector, aiming to decrease the overall GHG-based emissions, lower costs, and improve customer satisfaction. They developed a hybrid multi-objective Genetic-Prometee algorithm (MOGPA) and compared its performance with the Multi-Objective Imperialist Competitive Algorithm (MOICA) and NSGA-II.

4.3.2 Particle Swarm Optimization (PSO)

Particle Swarm Optimization (PSO) is a population-based metaheuristic algorithm inspired by the social behaviors observed in bird flocking and fish schooling. It optimizes a problem by iteratively improving a set of candidate solutions based on a predefined quality metric. Each individual in the swarm, referred to as a particle, represents a potential solution and explores the search space, influenced by its own best position and the best positions identified by other particles (Okulewicz & Mańdziuk, 2017). This collaborative process allows the swarm to progressively converge toward optimal or near-optimal solutions.

Kumar et al. (2016) integrated the PRP with the Production Routing Problem (PRP-I) for the first time. The researchers developed an innovative MILP formulation aimed at minimizing operational costs, including those associated with setup, manufacturing, inventory management, vehicle routing, and penalties for late deliveries, while also focusing on reducing fuel consumption. In order to address large-scale problem instances, they implemented a PSO-based solution approach and compared its performance to that of NSGA-II. During the computational experiments involving between 15 and 200 customer locations, the developed PSO-based solution approach showcased superior performance compared to NSGA-II in terms of efficiency and solution quality.

Building on this foundation, Norouzi et al. (2017) aimed to decrease the overall CO_2 emissions by minimizing travel times of vehicles, avoiding congested routes, and improving the service quality at the customer locations. Their approach accounted for diverse elements, such as alterations in the road slope, wind resistance, size of vehicles, and loading capacity. They developed an enhanced PSO algorithm to address the problem and evaluated its performance against the basic version of the PSO algorithm based on the commonly used problem instances from Solomon (1987), considering fixed and variable travel times. In another effort, Poonthalir and Nadarajan (2018) applied the goal programming method with the objective of minimizing the overall fuel consumption and travel costs. The authors adopted a Time-Varying Acceleration Coefficient PSO as a solution approach. Their research highlighted the impact of varying speeds on fuel consumption, with the goal of identifying fuel-efficient routes by incorporating flexible limits on vehicle speed.

In a later study by Cai et al. (2021), the authors presented a PSO-based solution approach combined with nonlinear mixed-integer programming in order to address the PRP, demonstrating the effectiveness of this hybrid approach. Lo (2022) explored the optimization of operations in cross-docking facilities to decrease the overall CO_2 emissions and applied a PSO-based solution approach. The developed algorithm was compared to the GA solution algorithm. The outcomes from the computational experiments indicated that PSO outperformed GA in many scenarios, further solidifying its efficacy in solving PRP-related challenges.

4.3.3 Tabu Search (TS)

TS is a metaheuristic optimization technique designed to guide a local search process beyond local optima by utilizing memory structures. Unlike conventional local search algorithms, which can become trapped in local optima, TS maintains a "tabu list" that records previously explored solutions to prevent the search from revisiting these states and falling into cycles. This tabu list enforces restrictions, known as "tabus", allowing the algorithm to explore new regions of the search space (Brandão, 2006).

Additionally, TS employs aspiration criteria, which allow for the acceptance of tabu moves if they lead to significant improvements in the objective function. The algorithm also incorporates intensification strategies to focus the search on promising areas and diversification mechanisms to explore less-explored regions of the solution space (Brandão, 2006). These features make TS a robust method for addressing combinatorial optimization problems with complex landscapes and multiple local optima. Its adaptability and success in various problem domains, such as scheduling, routing, and resource allocation, have contributed to its widespread application.

Zhang et al. (2015) introduced an MILP-based optimization model to address the low-carbon PRP, with a focus on minimizing CO_2 emissions and fuel consumption. The authors developed the Route Splitting Tabu Search (RS-TS) algorithm, which utilized three Neighborhood Search (NS) algorithms to encode and decode the model. The RS-TS algorithm evaluated all the feasible next stops to identify the most promising alternative to split the route at its endpoint. Similarly, Patoghi et al. (2017) addressed time-sensitive PRPs by proposing both an MILP model and a TS algorithm, targeting the reduction of toll costs, environmental impact, and travel time, with optimization driven by time sensitivity.

Li et al. (2018) tackled vehicle fleet assignment by considering both fuel consumption and CO_2 emissions. They proposed an MILP-based formulation along with a Split-based Adaptive TS (SA-TS) algorithm, integrating route-splitting techniques with the flexibility of the TS method. Koç et al. (2019) examined the impact of multiple trips, heterogeneous fleets, and route planning on CO_2 emissions and fuel consumption. They optimized routes using Geographic Information Systems (GIS) in conjunction with a TS algorithm, achieving a significant reduction in emissions. In a separate study, Kim et al. (2019) developed an MILP optimization model paired with a TS solution approach to balance service needs while reducing CO_2 emissions across different time periods. Their findings showed that multi-period route planning could reduce emissions without raising costs.

Lastly, Shi et al. (2020) addressed uncertainty in their model, incorporating synchronized visits and reductions in greenhouse gas (GHG) emissions. They utilized a hybrid approach combining Tabu Search with Simulated Annealing (SA), demonstrating that strategic flexibility and adaptability could significantly enhance environmental performance.

4.3.4 Simulated Annealing

SA is a metaheuristic optimization technique inspired by the annealing process in metallurgy, where materials are heated and gradually cooled to minimize energy and remove defects. SA mimics this process by exploring the solution space and occasionally accepting worse solutions to avoid becoming trapped in local optima (Kuo, 2010). The algorithm begins with a high "temperature', which permits a higher probability of accepting suboptimal solutions, and progressively lowers the temperature, reducing the likelihood of such moves. As the temperature decreases, the search becomes more focused on improving solutions, with the aim of converging to a near-optimal or optimal result.

Xiao et al. (2012) applied the SA-based solution approach to address the PRP variant, concentrating on minimizing fuel consumption and fixed vehicle costs. The performed numerical experiments demonstrated interesting outcomes, with the SA algorithm reducing fuel consumption by approximately 5%. Building on this work, Suzuki (2016) designed an SA-based solution approach and enhanced the obtained solutions using a TS algorithm, achieving similar fuel savings of around 5%.

In a subsequent study, Feng et al. (2017) extended the analysis to account for fuel consumption and fixed vehicle costs, under the assumption that the mean speeds on each route followed a normal distribution. The authors proposed an MILP model coupled with an Improved SA (ISA) solution algorithm. The obtained results demonstrated that variable vehicle speeds generally increased the overall fuel consumption, while employing a heterogeneous fleet substantially reduced overall costs. Pollution-related costs were reduced by between 0.82 and 18.32%, underscoring the environmental and operational benefits of using a diverse fleet.

4.3.5 Local Search (LS)

LS is an optimization method that iteratively explores the neighborhood of a given solution to find an improved one. Starting with an initial candidate solution, the algorithm moves to a neighboring solution if it enhances the objective value. This process continues until no further improvements can be identified, at which point the algorithm is considered to have reached a local optimum. While LS techniques are generally simple and computationally efficient, particularly for large-scale optimization problems, they often face the challenge of getting trapped in local optima, where no further improvements are possible, even though the global optimum has not been achieved (Arnold & Sörensen, 2019). To overcome this limitation, LS is often combined with other strategies, such as diversification techniques, restarts, or metaheuristics like Tabu Search and Simulated Annealing, to enable the algorithm to escape local optima and explore different regions of the solution space.

Kramer et al. (2015a) introduced an LS-based approach to solve the PRP decision problem. The presented solution methodology was evaluated on modified instances of the VRP benchmarks, focusing on fuel consumption, energy minimization, and time windows. The results demonstrated the algorithm's capability to deliver high-quality solutions. In addition, they proposed the Speed and Departure Time Optimization Algorithm (SD-TOA). This solution algorithm relied on a divide-and-conquer strategy to tackle the Sequential Ordering Problem (SOP) by removing time windows for some of the intermediate stops.

In a separate study, Rao et al. (2016) investigated the effect of road gradients on fuel consumption using an MILP-based formulation and a Two-Objective Hybrid Local Search (TOHLS) algorithm. This method was an extension of the Hybrid Local Search (HLS) approach, improving the algorithmic search capabilities and yielding fuel consumption reductions between 2 and 13%. Saka et al. (2017) presented an

MILP-based model alongside four heuristic solution approaches to address the PRP, incorporating a heterogeneous vehicle fleet and delivery deadlines. Their research revealed that the third heuristic approach, though slower than the first one, was the most effective and cost-efficient. When comparing the MILP-based formulation with the third algorithm for ten separate problem instances, the heuristic successfully handled seven cases but encountered local optimal solutions in the remaining three.

4.3.6 Neighborhood Search (NS)

NS is a fundamental concept in optimization algorithms, particularly within local search methods. The core idea is to explore the "neighborhood" of a current solution by making small, allowable modifications, such as swapping elements, changing assignments, or adjusting parameters, in search of better alternatives (Gu et al., 2019). A solution's neighborhood consists of the set of solutions reachable through these changes. The algorithm moves to a neighboring solution if it improves the objective function. This systematic exploration of nearby solutions helps refine the search process and often leads to improved outcomes.

Demir et al. (2012) developed an Adaptive Large NS (ALNS) algorithm with two stages to solve the PRP. In the initial stage, an enhanced ALNS was used to remove and add routes in pursuit of superior solutions. In the following stage of the solution procedure, the Speed Optimization Algorithm (SOA) adjusted the speeds of vehicles, improving the overall performance of the ALNS method. In a follow-up study, Demir et al. (2014b) integrated SOA and ALNS to find promising solutions using various alternatives, including the normalized weighting method, standard weighting method, ε-constraint-based method, along with a novel Hybrid Method that combined multiple approaches. Koç et al. (2014) developed the HEA++ algorithm, which incorporated a novel Heterogeneous ALNS (HALNS) and the SOA approach to fine-tune vehicle speeds on various transportation routes. Koç et al. (2016) later investigated the impact of depot and route selection on emissions in urban deliveries. Their research found that in cities with strict speed limits, optimizing routes from a central warehouse significantly reduced fuel consumption and CO_2 emissions. Their ALNS algorithm achieved a 34% decrease in the overall fuel consumption, while the use of a heterogeneous vehicle fleet yielded the mean cost savings of 17%.

Franceschetti et al. (2017) adapted ALNS to optimize delivery routes and speeds while accounting for rush-hour traffic. Wang et al. (2017) developed a new MILP-based optimization model and a Variable Neighborhood Search (VNS) algorithm to reduce the costs associated with driver wages and fuel consumption, achieving fuel savings of 6.37%. Majidi et al. (2018) proposed an approach for simultaneous pickup and delivery aimed at reducing CO_2 emissions using an ALNS algorithm integrated with SA to generate promising solutions for the problem. Kancharla and Ramadurai (2019) concentrated on reducing the overall fuel consumption of vehicles by taking into account the load weights across various vehicle types and depots, accomplishing the mean fuel consumption savings of around 13%.

Karakostas et al. (2020) addressed pollution, inventory, and routing management by applying an MILP formulation for small-scale instances and VNS for large-size problem cases. Qiu et al. (2020) integrated the carbon pricing principle in the PRP, significantly decreasing the overall truck delivery emissions. The authors used a combination of Fuzzy Logic Controlled PSO (FLCPSO) and a modified version of the ALNS algorithm to handle varying scenario sizes effectively. Finally, Zhang et al. (2024) presented a decision-making model for both non-collaborative and collaborative order assignments between various logistics service companies. Their model was assessed using an Improved ALNS (IALNS) algorithm, TS method, and the Oriented Ant Colony Algorithm, offering efficient solutions for order assignment and route optimization.

4.3.7 Clarke and Wright (CW)

The Clarke and Wright Savings Algorithm is a heuristic method used to solve the VRP by optimizing the number of routes and minimizing total travel distance or costs. Initially, the algorithm treats each customer as having a separate route connected directly to the depot. It then calculates the "savings" that can be achieved by merging two routes into one, where savings are defined by the reduction in total distance when two customers are served on the same route instead of being served individually (Caccetta et al., 2013). The algorithm prioritizes merging routes with the highest savings, while ensuring that vehicle capacity constraints are not violated. This process continues until no further savings can be made.

Li et al. (2015) focused on the many-to-many demand component of the tractor-trailer routing decision problem with the goal of minimizing CO_2 emissions. To address this, they implemented an enhanced version of the CW solution method, which was improved through the LS step and the refinement phase. This approach proved being effective in balancing the routing efficiency with environmental considerations. Li et al. (2016) analyzed CO_2 emissions generated in a two-echelon linehaul-delivery transportation system. The authors designed an MILP-based formulation and employed the CW solution algorithm to address the resulting routing problem. Their outcomes highlighted the importance of the central depot's strategic location and the movements of associated loaded semi-trailers in influencing routing decisions and impacting CO_2 emissions.

4.3.8 Ant Colony Optimization (ACO)

ACO is a bio-inspired metaheuristic algorithm designed to solve complex combinatorial optimization problems by imitating the foraging behavior of ants. In nature, ants communicate and find the shortest path to food sources by depositing pheromones along their paths, which guides other ants toward the most promising routes (Yu et al., 2009). Similarly, in ACO, artificial ants explore the solution space of a problem by building paths based on a probabilistic decision rule, influenced by pheromone levels and heuristic information. As better solutions are discovered, the corresponding paths are reinforced with more pheromones, encouraging future ants to follow those routes. To prevent premature convergence to suboptimal solutions, pheromone trails gradually evaporate over time, which ensures continuous exploration of the search space.

Androutsopoulos and Zografos (2017) proposed an optimization model that incorporated variables such as cargo weight, time, along with delivery deadlines. The authors presented an ACO-based solution approach to simultaneously plan schedules and routes, identifying two routes that reduced travel time by 7.35% and fuel consumption by 5.34%. Similarly, Islam and Gajpal (2021) developed a mathematical formulation aimed at optimizing the use of standard and eco-friendly vehicles to minimize CO_2 emissions. They utilized a hybrid ACO-Variable NS (ACO-VNS) algorithm, which outperformed other optimization techniques. Their experiments demonstrated that using a heterogeneous fleet led to a 6.90% reduction in CO_2 emissions when comparing with a homogeneous fleet.

4.3.9 Others

In our comprehensive literature review, we explored various approaches for addressing the PRP, with a particular focus in this section on studies utilizing unconventional methods. Franceschetti et al. (2013) formulated an MILP model aimed at reducing both driver wages and pollution by factoring in the effects of rush-hour traffic on emissions. Their research suggested that strategically stopping vehicles during traffic congestion could help reduce pollution. The authors presented the Divide and Solve Optimization Algorithm for large-scale instances, which performed well during the computational experiments. Oberscheider et al. (2013) presented an MILP formulation along with a heuristic algorithm, demonstrating that focusing on fuel consumption reduction leads to greater CO_2 savings than simply targeting travel costs. Pradenas et al. (2013) addressed energy consumption on various routes by incorporating cargo weight and traveled distance using a scatter search method to solve the decision problem at hand.

Hwang and Ouyang (2015) addressed the vehicle assignment shortest path problem by developing a heuristic algorithm, which effectively reduced the total cost, particularly those cost components associated with the pollution. Fukasawa et al. (2016) introduced two innovative methods for the PRP: one based on Mixed-Integer Second-Order Cone Programming and another one utilizing MILP. Yu et al. (2016) focused on reducing CO_2 emissions through improved schedule planning, employing a set-partitioning formulation that led to a 10% decrease in emissions. Guo and Liu (2017) optimized pick-up and drop-off schedules to minimize both driver wages and fuel costs using a heuristic algorithm within a set-partitioning formulation to improve the quality of solutions.

Masmoudi et al. (2018) concentrated on reducing CO_2 emissions by investigating the use of alternative energy sources for vehicles and designed various versions of the Artificial Bee Colony algorithm. Their study highlighted the trade-offs between lowering emissions and managing the wages of drivers. In a separate research effort, Karami and Ghezavati (2018) implemented a differential evolution solution method, which demonstrated the capability to provide satisfactory solutions within reasonable timeframes. Raeesi and Zografos (2019) proposed an MILP optimization model combined with the Path Elimination Procedure to find optimal solutions. Ouertani et al. (2022) introduced two novel approaches, the Vector Evaluated Evolutionary Algorithm and the improved Adaptive Local Search algorithm, both of which proved to be effective for tackling the PRP.

5 Future Research

Based on the review of the collected studies, this section discusses potential directions for the future research. We classify the identified gaps and potential research avenues into three main areas: problem classification, solution approaches, and objective functions, as outlined in the next sections.

5.1 Problem Classification

This section identifies several promising research avenues related to the classification of problems, particularly the integration of the PRP with other VRP variants, while considering taxonomic elements. Key prospects for future research categorization include the following.

5.1.1 New Modeling Approaches

Researchers can explore innovative modeling techniques by combining traditional VRP variations with the PRP. Although some studies have integrated VRP variants—such as VRP with pick-up and delivery tasks (Tajik et al., 2014), Production and PRP (Kumar et al., 2016), along with food transportation (Soysal et al., 2014)—there remains considerable room for further investigation. Promising VRP variants include periodic VRP, site-dependent VRP, and multi-compartment VRP, which could introduce new dimensions of complexity and insight into PRP modeling. Moreover, the emerging flexible transportation types, including Flex-Route Transit, have never been addressed with respect to sustainability indicators (Shahin et al., 2024).

5.1.2 Static Versus Dynamic Problem Variants

Currently, approximately 82% of research in the field addresses static and deterministic versions of the PRP, thereby oversimplifying the complexities of real-world scenarios. To address this gap, future research should focus on developing actionable methodologies for dynamic environments. Specifically, researchers can design frameworks that incorporate real-time re-optimization by leveraging technologies such as the Internet of Things (IoT) and advanced sensors to gather live data on traffic conditions, vehicle performance, and demand fluctuations. This data can then be processed using machine learning algorithms to predict and respond to changes in real-time, enabling adaptive route adjustments that minimize emissions and operational costs.

5.1.3 Electric Vehicles and Sustainability

The increasing adoption of electric vehicles (EVs) and autonomous systems presents significant opportunities and challenges for the future development of PRP research. EVs, while reducing emissions compared to conventional vehicles, still have environmental impacts, including energy consumption and the implications of electricity generation. Moreover, integrating EVs into PRP models requires addressing critical challenges such as the availability and optimization of charging infrastructure, variability in charging times, and limitations in vehicle range. These factors add a layer of complexity to PRP solutions and necessitate the development of advanced routing and scheduling algorithms tailored to EV operations.

In addition to EVs, the integration of autonomous vehicles (AVs) into PRP models could transform the field, offering potential benefits such as enhanced efficiency, reduced human error, and improved safety. However, the deployment of AVs introduces regulatory, ethical, and technological challenges. For instance, considerations around data privacy, vehicle-to-infrastructure communication, and compliance with evolving legal frameworks must be incorporated into model designs. Future research should explore how these challenges can be addressed while maximizing the environmental and operational benefits of AV integration.

Furthermore, beyond emissions, other environmental factors, including noise pollution, impacts on wildlife, and emissions from battery recycling, remain underexplored. Investigating these sustainability considerations within the context of EV and AV integration can provide a more holistic understanding of the long-term environmental impact of PRP solutions. Expanding PRP research in these directions could significantly enhance its relevance in addressing real-world transportation and sustainability challenges.

5.2 Solution Methods

Determining the links between different solution approaches and the PRP, as illustrated in Fig. 3, uncovers several avenues for further exploration. Metaheuristic algorithms—especially NS, GA, and NSGA-II—have been the most commonly employed strategies, underscoring their dominance in tackling the PRP compared to other methods. However, exact methods have been underutilized. To provide clarity, all solution approaches can be categorized into two main groups: approximation methods (including heuristic, metaheuristic, and hybrid approaches) and exact methods (comprising exact techniques and solvers). According to Fig. 3, approximately 70% (a total 52 studies out of 75) of the studies employed approximate solution methods, while about 30% (23 out of 75) used exact methods or solvers to find optimal solutions. Based on these findings, we propose the following research directions.

Although some research has explored stochastic versions of the PRP, the methodologies applied remain limited, highlighting a significant research gap in leveraging advanced stochastic programming techniques. For instance, the Sample Average Approximation (SAA) method can be a powerful tool, where "here-and-now" decisions are optimized across multiple scenarios, and "wait-and-see" decisions adapt to specific outcomes, yielding robust solutions. To implement this, researchers can model key uncertainties such as fuel consumption, traffic delays, and customer demand using probabilistic distributions, generating a set of realistic scenarios. These scenarios can be optimized through metaheuristic algorithms, such as genetic algorithms or particle swarm optimization, to handle the computational complexity of large-scale problems. Additionally, integrating scenario-based optimization with real-time data sources, such as IoT-enabled traffic monitoring systems, can ensure dynamic updates to these models, bridging the gap between theoretical frameworks and their practical application in transportation planning.

5.3 Objective Function

Objective functions are critical components of mathematical models and play a key role in shaping directions for the future research. A significant portion of the recent studies on the PRP has primarily focused on optimization models with a single objective function, which tends to be simpler compared to the complexities involved in multi-objective optimization. Among these, fuel consumption has been the most frequently addressed factor in single-objective and multi-objective alternatives. Nevertheless, other sustainability-related factors, such as customer satisfaction and workload of drivers, have been largely overlooked, representing potential areas for further exploration. In optimization models with multiple objectives, specific aspects received more attention in the broader VRP group of studies, such as the overall system reliability (Rabbani et al., 2018), route balancing (Tunga et al., 2017), customer satisfaction (Ghannadpour & Zarrabi, 2019), and the overall social cost (Ramos et al., 2014). The aforementioned objectives are significant and could be extended into PRP research to diversify the range of optimization objectives. Based on these insights, we suggest the following research directions.

In real-world scenarios, balancing multiple objectives is crucial, making the integration of multi-objective optimization in PRP research imperative. Of the 75 reviewed articles, only 11 (approximately 14.67%) employed this approach, signaling a significant opportunity for broader applications. Multi-objective optimization can provide more realistic and holistic solutions by simultaneously considering factors like fuel consumption, emissions, delivery time, and operational costs. To implement this, researchers can employ Pareto optimization techniques, where algorithms like NSGA-II or MOEA/D are used to generate a diverse set of trade-off solutions along a Pareto front. These solutions allow decision-makers to evaluate and prioritize objectives based on specific operational or environmental considerations. Practical implementation involves defining objective functions for each key factor (e.g., minimize emissions, minimize costs) and constraints (e.g., vehicle capacity, time windows) and using evolutionary algorithms to iteratively optimize the solutions. Integrating real-world data, such as live traffic conditions and fuel consumption rates through IoT devices, can enhance model realism and adaptability. Additionally, decision-support tools that visualize Pareto fronts and allow stakeholders to interactively explore trade-offs can facilitate practical adoption of these methods in transportation planning.

6 Conclusion

The VRP focuses on determining the most efficient vehicle routes and has garnered significant interest due to its practical applications in logistics and transportation. Numerous studies have explored various solutions to this problem. However, with growing global concerns about climate change, there has been a shift in focus toward environmentally sustainable solutions. As a result, the Pollution-Routing Problem (PRP), an advanced variant of the VRPTW, has gained prominence. The PRP emphasizes eco-friendly vehicle routing by integrating fuel consumption and CO_2 emissions as the objective components in the optimization model. Bektaş and Laporte (2011) were the first to introduce the PRP into academic literature, and since then, it has attracted substantial attention, especially in the area of methodological developments.

This study provided a comprehensive review of the current research on the PRP, starting with an introduction to the problem and an exploration of various methodologies designed to solve it. The studies were classified based on their solution approaches, which were categorized as either exact or metaheuristic methods. The review also offered valuable insights to enhance the understanding of the PRP research field and its potential applications. While this study provided a comprehensive review of the literature focusing on applied methodologies, it does have certain limitations. Specifically, its scope is restricted to classifying studies based on the methodologies employed. A more detailed analysis could be conducted by examining the papers with respect to objective function components, single versus multiple objectives, the inclusion of case studies, and a critical evaluation of algorithmic performance.

Several future research directions for the PRP were proposed. One key recommendation is the design of hybrid solution algorithms that combine different solution approaches, such as heuristic and exact methods, to leverage their respective strengths and produce more robust solutions. Another important suggestion is the incorporation of real-time optimization, which would enhance the responsiveness and accuracy of vehicle routing solutions by utilizing real-time traffic data. The integration of real-time data into the modeling framework is expected to significantly improve the efficiency and effectiveness of logistics procedures.

In summary, this chapter not only highlights the significance of the PRP as a critical extension of the traditional VRP but also underscores its potential to drive environmentally sustainable logistics practices. By offering a structured review of current methodologies and identifying key areas for improvement, this study provides a foundation for advancing both theoretical research and practical applications in this field. As the urgency to combat climate change intensifies, the PRP represents an opportunity for the logistics and transportation industries to align operational efficiency with environmental stewardship. Researchers and practitioners are encouraged to prioritize the development of innovative, scalable, and sustainable routing solutions that address the complexities of real-world scenarios. This collective effort will be instrumental in shaping a greener future for global transportation systems.

References

Abad, H. K. E., Vahdani, B., Sharifi, M., & Etebari, F. (2018). A bi-objective model for pickup and delivery pollution-routing problem with integration and consolidation shipments in cross-docking system. *Journal of Cleaner Production, 193*, 784–801.

Allen, J., Browne, M., & Piecyk, M. (2017). *Assessing the European Commission's target of essentially CO_2-free city logistics in urban centres by 2030*. University of Westminster.

Androutsopoulos, K. N., & Zografos, K. G. (2017). An integrated modelling approach for the bicriterion vehicle routing and scheduling problem with environmental considerations. *Transportation Research Part C: Emerging Technologies, 82*, 180–209.

Arnold, F., & Sörensen, K. (2019). Knowledge-guided local search for the vehicle routing problem. *Computers & Operations Research, 105*, 32–46.

Asghari, M., Al-e, S. M. J. M., et al. (2021). Green vehicle routing problem: A state-of-the-art review. *International Journal of Production Economics, 231*, Article 107899.

Baker, B. M., & Ayechew, M. (2003). A genetic algorithm for the vehicle routing problem. *Computers & Operations Research, 30*(5), 787–800.

Behnke, M., & Kirschstein, T. (2017). The impact of path selection on GHG emissions in city logistics. *Transportation Research Part E: Logistics and Transportation Review, 106*, 320–336.

Behnke, M., Kirschstein, T., & Bierwirth, C. (2021). A column generation approach for an emission-oriented vehicle routing problem on a multigraph. *European Journal of Operational Research, 288*(3), 794–809.

Bektaş, T., & Laporte, G. (2011). The pollution-routing problem. *Transportation Research Part B: Methodological, 45*(8), 1232–1250.

Boru İpek, A. (2022). Multi-objective simulation optimization integrated with analytic hierarchy process and technique for order preference by similarity to ideal solution for pollution routing problem. *Transportation Research Record,* 03611981221105503.

Brandão, J. (2006). A new Tabu search algorithm for the vehicle routing problem with backhauls. *European Journal of Operational Research, 173*(2), 540–555.

Bravo, M., Rojas, L. P., & Parada, V. (2019). An evolutionary algorithm for the multi-objective pick-up and delivery pollution-routing problem. *International Transactions in Operational Research, 26*(1), 302–317.

Caccetta, L., Alameen, M., & Abdul-Niby, M. (2013). An improved Clarke and Wright algorithm to solve the capacitated vehicle routing problem. *Engineering, Technology & Applied Science Research, 3*(2), 413–415.

Cai, L., Lv, W., Xiao, L., & Xu, Z. (2021). Total carbon emissions minimization in connected and automated vehicle routing problem with speed variables. *Expert Systems with Applications, 165*, Article 113910.

Cherkesly, M., Desaulniers, G., & Laporte, G. (2015). Branch-price-and-cut algorithms for the pickup and delivery problem with time windows and last-in-first-out loading. *Transportation Science, 49*(4), 752–766.

Dabia, S., Demir, E., & Van Woensel, T. (2017). An exact approach for a variant of the pollution-routing problem. *Transportation Science, 51*(2), 607–628.

Dell'Amico, M., Righini, G., & Salani, M. (2006). A branch-and-price approach to the vehicle routing problem with simultaneous distribution and collection. *Transportation Science, 40*(2), 235–247.

Demir, E. (2018). Value creation through green vehicle routing. In *Sustainable freight transport* (pp. 63–78). Springer.

Demir, E., Bektaş, T., & Laporte, G. (2011). A comparative analysis of several vehicle emission models for road freight transportation. *Transportation Research Part D: Transport and Environment, 16*(5), 347–357.

Demir, E., Bektaş, T., & Laporte, G. (2012). An adaptive large neighborhood search heuristic for the pollution-routing problem. *European Journal of Operational Research, 223*(2), 346–359.

Demir, E., Bektaş, T., & Laporte, G. (2014a). A review of recent research on green road freight transportation. *European Journal of Operational Research, 237*(3), 775–793.

Demir, E., Bektaş, T., & Laporte, G. (2014b). The bi-objective pollution-routing problem. *European Journal of Operational Research, 232*(3), 464–478.

Desaulniers, G., Desrosiers, J., & Solomon, M. M. (eds.). (2006). *Column generation* (Vol. 5). Springer Science & Business Media.

Dündar, H., Soysal, M., Ömürgönülşen, M., & Kanellopoulos, A. (2022). A green dynamic TSP with detailed road gradient dependent fuel consumption estimation. *Computers & Industrial Engineering, 168*, Article 108024.

Eguia, I., Racero, J., Molina, J. C., Guerrero, F. (2013). Environmental issues in vehicle routing problems. In *Sustainability appraisal: Quantitative methods and mathematical techniques for environmental performance evaluation* (pp. 215–241).

Eshtehadi, R., Fathian, M., & Demir, E. (2017). Robust solutions to the pollution-routing problem with demand and travel time uncertainty. *Transportation Research Part D: Transport and Environment, 51*, 351–363.

Feng, Y., Zhang, R.-Q., & Jia, G. (2017). Vehicle routing problems with fuel consumption and stochastic travel speeds. *Mathematical Problems in Engineering, 2017*, 1–10.

Franceschetti, A., Demir, E., Honhon, D., Van Woensel, T., Laporte, G., & Stobbe, M. (2017). A metaheuristic for the time-dependent pollution-routing problem. *European Journal of Operational Research, 259*(3), 972–991.

Franceschetti, A., Honhon, D., Van Woensel, T., Bektaş, T., & Laporte, G. (2013). The time-dependent pollution-routing problem. *Transportation Research Part B: Methodological, 56*, 265–293.

Frölicher, T. L. (2016). Strong warming at high emissions. *Nature Climate Change, 6*(9), 823–824.

Fukasawa, R., He, Q., & Song, Y. (2016). A disjunctive convex programming approach to the pollution-routing problem. *Transportation Research Part B: Methodological, 94*, 61–79.

Gao, W., Luo, Z., & Shen, H. (2023). A branch-and-price-and-cut algorithm for time-dependent pollution-routing problem. *Transportation Research Part C: Emerging Technologies, 156*, Article 104339.

Ge, X., Ge, X., & Wang, W. (2020). A path-based selection solution approach for the low carbon vehicle routing problem with a time-window constraint. *Applied Sciences, 10*(4), 1489.

Ghannadpour, S. F., & Zarrabi, A. (2019). Multi-objective heterogeneous vehicle routing and scheduling problem with energy minimizing. *Swarm and Evolutionary Computation, 44*, 728–747.

Govindan, K., Salehian, F., Kian, H., Hosseini, S. T., & Mina, H. (2023). A location-inventory-routing problem to design a circular closed-loop supply chain network with carbon tax policy for achieving circular economy: An augmented epsilon-constraint approach. *International Journal of Production Economics, 257*, Article 108771.

Gu, W., Cattaruzza, D., Ogier, M., & Semet, F. (2019). Adaptive large neighborhood search for the commodity constrained split delivery VRP. *Computers & Operations Research, 112*, Article 104761.

Guo, J., & Liu, C. (2017). Time-dependent vehicle routing of free pickup and delivery service in flight ticket sales companies based on carbon emissions. *Journal of Advanced Transportation, 2017*, 1–10.

Hsiang, S., et al. (2017). Estimating economic damage from climate change in the United States. *Science, 356*(6345), 1362–1369.

Hwang, T., & Ouyang, Y. (2015). Urban freight truck routing under stochastic congestion and emission considerations. *Sustainability, 7*(6), 6610–6625.

Islam, M. A., & Gajpal, Y. (2021). Optimization of conventional and green vehicles composition under carbon emission cap. *Sustainability, 13*(12), 6940.

Kancharla, S. R., & Ramadurai, G. (2019). Multi-depot two-echelon fuel minimizing routing problem with heterogeneous fleets: Model and heuristic. *Networks and Spatial Economics, 19*(3), 969–1005.

Kanniah, K. D., Zaman, N. A. F. K., Kaskaoutis, D. G., & Latif, M. T. (2020). Covid-19's impact on the atmospheric environment in the Southeast Asia region. *Science of the Total Environment, 736*, Article 139658.

Karakostas, P., Sifaleras, A., & Georgiadis, M. C. (2020). Variable neighborhood search-based solution methods for the pollution location-inventory-routing problem. *Optimization Letters*, 1–25.

Karami, M., & Ghezavati, V. (2018). A differential evolution algorithm to solve new green VRP model by optimizing fuel consumption considering traffic limitations for collection of expired products. *Journal of Industrial and Systems Engineering, 11*(2), 251–269.

Kim, H.-W., Joo, G.-H., & Lee, D.-H. (2019). Multi-period heterogeneous vehicle routing considering carbon emission trading. *International Journal of Sustainable Transportation, 13*(5), 340–349.

Koç, Ç., Bektaş, T., Jabali, O., & Laporte, G. (2014). The fleet size and mix pollution-routing problem. *Transportation Research Part B: Methodological, 70*, 239–254.

Koç, Ç., Bektaş, T., Jabali, O., & Laporte, G. (2016). The impact of depot location, fleet composition and routing on emissions in city logistics. *Transportation Research Part B: Methodological, 84*, 81–102.

Koç, Ç., Erbaş, M., & Özceylan, E. (2019). The impact of routing on CO_2 emissions at a retail grocery store chain: A GIS-based solution approach. In *Lean and green supply chain management* (pp.143–160). Springer.

Kopfer, H. W., Schönberger, J., & Kopfer, H. (2014). Reducing greenhouse gas emissions of a heterogeneous vehicle fleet. *Flexible Services and Manufacturing Journal, 26*(1), 221–248.

Kramer, R., Maculan, N., Subramanian, A., & Vidal, T. (2015a). A speed and departure time optimization algorithm for the pollution-routing problem. *European Journal of Operational Research, 247*(3), 782–787.

Kramer, R., Vidal, T., & Lucıdio dos Anjos, F. C. (2015b). A matheuristic approach for the pollution-routing problem. *European Journal of Operational Research, 243*(2), 523–539.

Kumar, R. S., et al. (2016). Multi-objective modeling of production and pollution routing problem with time window: A self-learning particle swarm optimization approach. *Computers & Industrial Engineering, 99*, 29–40.

Kuo, Y. (2010). Using simulated annealing to minimize fuel consumption for the time-dependent vehicle routing problem. *Computers & Industrial Engineering, 59*(1), 157–165.

Lai, D., Costa, Y., Demir, E., Florio, A., & Van Woensel, T. (2021). *The pollution-routing problem with speed optimization and uneven topography*. arXiv:2105.09229

Lai, D., Costa, Y., Demir, E., Florio, A. M., & Van Woensel, T. (2024). The pollution-routing problem with speed optimization and uneven topography. *Computers & Operations Research, 164*, Article 106557.

Laporte, G. (2016). Scheduling issues in vehicle routing. *Annals of Operations Research, 236*(2), 463–474.

Li, H., Lv, T., & Li, Y. (2015). The tractor and semitrailer routing problem with many-to-many demand considering carbon dioxide emissions. *Transportation Research Part D: Transport and Environment, 34*, 68–82.

Li, H., Zhang, L., Lv, T., & Chang, X. (2016). The two-echelon time-constrained vehicle routing problem in linehaul-delivery systems. *Transportation Research Part B: Methodological, 94*, 169–188.

Li, J., Wang, D., & Zhang, J. (2018). Heterogeneous fixed fleet vehicle routing problem based on fuel and carbon emissions. *Journal of Cleaner Production, 201*, 896–908.

Lin, C., Choy, K. L., Ho, G. T., Chung, S. H., & Lam, H. (2014). Survey of green vehicle routing problem: Past and future trends. *Expert Systems with Applications, 41*(4), 1118–1138.

Lo, S.-C. (2022). A particle swarm optimization approach to solve the vehicle routing problem with cross-docking and carbon emissions reduction in logistics management. *Logistics, 6*(3), 62.

Lo, S.-C., & Shih, Y.-C. (2021). A genetic algorithm with quantum random number generator for solving the pollution-routing problem in sustainable logistics management. *Sustainability, 13*(15), 8381.

Majidi, S., Hosseini-Motlagh, S.-M., & Ignatius, J. (2018). Adaptive large neighborhood search heuristic for pollution-routing problem with simultaneous pickup and delivery. *Soft Computing, 22*(9), 2851–2865.

Masmoudi, M. A., Hosny, M., Demir, E., & Cheikhrouhou, N. (2018). A study on the heterogeneous fleet of alternative fuel vehicles: Reducing CO_2 emissions by means of biodiesel fuel. *Transportation Research Part D: Transport and Environment, 63*, 137–155.

Mazinani, M., Tavakkoli-Moghaddam, R., & Bozorgi-Amiri, A. (2023). A multi-objective cash-in-transit pollution-location-routing problem based on urban traffic conditions. *International Journal of Engineering, 36*(2), 299–310.

Meng, F., Ding, Y., Li, W., & Guo, R. (2019). Customer-oriented vehicle routing problem with environment consideration: Two-phase optimization approach and heuristic solution. *Mathematical Problems in Engineering, 2019*, 1–11.

Moghdani, R., Salimifard, K., Demir, E., & Benyettou, A. (2021). The green vehicle routing problem: A systematic literature review. *Journal of Cleaner Production, 279*, Article 123691.

Mohsenizadeh, M., Tural, M. K., & Kentel, E. (2020). Municipal solid waste management with cost minimization and emission control objectives: A case study of Ankara. *Sustainable Cities and Society, 52*, Article 101807.

Norouzi, N., Sadegh-Amalnick, M., & Tavakkoli-Moghaddam, R. (2017). Modified particle swarm optimization in a time-dependent vehicle routing problem: Minimizing fuel consumption. *Optimization Letters, 11*(1), 121–134.

Oberscheider, M., Zazgornik, J., Henriksen, C. B., Gronalt, M., & Hirsch, P. (2013). Minimizing driving times and greenhouse gas emissions in timber transport with a near-exact solution approach. *Scandinavian Journal of Forest Research, 28*(5), 493–506.

Okulewicz, M., & Mańdziuk, J. (2017). The impact of particular components of the PSO-based algorithm solving the dynamic vehicle routing problem. *Applied Soft Computing, 58*, 586–604.

Ouertani, N., Ben-Romdhane, H., Krichen, S., & Nouaouri, I. (2022). A vector evaluated evolutionary algorithm with exploitation reinforcement for the dynamic pollution routing problem. *Journal of Combinatorial Optimization,* 1–28.

Park, Y., & Chae, J. (2014). A review of the solution approaches used in recent G-VRP (green vehicle routing problem). *International Journal of Advanced Logistics, 3*(1–2), 27–37.

Patoghi, A., Shakeri, Z., & Setak, M. (2017). A time dependent pollution routing problem in multi-graph. *International Journal of Engineering, 30*(2), 234–242.

Poonthalir, G., & Nadarajan, R. (2018). A fuel efficient green vehicle routing problem with varying speed constraint (F-GVRP). *Expert Systems with Applications, 100*, 131–144.

Pradenas, L., Oportus, B., & Parada, V. (2013). Mitigation of greenhouse gas emissions in vehicle routing problems with backhauling. *Expert Systems with Applications, 40*(8), 2985–2991.

Qian, J., & Eglese, R. (2016). Fuel emissions optimization in vehicle routing problems with time-varying speeds. *European Journal of Operational Research, 248*(3), 840–848.

Qiu, R., Xu, J., Ke, R., Zeng, Z., & Wang, Y. (2020). Carbon pricing initiatives-based bi-level pollution routing problem. *European Journal of Operational Research, 286*(1), 203–217.

Rabbani, M., Bosjin, S., Yazdanparast, R., & Saravi, N. (2018). A stochastic time-dependent green capacitated vehicle routing and scheduling problem with time window, resiliency and reliability: A case study. *Decision Science Letters, 7*(4), 381–394.

Raeesi, R., & Zografos, K. G. (2019). The multi-objective Steiner pollution-routing problem on congested urban road networks. *Transportation Research Part B: Methodological, 122*, 457–485.

Ramos, T. R. P., Gomes, M. I., & Barbosa-Póvoa, A. P. (2014). Planning a sustainable reverse logistics system: Balancing costs with environmental and social concerns. *Omega, 48*, 60–74.

Rao, W., Liu, F., & Wang, S. (2016). An efficient two-objective hybrid local search algorithm for solving the fuel consumption vehicle routing problem. *Applied Computational Intelligence and Soft Computing, 2016*, 1–11.

Rauniyar, A., Nath, R., & Muhuri, P. K. (2019). Multi-factorial evolutionary algorithm based novel solution approach for multi-objective pollution-routing problem. *Computers & Industrial Engineering, 130*, 757–771.

Saka, O. C., Gürel, S., & Van Woensel, T. (2017). Using cost change estimates in a local search heuristic for the pollution-routing problem. *OR Spectrum, 39*(2), 557–587.

Salimifard, K., & Raeesi, R. (2014). A green routing problem: Optimising CO_2 emissions and costs from a bi-fuel vehicle fleet. *International Journal of Advanced Operations Management, 6*(1), 27–57.

Shahin, R., Hosteins, P., Pellegrini, P., Vandanjon, P.-O., & Quadrifoglio, L. (2024). A survey of flex-route transit problem and its link with vehicle routing problem. *Transportation Research Part C: Emerging Technologies, 158*, Article 104437.

Shamsi, A. A., Raisi, A. A., & Aftab, M. (2014). Pollution-inventory routing problem with perishable goods. In *Logistics operations, supply chain management and sustainability* (pp. 585–596). Springer.

Shi, Y., Zhou, Y., Ye, W., & Zhao, Q. Q. (2020). A relative robust optimization for a vehicle routing problem with time-window and synchronized visits considering greenhouse gas emissions. *Journal of Cleaner Production, 275*, Article 124112.

Shuang, Y., Diabat, A., & Liao, Y. (2019). A stochastic reverse logistics production routing model with emissions control policy selection. *International Journal of Production Economics, 213*, 201–216.

Solomon, M. M. (1987). Algorithms for the vehicle routing and scheduling problems with time window constraints. *Operations Research, 35*(2), 254–265.

Soysal, M., Bloemhof-Ruwaard, J. M., & Van Der Vorst, J. (2014). Modelling food logistics networks with emission considerations: The case of an international beef supply chain. *International Journal of Production Economics, 152*, 57–70.

Srivastava, G., Singh, A., & Mallipeddi, R. (2021). NSGA-II with objective-specific variation operators for multiobjective vehicle routing problem with time windows. *Expert Systems with Applications, 176*, Article 114779.

Sung, I., & Nielsen, P. (2020). Speed optimization algorithm with routing to minimize fuel consumption under time-dependent travel conditions. *Production & Manufacturing Research, 8*(1), 1–19.

Suzuki, Y. (2016). A dual-objective metaheuristic approach to solve practical pollution-routing problem. *International Journal of Production Economics, 176*, 143–153.

Tajik, N., Tavakkoli-Moghaddam, R., Vahdani, B., & Mousavi, S. M. (2014). A robust optimization approach for pollution routing problem with pickup and delivery under uncertainty. *Journal of Manufacturing Systems, 33*(2), 277–286.

Tirkolaee, E. B., Goli, A., Faridnia, A., Soltani, M., & Weber, G.-W. (2020). Multi-objective optimization for the reliable pollution-routing problem with cross-dock selection using Pareto-based algorithms. *Journal of Cleaner Production, 276*, Article 122927.

Tunga, H., Bhaumik, A., & Kar, S. (2017). A method for solving bi-objective green vehicle routing problem (G-VRP) through genetic algorithm. *Journal of the Association of Engineers, 1*(2), 33–48.

Wang, K., Shao, Y., & Zhou, W. (2017). Matheuristic for a two-echelon capacitated vehicle routing problem with environmental considerations in city logistics service. *Transportation Research Part D: Transport and Environment, 57*, 262–276.

Wang, S., et al. (2022). Reducing carbon emissions for the vehicle routing problem by utilizing multiple depots. *Sustainability, 14*(3), 1264.

Wu, F., et al. (2021). Optimal driving for vehicle fuel economy under traffic speed uncertainty. *Transportation Research Part B: Methodological, 154*, 175–206.

Xiao, Y., et al. (2020). The continuous pollution-routing problem. *Applied Mathematics and Computation, 387*, Article 125072.

Xiao, Y., Zhao, Q., Kaku, I., & Xu, Y. (2012). Development of a fuel consumption optimization model for the capacitated vehicle routing problem. *Computers & Operations Research, 39*(7), 1419–1431.

Yu, B., Yang, Z. Z., & Yao, B. (2009). An improved ant colony optimization for vehicle routing problem. *European Journal of Operational Research, 196*(1), 171–176.

Yu, Y., et al. (2016). Reducing carbon emission of pickup and delivery using integrated scheduling. *Transportation Research Part D: Transport and Environment, 47*, 237–250.

Zhang, J., Zhao, Y., Xue, W., & Li, J. (2015). Vehicle routing problem with fuel consumption and carbon emission. *International Journal of Production Economics, 170*, 234–242.

Zhang, M., et al. (2024). A multi-depot pollution-routing problem with time windows in e-commerce logistics coordination. *Industrial Management & Data Systems, 124*(1), 85–119.

Zhou, Y., & Lee, G. M. (2017). A Lagrangian relaxation-based solution method for a green vehicle routing problem to minimize greenhouse gas emissions. *Sustainability, 9*(5), 776.

Zhu, L., & Hu, D. (2019). Study on the vehicle routing problem considering congestion and emission factors. *International Journal of Production Research, 57*(19), 6115–6129.

Reza Shahin has a Ph.D. in Computer Science with a focus on Operations Research and specializes in developing and implementing advanced optimization models, including Mixed-Integer Linear Programming (MILP) and machine learning techniques, for complex industrial and supply chain systems. He completed his Ph.D. program at University of Gustave Eiffel in France on the Flex-Route Transit problem. His Ph.D. focus was on integrating ad hoc stops into public transport in which the flexibility of demand responsive transit was merged with the low cost of public transportation, providing a door-to-door pick-up and drop-off to customers. Currently, he is a postdoctoral researcher at University of Polytechnique Montreal, where he works on scheduling tasks in a Canadian underground mine. His expertise includes applying machine learning methods, such as reinforcement learning and data-driven decision-making, to enhance operations in logistics, scheduling, and intelligent manufacturing.

Adoption of Electric Taxis in Hong Kong: Challenges and Opportunities

Yui-Yip Lau and Yang Andrew Wu

Abstract Taxis are one of the main modes of public transportation in Hong Kong. The HKSAR government published *Hong Kong's Climate Action Plan 2050* in October 2021. As such, Hong Kong has a series of climate change initiatives, including net-zero electricity generation, reducing carbon footprint, and promoting energy savings. To this end, these initiatives may achieve the dual goals of carbon peak and carbon neutralization in 2030 and 2060, respectively. At the beginning, this chapter points out the key challenges and opportunities of electric vehicle development in Hong Kong. Then, electric taxis are used as an illustrative case study to summarize the roadmap for the penetration of electric taxis in Hong Kong. Furthermore, the present research aims to identify licensed taxi owner intentions to purchase electric vehicles for operational use and to investigate the driving behavior of taxi drivers in Hong Kong. To address the research objectives, we carried out in-street interviews with 250 taxi drivers in Hong Kong. It was found that the waiting time to charge (i.e., time of queuing) and the time of charging are the key concerns for taxi drivers to determine whether to change from diesel to electric taxis. The government may provide an operating subsidy or other incentives for the taxi drivers to operate electric taxis. Based on the research findings, the implications for public policies in the future are further discussed. A step-by-step and continual promotion strategy is suggested to align with emission reduction goals and facilitate the shift from conventional taxis to electric taxis in Hong Kong. The examinations of numerous factors and the interest of various stakeholders in the local community are recommended as well.

Y.-Y. Lau (✉)
Division of Business and Hospitality Management, School of Professional Education and Executive Development, The Hong Kong Polytechnic University, Kowloon, Hong Kong
e-mail: yuiyip.lau@cpce-polyu.edu.hk

Y. A. Wu
Senior Lecturer, Division of Science, Engineering and Health Studies, School of Professional Education and Executive Development, The Hong Kong Polytechnic University, Kowloon, Hong Kong
e-mail: andrew.wu@cpce-polyu.edu.hk

M. A. Dulebenets et al. (eds.), *Transport and Climate Change*,
https://doi.org/10.1007/978-3-032-00563-2_2

Keywords Conventional taxis · Electric taxis · Carbon peak · Carbon neutralization · Electric vehicles

1 Introduction

Climate change is defined as the change in climate because of a series of human activities that induce a change in the atmospheric formation, along with natural climate changeability. Human activities generate rises in the atmospheric absorption of greenhouse gases, generally as a result of the burning of fossil fuels. The effect of human activities has induced the climate system to warm and increased the impact of greenhouse gases. Such a phenomenon is generally called 'global warming'. Recently, Hong Kong has been facing a severe climate change problem, including an increasing number of very hot days and hot nights, decreasing cold days, increasing temperature, more frequent heavy precipitation, stronger power of tropical cyclones, increasing mean sea level, and rising storm surge risk (Hong Kong Observatory, 2024). According to Climate Central (2024), some Hong Kong locations (e.g., Lai Chi Kok, Kai Tak, West Kowloon Cultural District, Tsing Yi, Central) will face a threat of sea level rise by 2050. Such locations are situated near the coastal area or converted from a reclamation area. Hazards in the coastal context, such as erosion and coastal storms, bring remarkable threats to coastal economic, social, and physical systems (Lau et al., 2022).

Storm Mangkhut, which occurred in 2018 and was followed by extreme rainfall events, imposed significant implications. First, a rise in the water flow because of heavy rain during tropical cyclones is likely to increase the trend of flooding incidents. Extreme rainfall could gradually create more floods. Flooding imposes broad and significant impacts on health, causing injuries and mental health problems. The transmission of several infectious diseases may create the risk for diffusion of arising pandemics. Moreover, the flooding of facilities may severely damage electrical power infrastructures because of heavy rainfall, notably in low-lying areas. In addition, high-intensity storms will decrease transportation efficiency, reduce accessibility to transportation networks, raise transportation costs, and negatively affect the development of green transportation infrastructures. However, densely populated and highly urbanized areas with infrastructure adjacent to coastal areas are more proactive to natural disasters (Lau et al., 2022).

In response, the HKSAR government proposed *Hong Kong's Climate Action Plan 2050* to foster net-zero electricity generation, reduce 'carbon footprint', and promote energy-saving. The ultimate goals are to attain carbon neutrality and mitigate climate change (HKSAR, 2021). As expected, the HKSAR government identified electric vehicles are the main vehicle technology in the forthcoming years. Due to the urgent demand for replacement of the internal combustion engines and the rising damage to the environment, there is an emerging trend in the use of electric vehicle technology worldwide (Lau et al., 2024). There is an emerging global trend of launching electric vehicles in different parts of the world. The Chinese government now subsidizes the development and deployment of electric vehicles. Also, China has recorded a 16% electric vehicle 4 W penetration. In addition, Norway constructed 10,000 charging stations before 2015. Indeed, Norway offered a 100% subsidy for public charger installation from 2009 to 2010. Moreover, the European Union has imposed a goal of one public charger for every ten electric vehicles. Furthermore, the United States strives toward setting up 500,000 public charges before 2030 and provided the financial support of USD 7.5 billion to this public charger project (KPMG, 2022).

According to Science Daily (2020), the average lifetime emissions from electric vehicles constitute 70% lower values than petrol vehicles since they are more efficient than petrol vehicles. Statistics indicated that in almost 90% of regions that comprise 95% of carbon emissions, electric vehicles are less emission-intensive than fossil fuel choices. We focus our attention on Asia because of its vital role in climate change. Political temperatures have always been boosted since political leaders support the public interest and encourage the nation to employ electric vehicles (Lau et al., 2024). Recently, the HKSAR government actively introduced electric vehicles into the public transport sector. Taxis will be considered as the large-scale electric vehicle implementation plan in the public transport sector in 2025.

Nowadays, carbon neutralization has arisen as the ‘emerging’ topic of the taxi industry because of climate change. To minimize the generation of carbon dioxide, the taxi industry has performed remarkable changes. Some taxis will be substituted by electric vehicles, enhancing environmental protection and traffic safety. Electric taxis normally contain energy storage devices like ultra-capacitors, fuel cells, and different kinds of batteries together with electric propulsion. Moreover, their energy management systems are based on diverse technologies manufactured using the principles of mechanical, chemical, automotive, electronic, and electrical engineering (Wu et al., 2023).

Hereto, the study area of electric vehicles has been commonly explored by some academic journals and research efforts. Nevertheless, the previous research studies mainly address operations management, econometric modelling, simulation, modelling, and experiments. To the best of the authors’ knowledge, Hong Kong is identified as a megalopolis, and detailed research studies on driver behavior and economic barriers with respect to the electric taxi deployment in Hong Kong are still needed (Lau et al., 2024). This research aims to explore licensed taxi owner intentions to purchase electric vehicles for operational use and investigate the driving behavior of taxi drivers in Hong Kong. In general, this book chapter is divided into five main sections. Section 1 mainly gives an overview of the research context, settings, and the development of the taxi industry in Hong Kong. Then, various studies related to electric vehicle progress, as well as electric taxi development, challenges, and opportunities in Hong Kong are discussed in Sect. 2. After that, methodology and data collection are described in Sect. 3. Then, findings and discussions are presented in Sect. 4. The conclusion is summarized in Sect. 5.

2 Literature Review

2.1 Overview of Electric Vehicle Progress in Hong Kong

A smart city is a city that has applied different types of technology to improve the city's policies and operations, as well as to improve the quality of life. A smart city not only improves convenience and quality of life but also leads to a more sustainable environment. A smart environment, to which people have been recently paying attention to, is one of the major factors in achieving the smart city settings. Indeed, electric vehicles bring fewer negative impacts towards climate change and air quality when compared to petrol or diesel vehicles (Environmental Protection Department, 2022). This is why electric vehicles gradually become a major trend over the past few years. In Hong Kong, the transportation sector has always been the second-largest source of carbon emissions (Environmental Protection Department, 2022). This is mainly caused by diesel engines, which are a major component of diesel vehicles, discharge different kinds of particulates, including nitrogen oxides (NOx), volatile organic compounds (VOC), benzene, as well as carbon monoxide (Choi, 2021). In 2018, the road transport in Hong Kong accounted for 18%, 21%, and 50% of the total emissions of NOx, VOC, and CO, respectively (Environmental Protection Department, 2022).

In addition to environmental impacts, the International Agency for Research and Cancer also found that diesel fumes increase the chance of lung cancer, acute respiratory illnesses, as well as respiratory symptoms and infections (International Agency for Research on Cancer, 2022). We can see how petrol and diesel cars severely affect the environment, as well as human beings. Contrarily, the implementation of electric vehicles can effectively mitigate the negative impacts brought by petrol and diesel cars, as the amount of carbon dioxide emission is 33% and 47% less on average compared to hybrid vehicles and petrol vehicles, respectively (CLP Power Hong Kong, 2022). Therefore, to effectively improve air quality and reduce air particulates, the Hong Kong government has vigorously promoted electric vehicles in recent years. The goal of the government is to achieve carbon neutrality by 2050, which means attaining the vision of a "Zero Carbon Emission Clean Air Smart City" (HKSAR Information Services, 2022). By continuously promoting electric vehicles and adopting other policies, the major roadside air pollutants, including nitrogen dioxide (NO_2) and respirable suspended particulates (PM_{10}), have gradually decreased by 41% and 38% from 2012 to 2021, respectively (Environmental Protection Department, 2022). As a result, the development of electric vehicles can pursue zero carbon emission targets, enhance air quality, and pursue a smart environment, as well as the smart city development.

Additionally, the government also plans to carry out trials for electric buses and minibuses (RTHK, 2022). By the end of April 2022, there are already 256 approved applications for the trials of green innovative transport technologies, with a total allowance of $209 million (HKSAR Information Services, 2022). Apart from promoting electric vehicles, the government also plans to expand the charging network for both electric private vehicles and commercial vehicles. For private vehicles, home charging and office charging are strongly encouraged, whereas dedicated fast charging facilities will be installed for electric commercial vehicles, for example, e-taxis (HKSAR Inland Revenue Department, 2022). The Hong Kong government has awarded a five-year contract to Smart Charge (HK) Limited, based on which the company will provide fast charging services to act in concert with the trial of electric taxis on Lantau Island (Environmental Protection Department, 2022).

Besides, the government will also conduct trials on e-Truck and e-public Light Bus in the New Territories, including Tuen Mun and Tseung Kwan O (Hong Kong Productivity Council, 2022). They will collect the trial results and the opinions of the drivers to evaluate and optimize the performance of electric vehicles. By making improvements in the pilot scheme, the government has dedicated itself to shortening the charging time of electric commercial vehicles, enhancing the usage time of fully charged vehicles, as well as developing a comprehensive charging network for different types of electric vehicles. The development of electric vehicles and the corresponding charging infrastructure can facilitate smart transportation.

Over the past few years, people have started paying attention to the environment, especially, the air pollution caused by petrol vehicles. They tend to purchase electric vehicles to reduce the emission of carbon dioxide and other air pollutants. However, for every single electric vehicle charger, there will be 7.4 electric vehicles vying for it generally. In other words, there are insufficient charging facilities nowadays. Although the government provides purchasing allowances to electric vehicle drivers, they have put less focus on providing electric vehicle charging services over the past years. Also, at the present stage, since the usage of electric vehicle chargers is free of charge, private companies may not find the electric vehicle charging market attractive as there are no profits to earn. Besides, as private companies may not receive any profits, it is hard to break even the operating and maintenance costs (Hong Kong Productivity Council, 2022). As such, the inadequacy of charging facilities may deflate people's willingness to purchase electric vehicles.

To encourage people to use electric vehicles, the rollout of government policies mainly focuses on purchasing allowance, profits tax deduction, increasing the number of charging facilities, etc. By the end of September 2022, there was a total of 39,114 electric vehicles, which accounts for approximately 4.3% of all the vehicles in Hong Kong. Out of 203 approved electric vehicle models, 45 of them are suitable for public transport as well as commercial vehicles. To cope with the development of electric vehicles, the government and other private companies have built 5,283 electric vehicle chargers for public use (Environmental Protection Department, 2022), which means that around 7.4 electric cars are still vying for an electric vehicle charger on average.

2.2 Electric Taxi Development, Challenges, and Opportunities in Hong Kong

Apart from the smart environment, a technologically advanced transport system may also be crucial for a smart city. So, to facilitate the intelligent transport system and traffic management, the Hong Kong government did not just promote electric private vehicles but also public transport, including electric buses, vans, as well as electric taxis. Take electric taxis as an example, the government has launched several trials in recent years. Although the pilot scheme launched in 2013 failed, the government planned to conduct an e-taxi trial again in 2022, with the support of the New Energy Transport Fund (Environmental Protection Department, 2022). The fund mainly focuses on shortening the charging cycle, increasing the reliability of electric vehicles, improving the cost and lifespan of electric vehicle batteries, and training qualified electric vehicle maintenance professionals (HKPC, 2014).

Taxis are commonly used in urban areas. In the past few decades, there has been a widespread number of electric vehicles. In the field of the public transportation sector, taxis, one of the key public transport service providers, are shifting to electric taxis to attain sustainable development. As of 2022, there were a sum of 18,163 taxis in Hong Kong, about 2,838 taxis in the New Territories (green), 75 taxis in Lantau (blue), and 15,250 urban taxis (red). Nearly all the taxis are rented out to self-employed drivers on the shifting principle and owned by different independent taxi firms. Since taxis are a very popular public transport mode in Hong Kong and provide a point-to-point convenient service to passengers, the daily taxi patronage is around 15 million on average (HKSAR Transport Department, 2024). Since nearly all inherent structures of taxis use internal combustion engines, they generate a considerable amount of carbon footprint emissions, notably for high mileage services. In other words, one taxi can consume as much energy and create as much pollution as ten private cars. Besides, a significant level of particulate pollution, especially PM2.5, is currently a serious Hong Kong health concern (Wu et al., 2023).

One of the major challenges in the transition from internal combustion engines to electric taxis is influencing in what way to design the development of charging stations to fulfill the changing travel patterns of electric taxis. Recent research studies were carried out on the distribution of gasoline, liquefied petroleum gas, and natural gas vehicle stations to decide optimal sites on the road networks. Also, electric taxis may face sustainability challenges. The major hesitation on sustainability is because of the various raw materials employed for the availability and production of batteries. A range of the components employed to manufacture electric taxis contain cobalt, lithium, and nickel. Most of the approaches used to source such materials are identified as unsustainable. In addition, the public is still concerned about the safety of electric taxis. Generally, electric taxis employ lithium-ion batteries as the key origin of power. Similar to all batteries, lithium-ion batteries may create inflammation and short circuits for electric taxis. Furthermore, the life of batteries is another challenge for electric taxis. To the best of the author's knowledge, the usage of electric taxis is relatively high. In other words, the batteries may require charging daily or two times per day. The life of the batteries may only last a few years. Also, it is not considerably appropriate to travel long distances per day (Wu et al., 2023).

Another challenge is that commercial electric vehicles will encounter ‘dead mileage’ during charging. To illustrate, as electric taxi drivers work on shifts, if they need to charge their electric taxis during the shift, they may not be able to carry passengers. This may affect their income, as they need to recover the cost of renting a taxi and fuel expenses. To this end, since dead mileage will bring severe impacts to electric taxi drivers, they may not be willing to change from petrol taxis to electric taxis. Moreover, BYD Co. Ltd entered the Hong Kong electric taxi market two years ago, but a fully charged BYD electric taxi may need around four hours (South China Morning Post, 2015), which may not be practical for taxi operations as well as other commercial vehicles.

Furthermore, there are only a limited number of electric vehicle models for commercial vehicles. Among 203 electric vehicle models, only 45 of them are accepted for commercial vehicles. Compared with electric private vehicles (PVs) and motorcycles, there are 158 electric vehicle models which provide a broad range of varieties for PV drivers to choose from. The common commercial and private electric vehicle models approved in Hong Kong are provided in Table 1. For light and medium-sized vehicles, only 15 electric vehicle models are available; and 13 electric vehicle models are available for buses and light buses. However, there is only one model for electric taxis, which means taxi drivers have no choice if they want to change to electric taxis. There is no other model for them to compare the pros and cons, and they must accept the drawbacks brought by the only model.

When it comes to electric PVs, apart from the shortage of electric vehicle chargers, the difficulty of electric vehicle maintenance is also a major concern for electric vehicle drivers. With the rapid development of electric vehicles, the Hong Kong government already tried its best to catch up with the pace. However, to successfully promote electric vehicles and achieve the goal of a smart city, an all-rounded charging network and professional and up-to-date repair and maintenance are indispensable. So far, the government has continuously expanded the charging network in Hong Kong, but there is still an inadequacy of professional electric vehicle repair and maintenance. Although the government plans to provide post-secondary training, retraining, and educational opportunities to cultivate more professionals and mechanical engineers, this might be a major concern to citizens about whether they should purchase an electric vehicle now. As the repair and maintenance of electric vehicles may require the knowledge and skills of electric engineering, the realization of timely maintenance may be a long-term goal at this moment. Even if post-secondary institutions push ahead to offer comprehensive training programs, it may take up to a couple of years to educate qualified engineers. In other words, it is hard for electric vehicle drivers to receive timely repair services if they encounter some problems now.

To facilitate commercial electric vehicle development, dedicated electric vehicle charging infrastructure can be built for commercial electric vehicles. Take electric taxis as an example, this year, the government is going to install at least 10 dedicated quick chargers for electric taxis in the New Territories, including Lantau Island and

Table 1 Common electric vehicle models approved in Hong Kong

Model	Category	Mileage (km)	Battery capacity (kWh)
FDG6102EVG	Commercial vehicle	280	324
Wuzhoulong FDG6700EVG	Commercial vehicle	180	101
JOYLONG EW4	Commercial vehicle	220	57.6
DFSK EC35	Commercial vehicle	233	41.4
Nissan E-NV200 Full Panel Van	Commercial vehicle	200–300	40
Mitsubishi Minicab-MiEV	Commercial vehicle	100–150	10.5–16
Renault Kangoo Van Z.E.	Commercial vehicle	170	22–33
KIA Niro EV+	Private car	288–455	39.2–64
Jaguar I-PACE EV400	Private car	396–470	90
Hyundai Kona Electric	Private car	303–484	39.2–64
Hyundai Ioniq Electric	Private car	200–274	28–38.3
Volkswagen e-GOLF	Private car	129–190	24.2
Smart Forfour	Private car	130	17.6
BYD e6	Private car	300–400	61–80
Renault ZOE	Private car	210–395	22–52
Renault Fluence Z.E.	Private car	145–233	22–36
BMW i3	Private car	130–322	18.2–37.9
Tesla Model X	Private car	383–523	60–100
Tesla Roadster	Private car	393	53
Tesla Model S	Private car	400–650	75–100
Tesla Model 3	Private car	354–504	54–82
Nissan LEAF	Private car	243–364	40–60
Mitsubishi iMiEV	Private car	160	16
Euauto MyCar	Private car	97	9.6

Source HKSAR Transport Department (2021)

Sai Kung (Environmental Protection Department, 2022). If this plan is successful, the government can build more dedicated electric vehicle chargers in all 18 districts to enhance the convenience of electric vehicle taxi drivers. The government also identifies appropriate charging locations, such as taxi stands, and transport interchanges, where taxi drivers can charge their electric taxis during parking or waiting for the next shift. By building a large e-taxi charging network, it is easier for taxi drivers to access electric vehicle chargers. Most importantly, taxi drivers do not need to share electric vehicle charging facilities with other types of electric vehicles. Moreover, if all the dedicated chargers apply fast-charging techniques, this may effectively shorten the charging time of electric taxis, thus bringing fewer negative impacts to the income of taxi drivers.

For electric PVs, one of the potential opportunities would be the development of home charging and office charging. Although there are already 5,283 electric vehicle chargers for public use, more electric vehicle chargers can be installed in different residential and commercial buildings to enhance the convenience of electric vehicle drivers. To accelerate the development of home charging, the government has introduced the "EV-charging at Home Subsidy Scheme" ("EHSS") with a total amount of $3.5 billion. This scheme is aimed at encouraging the installation of electric vehicle chargers in private residential buildings, and it is expected to cover approximately 140,000 private parking spaces in 700 existing car parks. For commercial buildings, some property management companies have already installed electric vehicle chargers at their commercial buildings and shopping malls to cope with the needs of their lessees and customers (Environmental Protection Department, 2022), which allows them to charge their electric vehicles during their working time or shopping time. As a result, by enlarging the electric vehicle charging network, electric vehicle drivers can charge their private vehicles during their work or when they are at home. The supply of electric vehicle chargers will gradually increase in the near future, and people may not even need to consider the issue of insufficient electric vehicle chargers anymore when they plan to purchase electric vehicles.

3 Methodology and Data Collection

We carried out in-street interviews with Hong Kong taxi drivers to explore the licensed taxi owner intentions to buy electric vehicles for operations and to investigate the driving behavior of Hong Kong taxi drivers. In general, Hong Kong has recorded 40,000 taxi drivers, and we chose 250 of them randomly for our research study via a convenience sampling method. As such, the drivers were selected randomly across Hong Kong. To improve the efficiency of survey data collection, our research team decided to collect survey data at popular taxi stations where many taxi drivers are waiting for passengers in order to collect the respondents' answers in a sufficient time. We selected the participants expecting they had at least 2 years of taxi driver's experience to provide constructive feedback and valid responses in our study. To boost the response rate and show appreciation, each survey respondent was offered a supermarket coupon after the completion of the survey. However, our research team faced several challenges that were encountered during the data collection. The main challenge consisted in the fact that some of the survey respondents were not willing to engage in the survey and failed to complete the survey due to the passengers taking on the taxi during the survey data collection process.

In general, the survey questionnaire was split into two key sections. In Section A, the survey respondents were expected to provide their profiles associated with gender, age, and driving experiences. In Section B, the survey respondents needed to indicate detailed information about their intention to buy electric taxis for operational use and their driving purposes. The main survey's questions mainly covered the following:

- How many kilometers does a taxi driver drive every day?
- What is the mileage of the electric vehicle that the taxi driver intends to drive?
- What is the acceptable number of charging times during a one-day operation?
- What is the acceptable driving time to the nearest available charging station?
- What kind of incentives would you like to have for getting an electric taxi?
- Which of the following location(s) is (are) suitable for installing charging facilities?
- What is (are) the main factor(s) for the main obstacle(s) to the development of electric taxis?

Before starting the data collection survey, we obtained the institutional ethical approval. The survey respondent answers were only used for academic purposes with high confidentiality. The collected responses were reviewed in an aggregate manner only. Also, the survey design and questions were based on the previous research studies and in-depth semi-structured discussions with various related parties involving academic scholars, taxi drivers, logistics and transport associations, and industrial practitioners. To validate and align with the key research study objectives, the survey questionnaire was double-checked to ensure that there are no unclear items and complicated wording. It is vital to ensure the validity of the content, which is a crucial measure of the survey instrument correctness. A series of discussions with the appropriate representatives resulted in small modifications in the wording and examples provided for some measurement items, which were eventually adopted as relevant and retained the content validity. The revised measurement items were integrated into the latest survey questionnaire (Nunnally, 1967).

The profile of survey respondents is summarized in Table 2. Overall, 97.6% of survey respondents were male taxi drivers, and 2.4% were female taxi drivers. In other words, the taxi industry is male-orientated. Also, 46.4% of survey respondents were between 51 and 60 years old, and 44.8% of survey respondents were more than 60 years old. In other words, there is a lack of young individuals to join the workforce in the taxi industry. In addition, 84% of the survey respondents indicated that they accumulated over 10 years of driving taxi experience.

4 Findings and Discussions

In terms of the travel distance of taxi drivers on a daily basis, over a quarter of survey respondents expressed that their travel distance was about 200 km, 18.4% of survey respondents pointed out that their travel distance was approximately 250 km, and 13.6% of survey respondents reflected that their travel distance was around 180 km. According to the results of the survey, we suppose that the daily travel distance of a taxi driver may range between 80 and 250 km on average.

Regarding the electric taxi's optimal travel distance, over 26% of survey respondents proposed about 300 km, 23.3% of survey respondents recommended 400 km, and 15.7% of survey respondents expressed 350 km. Hence, based on the collected

Table 2 Summary of survey respondent characteristics

Item	Percentage
Gender	
Male	97.6
Female	2.4
Age	
18–30	0
31–40	0.8
41–50	8
51–60	46.4
Over 60 years old	44.8
Driving experience	
Less than 2 years	0
2–5 years	8
6–10 years	8
Over 10 years	84

data, more than 65% of survey respondents expected that the electric taxi's optimal travel distance should reach 300 km at a minimum.

For the frequency of charging times, 99.2% of survey respondents would accept to charge once at the time of their shift only. The duration of charging electric taxis was found to be an important factor. Around 39% of survey respondents preferred that the charging duration would not exceed 5 min, over a quarter of survey respondents would agree with the charging duration of under 10 min, and 17.7% of survey respondents would be comfortable with the charging duration of 15 min. The survey outcomes indicated that certain respondents were unwilling to spend most of the time charging the electric taxis due to non-revenue activities. From the perspective of taxi drivers, the desirable plan is to charge once at the time of their shift only, and the acceptable charging duration is 15 min. However, the current technologies fail to fulfill these expectations (Lau et al., 2024).

Concerning the motivation for employing electric taxis, 82.8% of survey respondents highlighted that an operating subsidy was preferable, whereas 17.2% of survey respondents expressed that they would prefer to get subsidies at the time of buying electric taxis. These findings reflect that many respondents might not have sufficient funds to purchase an electric taxi. They may prefer to get an operating subsidy at a later stage (Wu et al., 2023).

For the places of setting up electric taxi charging stations, over 85% of survey respondents reflected that charging stations were appropriate to be set on the streets, over half of the survey respondents indicated that public car parks could be considered for establishing charging stations, and over 40% of survey respondents reflected that shopping malls were also good locations for building charging stations. The survey results highlighted that almost 90% of survey respondents expected charging stations to be installed in some public venues like public car parks and street parking, where it would be convenient for the taxi drivers to engage in electric taxi charging practice (Wu et al., 2023).

Regarding the attitudes in the direction of difficulties of electric taxi evolution, over 90% of survey respondents expressed that the waiting time to charge (i.e., queuing time) and the charging frequency are the determining factors in whether to shift from diesel to electric taxis. A total of 65.2% of survey respondents indicated that they often experience long queues between taxis loitering points and charging stations, and they are unwilling to search for a charging station located farther away. Moreover, some places like shopping malls, car parks, streets, etc. face inherent vulnerabilities (e.g., electricity supply, near the coastal area) that lead to difficulty in setting up electric taxi charging stations in a timely manner (Lau et al., 2024). Over 60% of survey respondents indicated that insufficient incentives were one of the key barriers to electric taxi development. Thus, there is a large number of taxi drivers who are inclined towards the continuous usage of the existing diesel taxis (Wu et al., 2023).

5 Conclusion

Currently, the HKSAR government strives toward introducing electric taxis to develop a smart city, achieve sustainable development goals, mitigate the negative impacts of climate change, and attain carbon neutralization. Nevertheless, the progress of electric taxis has faced unfolded challenges, including sustainability, battery life, and charging station design (Lau et al., 2024). The current study provides valuable insights into licensed taxi owner intentions to purchase electric vehicles for operational use and investigates the driving behavior of taxi drivers in Hong Kong. The survey results may help us to create an electric taxi driver behavioral model and design the optimal location of the charging facilities in the next research. In addition, the survey findings improve our understanding of the unfolded challenges in the adoption of electric taxis in Hong Kong and propose possible measures to respond to these challenges. The following measures can be considered to deal with the existing challenges in the forthcoming years.

- The HKSAR government may provide more resources and research funding to develop an electric taxi research center. The center's research area may focus on the optimal location of electric taxi charging stations, upgrading the infrastructure on the installation of fast-charging stations, passenger predictive behavior, green finance for a low-carbon economy, and sustainable energy. These measures may ease the anxiety of taxi drivers in terms of the frequency of charging times and the charging duration of electric taxis. Indeed, the HKSAR government may provide an operating subsidy or sufficient incentives to taxi drivers to increase the motivation of taxi drivers to support the implementation of electric taxis in the forthcoming years. As expected, electric taxis are likely to keep the sustainable development in the community.

- In general, teaching materials related to electric taxis are lacking in the current educational transport programs. As such, there is a lack of young people and talent engaged in the electric taxi sector. Also, the existing taxi drivers expressed that they are not familiar with certain aspects of driving electric taxis due to the lack of electric vehicle training programs. The sustainability of electric taxis is affected by these issues. In response, higher educational institutions may develop and implement vocational training programs related to electric vehicles. The training programs may concentrate on electric taxi areas covering operations, engineering, and facilities management.
- According to the survey results, over half of the survey respondents indicated that they often experience long queues between taxis loitering points and charging stations. The HKSAR government may collaborate with stakeholders to conduct city planning focusing on electric taxi charging facilities. The HKSAR government may provide the startup fund or matching fund for the industrial practitioners and associations to set up the optimal location and upgrade the electric taxi charging facilities in different points of Hong Kong. Moreover, the majority of survey respondents reflected that the charging facilities may be placed at the convenience points covering shopping malls, petroleum stations, public car parks, and streets. World Bank underlines that investing in charging infrastructures is 4–7 times more effective for electric vehicle adoption than giving electric taxi operating subsidies (KPMG, 2022). In doing so, it may reduce the demotivation of taxi drivers to support electric taxi development. In the long term, conducting a large-scale survey is crucial to understanding the taxi drivers' perceptions and operational behavior towards the electric taxi charging station locations and capacities.
- To a large extent, over 90% of survey respondents worried about the transition from diesel to electric taxis. The HKSAR government may design innovative promotional campaigns like Virtual Reality, Key Opinion Leaders, and social media related to the deployment of electric taxis in the community. In doing so, it may change the taxi drivers' and citizens' attitudes towards electric taxis in the community.

Along with the aforementioned findings and contributions, this study also has some limitations, which may bring future research directions. Firstly, this study only focuses on the Hong Kong study area which lacks generalization. As such, we may conduct a comparative study for other countries or cities like China, Norway, and Singapore, which largely adopt electric taxis, to generate a comparative electric taxi study. Secondly, this study mainly uses the quantitative research approach via the collection of survey questionnaires. To offset the limitations of the quantitative research method, we may consider including the qualitative research approach via semi-structured in-depth interviews with the relevant stakeholders, including industry practitioners, transport associations, government bodies, and the local community. Thirdly, we may incorporate the Theory of Planned Behavior to investigate the taxi drivers' behavior towards the adoption of electric taxis in Hong Kong. In doing so, we may create a theoretical framework to provide the academic contribution to the next research study. Fourthly, we cannot gather the percentage of drivers open to electric taxis due to the lack of available data and information that can be accessed by the public at this stage. Therefore, an electric taxi driving behavioral model can be developed via the baseline measurement method in the next research study.

Acknowledgements The research work discussed in this paper is funded by the Research Grant Council of Hong Kong SAR, project reference number UGC/FDS24/E08/21.

References

Choi, H. S. (2021). The HK Smart City Blueprint: An assessment in relation to international smart city performance standards. *HKIUD Journal, 5*, 91–102.

Climate Central. (2024). Retrieved November 9, 2024 from https://sealevel.climatecentral.org/

CLP Power Hong Kong Limited. (2022). Retrieved November 5, 2022 from https://www.clp.com.hk/en/business/low-carbon-solutions/electric-vehicle-charging

Environmental Protection Department. (2022). *Hong Kong air pollutant emission inventory—Road transport*. Retrieved December 3, 2022 from https://www.epd.gov.hk/epd/english/environmentinhk/air/data/emission_inve_transport.html

HKPC. (2014). An industry study on electric vehicle adoption in Hong Kong, Hong Kong.

HKSAR. (2021). *Government announces Hong Kong's climate action.* Retrieved November 10, 2024 from https://www.info.gov.hk/gia/general/202110/08/P2021100800588.htm

HKSAR Information Services Department. (2022). Retrieved October 8, 2022 from https://www.isd.gov.hk/eng/

HKSAR Inland Revenue Department. (2022). Retrieved August 6, 2022 from https://www.ird.gov.hk/eng/ppr/archives/22050401.htm

HKSAR Transport Department. (2024). Retrieved September 7, 2022 from https://www.td.gov.hk/en/transport_in_hong_kong/public_transport/taxi/index.html

Hong Kong Observatory. (2024). Retrieved November 9, 2024 from https://www.hko.gov.hk/tc/index.html

Hong Kong Productivity Council. (2022). *APAS Bolsters "Hong Kong roadmap on popularisation of electric vehicles" with all-round support for Hong Kong's EV development and to build a smart city with R&D.* Retrieved July 9, 2022 from https://www.hkpc.org/en/about-us/media-centre/press-releases/2021/apas-hkroadmap

International Agency for Research on Cancer. (2022). Retrieved December 3, 2022 from https://www.iarc.who.int/

KPMG. (2022). *Electric vehicle charging—The next big opportunity*. KPMG Assurance and Consulting Services LLP.

Lau, Y. Y., Wu, Y. A., Wong, L. M., Wu, J., Dong, Z., Yip, C., Lee, S. W., & Chan, J. K. Y. (2024). Optimizing electric vehicle charging station locations: A study on a small outlying island in Hong Kong. *Urban Science, 8*, 134.

Lau, Y. Y., Yip, T. L., Dulebenets, M. A., Tang, Y. M., & Kawasaki, T. (2022). A review of historical changes of tropical and extra-tropical cyclones: A comparative analysis of the United States, Europe, and Asia. *International Journal of Environmental Research and Public Health, 19*, 4499.

Nunnally, J. C. (1967). *Psychometric theory*. McGraw-Hill.

RTHK. (2022). *Govt making preparations for E-taxi trial runs.* Retrieved September 5, 2022 from https://news.rthk.hk/rthk/en/component/k2/1646991-20220504.htm

Science Daily. (2020). *Electric cars better for climate in 95% of the world*. Retrieved November 11, 2024 from https://www.sciencedaily.com/releases/2020/03/200323125602.htm

South China Morning Post. (2015). *Electric taxi project in Hong Kong goes belly up: China's BYD Brands 2-year campaign a 'failure'*. Retrieved October 9, 2022 from https://www.scmp.com/tech/innovation/article/1876321/electric-taxi-project-hong-kong-goes-belly-chinas-byd-brands-2-year

Wu, A. Y., Lau, Y. Y., Wong, L. M., & Wu, J. (2023). A preliminary feasibility study of electric taxi promotion in Hong Kong—Behavior modellling of driving patterns and preferences. *Applied Sciences, 13*, 1491.

Yui-Yip Lau is a Head of Student Affairs and Senior Lecturer from the School of Professional Education and Executive Development, The Hong Kong Polytechnic University. Until now, he has published more than 400 research papers in international journals and professional magazines, contributed over 25 book chapters, 8 books, and presented numerous papers at international conferences. He has collaborated with scholars from more than 20 countries and regions spreading over five continents on research projects. He has also secured over HK$ 10 million in research grants. Recently, he has been awarded a Certificate of Appreciation by the Institute of Seatransport in recognition of his outstanding performance on research and the Best Paper Award in leading international conferences. Moreover, he has participated in different consultancy projects with various (inter-)governmental organizations, academic institutions, and industrial associations. His research interests are cruise ships, ferries, impacts of climate change, shipping education and training, transport history, sustainability issues, resilient supply chain management, health logistics, human remains and regional development.

Yang Andrew Wu is a lecturer in the Division of Science, Engineering and Health Studies. He has taught various subjects covering electric vehicles, public policy, business management, smart engineering, and advanced mathematics. His research interests include energy markets, public policy, electric vehicles, and renewable energy sources. Dr. Wu received his Master of Science degree and Doctor of Philosophy degree in Electrical Engineering from the Hong Kong Polytechnic University and the Master of Science degree in Global Finance jointly conferred by the Hong Kong University of Science and Technology and New York University. He is a chartered Financial Risk Manager, a full member of Hong Kong Securities and Investment Institute, Mensa and the Institution of Engineering and Technology.

Electromobility of Urban Public Transport in Poland: Implementation Issues in the Context of Climate Change Challenges

Arkadiusz Kołoś, Łukasz Fiedeń, Krzysztof Gwosdz, Robert Guzik, Jakub Taczanowski, and Adam R. Parol

Abstract As urban areas and the global urban population grow, transport remains a major consumer of environmental resources and a significant contributor to greenhouse gas emissions. Developing efficient, eco-friendly urban transport is thus a key challenge for future cities. Electrification of public transport presents a promising solution, offering high efficiency with low environmental impact. Poland presents an intriguing case in urban electromobility, with its established electric public transport networks (trams, trolleybuses, and urban rail) alongside rapid deployment of modern battery electric buses across many cities. As a notable producer of low-emission vehicles, Poland demonstrates varied levels of electromobility adoption across its cities. Since around 2015, Polish cities have taken diverse approaches: some cities with extensive tram systems, including Wrocław and Łódź, have been cautious about electric buses, while others, such as Warsaw and Kraków, embraced large-scale electric bus programs. Additionally, smaller towns like Miechów and Ząbkowice Śląskie adopted electric buses as they introduced public transport for the first time. Olsztyn

A. Kołoś (✉) · Ł. Fiedeń · K. Gwosdz · R. Guzik · J. Taczanowski · A. R. Parol
Wydział Geografii i Geologii, Instytut Geografii i Gospodarki Przestrzennej, Uniwersytet Jagielloński w Krakowie, Kraków, Poland
e-mail: arkadiusz.kolos@uj.edu.pl

Ł. Fiedeń
e-mail: lukasz.fieden@uj.edu.pl

K. Gwosdz
e-mail: krzysztof.gwosdz@uj.edu.pl

R. Guzik
e-mail: robert.guzik@uj.edu.pl

J. Taczanowski
e-mail: jakub.taczanowski@uj.edu.pl

A. R. Parol
e-mail: ar.parol@doctoral.uj.edu.pl

A. R. Parol
Szkoła Doktorska Nauk Społecznych, Uniwersytet Jagielloński w Krakowie, Kraków, Poland

M. A. Dulebenets et al. (eds.), *Transport and Climate Change*,
https://doi.org/10.1007/978-3-032-00563-2_3

stands out as the only Central European city to reintroduce trams after dismantling its network fifty years ago. In general, urban electromobility adoption correlates with high socioeconomic development and supportive environmental and technological factors, reflecting a commitment to making cities more sustainable and attractive.

Keywords Electromobility · Urban transport · Electric bus · Tram · Trolleybus

1 Introduction

As climate change intensifies, it demands urgent actions to reduce greenhouse gas emissions and mitigate its adverse impacts. One of the most impactful areas for intervention in large urban centres is transportation. Shifting toward sustainable mobility, particularly through the adoption of electric transport or "electromobility", presents a promising path forward. This approach is widely recognized in current literature (Altenburg et al., 2016; Bartłomiejczyk & Kołacz, 2020; May, 2018; Pietrzak & Pietrzak, 2019; Połom & Wiśniewski, 2021; Yigitcanlar, 2022; Zhao, 2018) as a vital step in creating environmentally resilient cities. Furthermore, the electrification of transport serves objectives that extend beyond the reduction of greenhouse gas emissions. Its implementation also seeks to enhance air quality and stimulate economic growth by driving increased demand within the industrial sector.

Electric transport in cities is not a new concept—one might even categorize it into the first- and second-generation electromobility (Kołoś et al., 2023). The first-generation electromobility relies on vehicles powered by catenary or third-rail systems, requiring a continuous connection to the electricity grid. This type of electric transport was developed primarily before the First World War, although we are witnessing its renaissance in the twenty-first century. The second-generation electromobility encompasses vehicles that store energy onboard, requiring only temporary access to an energy source. This category includes battery electric vehicles (both plug-in and mains-charged), hydrogen-cell vehicles, and in some cases, hybrid powertrains, which are seen as transitional technologies until newer solutions are more widely established.

The primary objective of this chapter is to present and examine the development of electromobility in urban public transport (UPT) in Poland as a strategy for mitigating both the causes and effects of climate change. To achieve this objective, the following specific research questions have been formulated:

1. What are the main interdependencies between the adoption of electromobility in urban transport in Poland and the growth of domestic manufacturers of low-emission vehicles?
2. How has the development of the first- and second-generation electromobility in UPT progressed in Poland?
3. What are the distinctive characteristics of electric UPT distribution in Poland?

4. What historical, geographical, and economic factors have influenced the development of electric UPT vehicles in Poland?

The present study provides a comprehensive monographic approach, highlighting the changes in the structure of Poland's public transport fleet, driven primarily by the need to alleviate the transport sector's contribution to climate change. In contrast to other studies, the analysis demonstrates how various factors, such as the presence of manufacturers and historical developments, significantly influence the composition and evolution of the fleet.

This chapter is part of a broader research program conducted by the authors for over a decade (Gwosdz et al., 2011; Kołoś et al., 2023; Taczanowski et al., 2018). The program's core objective is to monitor the development of electromobility in Poland in real time, maintaining continuous iteration with this process, which is treated as a specific case of innovation diffusion. This approach enables the identification of trends, patterns, and turning points at an early stage, while avoiding many pitfalls of retrospective analysis (Rogers, 2003), including post hoc rationalization and challenges related to the availability of data sources, which is often highly fragmented. The fragmentation of data in this field means that on-going efforts to collect and analyze information yield more reliable results than attempts to gather data retrospectively at a later stage.

The chapter consists of 7 sections. Following this introduction, Sect. 2 discusses the research methodology. Then in Sect. 3, we have reviewed the literature to indicate the role of electric public transport in shaping environmentally friendly transport systems and to signal what challenges and potential negative impacts arise from the implementation of electromobility in urban public transport and how these can be mitigated. Section 4 discusses the coexistence of UPT manufacturers, organizers, and operators. Section 5 provides a detailed review of the development of electromobility in Poland since the nineteenth century. Section 6 discusses the spatial variation of electromobility in cities. Section 7 summarises this research and presents important concluding remarks.

2 Methodology

A number of different methods and techniques were used to find answers to the four specific research questions identified. The overall research methodology is summarized in Table 1.

To answer research question 1, an extensive analysis of secondary materials was conducted, which specifically included annual reports from manufacturers, industry reports, and statistics on the production and implementation of electric vehicles. Moreover, governmental documents, policies, and strategies related to public transport and sustainable mobility were reviewed as well. Additionally, the results of public transport tenders were analyzed. The analysis of secondary materials was supplemented with the interviews from industry and UPT experts. They were crucial

Table 1 Methodology of research

Research question	Research methodology and technique	Section
1. What are the main interdependencies between the adoption of electromobility in urban transport in Poland and the growth of domestic manufacturers of low-emission vehicles?	Secondary data analysis (reviewing: industry reports, statistics on the production and implementation of electric vehicles, public transport sector results, company annual reports); expert interviews	4
2. How has the development of the first- and second-generation electromobility in UPT progressed in Poland?	Preparation and analysis of a database and maps on electromobility in UPT containing historical data	5
3. What are the distinctive characteristics of electric UPT distribution in Poland?	Preparation and analysis of a database and maps on electromobility in UPT in cities with over 10,000 inhabitants	6
4. What historical, geographical, and economic factors have influenced the development of electric UPT vehicles in Poland?	Desk research, secondary data analysis	4, 5, 6

Source Own elaboration

for evaluation of electromobility implementation from two different points of view: of the producer and of the consumer. Consequently, the interviews enabled us to evaluate the importance of the co-existence of electric vehicle manufacturers and operators.

To conduct analyses addressing research questions 2 and 3, a database on urban public transport (UPT) electromobility in Poland was created. The following assumptions were made:

- The database included cities and towns with public transport, provided they had a population exceeding 10,000 residents (as of the end of 2023, there were 389 cities and towns of this size in Poland). Cities or towns that did not independently organize UPT but utilized existing systems were not considered separately (46 such locations were noted). Public transport was not organized in 96 cities or towns. Ultimately, 247 urban centres were included in the database.
- The Upper Silesian and Zagłębie Metropolitan Area (GZM), which encompasses 41 municipalities (including 24 cities with populations above 10,000), was treated as a single entity as it collectively organized public transport.
- Public transport organized by rural municipalities or counties was excluded from the database if it did not function as urban transport. Additionally, vehicles owned by cities that served solely other transport needs, such as school transport, were omitted.
- The research also utilized proprietary databases developed during previous studies for the years of 2016, 2017, 2020, and 2022.
- The databases documented the state of UPT vehicles, categorized into first-generation vehicles (urban rail, metro, trams, trolleybuses) and second-generation

vehicles (battery and hydrogen electric buses). Due to the specificity of Polish regulations allowing low-emission vehicles to temporarily substitute zero-emission vehicles, hybrid vehicles and those powered by alternative fuels (e.g., LNG) were also included.

- In the databases for 2016 and 2017, only information on hybrid and electric buses was collected.
- The databases recorded the introduction dates for the first and second-generation electromobility stages.

The main database reflects the situation as of the end of August 2024 and included a total of just over 17,000 vehicles, of which approximately 42% were zero- or low-emission vehicles. It is important to note that a challenge in building the database consisted in access to the information regarding the transport assets held by cities. While some organizers (e.g., ZKM Gdynia, ZTM Warsaw) and carriers (e.g., MPK Kraków, KM Rybnik) published very detailed and up-to-date data, others provided no information or frequently misled with outdated data. The authors have been collecting the information for 10 years (from government or municipal sources related to the EU fund reporting, dedicated industry websites such as www.transport-publiczny.pl and www.transinfo.pl or press releases) regarding the purchase of zero/low-emission vehicles, ensuring the accuracy of the count of such vehicles in the study. However, determining the exact number of diesel buses was more challenging. Occasionally, we relied on a fairly accurate nationwide database of public transport vehicles (www.phototrans.pl). Even this resource was sometimes inadequate for large private consortia operating in multiple cities (e.g., Kłosok Żory, PKS Grodzisk Mazowiecki). In such cases, the issue was not a lack of precise information about the fleet but rather a failure to assign it to specific contracts. Consequently, the final total of diesel vehicles was estimated with an error margin of less than 2%.

In addition, this analysis required the preparation of a set of maps. On the maps, the location of the cities or towns was shown as the centroid of their area. In the case of the GZM, the centroid of the main city of the metropolis, Katowice, was adopted as the representative point. Data was presented in the form of simple or complex cartodiagrams showing the number of vehicles in the cities under analysis. ArcGIS Pro software was used to prepare the maps.

As for question 4, desk research and secondary data analysis were utilized. They enabled us to identify the most important factors of historical, geographical, and economic nature, which have influenced the implementation of the first- and second-generation electromobility in Polish cities and towns characterized in our database.

3 Advantages and Challenges of Electric Public Transport

Electromobility is one of crucial instruments of mitigating climate change, in particular in urban areas. In fact, the role of electric energy, which provides zero-emission powertrains, cannot be overstated in the light of endeavors to place limits on greenhouse gas emissions. Although an electric vehicle itself does not contribute directly to mitigating climate changes, it is certainly an important instrument, which assists in alleviating the transport pressure leading to greenhouse effects. The Paris Agreement signed in December 2015 foresees a substantial reduction in these emissions in order to limit the global temperature increase to 2 °C above pre-industrial levels, striving to lower this value even more, that is to 1.5 °C above pre-industrial levels, recognizing that this would significantly reduce the negative impacts of climate change (Paris Agreement, 2015). Electric engines being energy-efficient, not causing local emissions of CO_2, NO_x, particulate matter (PM), and other harmful substances are a proper answer to these goals (Figenbaum et al., 2015; Longo et al., 2018). In fact, electric vehicles are 3.4 more energy-efficient than internal combustion engine (ICE) vehicles, and CO_2 emissions are even 4.5 lower if electric energy is produced from renewable sources (Holmberg & Erdemir, 2019). The shift towards electric energy is particularly important because passenger transport is one of the main sources of greenhouse gas emissions (Thaller et al., 2021).

Electromobility is crucial for mitigating a broad range of climate change effects. According to Jacob and Winner (2009), climate change will largely increase summertime surface ozone, in particular in urban areas. Consequently, stronger emission controls will be needed in order to meet air quality standards. This is true especially in cities which seem particularly vulnerable to climate change being exposed to numerous risks, including heat-related mortality and illness, water and nutrition shortage, energy limitations, danger of infrastructure damage, and economic loss (Gasper et al., 2011). In such conditions, it goes without saying that the growing need for efficient and reliable mobility in urban areas has increased the importance placed on resilience of municipal transport to climate conditions (Jaroszweski et al., 2014).

Moreover, as noted by Biresselioglu et al. (2018), transport electrification helps reduce the overall energy consumption, thereby strengthening Europe's energy security. This is because transport based on fossil fuels apart from numerous negative environmental impacts, such as air pollution, noise, and resource use, also contributes to growing fossil fuel dependency (Biresselioglu et al., 2018). The switch from fossil-fuel-based transport to electric one makes it possible to limit importing oil from abroad—often from politically unstable regions of the world—providing the opportunity for increased use of renewable energy sources found locally. The situation in the transport sector is more serious because in the EU, which aims to become a leader in zero-emission economy, the transport sector has so far failed to meet the previous emission requirements (Torkayesh & Venghaus, 2024).

Public transport has a crucial place in this shift towards electric powertrains. This is due to its numerous advantages over private cars. Along with lower greenhouse gas (GHG) emissions per passenger-kilometer, they include higher energy efficiency and better road space usage, which result in higher transport capacity, reduction of congestion, and consequently in better livability (Beirão & Cabral, 2007; Gallet et al., 2018; Gershon, 2005; Xylia & Silveira, 2018). It goes without saying that public transport is among crucial instruments in finding a solution for sustainable transport that would meet emission targets (Banister, 2011; Logan et al., 2020).

Urban areas seem to be crucial actors in responding to climate change challenges. As Bulkeley (2013) states, "cities (…) are central to the ways in which the vulnerabilities and risks of climate change are produced, and to the possibilities and challenges of responding to these issues". It is worth looking deeper into why it is in the cities that both interest in climate issues and the implementation of electromobility are concentrated. On the one hand, it is usually in the larger centres that a vanguard group of actors is concentrated, pushing with certainty for changes (Bayat & Biekart, 2009). On the other hand, in highly urbanized areas, phenomena, which are generally regarded as negative from transport and climate point of view, are very highly concentrated. It is essential to keep in mind that whereas more than half world's population lives in cities, they consume three quarters of the world's energy (Mi et al., 2019).

Just as the implementation of electromobility is attempting to reshape urban transport policy (lower emissions, promotion of public transport), cities are the place for mitigation and adaptation to climate change, which are absent or present on a much smaller scale outside them (heat island, air pollution, threats to biodiversity, reduction of green spaces, etc.). Hence, at both local and higher levels, cities are placed in the role of being the actors to deal with the above issues in the first place (Bulkeley, 2013). This is because many geographical spheres intersect with this issue, which are natural determinants of changes in transport. In fact, at present, the local governments have assumed a key role in climate policy (Fuhr et al., 2018). This is true especially in leading European cities that have stated even more ambitious goals than the EU as a whole and its single members.

However, as Kern (2019) emphasizes, the goals of 2015 Paris agreement can be met only if actions aiming at mitigating climate changes are not limited to the most important metropolitan regions but include also most medium-sized and small cities. In order to achieve that, it is indispensable that energy-efficient transport modes are introduced and a shift towards public transport takes place (Gambhir et al., 2019; Wimbadi et al., 2021). As reported by Christiadis et al. (2024), the majority of cities, which are active in implementing the EU Mission on Climate Neutral and Smart Cities, see "urban transport as a main sector and combine the introduction of new technologies with the promotion of public transport and active mobility".

It goes without saying that a success of such a policy cannot be achieved without the support of society. In this context, Mouratidis and Næss (2024) demonstrate that people who are concerned about climate change tend to use conventional cars less frequently and are ready to live without them, preferring denser neighborhoods, areas which are better served by public transport. This reminds us that a shift towards zero-emission economy cannot be achieved without an overall transport and spatial policy of a given city or rather functional urban area. Thaller et al. (2021) showed that so-called disruptive policy is required in order to drastically reduce passenger transport emissions. The importance of public transport organization and approach of local administration on the example of Poland was demonstrated by Taczanowski et al. (2018).

It is essential to emphasize the significant impact of legal regulations pertaining to electromobility. Połom (2015) points out that the EU policy has played a crucial role in this area, as outlined in a series of regulations, such as the "Green Paper on the Urban Environment", the "Plan for Urban Mobility and Transport", and the "White Paper on Transport", as well as in the methods of allocating grants from EU funds. An inspiring example to follow is Norway as one of world's leaders in electromobility implementation. According to the Norwegian National Transport Plan, all new city buses have to be zero-emission vehicles or use biogas by 2025 (The Ministry's Digitization Organization, 2024).

As Thorne et al. (2021) demonstrate, in Oslo, a pressure from the city's policymakers on public transport authorities to reduce GHG emissions has resulted in changes to existing contracts, enabling a fast purchase of electric buses. The authors also suggest other possible instruments, including setting high importance of environment/GHG criteria for winning the tender as well as zero-emission solutions requirements. In fact, the economic side of public transport electrification is critical due to high costs of electric vehicles and infrastructure purchase. That is why recently the idea of eco-rational planning has gained importance. It promotes an approach which links financial efficiency, rationality and efficiency for the transport system, sustainability for people's health and for the environment, and which is acceptable by the stakeholders, being coherent with the development of new transport technologies (Carteni, 2017).

In the Polish context, the pivotal role of the Act on Electromobility and Alternative Fuels (2018) has been crucial, as it established mandatory targets, particularly for large cities. These regulations are subject to dynamic changes. For instance, the initial version of the 2018 Act stipulated that in public transportation systems organized by municipalities with a population exceeding 50,000, the proportion of zero-emission vehicles had to be at least 30% in 2028 (5% in 2021, 10% in 2023, and 20% in 2025), although exceptions were permitted (Mercik, 2020). As noted by Zawieska (2019), the Electromobility Development Plan adopted in Poland in 2017 aimed to achieve a total of 1,000 electric buses in public transportation by the year 2025. It is also noteworthy that initially, the category of zero-emission vehicles included low-emission vehicles (e.g., those powered by natural gas), but excluded first-generation vehicles (such as trams or trolleybuses).

However, we should not take it for granted that the influence of electromobility in urban public transport on climate change challenges is always only positive. In fact, some—at least potentially—negative mechanisms can be observed too. The first is the energy issue. In fact, there is a risk that a large-scale introduction of electric vehicles may cause growing demand for electric energy and functioning of the grid (Biresselioglu et al., 2018; Clairand et al., 2019). This can result in another challenge: how to reconcile increased energy production and distribution with reduced environmental impacts. This is problematic because electric power generation produces large amounts of greenhouse gas emissions. Consequently, electricity production from primary energy sources combined with large electricity transmission to supply electric vehicles can make electromobility not particularly environmentally friendly and even disadvantageous for the environment to some extent (Kabeyi & Olanrewaju, 2022; Skrúcaný et al., 2019).

A proper solution is renewable energy. As stated by Bhatti et al. (2016), a combination of electric vehicles and low carbon power generation is the way to reduce greenhouse gas emissions. Wind and solar technologies seem to be the best solutions here (Nadolny et al., 2022; Nazari et al., 2024). On the other hand, a shift toward electromobility can be a factor enhancing a shift towards green energy production and promotion of large-scale renewable energy integration (Yuan et al., 2021). This is connected with the resilience of the electromobility supply chain as observed by Torres-Rivera et al. (2013). Another environmental concern connected with electromobility implementation is the management of end-of-life battery systems (Hertel et al., 2024). This is crucial because—as predicted in the EU's Industrial Policy on Batteries—production capacity of lithium-ion battery cells could rise from 44 GWh in 2020 to approximately 1,200 by 2030. A question arises also about the sustainability of the production process, where critical raw materials like cobalt, lithium, natural graphite, and rare earth elements are indispensable (Lehtimäki et al., 2024).

Finally, there is a risk that operational costs of electric public transport may become too high, which could consequently contribute to scaling down of public transport and increase car use. This threat was analyzed by Połom (2021) on the example of towns and cities in Poland, which were characterized by low experience in managing and maintaining electric vehicles. A scheme of positive and negative environmental consequences of electric vehicle shift in public transport is shown in Fig. 1.

4 Development of Electric Vehicle Manufacturers in Poland

Regulatory frameworks, particularly those influencing bus manufacturers, combined with technology development and evolving customer preferences—such as increasing mobility and heightened awareness of reducing carbon footprints—are the key factors driving the evolution of manufacturers' strategies. In a context of growing urbanization and mobility, the structural transformation of leading companies in this sector has been rapid and profound. For example, the leading Polish company Solaris Bus & Coach saw the share of low-emission vehicles in its production portfolio rise from 29 to 82% within just five years (2018–2023), while the total number of buses produced grew from 1,250 to 1,411 (Solaris, 2024).

Poland holds a significant position among European countries in the production of both first- and second-generation e-mobility solutions. Since 2021, Poland has been the largest producer and exporter of electric buses in the European Union, with exports valued at over 1.3 billion euros in 2023, thanks in part to companies like Solaris, which sells its products in many European countries and globally. Moreover,

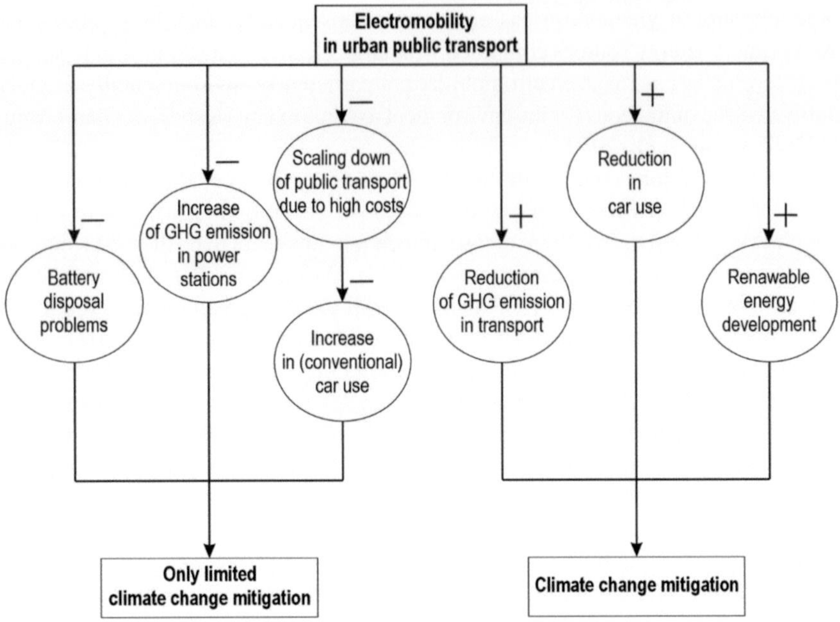

Fig. 1 Positive and negative environmental consequences of electromobility implementation in public urban transport

Poland is also a notable player in the electric train sector with important domestic producers like Newag and Pesa, which accounted in 2023 for approximately 15% of the total EU production of passenger trains. Due to investments from Chinese and Korean manufacturers, Poland is also now the world's largest producer and exporter of electric vehicle batteries after China, with exports exceeding 9 billion euros in 2023. This section of the study mainly discusses the development of major manufacturers and their role in shaping the electromobility landscape in Poland, focusing exclusively on final producers of low-emission vehicles for urban public transportation. A recent broader analysis of the entire supply chain of the e-mobility segment in Poland is covered by Domański et al. (2024), while manufacturers of rail vehicles are discussed in Graff (2021).

Currently, companies producing rail vehicles in Poland (first-generation electromobility) can be categorized into two main types. The first consists of global corporations that, in the 1990s, either acquired state-owned manufacturers (e.g. Alstom in Chorzów, Bombardier in Wrocław) or established new greenfield sites (e.g. Stadler in Siedlce). The second type includes companies with Polish capital that evolved from rolling stock repair facilities, such as PESA in Bydgoszcz, Newag in Nowy Sącz, and Modertrans in Poznań. For both types of companies, the Polish market plays a significant role (see Table 2).

The Konstal plant in Chorzów has become Alstom's centre of expertise for tram and metro car production, with additional significant non-production functions, such

Table 2 Manufacturers of electromobility products in Poland

Producer	Main products	Origin of capital	Employment (2023)	Revenues PLN (2023)	Export share (2023)
Electromobility 1 generation					
Alstom Polska	Trains, metro cars, trams, signalling systems	Foreign (domestic acquired in 1990s by Alstom)	3155	2,871,674	n.d. (but majority of output)
Modertrans Poznań	Trams	Domestic	285	295,259	0%
Newag	Trains, trams	Domestic	1194	1,231,096,000	n.d. (but minority of output)
Pojazdy Szynowe Pesa Bydgoszcz	Trains, trams	Domestic	2058 (whole company)	1,924,178,012	37%
Stadler	Trams, metro trains	Foreign (greenfield)	1421	2,058,091,692	48%
Electromobility 2 generation					
ARP e-vehicles (PILEA brand)	Buses	Domestic	89	24,135,937	0%
MAN Bus	Buses	Foreign (domestic acquired by MAN in 1999)	2792	3,054,350,811	96%
PAK-PCE Polski Autobus Wodorowy (Nesobus brand)	Buses	Domestic	113	53,780.2	0%
Scania Production Słupsk	Buses (until 2024), bus chassis	Foreign (greenfield)	977 (2022)	191,280,000	100%

(continued)

Table 2 (continued)

Producer	Main products	Origin of capital	Employment (2023)	Revenues PLN (2023)	Export share (2023)
Solaris Bus & Coach	Buses, trolleybuses; tramways	Domestic, acquired by Spanish CAF in 2018	2344	3,737,022	71%
Volvo	Buses (until 2023)	Foreign (greenfield)	3,315 (whole company volvo polska)	2,094,639 (bus division only) Whole company: 6,671,308	95%

Source Authors' compilation based on the company data (yearly reports)
Notes n.d.—no data

as a large engineering office in Katowice. Stadler, based in Siedlce, manufactures metro cars, and since acquiring a factory from Solaris in Środa Wielkopolska in 2018, it has also been producing trams. Two notable examples of product expansion and competitive growth are the Polish manufacturers PESA and Newag. The portfolios of these two rolling stock producers include also trams, a particularly important segment for PESA. In addition to these four major manufacturers, Modertrans, a medium-sized tram producer based in Poznań, is also noteworthy. A similar company, Poltrans, operated in Wrocław until 2016, also evolved from a tram repair facility, but it was unable to remain competitive in the market.

In the area of "second-generation electromobility" products, Poland has established itself as a major producer of buses and coaches within the European Union. In 2023, Poland's total bus production, including all types of buses, reached 9,128 units, accounting for 32.5% of the EU's total output. The country's particular strength lies in the production of city buses, especially electric models. By 2023, four major manufacturers—MAN in Starachowice, Solaris in Bolechowo, Scania in Słupsk, and Volvo in Wrocław—had the capacity to produce more than 1,500 vehicles annually. These Polish-based facilities serve as significant production hubs for these companies, with export shares in 2023 ranging from 71% for Solaris to 100% for Scania (see Table 2).

However, the post-pandemic period has introduced challenges. Both Volvo and Scania announced plans to discontinue bus manufacturing in Europe, with Scania retaining only chassis production in Słupsk. In 2024, the Volvo factory in Wrocław was sold to a Swedish heat pump manufacturer. Solaris deserves special mention, as its development over the past three decades has made it one of the most prominent success stories in the European transport industry (Gwosdz et al., 2011; Micek et al., 2021; Taczanowski et al., 2018). Founded by Krzysztof and Solange Olszewski in 1996 and headquartered in Bolechowo near Poznań, Solaris has been a pioneer in battery electromobility. Originally a Polish-owned company, Solaris was acquired by the Spanish company CAF in 2018. Over the years, it has introduced various zero-emission vehicles, including trolleybuses in 2001, hybrid buses in 2006, battery electric buses in 2011, and hydrogen-powered buses in 2014. By the end of 2023, Solaris had delivered or contracted over 3,000 zero-emission battery vehicles, securing its position as a leader in the European electric bus market (BEV + FCEV) with a cumulative market share of 14.5% over the past decade. In 2023, Solaris also ranked first in Europe in the zero-emission bus segment, encompassing both battery and hydrogen buses, with a market share of 15.2% (Solaris, 2024).

The imperative shift toward electric propulsion, coupled with the loss of traditional high-carbon markets, has spurred the emergence of several companies in Poland aspiring to manufacture low-emission vehicles. In 2017, Rafako, the manufacturer of conventional and CFB boilers for power generation, entered the e-mobility segment. Later, the company was acquired by the state-owned fund ARP and began operating under the brand Pilea, completing its first contracts in 2023. In the same year, a new hydrogen-powered bus factory, NesoBus, was established in Świdnik by a consortium of private Polish investors, including a lignite mining and power-generation company, with a production capacity of 100 vehicles. Unfortunately, Autosan from Sanok, a leading bus manufacturer before 1989, was unable to capitalize on the shift towards e-mobility. The company attempted to compete through pricing while simultaneously presenting designs for electric and LNG-powered buses. It also developed a prototype of hydrogen bus. However, struggling with ongoing difficulties, Autosan has to withdraw from the bus market and was eventually acquired by a producer of military equipment.

Another failed venture was Ursus Bus (2015–2019), which was intended to be a strategic lifeline for the struggling agricultural machinery manufacturer. Ursus's brief foray into producing electric buses and trolleybuses ended in 2021. Despite signing contracts for electric vehicles, the company failed to fulfill its orders. However, it did produce trolleybuses for Lublin before its exit from the market. It is also worth noting that Poland ranks among the top European countries in the production of high-power charging infrastructure. A key player, Ekoenergetyka, has supplied fast-charging systems to 28 countries, including major cities such as Berlin, Munich, Paris, and Barcelona. Other important manufacturers in this sector include Enelion, EC Engineering, and Garo, all contributing to the country's strong position in this rapidly growing industry.

All in all, the rapid evolution of the industry, driven by regulatory changes and the growing demand for sustainable transport, has led to profound transformations, the emergence of new players, but also the withdrawal of some established manufacturers from production in Poland. Currently, both in the "first-generation electromobility" and the "second wave", Poland is home to the key companies within the European landscape, including subsidiaries of global corporations as well as firms with local capital. This highlights, on the one hand, the significant upgrading of Poland's public transport manufacturing sector, and on the other, the mutually reinforcing relationship between producers and consumers, which is accelerating the development of electromobility in the country. This self-reinforcing dynamic rests on several key pillars.

Firstly, Poland represents a significant market on a European scale for both first- and second-generation electromobility products. This market is largely driven by public support, primarily through the EU structural funds, which have enabled substantial investment in both infrastructure and vehicle procurement. Secondly, partnerships between local governments and the transport companies they manage establish a framework of mutual, bilateral benefits. Within this framework, businesses gain access to a real-world testing ground for innovative solutions, while clients benefit from products that are flexibly tailored to their operational and contextual needs. A critical element of this dynamics is the ability of manufacturers to respond rapidly to emerging challenges, facilitated by both organizational (including trust) and geographical proximity. This agility not only strengthens relationships between stakeholders but also enhances the competitiveness of Polish manufacturers on the global stage. Lastly, the experience accumulated within the local market, which is both highly diverse and fragmented, provides a robust foundation for international expansion.

5 Electromobility in Polish Urban Transport

5.1 The First-Generation Electromobility Development

The first generation of urban electromobility (trams) developed on the present-day territory of Poland in the second half of the nineteenth century (see Table 3 and Fig. 2). This development was not uniform: the lands then belonging to Germany developed best economically, and those belonging to Russia were less developed. Hence, to this day—despite the investments made in the 1920s in the area of Zagłębie Dąbrowskie and in 1959 in Częstochowa—the largest Polish cities without urban rail transport are located in the eastern part of the country (Lublin, Białystok—over 300,000 inhabitants). A particular case is the one of Olsztyn which is the first and so far the only city in Central-Eastern Europe to have a reintroduced tram system after having closed the old network in the 1960s. There are currently 15 tram networks in Poland, including two extensive metropolitan networks (in the Upper Silesian and Zagłębie Metropolitan Area (GZM) and in Łódź). In total, they cover 29 cities and towns. In addition, there are systems with urban railway parameters in Poland, but they are part of urban transport only in Warsaw. In addition, as the only city in Poland, the capital has metro lines.

From the beginning of the twentieth century, trolleybuses also appeared, although they have never been very developed in Poland (see Fig. 2). Currently, 3 cities have trolleybus networks (Lublin, Gdynia, and Tychy—within the polycentric GZM Metropolis). In 2024, the cities in Poland operated 4,200 first-generation electromobility vehicles. This number has increased slightly in recent years, mainly due to the opening of the second metro line and the expansion of the urban rail network (SKM) in Warsaw (see Table 4).

Table 3 Cities and towns[1] in Poland where the electric trams were in operation between 1893 and 1912 according to the partitions

Cities or town within		Population of the city in 1910 [thousand]			
		>100	50–100	20–50	Total
Germany	Total in 1910	5	8	24	37
	With tram in 1910	5	8	13	26
		100%	100%	54%	70%
	With tram in 2024	5	5	6	16
Austria	Total in 1910	1	1	6	8
	With tram in 1910	1	–	3	4
		100%	–	50%	50%
	With tram in 2024	1	–	–	1
Russia	Total in 1910	2	5	16	23
	With tram in 1910[a]	2	–	4	6
		100%	–	25%	26%
	With tram in 2024[b]	2	2	5	9
Total	Total in 1910	8	14	46	68
	With tram in 1910	8	8	20	36
		100%	57%	43%	53%
	With tram in 2024	8	7	11	26

Source Own elaboration based on Jelonek (1956), Lijewski (1977), Wojcieszak (1996)
[a] Including the towns in the Łódź network, two of them before acquiring city rights
[b] Including the towns of Zagłębie Dąbrowskie (since 1928) and Częstochowa (since 1959)

5.2 *The Second-Generation Electromobility Development*

The first electric buses in Poland were purchased in 2015 by two medium-sized cities, Jaworzno and Ostrołęka. A year earlier, the first bus line operated by battery electric buses was launched in the second largest city in Poland, Krakow. However, these were test vehicles only (Kołoś et al., 2023). In 2016, battery electric buses were used in only 5 cities, Warsaw and Krakow, which are viewed as larger ones, and Ostrołęka, Inowrocław, and Jaworzno, which are viewed as smaller ones (see Tables 5 and 6). In total, there were 23 vehicles which accounted for only 0.2% of all vehicles used in urban public transport in Poland. In 2017, the number of vehicles in question tripled (see Fig. 3). This was mainly due to the purchase of new vehicles in Warsaw, Krakow, and Jaworzno. They were also present in 2 new cities: Wągrowiec and Polkowice (see Fig. 4).

[1] The analysis concerns cities and towns with a population of 20,000 or more in 1910; towns currently divided by political borders are also included: Słubice/Frankfurt, Gubin/Guben, Zgorzelec/Gőrlitz, Cieszyn/Český Těšin, Kostrzyn/Kietz.

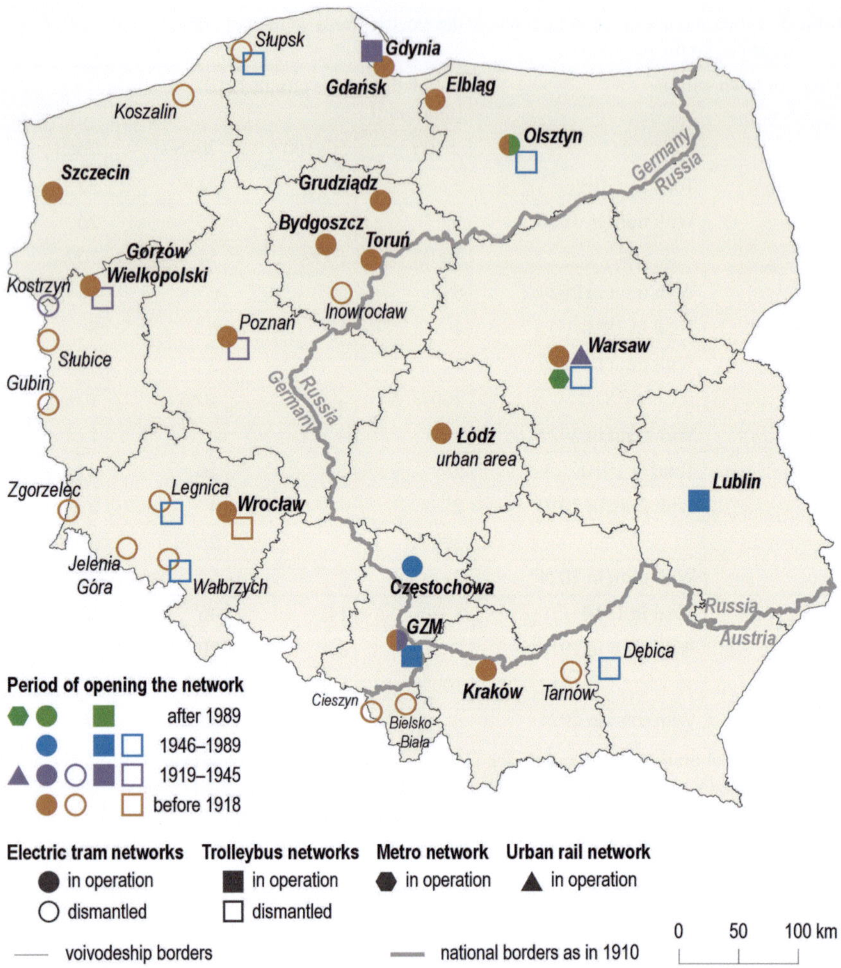

Fig. 2 Introduction and decommissioning of the first-generation electric transport in contemporary Poland. *Source* Own elaboration

Table 4 Number and shares of the first-generation electromobility vehicles in Poland

First-generation electromobility vehicles	2020	2022	2024
Urban rail carriages	160	222	247
Underground carriages	450	456	612
Trams	3099	3096	3097
Total urban rail transport vehicles	3709	3774	3956
Trolleybuses	246	270	244
Total	3955	4044	4200

Source Own elaboration based on own databases

Table 5 Number and shares of low- and zero-emission buses in urban public transport in Poland

Low- and zero-emission buses		2016	2017	2018	2020	2022	2024
Zero-emission buses		23	71	200	426	768	1263
Including:	Hydrogen buses	–	–	–	–	–	50
	Electric buses	23	71	200	426	768	1213
Low-emission buses		396	n.d.	895	1214	1494	1716
Including:	Hybrid buses	96	138	326	497	621	737
	CNG/LNG buses	300	n.d.	569	717	873	979
Other buses (mainly diesel b.)		11,554	11,791	11,405	11,119	10,619	10,097
Total		11,973	12,000	12,500	12,759	12,881	13,076
Share of zero-emission buses in the fleet (%)		0.2	0.6	1.6	3.3	6.0	9.7
Share of low-emission buses in the fleet		3.3%	n.d.	7.2%	9.5%	11.6%	13.1%

Source Own elaboration based on own databases and ZDG Tor (2018)
Notes n.d.—no data

Table 6 Number and share of cities and towns with zero-emission and low-emissions buses

Low and zero-emission buses		Number of cities and towns with				
		2016	2017	2020	2022	2024
Zero-emission buses		5	7	34	60	92
Including:	Hydrogen buses	0	0	0	0	4
	Electric buses	5	7	34	60	92
Low-emission buses		n.d.	n.d.	63	61	68
Including:	Hybrid buses	13	14	40	44	49
	CNG/LNG buses	n.d.	n.d.	28	26	30
Cities/towns with urban public transport		240	240	241	246	247
		Share of cities and towns with				
Zero-emission buses (%)		2.1	2.9	14.1	24.4	37.2
Low-emiision buses		n.d.	n.d.	26.1%	24.8%	27.5%

Source Own elaboration based on own databases and ZDG Tor (2018)
Notes n.d.—no data

Since 2018, a steadily increasing share of zero-emission buses in the fleet has been observed in Poland—from less than 2% in 2018 to almost 10% in 2024 (see Fig. 3). At the same time, a decrease in the growth rate of low-emission bus share can be observed, which means that city authorities are more likely to decide to purchase electric than hybrid or CNG/LNG-powered buses. Initially, electric buses appeared mainly in the largest cities (Warsaw, Krakow) and in single units in smaller cities (the exception here is Jaworzno, which has already been intensively expanding its

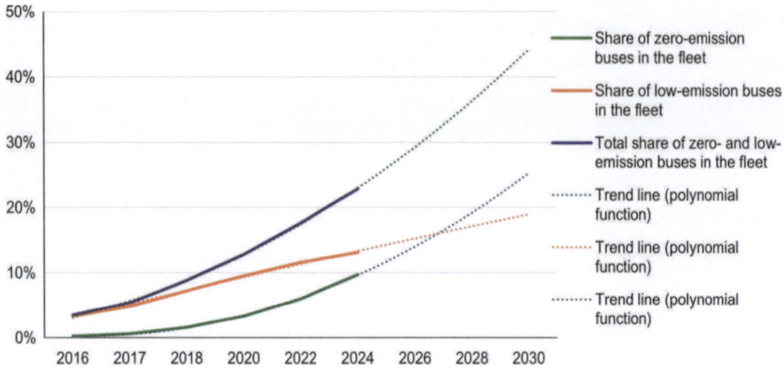

Fig. 3 Shares of low- and zero-emission buses in urban public transport in Poland. *Source* Own elaboration based on own databases and ZDG Tor (2018). *Note* The polynomial function found in the Microsoft 365 Excel spreadsheet was used to prepare the trend line

electric bus fleet since 2016). However, we can see that from 2020 onwards, such vehicles have been increasingly appearing in smaller cities albeit in small numbers (see Fig. 4). In 2022, electric buses were already present in 60 towns and cities, with a total of almost 800 buses, mostly in the largest cities, such as Warsaw, Krakow, Poznań, and GZM, but also in medium-sized cities, such as Jaworzno and Zielona Góra. Electromobility has been least developed in small towns, although it is worth noting that in some of them, such as Miechów and Ząbkowice Śląskie, electric (or hybrid) buses have been introduced when launching public transport for the first time.

Currently, electric buses are present in 92 Polish urban centres, with a total of almost 1,300 vehicles. This represents almost 10% of the vehicles in the fleet of public transport systems. The largest Polish cities have the largest number of such vehicles: Warsaw (161), Krakow (121), GZM (87), and Poznań (58), as well as Zielona Góra (69) and Jaworzno (50). Among the largest Polish cities, Wrocław (13 buses) and Łódź (25 buses) stand out negatively in this respect. As of 2024, Polish cities are using not only battery electric buses, but also hydrogen electric buses (we consider vehicles purchased by the cities, not just rented for testing). So far, there are only 50 of them in 4 cities: Poznań (25, Solaris), Rybnik (20, Nesobus), Gdańsk (4, Nesobus), and Konin (1, Solaris).

6 Spatial Differentiation of Electromobility in Cities and Towns

Only slightly less than half of the 247 centres organizing public transport in Poland have decided to operate any vehicles utilizing zero- or low-emission technology (see Table 7). Noteworthy, however, 95 have opted for electromobility development.

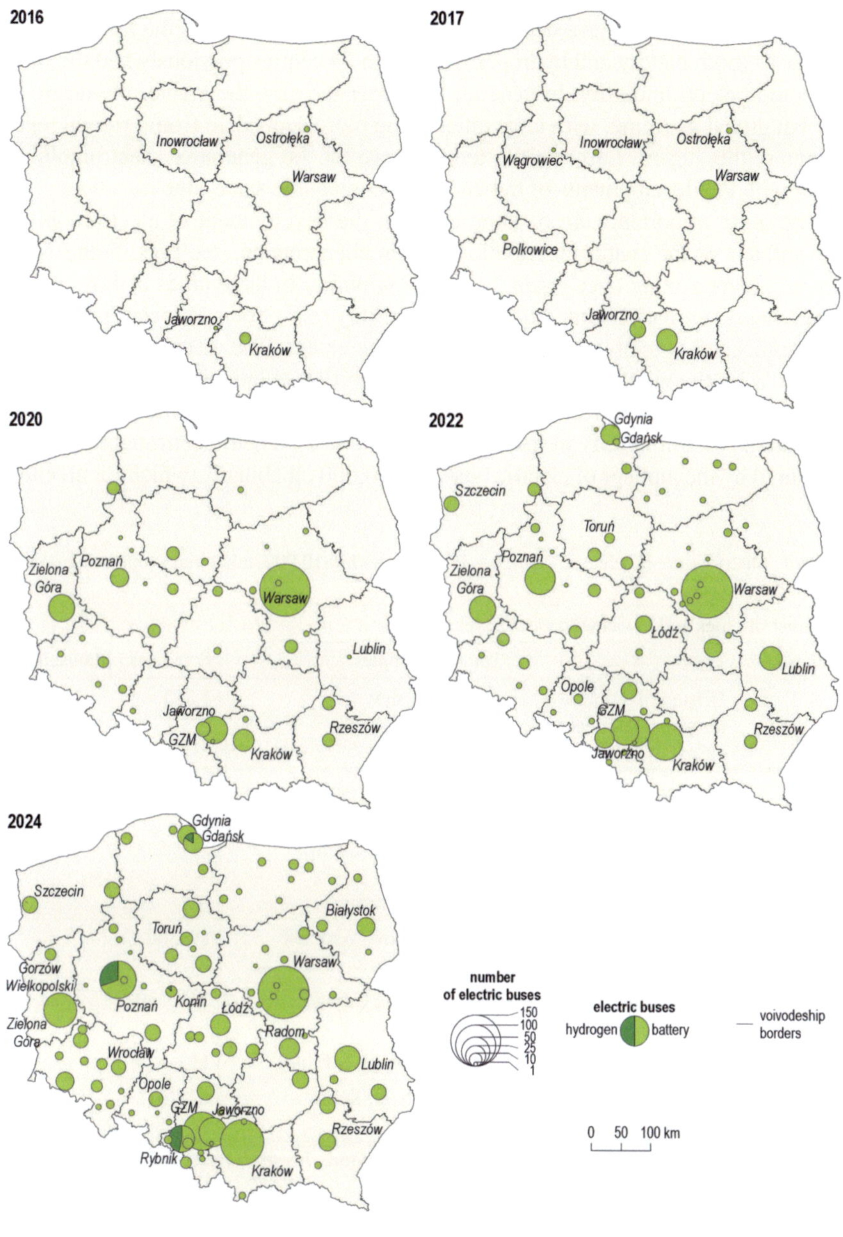

Fig. 4 Development of the second-generation electric urban public transport in Poland. *Source* Own elaboration

Among them, 92 cities have decided to introduce an innovation in the form of electric buses (both battery and hydrogen), of which 14 centres previously had the first-generation electromobility. In general, the larger the city, the greater the desire to combat global warming, with large cities being more dependent from external regulations in this regard. Larger cities tend to have the 1st generation electromobility along with the developments of the 2nd generation at the same time.

The more important role of large cities in the development of electromobility also influences the spatial distribution of this phenomenon (see Figs. 5 and 6). It is worth emphasizing once again that electromobility in large cities is, however, is largely based on the 1st generation electromobility (e.g., Łódź, Wrocław), although its significance is relatively decreasing to some extent. The electromobility index measured by the number of electric vehicles per 10,000 inhabitants ranges from 0 to 9.64, with the highest recorded in Warsaw, and there are seven cities with the 2nd generation electromobility in the top ten (see Table 8). Their electromobility index measured by the number of electric buses per 10,000 inhabitants (which for all cities

Table 7 Distribution of electric vehicles in cities with >10,000 inhabitants operating public transport in 2024

Number of cities and towns with electric vehicles				
Generation	Types	Total	Examples	Population [thousands]
1 + 2	Urban rail + metro + trams + battery electric buses	1	Warsaw	1,862
1 + 2	Trams + trolleybuses + battery electric buses	1	GZM	2,130
1 + 2	Trams + battery electric buses + hydrogen buses	2	Poznań, Gdańsk	487–538
1 + 2	Trams + battery electric buses	8	Kraków, Wrocław	89–806
1 + 2	Trolleybuses + battery electric buses	2	Lublin, Gdynia	241–330
1	Trams	3	Bydgoszcz, Elbląg	113–326
2	Battery electric buses + hydrogen buses	2	Rybnik, Konin	67–131
2	Battery electric buses	75	Zielona Góra, Jaworzno	10–292
Total		95		142 (average)
With zero- or low-emission vehicles		121		122 (average)
Without zero or low-emission vehicles		126		25 (average)
Total cities and towns with UPT		247		74 (average)

Source Own elaboration based on own databases

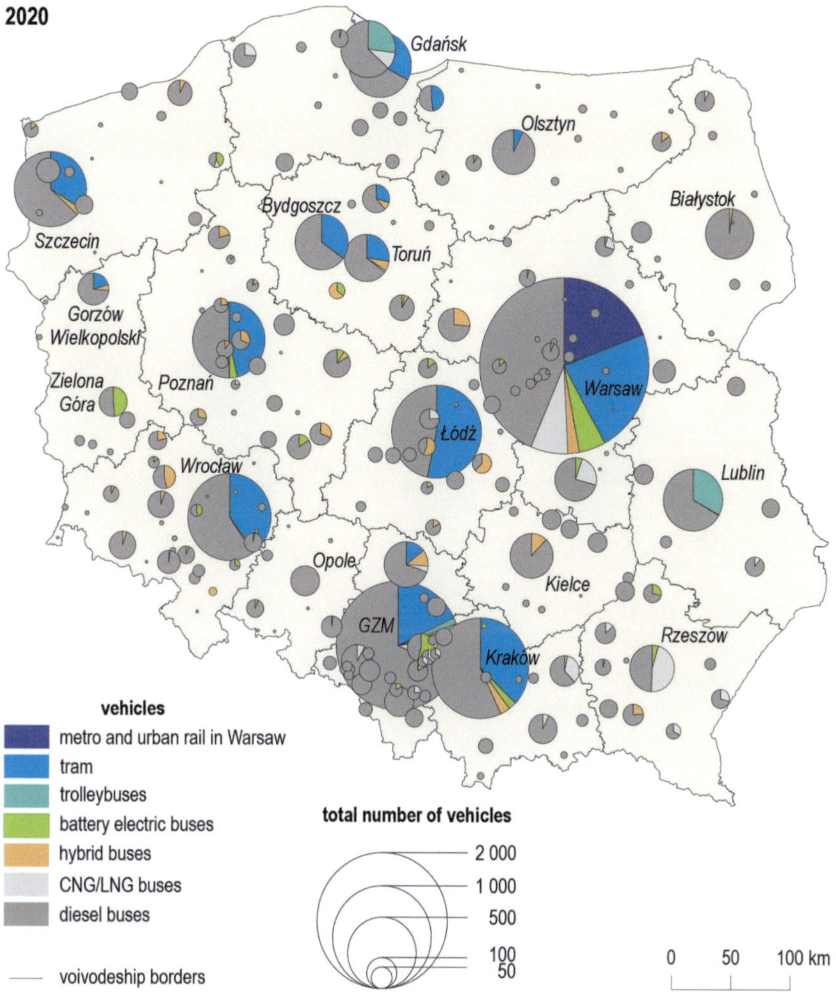

Fig. 5 Volume and structure of electromobility in public urban transport in Polish cities in 2020. *Source* Own elaboration

ranged from 0 to 6.56 for Polkowice) is sometimes very low. As an example, in the case of Wrocław, it is only 0.19.

Battery electromobility, as already indicated in the previous section, has developed most rapidly in medium-sized towns (Jaworzno, Zielona Góra). On the other hand, an interesting case is that some small towns, such as the aforementioned Polkowice with a population of 21.3 thousand or Opoczno with a population of 18.9 thousand, can potentially achieve very high electromobility rates. It can be pointed out that in Poland there were 8 urban centres in 2024 basing all their transport on electromobility. However, these were small towns, often with only a few vehicles at their disposal.

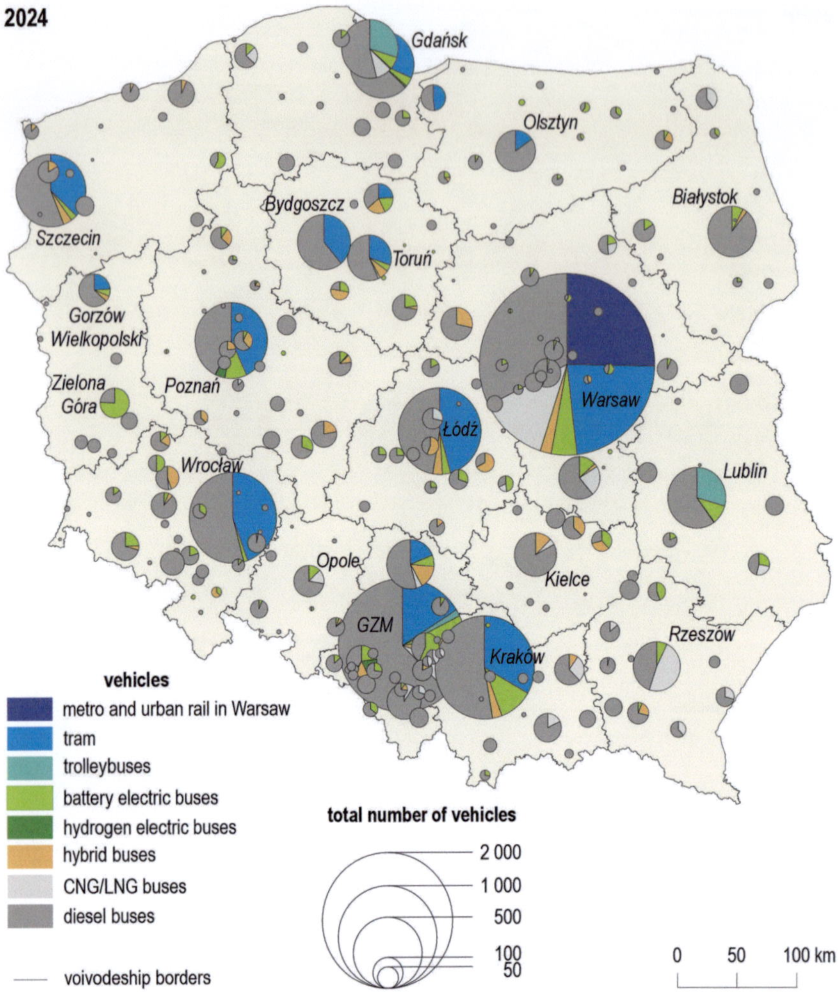

Fig. 6 Volume and structure of electromobility in public urban transport in Polish cities in 2024. *Source* Own elaboration

In terms of spatial distribution, in addition to the large cities mentioned, smaller centres, mainly located in Western Poland, play an important role. This may be related to the location of the most bus fleet manufacturers in this part of the country. Interestingly, the few cities with a significant proportion of bus fleets powered by gas (CNG/LNG) are located in the eastern part of the country (e.g. Rzeszów, Suwałki, Sanok, Przemyśl).

Table 8 Ranking of cities and towns with >10,000 inhabitants and public transport in 2024 in terms of electromobility

Cities and towns	Population	Number of electric vehicles (generation I + II) per 10,000 inhabitants (ranking)	Number of electric buses per 10,000 inhabitants (ranking)
		Indicator in 2024	
Warsaw	1,861,599	**9.64 (1)**	0.86 (59)
Polkowice	21,350	**6.56 (2)**	**6.56 (1)**
Opoczno	18,913	**6.34 (3)**	**6.34 (2)**
Poznań	538,439	**6.22 (4)**	1.54 (28)
Kraków	806,201	**6.00 (5)**	1.50 (31)
Wrocław	673,743	**5.79 (6)**	0.19 (91)
Łódź	652,015	5.78 (7)	0.38 (87)
Jaworzno	86,812	5.76 (8)	**5.76 (3)**
Szczecin	389,066	5.55 (9)	0.41 (83)
Gdynia	241,189	5.31 (10)	1.00 (49)
Zielona Góra	138,932	4.97 (11)	**4.97 (4)**
Lublin	329,565	4.52 (12)	1.18 (42)
Grudziądz	88,658	4.40 (13)	1.92 (17)
Szczecinek	37,557	3.99 (14)	**3.99 (5)**
Toruń	194,771	3.80 (15)	0.51 (77)
Bydgoszcz	326,434	3.71 (16)	0.00 (–)
Gdańsk	487,371	3.45 (17)	0.51 (78)
Rybnik	130,877	3.29 (18)	**3.29 (6)**
Częstochowa	205,969	3.25 (19)	0.92 (53)
Gorzów Wielkopolski	115,247	3.04 (20)	0.69 (67)

Source Own elaboration based on own databases

7 Discussion and Conclusions

The reviewed literature shows that electromobility contributes significantly to shaping environmentally friendly transport systems by reducing greenhouse gas emissions and promoting climate resilience in urban areas. Consequently, electric transport enables cities a transition to a low-carbon economy while improving air quality and public health. This is especially true in the case of electric public transport, which is distinguished by high energy efficiency and transport capacity accompanied by low resource consumption and which can serve as a particularly appropriate tool for creating sustainable and liveable cities. Therefore, electric public transport is a proper answer not only in terms of mitigating diverse effects of climate change. What is more, as public transport electrification requires an increase in electricity production and simultaneously enables to limit the role of mobility based on fossil fuel, it favours a growth of renewable energy provision while reducing dependence on energy imports and consequently promotes energy security.

The rapid development of electromobility in Poland has been driven by a combination of national policies, EU subsidies, and targeted local initiatives. Expanding such funding mechanisms, particularly for smaller municipalities, is critical to ensure equitable access to clean transport and to bridge the gap between economically developed and less developed regions. The implementation of electromobility faces numerous challenges, such as infrastructure requirements, environmental concerns associated with battery production and disposal, as well as with a potential increase in electricity production from fossil fuels, and finally high investment and operational costs. These high costs may result in a scaling-down of public transport, which would result in an increase in private vehicle use, including traditional diesel cars. In this case public transport electrification would be revealed as rather a superficial if not a counter-productive undertaking.

These challenges, however, can be mitigated through policies promoting renewable energy, sustainable resource management, and public–private partnerships. It goes without saying that a holistic view of transport, climate, and energy policies is needed here in order to make the best use of the potential offered by electric public transport development while minimizing its negative impacts. Unfortunately, it is difficult to relate this research about the diffusion of electromobility in public transport in Poland to other countries due to the lack of similar studies. In fact, the existing analyses of electromobility implementation, which apply to the national scale, mostly consider private electric cars (Cansino et al., 2018; Horváth et al., 2023; Łuszczyk et al., 2021; Tucki et al., 2019).

Research on electromobility implementation in public transport is much rarer. What is more, the existing relevant studies are focused mainly on technological issues applying to one country (Pencheva et al., 2024; Thorne et al., 2021) or—even more frequently—to local conditions in selected cities or on chosen lines (Bartłomiejczyk & Połom, 2019; Lin et al., 2019; May, 2018; Poprik, 2022). Future studies should explore the long-term impacts of electromobility on smaller municipalities, focusing on cost–benefit analyses and environmental outcomes. Additionally, comparative research on electromobility strategies across Central and Eastern Europe could provide valuable insights into region-specific barriers and opportunities.

This study addressed four key research questions concerning the development of electromobility in urban public transport (UPT) in Poland and its role in combating climate change. Regarding the coexistence of vehicle manufacturers and operators, the analysis shows that this synergy drives the diffusion of electromobility in Poland. This is an important factor of both the first- and second-generation electromobility development, as Poland is one of the main electric bus, trolleybus, and tram producers in Europe. Collaboration between manufacturers (representing both Polish and foreign capital) and operators has resulted in innovations tailored to urban needs, strengthening both the public transport sector and the domestic manufacturing industry. Thanks to well-functioning partnerships between municipal governments and urban transport companies, a framework of mutual and bilateral benefits could be established.

The research highlights distinct trajectories for the first- and second-generation electromobility. Poland is generally characterised by a well-developed system of 15 tram networks serving most of its large and some medium-sized cities. Nevertheless, their spatial distribution is strongly dependent on historical factors, in particular on political geography at the turn of the nineteenth and twentieth centuries. It is worth noting that—similarly to other Central-European countries but in contrast to most of Western and Southern Europe—tram networks have not been dismantled here on a large scale. What is more, one Polish city, Olsztyn, has seen a reintroduction of this means of electric urban transport, following many Western European examples but so far being the only case of tram revival in this part of the continent. By contrast, trolleybuses have only a limited role as there are only three networks in the entire country. Besides, in Warsaw, metro and urban rail are also in operation. At present, the Polish first-generation electromobility fleet sums up to 4,200 units with almost 3,100 trams and almost 250 trolleybuses (and the remaining vehicles being metro and urban rail carriages).

Electric and hybrid buses were first introduced in Poland for the first time in 2015. Quite surprisingly, among the first centres to purchase them apart from the capital Warsaw and the second largest city, Krakow, we could also find some medium-sized cities, some of which (like Jaworzno and Inowrocław) have decided to largely invest in electric buses treating electric public transport as part of the local city development strategies. In the next years, a quick diffusion of electric and hybrid buses has been observed all over the country with steadily increasing share of zero-emission buses in the fleet—from less than 2% in 2018 to almost 10% in 2024. Interestingly, from 2020 onwards, electric and hybrid buses have been increasingly appearing in smaller towns, in some cases (e.g., Miechów or Ząbkowice Śląskie) being introduced when launching public transport for the first time. It is worth noting, however, that the very rapid development of the second-generation electromobility is nevertheless accompanied by the stagnation of older solutions that could be more efficient after all (Bartłomiejczyk & Połom, 2019; Poprik, 2022; Tuszyński, 2015).

For the time being, in Poland there are almost 1,300 electric and 50 hydrogen vehicles, accompanied by over 700 hybrid and almost 1,000 CNG/LNG ones. Electric and hydrogen buses sum up to almost 10% of the fleet of Polish urban public transport systems. These vehicles are present in 92 Polish cities and towns, representing less than half of all centres organizing public transport in Poland. The largest Polish cities also have the largest number of electric buses with Warsaw (161) and Krakow (121) at the top of the list. Characteristically, some medium-sized centres like Zielona Góra and Jaworzno are also distinguished by relatively high numbers of such vehicles. On the other hand, we can find two large cities, Wrocław and Łódź, which have only started to implement this kind of vehicles to their fleets. An interesting but small niche is represented by 8 small towns which base their entire transport fleet on electromobility. This is a particularly interesting case because the introduction of electric buses in this kind of centres represents numerous challenges like budget limitations and stronger needs for the sense of community (Wang & González, 2013).

However, when taking into consideration also the first-generation electromobility vehicles (mainly trams), the general image of electromobility in Poland looks different. The importance of this kind of electric vehicles is particularly high in large cities (e.g. Wrocław and Łódź). Nevertheless, it can be stated that in general large cities tend to maintain the 1st generation electromobility assets and to develop the 2nd generation assets at the same time. Consequently, it seems that the larger the city, the greater the desire to combat global warming. The spatial distribution of cities and towns having electric transport is largely influenced by the distribution of large cities; however, a distinctive feature is also the presence of some small towns mainly located in Western Poland. Here, a relation to the location of the most important bus manufacturers seems to be—at least partially—the reason.

The development of electromobility in Poland reflects a complex urban landscape, with adoption rates varying across cities and towns due to the differences in their historical development connected with the location in Western or Central-Eastern part of the country, their economic capacity, policy support, and local technological resources. The importance of local authorities and their willingness to implement technological innovations cannot be overrated, as demonstrated by the examples of Jaworzno and Inowrocław. In fact, smaller towns, while in theory slower to adopt, can show great potential when supported by targeted environmental policies and incentives.

In order to fully realize the potential of electromobility, Poland should prioritize integrated urban planning that combines the first- and second-generation solutions. Specific measures include incentivizing the adoption of electric vehicles in smaller towns, expanding charging infrastructure, and implementing comprehensive training programs for municipal authorities to manage and maintain electric transport systems effectively. The rapid expansion of electromobility in Poland raises important questions about its long-term trajectory. Ensuring the sustainability of these efforts will require continuous monitoring and adaptation of policies to address emerging challenges, such as shifts in energy demand, evolving battery technologies, and the integration of renewable energy sources.

In conclusion, electromobility presents a viable solution to urban sustainability and climate change mitigation goals. We argue that electromobility should be treated in broad sense, including the one of the first generation as well. A co-existence of the two is strongly needed. This is because trams and trolleybuses offer established, stable solutions, while battery electric and hydrogen buses provide flexible options for diverse urban contexts. This flexibility also includes a possibility to link the features of the first and second generation of electromobility, as demonstrated by trolleybuses equipped with batteries to serve line sections without overhead cables (Bartłomiejczyk & Połom, 2019). These findings underscore the importance of targeted policies and infrastructure investments to optimize the unique strengths of both generations. However, its success depends on integrated strategies combining technological advancements, supportive policies, and continuous monitoring of environmental impacts. Future studies should focus on long-term assessments of environmental and economic benefits from electromobility, particularly for smaller municipalities, and explore innovative solutions to overcome identified challenges. This chapter demonstrates that electromobility is more than a technological advancement; it is a transformative force linking climate action with industrial development, driving progress toward sustainable and resilient urban futures.

Acknowledgements The publication has been supported by a grant from the Faculty of Geography and Geology under the Strategic Programme Excellence Initiative at Jagiellonian University.

References

Altenburg, T., Schamp, E. W., & Chaudhary, A. (2016). The emergence of electromobility: Comparing technological pathways in France, Germany, China, and India. *Science and Public Policy, 43*(4), 464–475. https://doi.org/10.1093/scipol/scv054

Banister, D. (2011). Cities, mobility, and climate change. *Journal of Transport Geography, 19*(6), 1538–1546. https://doi.org/10.1016/j.jtrangeo.2011.03.009

Bartłomiejczyk, M., & Kołacz, R. (2020). The reduction of auxiliaries power demand: The challenge for electromobility in public transportation. *Journal of Cleaner Production, 252,* Article 119776. https://doi.org/10.1016/j.jclepro.2019.119776

Bartłomiejczyk, M., & Połom, M. (2019). The road to the development of electromobility in the Czech's Prague: From electric buses to... trolleybuses? *AUTOBUSY – Technika Eksploatacja Systemy Transportowe, 24,* 22–28. https://doi.org/10.24136/atest.2019.119

Bayat, A., & Biekart, K. (2009). Cities of extremes. *Development and Change, 40*(5), 815–825. https://doi.org/10.1111/j.1467-7660.2009.01584.x

Beirão, G., & Cabral, J. A. S. (2007). Understanding attitudes towards public transport and private car: A qualitative study. *Transport Policy, 14*, 478–489. https://doi.org/10.1016/j.tranpol.2007.04.009

Bhatti, A. R., Salam, Z., Aziz, M. J. B. A., & Yee, K. P. (2016). A critical review of electric vehicle charging using solar photovoltaic. *International Journal of Energy Research, 40*, 439–461. https://doi.org/10.1002/er.3472

Biresselioglu, M. E., Kaplan, M. B., & Yilmaz, B. K. (2018). Electric mobility in Europe: A comprehensive review of motivators and barriers in decision-making processes. *Transportation Research Part A, 109*, 1–13. https://doi.org/10.1016/j.tra.2018.01.017

Bulkeley, H. (2013). *Cities and climate change.* Routledge. https://doi.org/10.4324/9780203077207

Cansino, J. M., Sánchez-Braza, A., & Sanz-Díaz, T. (2018). Policy instruments to promote electromobility in the EU28: A comprehensive review. *Sustainability, 10,* 2507. https://doi.org/10.3390/su10072507

Carteni, A. (2017). A plug-in hybrid electric bus fleet as a rational and sustainable urban transport policy: A real casa application in Italy. *International Journal of Transportation Systems, 2,* 46–53.

Christiadis, P., Ulpiani, G., Stepniak, M., & Vetters, N. (2024). Research and innovation paving the way for climate neutrality in urban transport: Analysis of 362 cities on their journey to zero emissions. *Transport Policy, 148*, 107–123. https://doi.org/10.1016/j.tranpol.2024.01.008

Clairand, J.-M., Guerra-Terán, P., Serrano-Guerrero, X., González-Rodríguez, M., & Escrivá-Escrivá, G. (2019). Electric vehicles for public transportation in power systems: A review of methodologies. *Energies, 12,* Article 3114. https://doi.org/10.3390/en12163114

Domański, B., Micek, G., Guzik, R., Gwosdz, K., & Kocaj, A. (2024). *The evolution of European manufacturing industries: The dynamics of core-periphery relationships.* Taylor & Francis. https://doi.org/10.4324/9781003430117

Figenbaum, E., Assum, T., & Kolbenstvedt, M. (2015). Electromobility in Norway: Experiences and opportunities. *Research in Transportation Economics, 50*, 29–38. https://doi.org/10.1016/j.retrec.2015.06.004

Fuhr, H., Hickmann, T., & Kern, K. (2018). The role of cities in multi-level climate governance: Local climate policies and the 1.5 °C target. *Sustainability, 30*, 1–6. https://doi.org/10.1016/j.cosust.2017.10.006

Gallet, M., Massier, T., & Hamacher, T. (2018). Estimation of the energy demand of electric buses based on real-world data for large-scale public transport networks. *Applied Energy, 230*, 344–356. https://doi.org/10.1016/j.apenergy.2018.08.086

Gambhir, A., Rogelj, J., Luderer, G., Few, S., & Napp, T. (2019). Energy system changes in 1.5 °C, well below 2 °C and 2 °C scenarios. *Energy Strategy Reviews, 23*, 69–80. https://doi.org/10.1016/j.esr.2018.12.006

Gasper, R., Blohm, A., & Ruth, M. (2011). Social and economic impacts of climate change on the urban environment. *Current Opinion in Environmental Sustainability, 3*(3), 150–157. https://doi.org/10.1016/j.cosust.2010.12.009

Gershon, R. R. M. (2005). Public transportation: Advantages and challenges. *Journal of Urban Health: Bulletin of the New York Academy of Medicine, 82*(1), 7–9. https://doi.org/10.1093/jurban/jti003

Graff, M. (2021). Producenci taboru kolejowego oraz nowe pojazdy szynowe w Polsce. *Problemy Kolejnictwa, Zeszyt, 192*. https://doi.org/10.36137/1922P

Gwosdz, K., Guzik, R., & Domański, B. (2011). Environmental pressure in fragmented markets: The rise and fall of bus-makers in Poland. *European Review of Industrial Economics and Policy, 3*. http://revel.unice.fr/eriep/index.html?id=3317

Hertel, D., Bräunig, G., & Thürer, M. (2024). Towards a green electromobility transition: A systematic review of electric vehicle battery systems disassembly. *Journal of Manufacturing Systems, 74*, 387–396. https://doi.org/10.1016/j.jmsy.2024.03.013

Holmberg, K., & Erdemir, A. (2019). The impact of tribology on energy use and CO_2 emission globally and in combustion engine and electric cars. *Tribology International, 135*, 389–396. https://doi.org/10.1016/j.triboint.2019.03.024

Horváth, G., Bai, A., Szegedi, S., Lázár, I., Máthé, C., Huzsvai, L., Zakar, M., Gabnai, Z., & Tóth, T. (2023). A comprehensive review of the distinctive tendencies of the diffusion of E-mobility in Central Europe. *Energies, 2023*(16), 5421. https://doi.org/10.3390/en16145421

Jacob, D. J., & Winner, D. A. (2009). Effect of climate change on air quality. *Atmospheric Environment, 43*, 51–63. https://doi.org/10.1016/j.atmosenv.2008.09.051

Jaroszweski, D., Hooper, E., & Chapman, L. (2014). The impact of climate change on urban transport resilience in a changing world. *Progress in Physical Geography: Earth and Environment, 38*(4), 448–463. https://doi.org/10.1177/0309133314538741

Jelonek, A. (1956). *Liczba ludności miast i osiedli w Polsce w latach 1810–1955*. Dokumentacja Geograficzna, 5, IG PAN.

Kabeyi, M. J. B., & Olanrewaju, O. A. (2022). The role of electrification of transport in the energy transition. In *Proceedings of the 5th European International Conference on Industrial Engineering and Operations Management*, Rome, Italy, July 26–28, 2022 (pp. 1–15). https://doi.org/10.46254/eu05.20220426

Kern, K. (2019). Cities as leaders in EU multilevel climate governance: Embedded upscaling of local experiments in Europe. *Environmental Politics, 28*(1), 125–145. https://doi.org/10.1080/09644016.2019.1521979

Kołoś, A., Fiedeń, Ł, Taczanowski, J., Parol, A. R., Gwosdz, K., Guzik, R., & Łodziński, J. (2023). Evolution of second-generation electromobility in public transport in Polish cities. *Prace Komisji Geografii Komunikacji PTG, 26*(1), 22–39. https://doi.org/10.4467/2543859xpkg.23.002.17399

Lehtimäki, H., Karhu, M., Kotilainen, J. M., Sairinen, R., Jokilaakso, A., Lassi, U., & Huttunen-Saarivirta, E. (2024). Sustainability of the use of critical raw materials in electric vehicle batteries: A transdisciplinary review. *Environmental Challenges, 16*, Article 100966. https://doi.org/10.1016/j.envc.2024.100966

Lijewski, T. (1977). *Geografia transportu Polski*. PWE.
Lin, Y., Zhang, K., Max Shen, Z.-J., & Miao, L. (2019). Charging network planning for electric bus cities: A case study of Shenzhen, China. *Sustainability, 11,* 4713. https://doi.org/10.3390/su11174713
Logan, K. G., Nelson, J. D., & Hastings, A. (2020). Electric and hydrogen buses: Shifting from conventionally fuelled cars in the UK. *Transportation Research Part D, 85,* Article 102350. https://doi.org/10.1016/j.trd.2020.102350
Longo, M., Foiadelli, F., & Yaïci, W. (2018). Electric vehicles integrated with renewable energy sources for sustainable mobility. In L. R. Martínez & M. D. Prieto (Eds.), *New trends in electrical vehicle powertrains* (pp. 203–223). Intech Open Limited. https://doi.org/10.5772/intechopen.76788
Łuszczyk, M., Sulich, A., Siuta-Tokarska, B., Zema, T., & Thier, A. (2021). The development of electromobility in the European Union: Evidence from Poland and cross-country comparisons. *Energies, 2021*(14), 8247. https://doi.org/10.3390/en14248247
May, N. (2018). Local environmental impact assessment as decision support for the introduction of electromobility in urban public transport systems. *Transportation Research Part D: Transport and Environment, 64*, 192–203. https://doi.org/10.1016/j.trd.2017.07.010
Mercik, A. (2020). Elektromobilność w autobusowym transporcie publicznym organizowanym przez Górnośląsko-Zagłębiowską Metropolię jako narzędzie realizacji idei zrównoważonej mobilności. *Prace Komisji Geografii Komunikacji PTG, 23*(5), 18–33. https://doi.org/10.4467/2543859xpkg.20.029.13246
Mi, Z., Guan, D., Liu, Z., Liu, J., Viguié, V., Fromer, N., & Wang, Y. (2019). Cities: The core of climate change mitigation. *Journal of Cleaner Production, 207*, 582–589. https://doi.org/10.1016/j.jclepro.2018.10.034
Micek, G., Guzik, R., Gwosdz, K., & Domański, B. (2021). Newcomers from the periphery: The international expansion of Polish automotive companies. *Energies, 14*(9), Article 2617. https://doi.org/10.3390/en14092617
Mouratidis, K., & Næss, P. (2024). Climate change concern as a driver of sustainable mobility and reduced car use. *Transportation Research Part D: Transport and Environment, 134,* Article 104345. https://doi.org/10.1016/j.trd.2024.104345
Nadolny, A., Cheng, C., Lu, B., Blakers, A., & Stocks, M. (2022). Fully electrified land transport in 100% renewable electricity networks dominated by variable generation. *Renewable Energy, 182*, 562–577. https://doi.org/10.1016/j.renene.2021.10.039
Nazari, M. A., Blazek, V., Prokop, L., Misak, S., & Prabaharan, N. (2024). Electric vehicle charging by use of renewable energy technologies: A comprehensive and updated review. *Computers and Electrical Engineering, 118,* Article 109401. https://doi.org/10.1016/j.compeleceng.2024.109401
Paris Agreement. (2015). *United Nations*. http://unfccc.int/files/essential_background/convention/application/pdf/english_paris_agreement.pdf
Pencheva, P., Asenov, A., Georgiev, I., Georgiev, A., & Stoyanov, P. (2024). Systems for planning and operation of electric buses for public transport of passengers in cities. *Studies in Systems, Decision and Control, 563*, 309–374. https://doi.org/10.1007/978-3-031-69487-5_6
Pietrzak, K., & Pietrzak, O. (2019). Environmental effects of electromobility in a sustainable urban public transport. *Sustainability, 12*(3), Article 1052. https://doi.org/10.3390/su12031052
Połom, M. (2015). European Union funds as a growth stimulant of electromobility on the example of electric public transport in Poland. *Regional Barometer, 13*(3), 89–96. https://doi.org/10.56583/br.739
Połom, M. (2021). E-revolution in a post-communist country? A critical review of electric public transport development in Poland. *Energy Research & Social Science, 80,* Article 102227. https://doi.org/10.1016/j.erss.2021.102227
Połom, M., & Wiśniewski, P. (2021). Implementing electromobility in public transport in Poland in 1990–2020: A review of experiences and evaluation of the current development directions. *Sustainability, 13*(7), Article 4009. https://doi.org/10.3390/su13074009

Poprik, A. (2022). *Prague bus fleet electrification*. Prague trolleybuses (presentation). DPD hl.m.Praha a.s. Retrieved January 24, 2025 from https://jaspers.eib.org/activitiesNP/2022/Webinars%20on%20bus%20fleet%20decarbonisation/Second%20day/06_DPP%20Prague.pdf

Rogers, E. M. (2003). *Diffusion of innovations* (5th ed., 551 pp.). Free Press.

Skrúcaný, T., Kendra, M., Stopka, O., Milojević, S., Figlus, T., & Csiszár, C. (2019). Impact of the electric mobility implementation on the greenhouse gases production in Central European countries. *Sustainability, 11*, 4948. https://doi.org/10.3390/su11184948

Solaris. (2024). *Sprawozdanie z działalności spółki Solaris Bus & Coach Sp z o.o. w 2023 roku*. Retrieved November 15, 2024 from https://ekrs.ms.gov.pl/rdf/pd/search_df

Taczanowski, J., Kołoś, A., Gwosdz, K., Domański, B., & Guzik, R. (2018). The development of low-emission public urban transport in Poland. *Bulletin of Geography. Socio-economic Series, 41*, 79–92. https://doi.org/10.2478/bog-2018-0027

Thaller, A., Posch, A., Dugan, A., & Steininger, K. (2021). How to design policy packages for sustainable transport: Balancing disruptiveness and implementability. *Transportation Research Part D, 91*, Article 102714. https://doi.org/10.1016/j.trd.2021.102714

The Ministry's Digitization Organization. (2024). *National transport plan 2022–2033*. Retrieved January 15, 2025 from https://www.regjeringen.no/

Thorne, R. J., Hovi, I. B., Figenbaum, E., Pinchasik, D. R., Amundsen, A. H., & Hagman, R. (2021). Facilitating adoption of electric buses through policy: Learnings from a trial in Norway. *Energy Policy, 155,* 112310. https://doi.org/10.1016/j.enpol.2021.112310

Torkayesh, A. E., & Venghaus, S. (2024). Decoding the transport policy maze towards climate neutrality: Cross-sectoral policy landscapes. *Renewable and Sustainable Energy Reviews, 206*, Article 114865. https://doi.org/10.1016/j.rser.2024.114865

Torres-Rivera, A. D., McNamara Valdes, A. J., & Florencio Da Silva, R. (2013). The resilience of the renewable energy electromobility supply chain: Review and trends. *Sustainability, 15,* 10838. https://doi.org/10.3390/su151410838

Tucki, K., Orynycz, O., Świć, A., & Mitoraj-Wojtanek, M. (2019). The development of electromobility in Poland and EU states as a tool for management of CO_2 emissions. *Energies, 12,* 2942. https://doi.org/10.3390/en12152942

Tuszyński, M. (2015). Rozwój proekologicznego transportu zbiorowego w Gdyni. *Transport Miejski i Regionalny, 8*(2015), 28–34.

Wang, X., & González, J. A. (2013). Assessing feasibility of electric buses in small and medium-sized communities. *International Journal of Sustainable Transportation, 6,* 431–448, https://doi.org/10.1080/15568318.2012.667864

Wimbadi, R. W., Djalante, R., & Mori, A. (2021). Urban experiments with public transport for low-carbon mobility transitions in cities: A systematic literature review (1990–2020). *Sustainable Cities and Society, 72*, Article 103023. https://doi.org/10.1016/j.scs.2021.103023

Wojcieszak, J. (1996). Dzieje komunikacji tramwajowej na świecie. *Technika Transportu Szynowego, 1*(96), 6–9.

Xylia, M., & Silveira, S. (2018). The role of charging technologies in upscaling the use of electric buses in public transport: Experiences from demonstration projects. *Transportation Research Part A: Policy and Practice, 118*, 399–415. https://doi.org/10.1016/j.tra.2018.09.011

Yigitcanlar, T. (2022). Towards smart and sustainable urban electromobility: An editorial commentary. *Sustainability, 14*, 2264. https://doi.org/10.3390/su14042264

Yuan, M., Thellufsen, J. Z., Lund, H., & Liang, Y. (2021). The electrification of transportation in energy transition. *Energy, 236*, Article 121564. https://doi.org/10.1016/j.energy.2021.121564

Zawieska, J. (2019). Rozwój rynku elektromobilności w Polsce. In J. Gajewski, W. Paprocki, & J. Pieriegud (Eds.), *Elektromobilność w Polsce na tle tendencji europejskich i globalnych* (pp. 9–38). CeDeWu. Retrieved November 15, 2024 from https://www.efcongress.com/wp-content/uploads/2020/02/publikacje09__Elektromobilno%C5%9B%C4%87-w-Polsce-na-tle-tendencji-europejskich-i-globalnych.pdf

ZDG Tor. (2018). *Polska na drodze do elektromobilności*. Retrieved November 15, 2024 from https://www.cire.pl/pliki/1/2018/zdgtor_polska_na_drodze_do_elektromobilnosci.pdf

Zhao, Q. (2018). Electromobility research in Germany and China: Structural differences. *Scientometrics, 117*, 473–493. https://doi.org/10.1007/s11192-018-2873-9

Arkadiusz Kołoś is an Assistant Professor at the Institute of Geography and Spatial Management of the Jagiellonian University in Krakow. His main areas of interest are transport geography, especially in cities, as well as spatial management and legal issues in spatial planning. He is the author of 70 scientific papers and several dozen conference presentations. In addition, he is the chairman of the Commission for the Geography of Transport of the Polish Geographical Society and the chairman of the team for large-scale infrastructure projects at the National Spatial Planning Committee of the Polish Academy of Sciences.

Łukasz Fiedeń is a Transport Geographer and Assistant Professor at the Institute of Geography and Spatial Management, Jagiellonian University. His research focuses on the spatial differentiation of mobility and the social consequences of poor public transport accessibility in rural areas remote from cities. He is also interested in the impact of large road investments on land-use changes and quality of life at both local and regional scales. Additionally, his academic interests include the implementation of electromobility in urban transport systems, the application of geographic information systems (GIS) in human geography, public perceptions of natural disasters in local communities, political geography, and the accessibility of modern business service centers. He is the author or co-author of over 60 scientific publications and serves as a reviewer for international journals, including Journal of Transport Geography, Sustainable Cities and Society, and Bulletin of Geography. He has also contributed as an author and co-author of scientific reports commissioned by institutions such as the Polish Ministry of Investment and Development, the Polish Ministry of Infrastructure, the Marshal's Office of the Pomeranian Voivodeship, the Association of Business Service Leaders (ABSL), the City of Kraków, and the Institute of Urban and Regional Development. He collaborates with the Center for Evaluation and Analysis of Public Policies at Jagiellonian University. In 2020, he was a recipient of the START scholarship from the Foundation for Polish Science. In 2022, he was awarded the Minister of Education and Science Scholarship for Outstanding Young Scientists.

Krzysztof Gwosdz is an Associate Professor in the Institute of Geography and Spatial Management of Jagiellonian University in Kraków. His research interests are focused on the issues of local and regional development, including restructuring of cities and industrial districts, urban regeneration, functioning of public spaces and geography of innovation. He is particularly interested in the processes of mechanisms of development and transformations of industrial cities and districts in Europe. He participated in several European projects financed by the European Union, British Economic and Social Council, and European Trade Union Institute in Brussels. He is an active member of international research and educational networks: GERPISA Réseau international de l'automobile, International Business 4.0, and CEEPUS-GEOREGNET. Currently, he coordinates a grant funded by the Polish National Science Center Digital entrepreneurs and Industry 4.0.

Robert Guzik is an Economic Geographer and an Adjunct Professor in the Institute of Geography and Spatial Management at Jagiellonian University. His research focuses on transport accessibility and transport exclusion, as well as the conditions and mechanisms of regional development processes. He is particularly interested in sustainable mobility, public transport, and their role in enhancing accessibility in both urban and rural areas.

Jakub Taczanowski is affiliated with the Jagiellonian University in Kraków, Poland and the Institute of Geography and Spatial Management, Department of Population, Settlement and Agricultural Geography. In 2014, he defended his Ph.D. thesis entitled "The influence of selected political and economic factors on the development of railway networks of selected successor states of

Austria-Hungary". He is an author or co-author of over 60 scientific articles and books and has presented the results of his research at over 50 scientific conferences. His research interests include transport, urban and historical geography. In the field of transport geography, Dr. Taczanowski focuses especially on questions of railway and municipal transport. He researches into problems of development and functioning of passenger rail transport systems as well as into transformation of electric public transport networks in changing urban contexts. The main area of interest is Central Europe with particular emphasis on Poland, Czechia, Slovakia, Hungary, and Austria as well as Italy. Dr. Taczanowski is active as a member of editorial boards of "Bulletin of Geography. Socio-economic series", "Review of Economic Perspectives", and "Transport Geography Paper of Polish Geographical Society". He is the deputy chairman of the Kraków branch of the Polish Geographical Society and the secretary of the Task Force for Large-Scale Infrastructure Projects of the Committee for Spatial Economy and Regional Planning of the Polish Academy of Sciences.

Adam R. Parol represents the Institute of Geography and Spatial Management of the Jagiellonian University in Kraków (Poland), where he is a Ph.D. Candidate in the Doctoral School of the Social Sciences. He graduated from the University of Warsaw, the Jagiellonian University, and the Kraków University of Economics with a degree in geography and economics. His research interests include a number of sub-disciplines of human geography—in particular transport and mobility geography, political geography and functional-spatial relations, as well as social geography, economic geography, settlement system, and spatial management. He is the author or co-author of more than a dozen scientific articles, two monographs, and numerous scientific expert reports.

Analyzing the Impact of the Co-evolution Policy of Land Use Allocation and Autonomous Vehicle Adoption on Carbon Dioxide Emission Reduction

Alireza Sahebgharani and Szymon Wiśniewski

Abstract Transportation and land use systems have environmental impacts on human settlements. Transportation emits pollutants through fuel combustion, while land use leaves ecological footprints. Addressing the exacerbation of climate change consequences requires diligent attention to policies aimed at reducing emissions. Meanwhile, the interplay between land use and transportation systems necessitates a coherent framework to guide future development and retrofit existing infrastructure. Furthermore, emerging autonomous vehicles (AVs) offer an opportunity to mitigate transportation-related environmental impacts that affect land use patterns. However, AV adoption rates (ARs) vary, and analyzing how AV usage together with land use allocation optimization could contribute to the CO_2 footprint reduction urges investigation. This matter, which has analytical and prescriptive implications, requires the development of optimization models that combine land use and transportation systems concerning the carbon footprint objectives. Taking the interconnection and hierarchy between land use allocation and traffic assignment decisions, this chapter presents a bi-level co-evolution model. The upper level models the emission of the whole system (land use together with transportation), and the lower level formulates mixed traffic assignment. Then, a solution framework, including a genetic algorithm and gradient projection method, is introduced and applied to analyze several scenarios. This study contributes to the literature by providing a bi-level co-evolution model for land use and AV adoption optimization and presenting ideas for solving the formulated problem. It also offers insights into mitigating carbon footprint when communities go toward adopting AVs and environmental-based optimal land use and transportation patterns. The obtained results show that utilizing AVs and optimized land use allocation could be effective in decreasing the CO_2 footprint.

A. Sahebgharani (✉)
Department of Transportation Engineering, Isfahan University of Technology,
Isfahan 84156-83111, Iran
e-mail: a_sahebgharani@iut.ac.ir

S. Wiśniewski
Faculty of Geographical Sciences, Institute of the Built Environment and Spatial Policy,
University of Lodz, Lodz, Poland
e-mail: szymon.wisniewski@geo.uni.lodz.pl

M. A. Dulebenets et al. (eds.), *Transport and Climate Change*,
https://doi.org/10.1007/978-3-032-00563-2_4

Keywords Climate change · Pollution emission · Autonomous vehicles · Adoption rate · Land use-transportation optimization

1 Introduction

According to the United Nations Framework Convention on Climate Change, climate change refers to long-term alterations in climate patterns caused by human activities that modify the atmosphere and lead to observable climatic variations over time (UNFCCC, 2022). The adverse effects of this phenomenon on both natural ecosystems and human settlements—such as flooding, wildfires, extreme heat, and inundation—make addressing climate change an unavoidable task at global, national, regional, and local levels (Norman, 2022). Two primary strategies for tackling climate change are mitigation and adaptation. Mitigation focuses on reducing greenhouse gas emissions or enhancing carbon sinks (UNFCCC, 2022), while adaptation involves adjusting systems to moderate harms or exploit opportunities associated with climate change (IPCC, 2022). These concepts are particularly relevant to the transportation sector, which accounts for 19.2% of global carbon emissions (Pal et al., 2023) and is a significant source of pollution (Borowska-Stefańska et al., 2022). Various strategies have been proposed to reduce the environmental impact of transportation, including promoting the use of low-carbon transportation modes (World Bank, 2006).

A notable recent development in the transportation sector is the emergence of autonomous vehicles (AVs), which have the potential to revolutionize urban transportation systems (Nikitas et al., 2021). AVs are designed to transport passengers and freight without human intervention, with autonomy levels ranging from no automation to full automation, indicating varying levels of driver dependence (Jones & Leibowicz, 2019). Alongside benefits such as reduced congestion (Zhao et al., 2023) and improved safety (Zhang et al., 2023), AVs have been identified as a potential climate change mitigation measure. Ahmed et al. (2023) found that AVs could reduce emissions by minimizing congestion and facilitating the use of electric vehicles (EVs). Furthermore, AVs, as part of adaptive strategies, can be integrated with intelligent systems to enhance resilience during disasters and extreme conditions (Hancock et al., 2020).

However, the environmental benefits of AVs depend on several factors, including the adoption rate—defined as the proportion of AVs in the overall vehicle fleet (Ryu & Kim, 2022)—and the type of fuel usage (Luca et al., 2023). For instance, Le Hong and Zimmerman (2021) reported that electric AVs could reduce CO_2 emissions by 6–20% in specific contexts. Similarly, Kopelias et al. (2020) highlighted that the environmental impact of AVs becomes more pronounced at higher penetration levels.

In addition to AV adoption, the spatial distribution of land use and activities plays a critical role in managing travel demand and addressing climate change (World Bank, 2006). Transportation and land use are deeply interconnected systems, with land use patterns influencing travel behavior and vice versa (Wegener & Fürst, 2004). Optimizing land-use distribution is crucial not only for regulating the trip generation and distribution but also for reducing CO_2 emissions and air pollution (Chen et al., 2023b). Factors such as land use intensity and location significantly affect these outcomes. As a result, researchers have increasingly focused on spatial land use optimization to achieve environmental and emission reduction objectives (Dong et al., 2022; Huang et al., 2021; Wang & Han, 2021; You & Wu, 2010; Zhang & Zhang, 2023).

Significant research has been conducted on individual aspects of AVs, such as their impact on the value of time (Singleton, 2019), road capacity (Martin-Gasulla et al., 2019), crash reduction (Ye & Yamamoto, 2019), parking demand (Zhang et al., 2015), technology acceptance (Detjen et al., 2021), and energy consumption (Chen et al., 2023a). Moreover, substantial efforts have been devoted toward land use distribution modeling and analysis, including bi-level land use transportation optimization (Lin & Feng, 2003; Zhao & Peng, 2010), environment-based land use allocation (Li et al., 2022, 2023b), equity-based land use distribution (Bilal & Giglio, 2023), and integrated model utilization (Cordera et al., 2021).

However, the combined effects of AV adoption and land use optimization within a co-evolution framework still require further investigation. Understanding this interplay is essential for evaluating how AV adoption rates, combined with optimized land use patterns, can reduce carbon emissions and provide actionable insights for policymakers in implementing adaptive and mitigating climate strategies. Therefore, the objective of this chapter is to introduce a co-evolution model that integrates land use optimization with traffic assignment simulations under varying AV adoption rates. Using a bi-level modeling approach, this study evaluates the effectiveness of a synthetic policy combining AV adoption and land use allocation in mitigating and adapting to climate change through CO_2 emission reductions.

The rest of this chapter is organized as follows. Section 2 reviews the literature to place this study in the body of existing knowledge. Section 3 describes the optimization model for analyzing the effect of co-evolution land use and AV adoption policy on the CO_2 emission reduction. Section 4 presents the solution procedure. Section 5 explains the model application and outputs and discusses the results. Section 6 illustrates conclusions and discusses the research limitations.

2 Literature Review

The literature on the interplay between land use and transportation systems, particularly in the context of AV adoption, spans a wide array of interdisciplinary studies. These studies explore various dimensions of the relationship, including the environmental implications of transportation, spatial land-use optimization, and emerging AV technologies. Transportation systems are major contributors to greenhouse gas emissions, while land-use patterns significantly influence travel behavior and environmental outcomes. The adoption of AVs introduces both challenges and opportunities for optimizing these interconnected systems. Given the breadth and complexity of the topic, it is crucial to systematically organize the review to ensure clarity and focus.

Therefore, the literature review is divided into two subsections. The first subsection examines the synergies of bi-level optimization models in transportation and land-use planning. This provides the theoretical and methodological basis for understanding the co-evolution nature of land-use allocation and transportation systems, emphasizing their role in addressing climate change. The second subsection delves into the environmental impacts of emerging vehicle technologies, specifically focusing on AVs, and their potential integration within sustainable urban frameworks. This division is informed by the dual objectives of the study: (1) to analyze the theoretical and methodological frameworks that underpin the integration of land use and transportation systems, and (2) to evaluate the practical implications of AV adoption as a climate change mitigation strategy. By separating these focal points, the review effectively contextualizes the study within existing research, while highlighting the novel contribution of the proposed co-evolution model in bridging these two domains.

2.1 Synergies in Bi-level Programming: Insights into Transportation and Infrastructure Management

Bi-level programming models have gained significant attention in addressing complex challenges in transportation and infrastructure management. These models effectively coordinate decision-making processes across hierarchical levels, enabling optimization in dynamic and resource-intensive systems. For example, Fathollahi-Fard et al. (2022) employed a bi-level programming framework for designing home healthcare supply chains, integrating metaheuristic algorithms like the Imperialist Competitive Algorithm to improve resource allocation and operational efficiency. This study underscores the applicability of bi-level models in the healthcare sector, where resource management and cost optimization are critical. In another effort, Zhang et al. (2024) proposed a bi-level model to optimize order acceptance and production planning in industrial robotics. By using an enhanced whale optimization algorithm, their study demonstrated efficiency in balancing profits and operational costs, providing valuable insights into supply chain coordination.

In the context of transportation systems, bi-level programming has been instrumental in addressing challenges associated with mixed vehicle fleets and infrastructure planning. Wang (2024) developed a mixed-integer nonlinear programming model to optimize ride-hailing networks with autonomous vehicles (AVs) and human-driven vehicles. Their results revealed a potential 20% increase in platform profits, especially when infrastructure costs for AVs were minimized. Similarly, Yan et al. (2024) integrated deep reinforcement learning with model predictive control in a bi-level traffic management model, achieving significant improvements in average vehicle speeds and traffic throughput. These findings highlight the potential of bi-level models to enhance transportation network efficiency, particularly in mixed-traffic environments.

The application of bi-level models extends to sustainable infrastructure development, including the allocation of electric vehicle charging stations (Li et al., 2023a; Piedra-de-la-Cuadra & Ortega, 2024). These studies demonstrated that heuristic methods could efficiently optimize charging station placement, ensuring backup coverage while maximizing user convenience. Najafi-Ghalelou et al. (2023) further expanded the scope by integrating stochastic elements into bi-level models for coordinating electric vehicles in distribution networks, achieving effective cost minimization under dynamic conditions. These advancements emphasize the role of bi-level programming in addressing sustainability challenges within transportation and infrastructure systems.

In urban environments, bi-level optimization has proven valuable for enhancing mobility and reducing environmental impacts. Huang et al. (2024) introduced a bi-level model to optimize vehicle relocation in Mobility-on-Demand systems, achieving reductions in empty vehicle trips and improved order fulfillment rates. Similarly, Liu et al. (2024) addressed airport de-icing operations using a bi-level framework, significantly reducing delays and operational inefficiencies. These studies exemplify the adaptability of bi-level programming in solving context-specific transportation challenges.

The literature demonstrates the extensive application of bi-level models across various domains of transportation and infrastructure management. These models leverage advanced optimization techniques, such as metaheuristics and dynamic programming, to improve resource allocation, reduce costs, and enhance system efficiency (Bakibillah et al., 2021; Burger et al., 2022). By integrating technological advancements, such as autonomous and electric vehicles, with spatial optimization strategies, bi-level programming continues to play a pivotal role in addressing the environmental and operational challenges of modern transportation systems.

2.2 *Balancing Benefits and Costs: Environmental Impacts of Emerging Vehicle Technologies*

The environmental impacts of vehicle technologies, particularly electric vehicles (EVs) and AVs, are multifaceted, with both opportunities and challenges emerging throughout their life cycles. EVs, for instance, offer significant reductions in operational emissions compared to internal combustion engine vehicles (ICEVs). However, their production—especially of lithium-ion batteries—generates high initial emissions (Moseman & Paltsev, 2022). Furthermore, challenges related to battery disposal and recycling highlight the need for robust circular economy practices to mitigate long-term environmental costs (Ahmed et al., 2023). This underscores the importance of clean energy infrastructure in maximizing the benefits of EV adoption.

AVs add an additional layer of complexity to the environmental discourse. While their efficient driving and traffic management capabilities can reduce emissions, their production—particularly of advanced components like sensors and cameras—increases resource consumption and carbon footprints (Silva et al., 2022). Moreover, AV deployment risks exacerbating urban sprawl, which is associated with increased land use, habitat disruption, and higher emissions (Liu et al., 2023). To address these challenges, AV adoption must be accompanied by policies that integrate land-use planning and infrastructure development, ensuring that environmental benefits outweigh potential drawbacks.

The potential of AVs to improve urban mobility and reduce emissions is significant but depends on effective implementation strategies. For instance, dynamic route planning and smoother driving styles enabled by AVs have been shown to reduce $PM_{2.5}$ and CO_2 emissions by 11.43% and 3.03%, respectively (Szűcs & Szűcs, 2024). Shared AVs (SAVs) further amplify these benefits by replacing multiple private vehicles, reducing congestion, and lowering energy consumption. However, their widespread deployment could unintentionally promote urban sprawl if not carefully regulated (Orieno et al., 2024). Policies focusing on urban density and sustainable mobility are essential to balance these competing outcomes.

Infrastructure considerations also play a critical role in maximizing AV benefits. Dedicated AV lanes and speed management strategies have been shown to reduce CO_2 emissions by over 16% in mixed-traffic scenarios (Kim, 2024). However, the construction and maintenance of such infrastructure require significant resources, potentially offsetting some environmental gains. Similarly, IoT-enabled AV systems, while facilitating adaptive route optimization and real-time environmental monitoring, demand high energy inputs for data processing and communication (Ferrari, 2022). Balancing these technological benefits with their resource demands is crucial for sustainable urban development.

Beyond urban applications, AVs have demonstrated potential in agricultural and industrial contexts. For example, autonomous electric vehicles used in precision agriculture reduce pollution through optimized resource use and energy-efficient operations (Zhou & Zhou, 2021). However, the production and maintenance of such vehicles involve energy-intensive processes, highlighting the need for sustainable manufacturing practices.

The environmental promise of AV and EV technologies lies in their ability to transform mobility while mitigating climate impacts. However, realizing this potential requires a nuanced approach that integrates renewable energy sources, innovative traffic management solutions, and comprehensive urban policies (Kopelias et al., 2020). Addressing challenges such as resource-intensive production, lifecycle emissions, and infrastructure investments is essential for achieving sustainable and equitable outcomes.

2.3 Research Gaps and Study Advancements

Despite extensive research on AVs and land-use optimization, several critical gaps remain in the literature. Existing studies have primarily focused on isolated aspects of either AV adoption or land-use distribution. For example, research on AVs has explored their potential to reduce emissions through improved traffic flow and congestion management (Kopelias et al., 2020; Zhao et al., 2023). Similarly, studies on land-use optimization have addressed issues such as carbon reduction through spatial reallocation (Dong et al., 2022) and equity-based land-use distribution (Bilal & Giglio, 2023). However, the combined effects of AV adoption and land-use optimization within a co-evolution framework remain underexplored.

Furthermore, while bi-level programming models have been widely used to address hierarchical decision-making in transportation and land use (Huang et al., 2024; Wang et al., 2019), few studies have integrated these systems in the context of dynamic AV adoption rates. The environmental and policy implications of synchronizing AV deployment with spatial optimization have not been comprehensively analyzed, leaving a gap in actionable insights for policymakers aiming to achieve sustainable urban development.

This study addresses these gaps by introducing a novel bi-level co-evolution model that integrates land-use optimization with traffic assignment simulations under varying AV adoption rates. The upper level focuses on minimizing the carbon footprint of land-use and transportation systems, while the lower-level models user equilibrium in mixed-traffic scenarios. This dual-level approach enables the simultaneous evaluation of spatial distribution and AV deployment strategies, providing a comprehensive framework for sustainable urban planning.

The advancements of this paper are threefold. First, it extends prior research by combining AV adoption and land-use optimization into a single framework, illustrating their synergistic effects on reducing CO_2 emissions. Second, it incorporates dynamic AV adoption rates, allowing for a more realistic representation of the gradual transition to autonomous mobility. Finally, the application of advanced optimization techniques, such as genetic algorithms and gradient projection methods, ensures the robustness and scalability of the proposed model, making it applicable to real-world urban scenarios.

By addressing the existing gaps, this study not only contributes to theoretical advancements in bi-level programming but also offers practical insights for policymakers and urban planners. The findings demonstrate how integrating AV adoption with optimized land-use allocation can lead to significant reductions in carbon emissions, paving the way for more sustainable and adaptive urban systems.

3 Model Development

Inspired by the previous efforts (Bilal & Giglio, 2021, 2023; Lin & Feng, 2003; Wang et al., 2019; Zhao & Peng, 2010), a bi-level problem is formulated to connect land use and transportation systems. The local government controls the upper level through the department of land use planning where spatial planners determine employment and population distributions in the spatial units with the objective of carbon footprint of land use and transportation systems minimization, taking the system's constraints and opportunities into account. The upper-level objective has two components. The first is related to land use and the second to the transportation systems. While the whole objective function calculates the CO_2 emission of the system, the latter component returns the traffic flow in the network given the population and employment arrangement. The lower-level problem is the representation of the transportation network user's behavior. At this level, the users attempt to minimize an objective function that models user equilibrium conditions in the network by having the land use arrangement determined in the upper-level problem.

In fact, in the lower-level problem, having the population and employment distributions obtained from the upper level, the flows in the networks are set, so that the equilibrium in the transportation component is fulfilled, which can be reflected through a relative gap minimization function. Figure 1 depicts the generalities of the bi-level optimization problem. The following sub-sections formulate the bi-level optimization problem, using these generalities. First, the spatial system is described. Second, the decision variables are explained. Finally, the upper- and lower-level problems alongside the constraints are modeled.

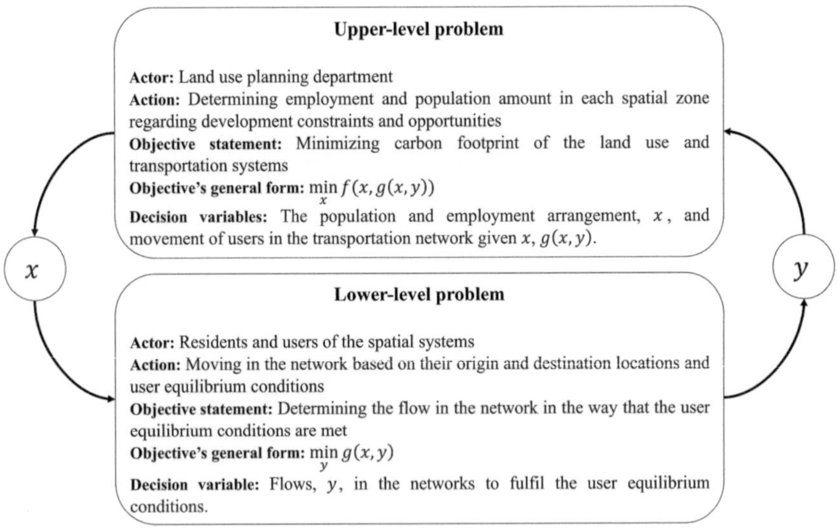

Fig. 1 The bi-level optimization problem generalities

3.1 Spatial System Description

Land use system: Let a spatial area be divided into zones denoted by Z that represent analysis or planning units. Each zone has developable land, which can be built with varying floor area ratios (FARs). These groups, which determine the development capacity of a zone, are members of $D = \{1, 2, \ldots, d, \ldots, T\}$. The amount of development capacity for each FAR group in zone $i \in Z$, denoted by F_i^d, could be derived from land suitability analysis, which outputs the maximum gross number of floors that can be constructed on a given land area. For instance, $F_{i=1}^{d=1} = 50$ and $F_{i=1}^{d=2} = 100$ indicate that the development capacity in zone 1 is 50 and 100 units, respectively, where the former is allocated to the first FAR category ($d = 1$) with 1 floor, and the latter to the second FAR category ($d = 2$) with 2 floors. Thus, the development capacity for $F_{i=1}^{d=2}$ is 100, while the land used is 50 units. This spatial system includes activities belonging to set A, and each activity $k \in A$ has a carbon footprint value in each development capacity group, denoted by CFP_d^k. CFP_d^k comprises two parts: the first includes material production, transportation, construction, and waste treatment, and the second relates to building use. The development level for each activity $k \in A$, DEV^k is also specified.

Transportation system: The above-mentioned zones are interconnected by a network represented by graph G consisting of nodes (N) and links (L). Each link $l \in L$ has a defined capacity for human-driven vehicles and AVs, and length, which are denoted by c_l^{HU}, c_l^{AV}, and g_l. The carbon emissions of human-driven vehicles and AVs are denoted by CE^{HU} and CE^{AV}, respectively. Furthermore, each link includes a performance function that incorporates capacity and volume (i.e., v_l^{HU} and v_l^{AV}), along with parameters β^l and γ^l, and free flow travel time (i.e., t_l^0) to calculate the link's travel time, t_l. Routes in the network, marked by their starting (r) and ending (s) nodes in N, belong to set RS. Multiple vehicle types, represented by $m \in M$, traverse the network, with route costs for vehicle type m on route $p \in rs \in RS$ represented by c_{pm}^{rs}. The adoption rate of AVs (AR) indicates the level of AV usage in the system.

3.2 Decision Variables

Two decision variables are included in the optimization model. The first, $x_{ij}^{k\acute{k},d}$, describes the flow of activity $k \in A$ in zone $i \in Z$, which consumes a development capacity in category $d \in D$ in i and moves to zone $j \in Z$ for activity $\acute{k}$ (e.g., if $k = 1$ represents residential activity, $k = 2$ represents working activity, and $d = 2$ signifies development capacity in 2-story buildings, then $x_{4,5}^{1,2,2}$ is the number of people living in 2-story buildings in zone 4 and commuting to zone 5 for work).

The second variable, f_{pm}^{rs}, represents the flow of vehicle type $m \in M$ assigned to route p connecting r to s in G. Note that "flow" refers to origin–destination pairs and connecting routes, while "volume" applies to link $l \in G$, representing the sum of flows associated with routes through a given link following traffic assignment. It should be noted that the concept of flow is used for origin–destination pairs and their connecting routes, while the concept of volume is utilized for $l \in G$. In other words, the volume of each link is the sum of flows associated with routes passing through that link after performing the traffic assignment. The model's condensed notations together with intermediate variables and functions applying during the optimization process are as follows.

Sets:

Z	Set of zones $i, j, r, s \in Z$
D	Set of development capacity groups
A	Set of activities $k, \acute{k} \in A$
N	Set of nodes in G $n \in N$
L	Set of links in G $l \in L$
M	Set of transportation modes $m \in M = \{HU, AV\}$
RS	Set of origin–destination pairs $rs \in RS$. r is the origin and s is the destination
P_m^{rs}	Set of paths (routes) between r and s traversed by m ($p \in P_m^{rs}$)

Parameters:

F_i^d	The amount of development capacity for d in zone i
CFP_d^k	The carbon footprint value of activity k in d, including material production, transportation, construction, waste treatment, and building use
DEV^k	The development level for activity k
c_l^{HU}	The capacity of link l for HU
c_l^{AV}	The capacity of link l for AV
g_l	The length of link l
CE^m	The carbon emission value of mode m
β^l and γ^l	Parameters of the intermediate function of t_l
t_l^0	The free flow time of link l
AR	The AV adoption rate
α^k	The converting rate of activity flow k to the FAR requirement
DEV^k	The total activity k in the system
ε	The acceptable relative gap used for controlling the number of calculations for optimizing the lower-level objective function. This parameter is a small number such as 10^{-3}

Decision variables:

Upper-level variable:

$x_{ij}^{k\acute{k},d}$ The flow of activity k in zone i, which consumes a development capacity in d in i and moves to zone j for conducting $\acute{k}$

Lower-level variable:

f_{pm}^{rs} The flow of m assigned to p connecting r to s

Infeasible solution repairing variable:

$y_{ij}^{k\acute{k},d}$ The amount of change in the flow of activity k in zone i, which consumes a development capacity in d in i and moves to zone j for conducting $\acute{k}$. This variable adjusts a given $x_{ij}^{k\acute{k},d}$ in a way $x_{ij}^{k\acute{k},d}$ undergoes minimum variation after combining with $y_{ij}^{k\acute{k},d}$ to satisfy the problem constraints

Intermediate functions:

t_l The travel time function of link l. This function calculates travel time of link l given v_l^m, c_l^m, β^l, γ^l, and t_l^0. This function is explained in Eq. 16

Intermediate variables:

v_l^m The volume of m on l. This variable is obtained by summing the flow of paths passing through l. This variable is obtained through $\sum_{rs\in RS}\sum_{p\in P_m^{rs}} f_{pm}^{rs}\delta_{plm}^{rs}$

c_{pm}^{rs} The travel cost of p by m from r to s. This variable is computed by summing t_l over l in p

$c_{\overline{p}m}^{rs}$ The travel cost of the shortest path $\overline{p}$ from r to s by m

$f_{\overline{p}m}^{rs}$ The flow of the shortest path $\overline{p}$ from r to s by m

δ_{plm}^{rs} A binary variable that is 1 when l is on p that connects r to s, and 0, otherwise

3.3 The Upper- and Lower-Level Problem Formulations

Considering the spatial system, transportation network, and the decision variables, the co-evolutionary land use and AV assignment is formulated as a bi-level optimization problem. The upper-level problem allocates activities in the system alongside movement costs in the network, while the lower-level problem assigns flows in the street network in alignment with spatial activity distribution. Detailed mathematical modeling and specifics are provided below.

In the upper-level problem, the objective (Eq. 1) is to minimize the carbon footprint across the entire system, including the land use-activity structure and transportation network. The first part of the objective and Eqs. 2–4 impose constraints on land use-activity allocation, which distributes activities across zones and development groups to reduce the total carbon footprint. Constraints in Eq. 2 ensure that the total development in i in category d does not exceed zone capacity based on the allocated activities. In essence, the floor area demand in a zone cannot exceed the supply. Constraints in Eq. 3 ensure that the total development for activity k matches DEV^k. The third group of constraints (Eq. 4) explains that the activity flows should be greater than or equal to 0.

The second part of the objective function formulates the CO_2 emission by the transportation system corresponding to $x_{ij}^{k\acute{k},d}$. This emission value is calculated via a traffic assignment model, detailed in the lower-level problem (Eq. 5), which balances network flows based on the AV adoption rate and origin–destination matrix. In the lower-level problem, the objective minimizes the relative gap between the total cost-flow product for routes and the lowest-cost routes within the network assignment process. This problem is subject to constraints, which outline a variational inequality problem that aims to identify the flow of feasible routes when Eqs. 6 and 7 are satisfied. Equation 6 ensures that the total route flow aligns with origin–destination pairs, and Eq. 7 restricts flows to non-negative values.

$$\min_{x_{ij}^{k\acute{k},d}} \left(\sum_{i\in Z}\sum_{d\in D}\sum_{k\in A}\sum_{j\in Z}\sum_{\acute{k}\in A} x_{ij}^{k\acute{k},d} \cdot \alpha^k \cdot CFP_d^k + \sum_{l\in L}\sum_{m\in M}\sum_{rs\in RS}\sum_{p\in P_m^{rs}} g_l \cdot CE^m \cdot f_{pm}^{rs} \cdot \delta_{plm}^{rs} \right) \tag{1}$$

Subject to:

$$\sum_{k\in A}\sum_{\acute{k}\in A}\sum_{j\in Z} x_{ij}^{k\acute{k},d}\alpha^k \le F_i^d, \forall i \in Z, d \in D \tag{2}$$

$$\sum_{i\in Z}\sum_{j\in Z}\sum_{\acute{k}\in A}\sum_{d\in D} x_{ij}^{k\acute{k},d}\alpha^k = DEV^k, \forall k \in A \tag{3}$$

$$x_{ij}^{k\acute{k},d} \ge 0, \forall i \in Z, j \in Z, k \in A, \acute{k} \in A, d \in D \tag{4}$$

$$\min_{f_{pm}^{rs}} \left(\frac{\sum_{m\in M}\sum_{rs\in RS}\sum_{p\in P_m^{rs}} c_{pm}^{rs} \cdot f_{pm}^{rs} - \sum_{m\in M}\sum_{rs\in RS}\sum_{p\in P_m^{rs}} c_{\overline{p}m}^{rs} \cdot f_{\overline{p}m}^{rs}}{\sum_{m\in M}\sum_{rs\in RS}\sum_{p\in P_m^{rs}} c_{pm}^{rs} \cdot f_{pm}^{rs}} \right) \tag{5}$$

Subject to:

$$\sum_{p\in P_m^{rs}} f_{pm}^{rs} = \sum_{k\in A}\sum_{\acute{k}\in A}\sum_{d\in D} x_{rs}^{k\acute{k},d}, \forall rs \in RS, m \in M \tag{6}$$

$$f_{pm}^{rs} \ge 0, \forall rs \in RS, m \in M, p \in P_m^{rs} \tag{7}$$

4 Solution Procedure

To solve the described problem, a two-stage procedure is followed. In line with the previous research (Bilal & Giglio, 2023; Sahebgharani & Wiśniewski, 2024; Wang et al., 2019), the first stage is managed by the genetic algorithm (GA), an algorithm that has demonstrated satisfactory results in solving non-convex and non-differentiable problems. The second stage employs the gradient projection algorithm (GPA) introduced by Jayakrishnan et al. (1994) and later adapted by Ryu and Kim (2022). Specifically, the GA determines feasible activity flows ($x_{ij}^{k\acute{k},d}$) between origin–destination pairs, optimizing them within the spatial system. The optimal traffic flows in the network, corresponding to the allocated activities, are then assigned by the GPA and fed back into the equation system. Figure 2 depicts the general solution procedure.

First, in the GA procedure, a set of feasible solutions (i.e., activity flow matrices) of size n are generated. Then, Eqs. 5–7 are solved for each solution to determine the load of each link after the traffic assignment, which is essential for calculating the upper-level objective function. For computing these equations, the route-based gradient projection algorithm for the multi-class mixed-mode assignment proposed by Ryu and Kim (2022) is used. In the initial step, the path set, and path and link flows are created. In subsequent steps, the route generation includes updating link costs, calculating the shortest paths, and updating the path set. Next, path costs, OD costs, and second derivatives are computed.

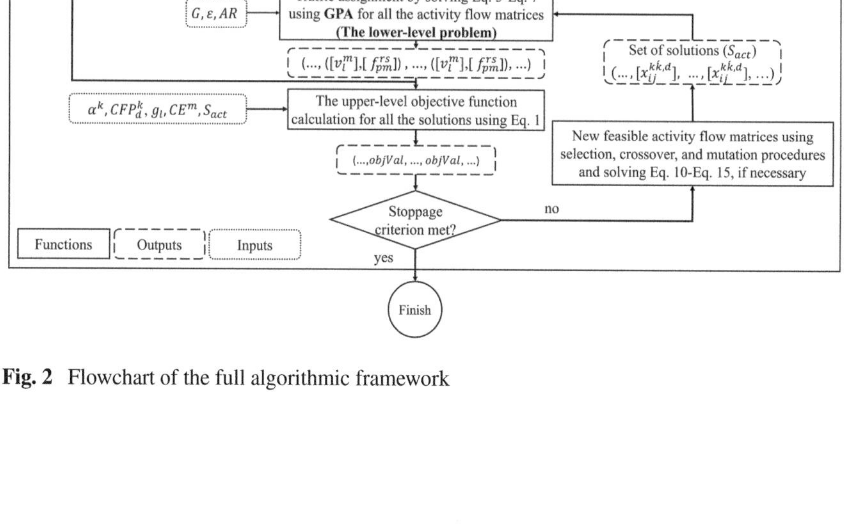

Fig. 2 Flowchart of the full algorithmic framework

In the final step, path flows and link flows are recalculated, and the convergence condition—indicated by the related gap in Eq. 5—is checked. If the GPA converges (i.e., the value of Eq. 5 is equal to or less than ε, a small number such as 10^{-3}), link flows, path flows, link travel times, and path travel times are output for use in other parts of the solution procedure. Otherwise, these steps iterate until the convergence criterion is met. Figure 3 shows the GPA procedure in detail, where L is the network, RS is the set of origin–destination pairs, RS_{Path} is the set of paths between origins and destinations, $RS_{Shortest}$ is the set that stores the shortest paths between origins and destinations, and Q is the demand between origins and destinations obtained from $x_{ij}^{k\acute{k},d}$.

Second, once the spatial distribution of activities and traffic volumes in the system, as well as the objective function values, are known, solutions are selected to generate new ones with the goal to obtain improved offspring in the next generation. Various selection procedures exist in the GA literature [e.g., random, ranking, elitist, scale, and tournament selection (Fattahi, 2011)]. This application uses the roulette wheel method, which has proven effective in previous studies (Li et al., 2016, 2019; Sahu, 2018; Wang et al., 2019; Zuo & Li, 2019). Crossover and mutation operators are employed to combine selected solutions. In the crossover operator, two solutions are synthesized based on certain logic [e.g., single-point, two-point, and multi-point (Fattahi, 2011)] to create a solution that inherits the quality characteristics of its parents. For this study, crossover is implemented via Eq. 8, where u is a uniform random number between 0 and 1, and $parent^{(1)}$ and $parent^{(2)}$ are the selected solutions chosen through the roulette wheel method. In a nutshell, a portion of the first parent is combined with the second randomly by applying the uniform distribution. For instance, the u part is obtained from $parent^{(1)}$ and $1-u$ from $parent^{(2)}$. Then, these values are summed to generate the new solution.

$$sol^{(new)} = u \times parent^{(1)} + (1-u) \times parent^{(2)} \tag{8}$$

In addition to the crossover operator, mutation is used to enhance exploitation within the solution space. Similar to the crossover and selection operators, there are several mutation procedures [e.g., flipping, interchanging, and replacement (Fattahi, 2011)]. This study uses the mutation operator presented in Eq. 9, where ub and lb represent the variable's upper and lower bounds. In brief, a random value between ub and lb is generated and added to $parent$ to change a part of the selected solution.

$$sol^{(new)} = parent + 0.1(ub - lb) \times u \tag{9}$$

Two rates, P_c and P_m, control the number of solutions generated by the crossover and mutation operators (Sahebgharani & Wiśniewski, 2024). These values designate the proportion of solutions produced by the operators and replaced within the population. Since these rates influence the GA algorithm's efficiency, values between 0.5–0.9 and 0.2–0.3 are recommended in previous studies (Bilal & Giglio, 2021, 2023; Carroll, 1996; Sahebgharani & Wiśniewski, 2024) for P_c and P_m, respectively.

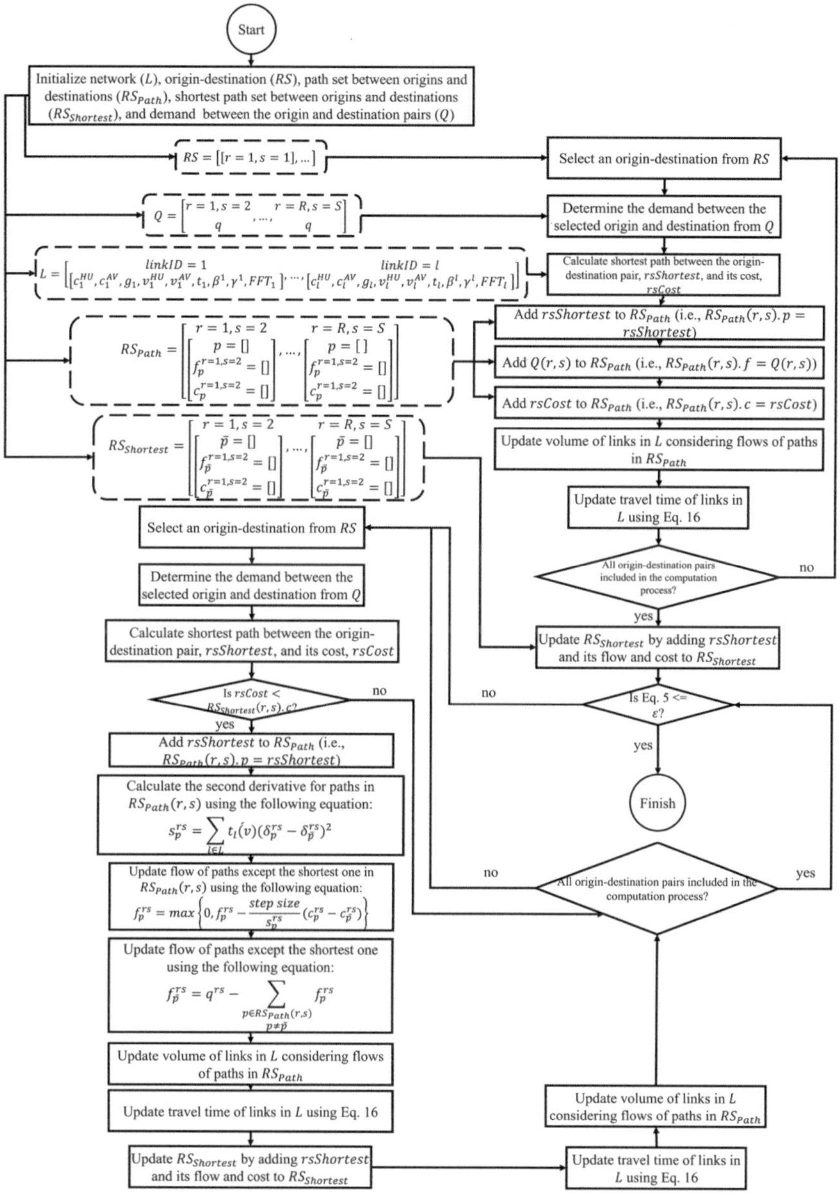

Fig. 3 Flowchart of the gradient projection algorithm (GPA)

After performing crossover and mutation, the output solution might violate problem constraints. Therefore, the infeasible solutions need repairing to hold the problem conditions. To keep the repaired solutions as close as the original ones and maintain the modifications carried out by the operators, a model is developed to find values that can be added to $x_{ij}^{k\acute{k},d}$ in the way the adjustments minimize the deviation of the repaired solution from the infeasible one. Equation 10 is a minimization function that models the difference between infeasible and adjusted values. This function is subject to constraints (Eqs. 11–15), ensuring the repaired solution satisfies development conditions (i.e., DEV^k and F_i^d).

$$\min_{\Delta x_{ij}^{k\acute{k},d}} \left(\sum_{i \in Z} \sum_{j \in Z} \sum_{k \in A} \sum_{\acute{k} \in A} \sum_{d \in D} x_{ij}^{k\acute{k},d} - y_{ij}^{k\acute{k},d} \right) \tag{10}$$

Subject to:

$$\sum_{i \in Z} \sum_{j \in Z} \sum_{\acute{k} \in A} \sum_{d \in D} x_{ij}^{k\acute{k},d} - y_{ij}^{k\acute{k},d} = DEV^k, \forall k \in A \tag{11}$$

$$\sum_{k \in A} \sum_{\acute{k} \in A} \sum_{j \in Z} x_{ij}^{k\acute{k},d} - y_{ij}^{k\acute{k},d} \le F_i^d, \forall i \in Z, d \in D \tag{12}$$

$$x_{ij}^{k\acute{k},d} \ge y_{ij}^{k\acute{k},d}, \forall i \in Z, j \in Z, k \in A, \acute{k} \in A, d \in D \tag{13}$$

$$y_{ij}^{k\acute{k},d} \in \mathbb{R}, \forall i \in Z, j \in Z, k \in A, \acute{k} \in A, d \in D \tag{14}$$

$$x_{ij}^{k\acute{k},d} - y_{ij}^{k\acute{k},d} \ge 0, \forall i \in Z, j \in Z, k \in A, \acute{k} \in A, d \in D \tag{15}$$

These processes continue iteratively until the stoppage criterion—such as the maximum number of iterations, run time, or objective value stability after a certain number of calculations (Alam Tabriz et al., 2013)—is achieved. The full GA process is presented in Fig. 4, where N_c and N_m are sets storing solutions generated by crossover and mutation operators, respectively, N is the set that stores solutions, P_m and P_c are the mutation and crossover rates, n_c is the number of solutions generated by the crossover operator, n_m is the number of solutions generated by the mutation operator, n is the number of solutions in a generation, S_{act} is a set of $x_{ij}^{k\acute{k},d}$ matrices, and $|.|$ determines the size of a set.

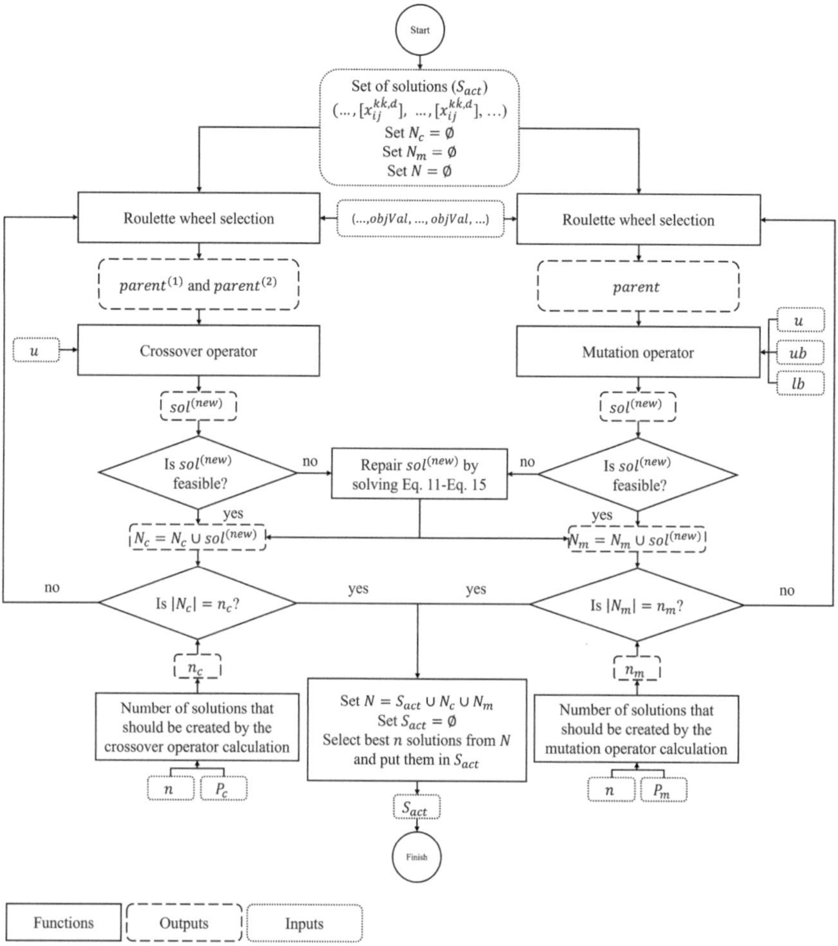

Fig. 4 Flowchart of the genetic algorithm (GA)

5 Model Application and Outputs

In this section, the case study and scenarios under consideration are described. Interesting managerial insights are uncovered using the proposed methodology, and supporting discussions are presented accordingly.

5.1 Experimental Case, Data, and Scenarios

To execute the co-evolution model, the city of Sioux Falls, South Dakota is considered (Chakirov & Fourie, 2014; Leblanc, 1975) (Fig. 5).

Two capacity categories are defined and randomly distributed among the zones. Since work-to-home trips are available in the study area, two activity types (i.e., living and working) are considered, and each can be assigned to the capacity categories. *F* is randomly generated, and *CFP* values are provided in Table 1. In the baseline scenario, 360,600 people are living and working in the study area, with 10% of this population designated to be added under future conditions (*DEV*). The flow-to-land transformation factors (α) are 30 m^2/unit of flow for living activities and 25 m^2/ unit of flow for working activities, respectively.

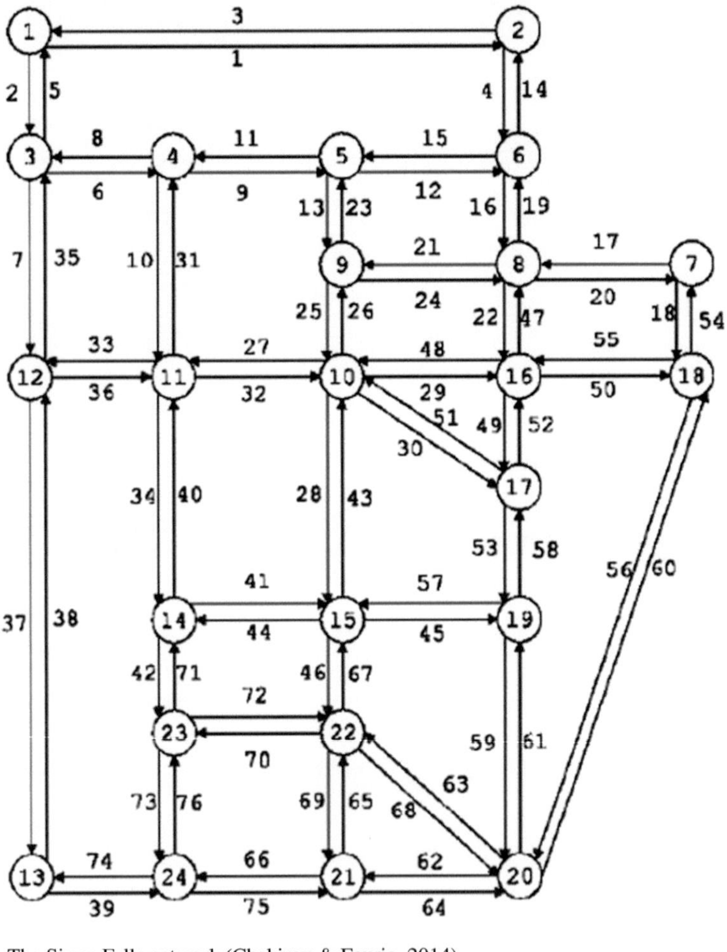

Fig. 5 The Sioux Falls network (Chakirov & Fourie, 2014)

Table 1 Carbon footprint of activities adapted from and inspired by Collings (2020) and Huang et al. (2017)

	CFP_d^k ($kgCO_2e/m^2/h$)-Building construction[a]		CFP_d^k ($kgCO_2e/m^2/h$)-Building use	
	$k = 1$ (residential buildings)	$k = 2$ (other buildings)	$k = 1$ (residential buildings)	$k = 2$ (other buildings)
$d = 1$	0.0020	0.0034	0.0084	0.0067
$d = 2$	0.0009	0.0045		

[a] *Note* The lifetime of a building is considered 30 years

The street network consists of 24 nodes (zones), 576 OD pairs, and 76 links. Following Ryu and Kim (2022), it is assumed that adding AVs to the streets increases link capacities by 50% of their initial values. The link performance function is provided in Eq. 16, where γ_l and β_l are set to 0.5 and 4, respectively. Additionally, AVs are assumed to operate using electric power. The carbon emission rates for human-driven vehicles (CE^{HU}) and AVs (CE^{AV}) are 0.0002 (kg/m) and 0.0001 (kg/m), respectively (Moseman & Paltsev, 2022).

$$t_l = t_l^0\left[1 + \gamma_l\left(\left(v_l^{AV} + v_l^{HU}\right)\left(\frac{v_l^{AV}}{v_l^{AV} + v_l^{HU}}\frac{1}{c_l^{AV}} + \frac{v_l^{HU}}{v_l^{AV} + v_l^{HU}}\frac{1}{c_l^{HU}}\right)\right)^{\beta_l}\right] \quad (16)$$

Given these general system conditions, six scenarios were defined to analyze the impact of co-evolution land use and transportation planning with AV implementation on carbon emissions in the study area. Explanations of the scenarios (S1–S6) are provided in Table 2.

Table 2 Details of the scenarios under consideration

Scenario ID	Scenario description
S1	In this scenario, the distribution of population and employment is in the same proportion as the status quo in the first capacity category, and the AV adoption rate is set to 0 ($AR = 0$). Based on these conditions, the traffic is assigned to the network, and the CO_2 emission and footprint are calculated
S2	In this scenario, the total activity in the system is reallocated regarding all development capacity categories and $AR = 0\%$
S3	Conditions of this scenario are the same as S2 but $AR = 25\%$
S4	Conditions of this scenario are the same as S2 but $AR = 50\%$
S5	Conditions of this scenario are the same as S2 but $AR = 75\%$
S6	Conditions of this scenario are the same as S2 but $AR = 100\%$

Table 3 GA parameter values

Parameters	Values
n	50, 100, 150
P_m	0.2, 0.25, 0.3
P_c	0.5, 0.7, 0.9
Number of iterations	50, 100, 150

* *Note* All combinations for these parameters were accounted for and run several times. Then, the best solution was reported in the findings and analyses

5.2 *Outputs, Analysis, and Discussion*

After defining the case study and scenarios, the adopted algorithms were programmed in MATLAB and implemented on a Core i5 CPU @ 2.5GHz laptop with 16 GB RAM and a 64-bit operating system. The algorithms were run multiple times with various parameter sets, and from these trials, the best solutions were identified and compared. Table 3 shows the values applied for conducting trials.

Figure 6 depicts the objective values (i.e., CO_2 footprints) for scenarios S1–S6. The results show that applying co-evolution planning (even when AV adoption is 0) reduces the CO_2 footprint. Moreover, as AV adoption increases, further reductions in CO_2 emissions are observed. A comparison of S2–S6 with S1 shows reductions in CO_2 footprints by approximately 15%, 28%, 37%, 47%, and 58%, respectively. Additionally, pairwise comparisons between successive scenarios reveal that spatial distribution of activities concerning for CO_2 emissions decreases by approximately 15.5% in cost, while combining land-use optimization with AV adoption gradually reduces the carbon footprint by 15.9% for S3 over S2, 12.2% for S4 over S3, 16% for S5 over S4, and 20% for S6 over S5.

To examine how co-evolution planning affects the carbon footprint of land use and transportation systems, traffic assignment was conducted for different AV adoption rates. Specifically, these experiments mirror S1 but with adoption rates of 25, 50, 75, and 100%. Figure 7 shows the gap between the co-evolution model (blue line) and cases that only use AVs at various adoption rates (red line). The results show that the utilization of AVs together with optimized land use allocation could be effective in decreasing the CO_2 footprint and contributing to city adaptation to climate change. In other words, replacing conventional vehicles with electric-powered AVs has the potential for reducing emissions and alleviating congestion. Also, these effects may be amplified by combining traffic assignment with land use optimization.

The results indicate that when only the introduction of AVs is considered at an adoption rate of 75%, the emission reduction could be promising. However, the co-evolution model yields competitive results even at lower adoption rates (i.e., 25, 50, and 75%). This finding also holds for an AR of 100%, where the co-evolutionary outputs outperform the results at lower adoption levels. These outputs signify that co-evolutionary planning is effective not only at higher AV adoption rates but also at lower and intermediate levels.

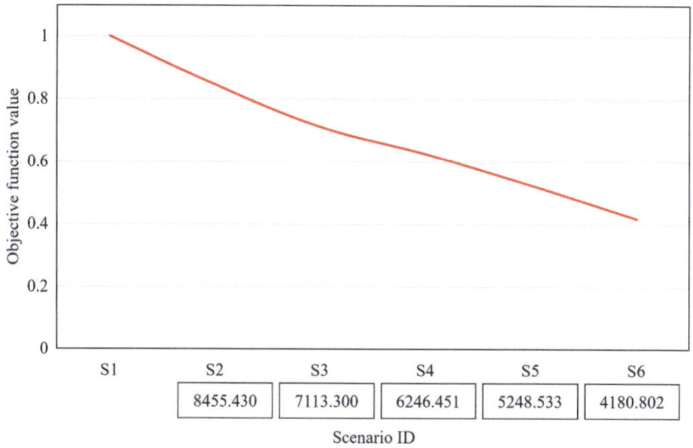

Fig. 6 The CO_2 footprint of the modeled scenarios (standardized values—lower values indicate better scenarios; boxes are the objective values multiplied by 0.001)

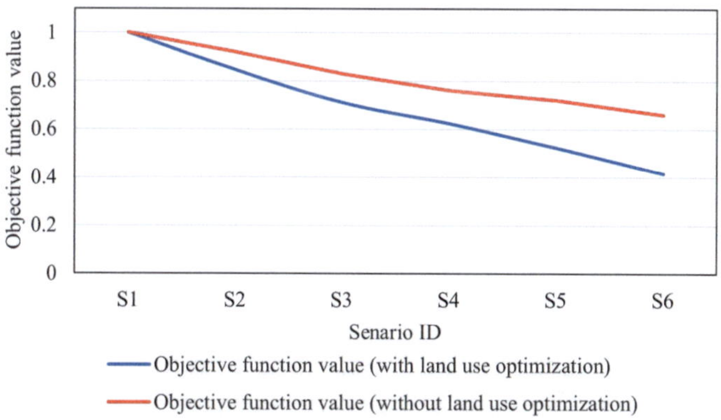

Fig. 7 Comparison of the co-evolution model and their counterpart that relies on AV applications only (standardized values—lower values indicate better scenarios)

The findings could be helpful in policymaking processes aiming to maximize the environmental benefits of AV technology. As shown in Figs. 6 and 7, high AV adoption rates combined with optimized spatial arrangements may deliver substantial emissions reductions, suggesting that policymakers should prioritize supporting AV infrastructure alongside spatial planning adjustments. For cities with limited resources, the application of the proposed model could assist in better resource management and carbon savings by promoting AV adoption in communities.

Aside from the above-mentioned findings, the proposed model and solution method have several perspectives for further investigation. First, the experiments depicted the positive effect of co-evolution modeling on the carbon footprint management of land use and transportation systems based on the AV adoption rates that were gradually and externally fed into the system. Although the utilized approach was along with the book chapter's research objective, the adoption rates can be obtained through the interaction of production and consumption in the market. For this purpose, previous research studies (Dubey et al., 2022; Talebian & Mishra, 2018) could provide a ground for reflecting such dynamism in the prototype model presented in this chapter.

Second, in this chapter, the externalities of the AV manufacturing and lifecycle like battery production, raw material extraction for the AV production, and disposal were not included in the framework of the proposed model. Formulating such elements in the objective functions can be pursued in future research. In line with this extension, other environmental items can be modeled, concerning the structure provided in this chapter, in the case of data availability. Besides, the results of this study could be extended to evaluate other indicators such as travel distance, fuel consumption, and more in its current form.

Third, this chapter focused on CO_2 emissions. However, it should be noted that AV adoption and land use distribution have socio-economic impacts, and such components are also drivers of land use and transportation systems. One aspect that could be included in future research is to model accessibility and equity since AV adoption affects the social gaps between different groups of society. Land use distribution and the rate of utilizing AVs concerning accessibility and equity objectives to control social gaps and evaluate how land use and transportation patterns affect social equity could be carried out in other studies. Another avenue for this study is to reinforce the proposed model by embracing land use market interactions, social attitudes, and behavioral mechanisms.

The fourth aspect is the presentation of time in the model. Adding a temporal dimension to the proposed model is a potential extension of this study. The static traffic assignment and land use allocation models have theoretical and practical implications; however, AV traffic assignment and land use evolution are dynamic processes that occur over time, which may involve both agglomerative and divisive effects. Although the static approach serves the purpose of this research (i.e., showing if co-evolution land use and transportation planning could impact the CO_2 emissions and analyzing the effect of AV adoption rates integrated with the land use optimization on the footprint of these spatial systems), utilizing temporal facets in the land use allocation and traffic assignment could be considered in other studies by illustrating sequential decisions and the dynamic assignment models, respectively.

Fifth, the computational procedure presented in this chapter is the GA with an embedded GPA. It should be noted that the purpose of the solution algorithm was to find an answer to the formulated problem, and introducing a versatile one was not the objective. Although the method was acceptable for the experimental case of Sioux Falls and the analytical objective of this study, the algorithm's efficiency and scalability require tests for large cases. It is worth noting that GPA was executed for a large-scale problem by Ryu and Kim (2022), and the functionality of GA in such a bi-level optimization problem was portrayed by several pieces of research (Bilal & Giglio, 2023; Wang et al., 2019). This way, responsiveness and applicability of these algorithms in real and practical cases could be expected. However, in large-size cases, computational approaches such as parallelization and developing efficient algorithms could be fruitful in tackling the problem's complexity. Another perspective to this study in the methodological line and for solving the presented mathematical model is checking the Karush–Kuhn–Tucker (KKT) conditions. If the Mangasarian-Fromovitz constraints hold for a problem alongside certain conditions (i.e., the objective functions and constraints are satisfactorily smooth, and the solution space is non-empty and finite), the KKT conditions could assist in transforming the bi-level problem into a single-objective one in cases when the lower-level problem is convex. Subsequently, the KKT can be presented through Lagrangian coefficients and complementary constraints.

Sixth, it was attempted to provide a baseline scenario (i.e., S1) to analyze the results of this study. The developed scenario could partly reflect a common ground for evaluating the outputs. However, since drawing such comparisons produces insights into the model application, setting a baseline by introducing a similar model or at least matching the ideas put forward in different modeling approaches (e.g., spatial interaction or discrete choice models) for land use distribution and can be pursued in other studies.

Finally, the formulated problem had some parameters that could be subject to the sensitivity analysis. Apart from the algorithm parameters set by trial and error to output the final land use arrangement and traffic volume distribution, F_i^d and DEV^k are substantial ones that can be investigated in future research to show how significant these parameters are in reducing carbon emissions in land use and transportation systems. Such analysis is expected to provide policy- and decision-makers insights into defining the bounds of development and spatial systems in confronting climate change and carbon footprint mitigation.

6 Conclusions

The objective of this chapter was to develop and test a co-evolutionary model of spatial allocation and the adoption of AVs as a strategy to reduce CO_2 emissions and the impact of the transportation system on climate change. The results confirmed the model's effectiveness, meeting the study's primary goal. The analysis shows that the proposed approach, combining AV implementation with land use optimization, achieves a more effective reduction in CO_2 emissions than scenarios involving only AV adoption without spatial allocation changes. The research demonstrated that integrating AVs with spatial optimization enables better emission reduction outcomes, improving the functioning of urban transport systems and effectively managing urban spatial development. Even at moderate AV adoption levels, the co-evolution model proved more effective than implementing AVs alone without adequate spatial support.

These findings suggest that combining high AV adoption rates with optimized urban activity allocation can lead to significant resource savings and substantial CO_2 reductions, thus supporting cities' climate goals. A policy that supports AV infrastructure development alongside concurrent spatial planning adjustments is a key tool for more sustainable urban management. The co-evolutionary model not only supports CO_2 emission reduction but also contributes to the efficient use of urban resources. Notably, cities with limited resources may benefit from this approach, as the integration of spatial and transport planning aids in achieving both environmental and economic goals. However, it is worth emphasizing that the full implementation of this model requires further studies that could better account for the full lifecycle of AVs, from production to disposal, enabling a more comprehensive assessment of this technology's environmental impact.

Based on the results and limitations of the present study, several directions for further research are proposed. Another important research area involves testing the model in diverse urban contexts and conducting comparative analyses to assess the model's scalability and adaptability to local conditions. Such studies could help identify factors influencing the model's effectiveness in various cities and guide further adjustments to meet specific needs. An additional valuable extension would be analyzing the impact of AVs and spatial allocation co-evolution on urban accessibility for different social groups, especially those with limited mobility. This would help determine whether AV integration and spatial optimization could reduce inequalities in access to urban spaces and services.

Finally, a cost–benefit analysis encompassing not only direct environmental effects but also long-term savings and impacts on the local economy would provide valuable feedback for local governments, supporting them in making informed investment decisions and optimally allocating resources. The study's findings not only confirm the effectiveness of the co-evolutionary model in reducing CO_2 emissions but also highlight the potential benefits for cities that implement AVs alongside spatial optimization. Future studies incorporating temporal dynamics, diverse urban contexts, and additional environmental objectives could further enhance the model's application in spatial planning and climate policy.

References

Ahmed, A. A., Nazzal, M. A., Darras, B. M., & Deiab, I. (2023). Global warming potential, water footprint, and energy demand of shared autonomous electric vehicles incorporating circular economy practices. *Sustainable Production and Consumption, 36*, 449–462.

Alam Tabriz, A., Zandieh, M., & Mohammad Rahimi, A. (2013). *Metaheuristic algorithms in combinatorial optimization*. Saffar Publications.

Bakibillah, A., Kamal, M. A. S., Tan, C. P., Susilawati, S., Hayakawa, T., & Imura, J.-I. (2021). Bi-level coordinated merging of connected and automated vehicles at roundabouts. *Sensors, 21*, 6533.

Bilal, M. T., & Giglio, D. (2021). Inequity evaluation for land use and transportation model on introduction of autonomous vehicles. In *2021 7th International Conference on Models and Technologies for Intelligent Transportation Systems (MT-ITS), 2021* (pp. 1–7). IEEE.

Bilal, M. T., & Giglio, D. (2023). Analysing inequity in land use and transportation models by genetic algorithm for realistically quantified penetration rate of advanced driving system equipped vehicles. *Transportation Research Interdisciplinary Perspectives, 20*, Article 100841.

Borowska-Stefańska, M., Kowalski, M., Kurzyk, P., Sahebgharani, A., & Wiśniewski, S. (2022). The effect of COVID-19 pandemic on emitted PM2. 5 in urban road networks: Using loop data and Kriging method for passenger cars in the central part of the City of Lodz. *Aerosol and Air Quality Research, 22,* 210313.

Burger, C., Fischer, J., Bieder, F., Taş, Ö. Ş., & Stiller, C. (2022). Interaction-aware game-theoretic motion planning for automated vehicles using bi-level optimization. In *2022 IEEE 25th International Conference on Intelligent Transportation Systems (ITSC), 2022* (pp. 3978–3985). IEEE.

Carroll, D. L. (1996). Genetic algorithms and optimizing chemical oxygen-iodine lasers. *Developments in Theoretical and Applied Mechanics, 18,* 411–424.

Chakirov, A., & Fourie, P. (2014). *Enriched Sioux Falls scenario with dynamic and disaggregate demand*. Working paper. Future Cities Laboratory, Singapore-ETH Centre (SEC), Singapore, Tech. Rep.

Chen, B., Chen, Y., Wu, Y., Xiu, Y., Fu, X., & Zhang, K. (2023a). The effects of autonomous vehicles on traffic efficiency and energy consumption. *Systems, 11,* 347.

Chen, N., Xin, C., Zhang, B., Xin, S., Tang, D., Chen, H., & Ma, X. (2023b). Contribution of multi-objective land use optimization to carbon neutrality: A case study of Northwest China. *Ecological Indicators, 157,* Article 111219.

Collings, D. (2020). Carbon footprint benchmarking data for buildings. *The Structural Engineer: Journal of the Institution of Structural Engineer, 98,* 10–13.

Cordera, R., Nogués, S., González-González, E., & Moura, J. L. (2021). Modeling the impacts of autonomous vehicles on land use using a LUTI model. *Sustainability, 13,* 1608.

Detjen, H., Faltaous, S., Pfleging, B., Geisler, S., & Schneegass, S. (2021). How to increase automated vehicles' acceptance through in-vehicle interaction design: A review. *International Journal of Human–Computer Interaction, 37,* 308–330.

Dong, H., Huang, Q., Zhang, F., Lu, X., Zhang, Q., Cao, J., Gen, L., & Li, N. (2022). Path of carbon emission reduction through land use pattern optimization under future scenario of multi-objective coordination. *Frontiers in Environmental Science, 10,* 1065140.

Dubey, S., Sharma, I., Mishra, S., Cats, O., & Bansal, P. (2022). A general framework to forecast the adoption of novel products: A case of autonomous vehicles. *Transportation Research Part B: Methodological, 165,* 63–95.

Fathollahi-Fard, A. M., Hajiaghaei-Keshteli, M., Tavakkoli-Moghaddam, R., & Smith, N. R. (2022). Bi-level programming for home health care supply chain considering outsourcing. *Journal of Industrial Information Integration, 25,* Article 100246.

Fattahi, P. (2011). *Metaheuristic algorithms*. Bu-Ali Sina University.

Ferrari, E. (2022). IoT-enabled environmental monitoring for autonomous vehicle safety. *Journal of AI-Assisted Scientific Discovery, 2,* 86–107.

Hancock, P. A., Kajaks, T., Caird, J. K., Chignell, M. H., Mizobuchi, S., Burns, P. C., Feng, J., Fernie, G. R., Lavallière, M., & Noy, I. Y. (2020). Challenges to human drivers in increasingly automated vehicles. *Human Factors, 62,* 310–328.

Huang, S., Xi, F., Chen, Y., Gao, M., Pan, X., & Ren, C. (2021). Land use optimization and simulation of low-carbon-oriented—A case study of Jinhua, China. *Land, 10,* 1020.

Huang, W., Li, F., Cui, S.-H., Huang, L., & Lin, J.-Y. (2017). Carbon footprint and carbon emission reduction of urban buildings: A case in Xiamen City, China. *Procedia Engineering, 198,* 1007–1017.

Huang, Y., Zhu, P., Zhong, R., & Geroliminis, N. (2024). A bi-level approach for optimal vehicle relocating in mobility-on-demand systems with approximate dynamic programming and coverage control. *Transportation Research Part E: Logistics and Transportation Review, 192,* Article 103754.

IPCC. (2022). *Climate change 2022: Impacts, adaptation and vulnerability.*

Jayakrishnan, R., Tsai, W. T., Prashker, J. N., & Rajadhyaksha, S. (1994). *A faster path-based algorithm for traffic assignment*. University of California Transportation Center.

Jones, E. C., & Leibowicz, B. D. (2019). Contributions of shared autonomous vehicles to climate change mitigation. *Transportation Research Part D: Transport and Environment, 72*, 279–298.

Kim, H. K. (2024). The environmental benefits of an automatic idling control system of connected and autonomous vehicles (CAVs). *Applied Sciences, 14*, 4338.

Kopelias, P., Demiridi, E., Vogiatzis, K., Skabardonis, A., & Zafiropoulou, V. (2020). Connected & autonomous vehicles—Environmental impacts—A review. *Science of the Total Environment, 712*, Article 135237.

Le Hong, Z., & Zimmerman, N. (2021). Air quality and greenhouse gas implications of autonomous vehicles in Vancouver, Canada. *Transportation Research Part D: Transport and Environment, 90*, Article 102676.

Leblanc, L. J. (1975). An algorithm for the discrete network design problem. *Transportation Science, 9*, 183–199.

Li, J., Liu, C., Wang, Y., Chen, R., & Xu, X. (2023a). Bi-level programming model approach for electric vehicle charging stations considering user charging costs. *Electric Power Systems Research, 214*, Article 108889.

Li, L., Chen, Z., & Wang, S. (2022). Optimization of spatial land use patterns with low carbon target: A case study of Sanmenxia, China. *International Journal of Environmental Research and Public Health, 19*, 14178.

Li, T., Sun, H., Wu, J., Gao, Z., Ge, Y.-E., & Ding, R. (2019). Optimal urban expressway system in a transportation and land use interaction equilibrium framework. *Transportmetrica A: Transport Science, 15*, 1247–1277.

Li, T., Wu, J., Sun, H., & Gao, Z. (2016). Integrated co-evolution model of land use and traffic network design. *Networks and Spatial Economics, 16*, 579–603.

Li, W., Chen, Z., Li, M., Zhang, H., Li, M., Qiu, X., & Zhou, C. (2023b). Carbon emission and economic development trade-offs for optimizing land-use allocation in the Yangtze River Delta, China. *Ecological Indicators, 147*, Article 109950.

Lin, J.-J., & Feng, C.-M. (2003). A bi-level programming model for the land use–network design problem. *The Annals of Regional Science, 37*, 93–105.

Liu, C.-J., Chen, K., Bao, Z., Ng, S. T., Zhang, C., & Jiang, Z.-H. (2023). Assessing the impacts of connected-and-autonomous vehicle management strategy on the environmental sustainability of urban expressway system. *Sustainable Cities and Society, 99*, Article 104904.

Liu, J., Huang, Q., Han, Y., Chen, S., Pan, N., & Xiao, R. (2024). Bi-level optimization for de-icing position allocation and unmanned de-icing vehicle fleet routing problem. *Biomimetics, 9*, 26.

Luca, O., Andrei, L., Iacoboaea, C., & Gaman, F. (2023). Unveiling the hidden effects of automated vehicles on "do no significant harm" components. *Sustainability, 15*, 11265.

Martin-Gasulla, M., Sukennik, P., & Lohmiller, J. (2019). Investigation of the impact on throughput of connected autonomous vehicles with headway based on the leading vehicle type. *Transportation Research Record, 2673*, 617–626.

Moseman, A., & Paltsev, S. (2022). *Are electric vehicles definitely better for the climate than gas-powered cars*. MIT Climate Portal. Retrieved August 30, 2023.

Najafi-Ghalelou, A., Khorasany, M., & Razzaghi, R. (2023). Stochastic two-stage coordination of electric vehicles in distribution networks: A multi-follower bi-level approach. *Journal of Cleaner Production, 414*, Article 137610.

Nikitas, A., Vitel, A.-E., & Cotet, C. (2021). Autonomous vehicles and employment: An urban futures revolution or catastrophe? *Cities, 114*, Article 103203.

Norman, B. (2022). *Urban planning for climate change*. Taylor & Francis.

Orieno, O. H., Ndubuisi, N. L., Ilojianya, V. I., Biu, P. W., & Odonkor, B. (2024). The future of autonomous vehicles in the US urban landscape: A review: Analyzing implications for traffic, urban planning, and the environment. *Engineering Science & Technology Journal, 5*, 43–64.

Pal, P., Gopal, P., & Ramkumar, M. (2023). Impact of transportation on climate change: An ecological modernization theoretical perspective. *Transport Policy, 130*, 167–183.

Piedra-De-La-Cuadra, R., & Ortega, F. A. (2024). Bilevel optimization for the deployment of refuelling stations for electric vehicles on road networks. *Computers & Operations Research, 162*, Article 106460.

Ryu, S., & Kim, M. (2022). Solving a multi-class traffic assignment model with mixed modes. *Applied Sciences, 12*, 3678.

Sahebgharani, A., & Wiśniewski, S. (2024). Developing a bi-level optimization model for the coupled street network and land subdivision design problem with various lot areas in irregular blocks. *ISPRS International Journal of Geo-Information, 13*, 224.

Sahu, A. (2018). A methodology to modify land uses in a transit oriented development scenario. *Journal of Environmental Management, 213*, 467–477.

Silva, Ó., Cordera, R., González-González, E., & Nogués, S. (2022). Environmental impacts of autonomous vehicles: A review of the scientific literature. *Science of the Total Environment, 830*, Article 154615.

Singleton, P. A. (2019). Discussing the "positive utilities" of autonomous vehicles: Will travellers really use their time productively? *Transport Reviews, 39*, 50–65.

Szűcs, H., & Szűcs, J. (2024). The environmental sustainability potential of autonomous vehicles: An overview. *Periodica Polytechnica Transportation Engineering, 52*, 246–256.

Talebian, A., & Mishra, S. (2018). Predicting the adoption of connected autonomous vehicles: A new approach based on the theory of diffusion of innovations. *Transportation Research Part C: Emerging Technologies, 95*, 363–380.

UNFCCC. (2022). *Glossary of climate change acronyms and terms* [Online]. Retrieved May 15, 2024 from https://unfccc.int/process-and-meetings/the-convention/glossary-of-climate-change-acronyms-and-terms

Wang, G., & Han, Q. (2021). The multi-objective spatial optimization of urban land use based on low-carbon city planning. *Ecological Indicators, 125*, Article 107540.

Wang, Q. (2024). Study on the impact of autonomous driving technology on the economy and society. In *SHS Web of Conferences, 2024*. EDP Sciences.

Wang, Z., Han, Q., & De Vries, B. (2019). Land use oriented bi-level discrete road network design. *Transportation Research Procedia, 37*, 35–42.

Wegener, M., & Fürst, F. (2004). *Land-use transport interaction: State of the art.* Available at SSRN 1434678.

World Bank. (2006). *Urban transport and climate change* [Online]. Retrieved May 15, 2024 from https://www.worldbank.org/en/news/feature/2006/05/03/urban-transport-and-climate-change

Yan, L., Liang, J., & Yang, K. (2024). *Bi-level control of weaving sections in mixed traffic environments with connected and automated vehicles.* arXiv:2403.16225.

Ye, L., & Yamamoto, T. (2019). Evaluating the impact of connected and autonomous vehicles on traffic safety. *Physica A: Statistical Mechanics and Its Applications, 526*, Article 121009.

You, H.-Y., & Wu, C.-F. (2010). Carbon emission efficiency and low carbon optimization of land use: Based on the perspective of energy consumption. *Journal of Natural Resources, 25*, 1875–1886.

Zhang, M., Kong, M., Shi, H., Tan, W., Fathollahi-Fard, A. M., & Yaseen, Z. M. (2024). Bi-level programming for joint order acceptance and production planning in industrial robot manufacturing enterprise. *Computers & Industrial Engineering, 196*, Article 110471.

Zhang, P., Zhu, B., Zhao, J., Fan, T., & Sun, Y. (2023). Safety evaluation method in multi-logical scenarios for automated vehicles based on naturalistic driving trajectory. *Accident Analysis & Prevention, 180*, Article 106926.

Zhang, W., Guhathakurta, S., Fang, J., & Zhang, G. (2015). Exploring the impact of shared autonomous vehicles on urban parking demand: An agent-based simulation approach. *Sustainable Cities and Society, 19*, 34–45.

Zhang, X., & Zhang, D. (2023). Urban carbon emission scenario prediction and multi-objective land use optimization strategy under carbon emission constraints. *Journal of Cleaner Production, 430*, Article 139684.

Zhao, C., Yu, H., & Molnar, T. G. (2023). Safety-critical traffic control by connected automated vehicles. *Transportation Research Part C: Emerging Technologies, 154*, Article 104230.

Zhao, L., & Peng, Z.-R. (2010). Integrated bilevel model to explore interaction between land use allocation and transportation. *Transportation Research Record, 2176*, 14–25.
Zhou, X., & Zhou, J. (2021). Optimization of autonomous driving state control of low energy consumption pure electric agricultural vehicles based on environmental friendliness. *Environmental Science and Pollution Research, 28*, 48767–48784.
Zuo, F., & Li, T. (2019). Study on integrated design method of transport and land use for junction center station of urban rail transit. In *CICTP 2019*.

Alireza Sahebgharani is an Assistant Professor in the Department of Transportation Engineering at Isfahan University of Technology in Isfahan, Iran. He has published several scientific articles, presented at conferences, and participated in research projects. His research interests encompass land use and transportation systems modeling and optimization, urban dynamics, city logistics, SDSS development, and public transit planning.

Szymon Wiśniewski is an Associate Professor at the University of Łódź, serving in the Institute of Built Environmental and Spatial Policy within the Faculty of Geographical Sciences. He holds the position of Vice Dean for Development and Promotion of the Discipline of Socio-Economic Geography and Spatial Management. Additionally, he directs the Research Centre for European Spatial Policy and Local Development. His research focuses on transport infrastructure planning and development, transport geography, urban logistics, transport accessibility, spatial mobility, intelligent transport systems, and traffic engineering. His studies have addressed various issues, including the role of low-emission transport zones in shaping urban residents' mobility behavior, the sensitivity of transport systems to urban floods, and changes in travel behavior in Łódź following the implementation of Sunday trade restrictions. As Vice Dean, he oversees the research activities of academic staff, ensuring alignment with institutional goals and disciplinary standards. He monitors research progress, assesses publication and project activities, and documents the societal and economic impact of scientific endeavors. He also supervises the preparation of reports demonstrating the societal relevance of research, assisting researchers in compiling the necessary evidence. Szymon Wiśniewski has authored and co-authored over 130 publications, including four academic monographs. He has participated in numerous research projects, serving as the principal investigator in grants funded by the National Science Centre (NCN) and the National Agency for Academic Exchange (NAWA). His academic journey includes a bachelor's degree in Spatial Management (2009) and Logistics (2011), a master's degree in Spatial Economy (2011), a Ph.D. in Geographical Sciences (2014), and a habilitation in Social Sciences (2022). He has also completed postgraduate studies in Traffic Engineering and Transport Planning (2019) and National Security (2023). As an educator, Szymon Wiśniewski delivers courses aligned with his research interests, equipping students with knowledge to address contemporary challenges in transport planning and spatial development.

Smart Parking as a Tool for Climate Change Mitigation and Preparedness

Lasse Berntzen, Marius Rohde Johannessen, and Jarle Løwe Sørensen

Abstract Rapid urbanization and climate change are pressing issues that require innovative, research-based methods for mitigation and preparedness. This chapter explores the potential of the smart city concept, specifically smart parking systems, in managing the environmental impact of the transportation sector and enhancing climate change preparedness. A conceptual, semi-structured literature review approach is applied to investigate the utilization of sensor networks and crowd-sensing and explore their role in informing policy decisions and fostering the creation of smart cities. Moreover, this chapter presents a conceptual model of smart parking, which provides a detailed examination of the main components, including information sensing, infrastructure, direction guidance, reservation handling, and payment. The chapter concludes by discussing how this conceptual model can be applied in crisis management, including disruptions caused by climate change.

Keywords Smart parking · Smart cities · Sustainability · Smart traffic · Climate change mitigation

---

L. Berntzen · M. R. Johannessen · J. L. Sørensen (✉)
USN School of Business, Department of Business, History and Social Sciences, University of South-Eastern Norway, Borre, Norway
e-mail: jarle.sorensen@usn.no

L. Berntzen
e-mail: lasse.berntzen@usn.no

M. R. Johannessen
e-mail: marius.johannessen@usn.no

L. Berntzen
USN School of Business, Department of Business, Marketing and Law, University of South-Eastern Norway, Borre, Norway

M. A. Dulebenets et al. (eds.), *Transport and Climate Change*,
https://doi.org/10.1007/978-3-032-00563-2_5

1 Introduction

Smart cities have been on the agenda of researchers and practitioners for the last decade, and while broad, the concept of smart cities contributes to structuring a wide range of sociotechnical efforts to improve urban spaces and quality of life (Gil-García et al., 2015; Townsend, 2014). Smart cities refer to "*places where information technology is combined with infrastructure, architecture, everyday objects, and even our bodies to address social, economic, and environmental problems*" (Townsend, 2014). Hence, technology is an important aspect of smart cities. For this reason, this chapter will explore the concept of smart parking and its potential for crisis management. Smart parking refers to the use of sensors, intelligent displays and data analytics to aid drivers in finding a parking space and to inform and aid cities in optimizing the placement and use of this limited resource (Winter & Goel, 2021). We present an overview and model of smart parking and discuss how smart parking can be useful not only for drivers, but also as part of the crisis management tool box. Our intention with the present chapter is to provide an overview of the concept and discuss some potential scenarios where smart parking can aid crisis management efforts, and we do this by presenting a conceptual model of smart parking and discussing how this model can be applied for increased preparedness in a crisis situation.

The escalating impacts of climate change are increasingly threatening nations and cities globally, as evidenced by the devastating floods experienced across various European countries, including Spain, Austria, Germany, Poland, and Italy, in 2024 (Center for Disaster Philanthropy, 2024). A particularly destructive incident was the storm "Boris" which precipitated a month's worth of rain within 24 h, leading to the loss of 26 lives, widespread evacuations, and substantial infrastructure damage. Particularly hard-hit were Ukrainian refugees residing in Poland, specifically in Lądek-Zdrój, where 90 mothers and their children were rendered temporarily homeless after the floods decimated three accommodation centers. This catastrophic event led EU crisis management commissioner Janez Lenarcic to underscore the pressing need for enhanced disaster preparedness and response efforts at both national and European levels, stating that "*the global reality of climate breakdown has moved into the everyday lives of Europeans*" (Lenarčič, 2024). The subsequent crisis in Spain, which resulted from heavy rainfall causing significant flooding and over 200 fatalities, further emphasized this requirement. The rising water levels resulted in extensive damage and public outcry over the government's response efforts (Keeley, 2024).

Contemporary data reveals that approximately 60% of cities with populations exceeding 5 million are situated within 100 km of the coastline, with a significant proportion of their inhabitants living below the annual coastal flood level. This figure is projected to rise dramatically, and comprise 340 million people by 2050 (Kulp & Strauss, 2019). Consequently, the urgent need to implement innovative, research-based methods to mitigate the impacts of rapid urbanization is evident (Ruan et al., 2024).

City resilience refers to the ability of "*building resilient cities that are able to resist and absorb threats and are capable of adapting to and recovering from shocks and stresses is vital for the wellbeing of society*" (Khatibi et al., 2022). In that context, this chapter will specifically delve into the role of smart city technology in advancing mitigation efforts and preparedness for crises. The central argument lies in the significant role that smart parking technology can play in decision-making for crisis management, facilitated by real-time data derived from integrated sensors and Internet of Things (IoT) devices. By aligning this technology with other smart city systems, it is possible to ensure a swift and efficient response to crises. Smart city technology is not only capable of facilitating efficient planning and resource allocation during emergencies, but it also contributes to sustainability by minimizing unnecessary energy consumption. Moreover, with the integration of advanced security features like surveillance cameras and sensors, smart city technology can provide immediate detection and response to safety threats during a crisis scenario.

The main purpose of this chapter is to explore and understand the application of smart city technology for climate change mitigation and preparedness in the transportation sector. We aim to investigate the utilization of sensor networks in smart parking systems to illustrate the potential of smart city technology in managing the environmental impact of the transportation sector, aiding in emission control measures, and reducing the frequency and severity of climate-related crises. The chapter will also examine the role of crowdsensing, and the value of the data collected for informing policy decisions and contributing to the creation of more sustainable, smart cities. Through this research, we intend to provide valuable insights that can guide the transformation of the transportation sector toward a more sustainable and resilient future.

2 Methods

The chapter is a conceptual exploration of smart parking systems and how they can contribute to climate change mitigation and preparedness. As such, we have approached this as a semi-systematic literature review (Snyder, 2019). This approach involves four steps: design, conduct, analyze, and write. Design and conduct refers to research theme, keywords and search strategies, while analysis refers to how the papers are sorted and findings are structured. As smart parking in the context of this chapter is not a well-researched topic, it is suitable for this kind of exploratory research (Snyder, 2019). We searched our institutional library for combinations of keywords: "smart parking, preparedness, climate, smart city, smart parking frameworks". Papers were screened by reading titles and abstracts. In addition to the literature review, we base the chapter on conversations and brainstorming sessions we have had with city officials in our local region of Vestfold, Norway. Snowballing was also employed, as we found some of the cited articles through the references section of papers initially identified.

The remainder of the chapter is structured as follows. In Sect. 3, we describe the concept of "smart city" as an umbrella term for the use of technology to improve cities and transition to smart parking as one aspect of smart cities, which can contribute to climate change mitigation and preparedness. Moreover, Sect. 3 provides more details on smart parking and presents a conceptual model of smart parking solutions. In Sect. 4, we discuss how smart parking can serve as a tool for preparedness and climate change mitigation. Finally in Sect. 5, we provide important concluding remarks and present some limitations along with future research directions.

3 Literature Review

3.1 Smart City

The "smart city" has no unified definition, but most proposals include the use of information and communication technologies, with the main objective of improving the quality of life for its citizens. Quality of life is a vague concept, but authors point to such things as better services, reduced environmental footprint, and citizen participation as part of the concept (Fybr, 2024; ITU-T Focus Group of Smart Sustainable Cities, 2014; Smart Parking Ltd., 2024; Vrabie, 2016). The "smart city" can also be seen as an umbrella for ideas and projects that contribute to making the city more sustainable, efficient, and citizen centric. For this chapter, we apply the broad definition of Townsend (2014), who defines smart cities as "*places where information technology is combined with infrastructure, architecture, everyday objects, and even our bodies to address social, economic, and environmental problems*".

According to Gil-García et al. (2015) and Lazaroiu and Roscia (2012), the smart city concept covers a wide range of application areas, including communication, culture, energy, emergency services, environment and climate, health, leisure, safety and security, tourism, transport, and work. Essentially, it encompasses nearly every aspect of urban life. The framework of smart cities proposed by Gil-García et al. (2015) illustrates this by showing how smart cities are made up by government institutions and services; societal dimensions such as the knowledge economy, human capital, and a collaborative mindset; and the physical environment, including natural, built, and infrastructure. Data, information and digital technologies can work together with these dimensions to reach the goal of improving life for citizens (see Fig. 1).

"Smart cities" is a popular research area, with over 2 million results on Google Scholar. The topics range from sustainability and urban planning to citizen inclusion and participation. The following studies can serve as examples, just to illustrate the diversity of smart city research. Carrozzino et al. (2014) explore how augmented reality can enhance tourist satisfaction in cities. Roche (2014) presents geographic information system-based solutions for citizen participation. Stratigea et al. (2015)

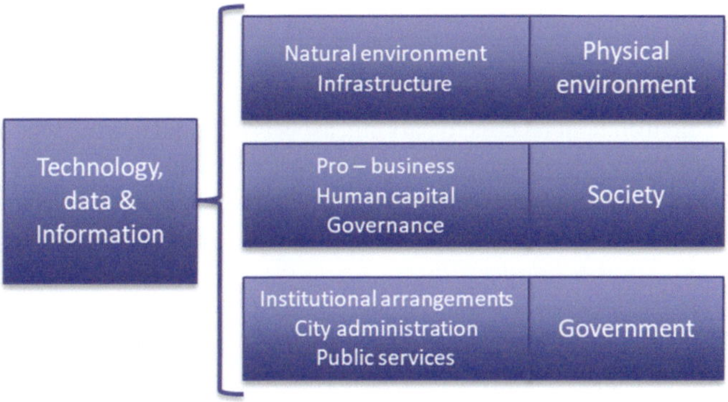

Fig. 1 Smart city framework

offer a framework for involving citizens in urban sustainability planning. Additionally, the research by Sáez-Martín et al. (2014) indicates that cities actively using social media or online communication are more successful in engaging citizens.

In terms of technologies being utilized, the Internet of Things (IoT), including smartphones and diverse sensors, as well as data analytics, are prominent topics in the literature. Bhoraskar et al. (2012) illustrate how smartphone sensors like GPS and accelerometers can be used for traffic monitoring, a feature we see in action with the traffic layer in Google Maps today. Further, Madni et al. (2017) demonstrate how collected data can be analyzed using various machine learning or statistical techniques to predict, recognize patterns, make forecasts, or aid in decision-making. Quercia and Saez (2014) show how social media data can be used to identify underdeveloped neighborhoods, while Appel et al. (2014) discuss how data mining and predictive analytics can forecast the number of vacant properties in a city. Massa and Campagna (2014) demonstrate how geotagged social media data can enhance urban planning projects.

The smart city literature also discusses climate change, mitigation and adaptation. Huang-Lachmann (2019), for example, performed a literature review of climate change adaptation in the smart city context and created a "model of synergetic coproduction" that illustrates how a combination of smart people, smart government, and technology can help with climate change adaptation in various ways. Leal Filho et al. (2022) argue for the use of artificial intelligence (AI) and show examples of how American cities have applied AI for climate change adaptation. Moreover, Chu et al. (2021) show that smart city development has led to pollution reduction in Chinese cities through the use of big data analytics for land allocation. Finally, Obringer and Nateghi (2021) demonstrate various ways to use big data for climate change mitigation and adaptation, which brings us to the purpose of the chapter, to examine one aspect of the smart city in more detail.

This chapter looks at one specific application of a smart city, i.e. smart parking. Perković et al. (2020) point out that smart parking solutions are essential to ensure smart and sustainable mobility. The idea of smart parking is to improve the parking situation by effectively directing vehicles to unoccupied parking spots and handling the payment of parking fees with minimal effort from the user. In terms of the smart city framework, smart parking is the technical solution addressing issues in the physical environment (access to parking, aspects of built environment in a crisis), supported by human capital (such as willingness to participate in crowdsensing and to use smart parking solutions) and institutional arrangements in local government (for example, effective use of data dashboards). We examine this in more detail in Sect. 4.

3.2 Smart Parking

In their book on smart parking, Winter and Goel (2021) indicate that smart parking builds on the premise that parking space is a limited resource that needs to be efficiently utilized to fulfill the goals of the smart city discussed earlier. The problem occurs when too many drivers are driving around to find a parking spot. These drivers are causing traffic problems since they are blocking other cars trying to get through traffic-dense areas. In their search for parking spots, the cars also contribute to the pollution of city space. A mix of geospatial technologies, the Internet of Things (IoT), computer vision, and data analytics can help alleviate these challenges.

According to Lee et al. (2008), the smart parking service normally includes on-road sensors along with intelligent displays that can effectively guide motorists and show the best path to the parking spaces in the city. Smart parking has numerous advantages. Effective allocation of parking spaces may reduce the total travel time of vehicles, emissions produced by these vehicles, and alleviate traffic congestion in busy metropolitan areas. The smart parking service can be seamlessly incorporated within the urban IoT infrastructure systems, as many European companies have already market products in place for smart parking applications. Moreover, the implementation of short-range communication systems, including Near Field Communication (NFC) and Radio Frequency Identification (RFID), will facilitate the electronic validation of parking tags in locations that are reserved for specific residents and individuals with disabilities. This will offer a better service to communities and will allow fast identification of parking violations. Their paper proposes a framework for smart parking solutions in diverse situations.

Our conceptual model consists of some of the components suggested by Lin et al. (2017) and also includes proactive components, as proposed by Ahmed et al. (2024) and Geng and Cassandras (2013), to create a parking management system, which collects information from sensors, the component that directs drivers to a free parking spot, and reservation handling where appropriate. As such, our model is a synthesis of the best practices from the existing literature on smart parking. Our conceptual model is shown in Fig. 2 and includes the following elements:

- **The parking management system** is the core of the model and communicates with various subsystems. The system receives information from the information

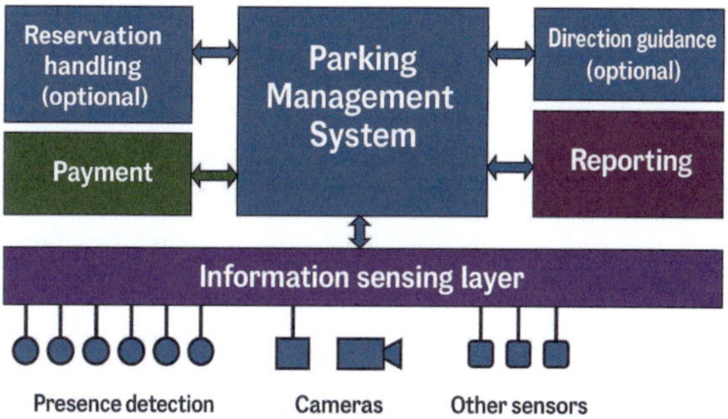

Fig. 2 Conceptual model of smart parking components

sensing layer, keeps track of occupied and unoccupied parking spots, and keeps track of vehicles and their parking periods.

- **The information-sensing layer** provides sensory information from each parking spot, from cameras collecting license plate information, and from other sensors, e.g. air quality and temperature.
- **The payment subsystem** keeps track of payments. Payments can be automatic based on license plate number and a contract between the car owner and the parking company. Payments can also be made through mobile phones or machines where the driver pays with a credit or debit card.
- **The reporting subsystem** provides managerial information on parking spot usage and can be important for planning additional parking spots or dynamically changing the pricing of parking.
- **The direction guidance subsystem** (optional) provides drivers with information on unoccupied parking spaces to reduce the time spent searching. Such guidance could be based on sensor information or crowdsensing.
- **The reservation handling subsystem** (optional) lets a user pre-reserve a parking space and has some mechanism to show that a specific parking space is reserved or physically blocked.

3.2.1 Parking Management System

The core component of "smart parking" is the parking management system. This component keeps track of the parking spots and their availability, communicates with sensors through a communication infrastructure, exchanges information with the payment handling system, provides information to the component directing the drivers looking for parking spots, and handles reservations made by drivers through a reservation subsystem.

3.2.2 Information Sensing

To direct drivers to unoccupied parking spaces, it is necessary to implement a system that keeps track of the occupancy of the parking spaces. Misra et al. (2019) state that early parking management systems for indoor parking were counting cars entering and leaving the parking area and showing the number of unoccupied spaces on a display board. On the other hand, current parking management systems utilize sensors to track occupied and unoccupied parking spots. The types of sensors used depend on environmental conditions. Outdoor parking spots imply more challenging conditions than indoor parking spots.

The common sensors for outdoor use are either sensors buried in the ground or cameras doing image recognition from above. These two types of sensors are the most typical applications. The most general approach is to put a sensor in a capsule embedded in the ground that detects the presence of a car by using ultrasound, light, or magnetism. Establishing wired infrastructure is expensive, and wireless communication is often a better option for outdoor use. An alternative is to use a camera from above to detect what parking spots are occupied. A camera can monitor several parking spots at the same time. For indoor use, the use of sensors is less demanding since the environmental conditions are better. Sensors can be mounted on walls, in ceilings, or on the floor. It is easier to establish wired infrastructure for power and communication. Sensors can be infrared, ultrasound, or magnetic.

The sensor platform typically has some wireless technology to communicate with network nodes. Urdain et al. (2012) described a wireless sensor network with a protocol to keep track of outdoor parking spaces. The unit used the Arduino Seeduino platform with the nRF24L01 low-power radio module and an infrared proximity sensor to detect the presence of a car. The authors reported to have made a cluster of ten units. It seems that the sensors were mounted next to each parking space. The main objective was to show the development of a feasible protocol for inter-unit communication. A somewhat similar approach was proposed by Jeffrey et al. (2012), using a wireless sensor network with an ambient light sensor used to detect occupied spaces. Wired sensors were found to be best suited for indoor use, as outdoor use is much more expensive due to the need to isolate wires for weather and put them in the ground. For wireless communications, several technologies have been proposed, such as ZigBee (smart home protocol), LoRa (low-range wireless), or Bluetooth. Wireless networks can be either centralized or mesh-based. Centralized networks rely on one single centralized hub for data transmission, while mesh-based networks are created from multiple nodes that share and distribute the signal. For example, in a Zigbee mesh network, each sensor node also acts as a repeater, and adding more units makes the network more robust.

3.2.3 Payment

An important part of smart parking solutions is the provision of payment services. In many cases, drivers need to pay parking fees, either because the owner (public or private) wants some profit from the land use or to restrict the use of a limited resource. A policy is necessary to decide on the amount to be paid. The parking fee can be progressive and vary according to the time of the day. Another policy issue is if the parking fee should be paid for a predefined period or if the user should pay for actual time used. The policy will also include penalties for not paying the parking fee or overstepping the period paid for. The use of technology is important to give the user the possibility to pay parking fees with minimum effort. Earlier, parking was paid for by putting coins in parking meters. Today, users pay by credit or debit cards, by mobile phone payment solutions such as Google/Apple Pay, via dedicated parking apps, or by dedicated devices. Figure 3 shows a screenshot from an app where the user selects the parking area on a map and then chooses the parking time. Another possibility is to use automated license plate recognition.

Huntsman et al. (2018) show that access to parking and reasonable parking solutions relates to property value. The potential financial gains can thus incentivize

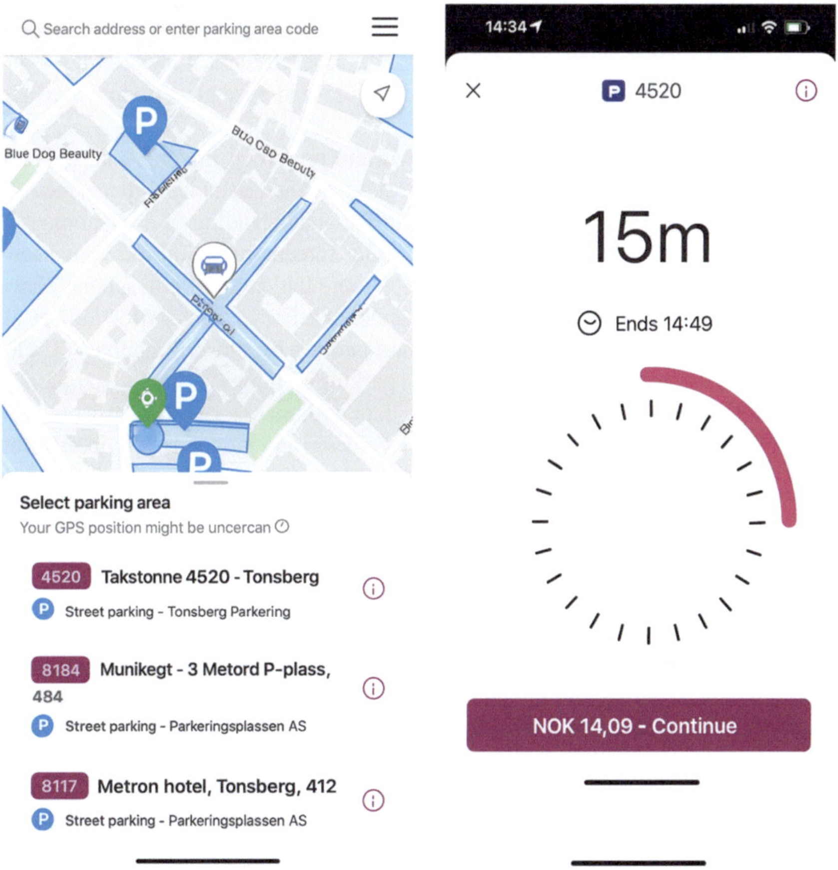

Fig. 3 Smartphone parking app

improved and smarter parking solutions. Companies such as Robotic Parking[1] and HL Mando[2] have developed robot technology that can park and stack cars more closely together, optimizing the use of indoor parking spaces and requiring less land for parking. Autonomous robots shuffle the cars around without customers having to park themselves. Such robots could lead to increased safety around parking spaces. Further, companies such as Volkswagen are also working on automated parking solutions, where the driver presses a button, and the car will find a parking spot and park itself.

3.2.4 Reporting

A combination of Internet-of-Things sensors, connected cars, and humans as sensors, perhaps combined with additional relevant data, provides possibilities for data analysis, statistics, and reporting. Atif et al. (2020) demonstrate how these data sources can be combined to create new data streams that offer valuable insights to decision-makers. There are examples of how time-series data can provide this type of insight. Piovesan et al. (2016), for instance, present an unsupervised machine-learning algorithm that uses statistical parking data to analyze and predict periods where the demand for parking is high.

Data-driven reporting is usually gathered and presented in some kind of data visualization dashboard. A data dashboard can be defined as a visual representation of data used to monitor conditions and facilitate understanding of a phenomenon. A typical dashboard will have several reporting screens based on different needs and offer the opportunity to drill down and explore the data. Several scholars have also examined how to design effective dashboards. For example, Mouchili et al. (2018) demonstrate data dashboards for smart parking by showing samples of relevant data output from smart traffic and smart parking solutions. Matheus et al. (2020) present ten design principles for smart city data dashboards, covering aspects such as data quality, timeliness, accuracy, types of views, and guidelines for interaction and presentation.

3.2.5 Providing Directions

Smart parking system can assist drivers to identify an available parking space in their vicinity. The data generated for the drivers may have many diverse options that range from public displays located on roadway segments to phone applications guiding the drivers to the available parking positions. Directions can be communicated through several channels. One option is to set up public displays through roadside boards showing the number of free parking spots in a location, a solution we see partially implemented in some cities through signs on parking garages showing the number

[1] https://www.roboticparking.com/.

[2] https://www.hlmando.com/en/solution/robot/parkingrobot.do.

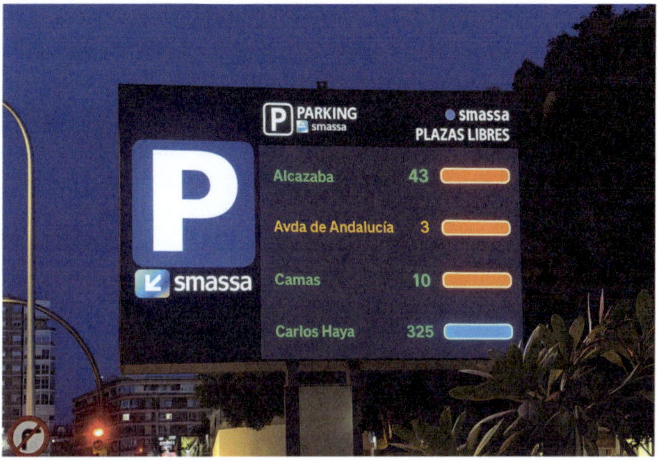

Fig. 4 Parking guidance system in Malaga, Spain

of available spots. This information is usually not centralized, but somewhat limited to single parking garages. Figure 4 shows a roadside board from Malaga, Spain, showing the availability of parking spaces in several locations. Directions can also be communicated via smartphone apps or the car's infotainment system, showing which area has the highest number of free parking spots and routing the user to the best option based on the destination input in the phone or car GPS.

An alternative, or supplement, to using technical sensors is to use "human sensors", i.e. letting people report (Berntzen et al., 2018). In this case, citizens could report unoccupied parking spaces, events, and similar issues that would have consequences for parking. For example, citizens could report if they observe an unforeseen and unwanted event, such as flooding or fire, which would then lead to a message being sent to people parked in the area that they need to move their cars. This type of applications already exists. Koster et al. (2014) developed "wePark", which is an Android application for smart parking based on observations of citizens. Available parking spots can be reported by citizens, and the application will eventually guide the drivers to the available parking locations. An earlier application developed by Google, which is called "Open Spot", relied on a similar methodology but was not successful. The authors suggested deploying the application to assess the motivations of users to report the available parking spots. Nevertheless, there was no follow-up study on this subject, and so far, this kind of reporting by fellow citizens has not been successful.

Even so, crowdsensing is an interesting approach to be considered, and it has the potential to provide valuable additional information along with the data supplied from technical sensors. The public can provide feedback on service, report problems with the parking solution, payment solution, or any issues related to the physical infrastructure or unforeseen events, to name a few examples. Earlier research from related fields shows that people want to participate if they see it as valuable to themselves or their local communities. Berntzen et al. (2018) made a study of crowdsensing in Germany and Norway using apps to report issues with local infrastructure (potholes, broken streetlights, waste, and other problems) and found that thousands of people made reports using these apps. An essential feature for increasing engagement was that citizens got feedback on the issues they reported and saw that reports led to actions or at least some kind of feedback. Lee et al. (2021) present a success story from crowdsensing of disease surveillance, where the key to success is the recruitment of a large panel of volunteers who report on the disease. Another study by Xu et al. (2022) of crowdsensing, in general, reveals that gamification vastly increases the chance of success and that monetary incentives could be another motivational factor in some cases.

3.2.6 Reservation

Another question is if car users should be able to reserve parking spots in advance. Reservation causes some additional challenges when it comes to showing that a parking space has already been booked by someone else. Some visual mechanism to clearly show that a specific parking space is reserved or that the space the car is currently in is booked by someone else from a particular time is needed. It is also a problem if someone does not move the car in time for the next customer to arrive. This visual mechanism must work regardless of weather and climate. For example, in countries where snowfall is regular, it is necessary to make sure that the visual mechanism is not blocked by snow. Also, sensors or cameras are needed that can recognize the car that has made the booking and register the visit. Further, reservations need to be connected to payment solutions, and probably the parking spot needs to be paid for even if it is not used or freed up if the car that has made the booking does not show up within a specific time window.

As reservations of individual parking spots offer some challenges, there is a need to discuss this need. Are parking spots allocated on a first-come first-served basis or is there a mechanism to pre-reserve a parking spot? Furthermore, in what circumstances would reservation of individual parking spots justify the additional cost of sensors needed in cold-weather climates?

4 Discussion

4.1 Smart Parking Applications in Crisis Management

Smart city technology and ingenious smart parking systems can be crucial in managing various disaster situations, not just flooding. Disaster situations include earthquakes, wildfires, and even man-made disasters. The use of sensor platforms, information sensing, and crowdsensing can contribute significantly to both crisis management and recovery. This is not explicitly discussed in the smart parking literature cited above. Hence, we discuss some potential use cases in this section.

In the case of an earthquake, for instance, smart parking systems can be integrated with seismic activity monitors to provide early warnings when an unusual seismic activity is detected. In the event of an impending wildfire, sensors in fire-prone areas can monitor temperature fluctuations and smoke levels, triggering alerts when threshold levels are exceeded. These alerts can be rapidly disseminated to citizens, allowing them to move their vehicles to safer areas. Infrastructure for city-wide environmental monitoring, for example, a central information hub, would increase the value of individual data points.

During a disaster, be it an earthquake, wildfire, or a man-made disaster such as a chemical spill, smart parking systems can guide emergency responses. Sensors can provide real-time data on the progression of the disaster, helping authorities assess the severity of the situation and strategize effective responses. If roads become inaccessible due to the disaster, the smart parking system can redirect drivers to alternative, safe parking locations. Such smart parking systems can help to prevent vehicles from becoming trapped or damaged.

In any disaster situation, citizens can play a crucial role through crowdsensing. By reporting incidents, road closures, or available parking spaces in safe locations, citizens can contribute to a more comprehensive understanding of the unfolding situation. While this reporting system has had limited success so far, in a disaster situation where timely information is of utmost importance, citizens may be more motivated to participate.

After a disaster, the data captured by the smart parking system can support post-disaster analysis. Such data can provide insights into the effectiveness of the disaster response, identify areas for improvement, and inform future disaster preparedness planning. The data can also inform decisions on infrastructure upgrades and repairs, such as the need for improved structural integrity in earthquake-prone areas or the construction of fire-resistant parking structures. The wireless technologies used in smart parking systems, such as ZigBee, LoRa, or Bluetooth, enhance the resilience of these systems even in disaster scenarios. Such technologies ensure that communication networks remain operational during a crisis, providing essential, up-to-date information. Revisiting our framework presented in Fig. 1, we summarize the discussion above in Table 1 to show how smart parking infrastructure can contribute to climate-related disaster management. Note that this is not a comprehensive list. Instead, selected examples of some of the possibilities of smart parking are presented for a crisis management context.

In summary, smart parking systems and smart city technologies can play a pivotal role in various disaster management contexts, including disasters caused by climate change. By providing real-time data, guiding emergency responses, encouraging citizen participation, and contributing to post-disaster learning, these systems can significantly enhance societal resilience in the face of diverse disaster situations.

Table 1 Conceptual model for smart parking and its contribution to climate-related disaster management

Conceptual model components	Contribution to climate-related disaster management	Example
Information sensing	Sensors monitor climate and natural environment variables. Crowdsensing or "citizens as sensors" supplements technological sensors	Early warning of seismic activity, wildfires, chemical spills, etc.
Direction guidance	Using information from the environmental monitoring system, guide citizens and emergency responses	Real-time data will help optimize routing of emergency services, route citizens to safe spaces, and help people avoid and circumvent dangerous areas
Reservation handling	Reservation of parking and ridesharing; crowdsensing through warnings of ongoing events	In a crisis, reservation handling can reserve critical parking areas for emergency services and send out messages to owners of parked cars that they need to move
Parking management system	Collects data from sensors and other data sources and acts as an information and analytics hub; post-disaster analysis	Post-crisis, the system would collect data from emergency services on how the event was handled, and data analysis could inform future decisions on similar situations

4.2 Policy Considerations

To successfully implement smart parking systems at scale, it is essential to consider several critical policy aspects, including data privacy, public–private partnerships, equitable access, and crisis management. Data privacy must be prioritized by adhering to regulations like the General Data Protection Regulation (GDPR), anonymizing data, securing data storage and transmission, and ensuring transparency and consent from users. Public–private partnerships are crucial, as they combine the strengths of both sectors, with clear roles and responsibilities, shared investment and risk, performance-based contracts, and robust data-sharing agreements. Equitable access should be ensured by addressing the digital divide, setting affordable pricing structures, designing systems accessible to disabled users, and involving communities in the planning process.

Additionally, incorporating crisis management policies is vital, as smart parking systems can provide real-time data, guide emergency responses, and support post-disaster analysis, thereby enhancing societal resilience in the face of disasters. To advance these initiatives, actionable recommendations for stakeholders include urban planners integrating AI and machine learning for predictive analytics in parking management, policymakers developing standardized frameworks for data sharing and privacy, and financial institutions supporting investments in smart infrastructure.

Emerging trends such as AI integration and IoT advancements should be explored in future research to optimize these systems further.

It is also crucial for central and local governments to collaborate with financial institutions and businesses to mitigate risks and ensure the widespread adoption of innovative solutions. By integrating these policy considerations, cities can leverage smart parking systems to improve urban mobility, reduce traffic congestion, enhance crisis management capabilities, and ultimately improve the overall quality of life while addressing the urgent need for climate change mitigation and resilience. By addressing these policy considerations, cities can create a robust governance framework that ensures the successful deployment and operation of smart parking systems. Urban planners should focus on integrating emerging technologies like artificial intelligence (AI) and machine learning to enhance predictive analytics and optimize parking management. Policymakers are encouraged to develop standardized frameworks that facilitate data sharing while maintaining stringent privacy protections, thus fostering an environment conducive to innovation and public trust.

Financial institutions and businesses play a pivotal role in this ecosystem by supporting investments in smart infrastructure, thereby driving economic growth and technological advancement. These stakeholders should also consider the potential for smart parking systems to contribute to broader urban resilience strategies, particularly in the context of climate-related disasters. For instance, smart parking technologies can be integrated with city-wide environmental monitoring systems to provide early warnings and guide emergency responses, helping to mitigate the impacts of natural disasters. Furthermore, local governments should prioritize community engagement to ensure that the benefits of smart parking systems are distributed equitably. This includes addressing the digital divide by providing alternative payment methods and information channels for those without access to smartphones or the internet and ensuring that systems are designed to be accessible to all users, including those with disabilities.

5 Conclusion and Future Work

Smart city technology, specifically smart parking systems, offers a promising approach for climate change mitigation and preparedness. The integration of sensor networks and crowdsensing allows for real-time monitoring and efficient resource allocation, contributing to societal resilience and environmental sustainability. The data collected from these systems is invaluable for policy-making and long-term climate change strategies. Despite some challenges, such as the unsuccessfulness of crowdsourced reporting of free parking spaces, these technologies hold significant promise.

This chapter provided a conceptual and literature-based overview of smart parking technology, highlighting its promising potential for climate change mitigation and preparedness. However, a notable limitation is the lack of empirical evidence to support our claims. Therefore, future research should focus on setting up empirical studies to validate these concepts, refining these systems, and examining how they can be integrated into broader preparedness policies. Additionally, exploring further applications of smart city technologies in climate change mitigation and resilience is crucial.

The findings underscore the urgent need for the continued development and implementation of smart, sustainable technologies in the face of our current climate crisis. By incorporating policy considerations related to data privacy, public–private partnerships, equitable access, and crisis management, and by providing actionable recommendations for various stakeholders, cities can leverage smart parking systems to improve urban mobility, reduce traffic congestion, and enhance societal resilience. Central and local governments, financial institutions, and businesses must collaborate to mitigate risks and ensure the successful adoption of these innovative solutions. This collaborative approach is essential for building a more sustainable and resilient urban future.

References

Ahmed, M., Hamad, M. S., Abdel-Khalik, A. S., et al. (2024). Improved utilization for 'smart parking systems' based on paging technique. *IEEE Transactions on Intelligent Transportation Systems, 25*(5), 3375–3388. Institute of Electrical and Electronics Engineers Inc.

Appel, S. U., Botti, D., Jamison, J., et al. (2014). Predictive analytics can facilitate proactive property vacancy policies for cities. *Technological Forecasting and Social Change, 89,* 161–173. North-Holland.

Atif, Y., Kharrazi, S., Jianguo, D., & Andler, S. F. (2020). Internet of things data analytics for parking availability prediction and guidance. *Transactions on Emerging Telecommunications Technologies, 31*(5), Article e3862.

Berntzen, L., Johannessen, M. R., Böhm, S., et al. (2018). Citizens as sensors: Human sensors as a smart city data source. In *SMART 2018: The Seventh International Conference on Smart Cities, Systems, Devices and Technologies, 2018.*

Bhoraskar, R., Vankadhara, N., Raman, B., et al. (2012). Wolverine: Traffic and road condition estimation using smartphone sensors. In *2012 4th International Conference on Communication Systems and Networks, COMSNETS 2012.* Epub ahead of print 2012. https://doi.org/10.1109/COMSNETS.2012.6151382

Carrozzino, M., Tecchia, F., Bergamasco, M., et al. (2014). From territory to smartphone: Smart fruition of cultural heritage for dynamic tourism development [online]. *Planning Practice and Research, 29*(3), 238–255. Routledge. https://doi.org/10.1080/02697459.2014.929837

Center for Disaster Philanthropy. (2024). *2024 Europe floods* [online]. https://disasterphilanthropy.org/disasters/2024-central-and-eastern-europe-floods/

Chu, Z., Cheng, M., & Yu, N. N. (2021). A smart city is a less polluted city. *Technological Forecasting and Social Change, 172,* 121037. North-Holland.

Fybr. (2024). *Making cities smarter with affordable, turnkey solutions* [online]. Retrieved November 19, 2024 from https://www.fybr.com/

Geng, Y., & Cassandras, C. G. (2013). New 'smart parking' system based on resource allocation and reservations. *IEEE Transactions on Intelligent Transportation Systems, 14*(3), 1129–1139.

Gil-García, J. R., Pardo, T. A., & Nam, T. (2015). What makes a city smart? Identifying core components and proposing an integrative and comprehensive conceptualization. *Information Polity, 20*(1), 61–87.

Huang-Lachmann, J. T. (2019). Systematic review of smart cities and climate change adaptation. *Sustainability Accounting, Management and Policy Journal, 10*(4), 745–772. Emerald Group Holdings Ltd.

Huntsman, B., Lasley, P., Metsker-Galarza, M., et al. (2018). *Finding the value of urban parking: An analysis of the impacts of smart parking systems on congestion and land values in Downtown Houston, final report*. Texas A&M Transportation Institute. Epub ahead of print 1 January 2018. https://doi.org/10.21949/1503647

ITU-T Focus Group of Smart Sustainable Cities. (2014). *Smart sustainable cities: An analysis of definitions*. Focus group technical report [online]. Retrieved November 19, 2024 from https://www.itu.int/en/ITU-T/focusgroups/ssc/Documents/website/web-fg-ssc-0100-r9-definitions_technical_report.docx

Jeffrey, J., Patil, R. G., Narahari, S. K. K., et al. (2012). Smart parking system using wireless sensor networks. In *SENSORCOMM 2012: The Sixth International Conference on Sensor Technologies and Applications*. IARIA.

Keeley, G. (2024). *Thousands protest over handling of Spanish flood disaster.* Reuters [online]. https://www.reuters.com/world/europe/thousands-protest-over-handling-spanish-flood-disaster-2024-11-09/

Khatibi, H., Wilkinson, S., Eriwata, G., Sweya, L. N., Baghersad, M., Dianat, H., & Javanmardi, A. (2022). An integrated framework for assessment of smart city resilience. *Environment and Planning B: Urban Analytics and City Science, 49*(5), 1556–1577.

Koster, A., Koch, F., & Bazzan, A. L. C. (2014). *Incentivising crowdsourced parking solutions*. Lecture notes in computer science (including subseries Lecture notes in artificial intelligence and Lecture notes in bioinformatics) (Vol. 8313).

Kulp, S. A., & Strauss, B. H. (2019). New elevation data triple estimates of global vulnerability to sea-level rise and coastal flooding. *Nature Communications, 10*(1), 1–12.

Lazaroiu, G. C., & Roscia, M. (2012). Definition methodology for the smart cities model. *Asia-Pacific Forum on Renewable Energy, 47*(1), 326–332.

Leal Filho, W., Wall, T., Rui Mucova, S. A., et al. (2022). Deploying artificial intelligence for climate change adaptation. *Technological Forecasting and Social Change, 180,* 121662. North-Holland.

Lee, L., Mukhi, S., & Bancej, C. (2021). FluWatchers a crowdsourcing approach: Crowdsourced disease surveillance success story: The FluWatchers program. *Canada Communicable Disease Report, 47*(9), 354.

Lee, S., Yoon, D., & Ghosh, A. (2008). Intelligent parking IoT application using wireless sensor networks. In *2008 International Symposium on Collaborative Technologies and Systems, CTS'08* (pp. 48–57).

Lenarčič, J. (2024). *Address by Commissioner Lenarčič to the European Parliament on the devastating floods in Central and Eastern Europe* [online]. https://ec.europa.eu/commission/presscorner/detail/en/speech_24_4743

Lin, T., Rivano, H., & Le Mouel, F. (2017). A survey of smart parking solutions. *IEEE Transactions on Intelligent Transportation Systems, 18*(12), 3229–3253. Institute of Electrical and Electronics Engineers Inc.

Madni, H. A., Anwar, Z., & Shah, M. A. (2017). Data mining techniques and applications—A decade review. In *ICAC 2017—2017 23rd IEEE International Conference on Automation and Computing: Addressing Global Challenges through Automation and Computing*. Institute of Electrical and Electronics Engineers Inc.

Massa, P., & Campagna, M. (2014). Social media geographic information: Recent findings and opportunities for smart spatial planning. *TeMA—Journal of Land Use, Mobility and Environment*. Epub ahead of print 15 May 2014. https://doi.org/10.6092/1970-9870/2500

Matheus, R., Janssen, M., & Maheshwari, D. (2020). Data science empowering the public: Data-driven dashboards for transparent and accountable decision-making in smart cities. *Government Information Quarterly, 37*(3), Article 101284.

Misra, P., Vasan, A., Krishnan, B., et al. (2019). The future of smart parking systems with parking 4.0. *GetMobile: Mobile Computing and Communications, 23*(1), 10–15.

Mouchili, M. N., Aljawarneh, S., & Tchouati, W. (2018). Smart city data analysis. In *Proceedings of the First International Conference on Data Science, E-learning and Information Systems* (pp. 1–6).
Obringer, R., & Nateghi, R. (2021). What makes a city 'smart' in the Anthropocene? A critical review of smart cities under climate change. *Sustainable Cities and Society, 75,* 103278. Elsevier.
Perković, T., Šolić, P., Zargariasl, H., et al. (2020). Smart parking sensors: State of the art and performance evaluation. *Journal of Cleaner Production, 262,* 121181. Elsevier.
Piovesan, N., Turi, L., Toigo, E., Martinez, B., & Rossi, M. (2016). Data analytics for smart parking applications. *Sensors, 16*(10), 1575.
Quercia, D., & Saez, D. (2014). Mining urban deprivation from foursquare: Implicit crowdsourcing of city land use. *IEEE Pervasive Computing, 13*(2), 30–36. Institute of Electrical and Electronics Engineers Inc.
Roche, S. (2014). Geographic information science I: Why does a smart city need to be spatially enabled? *Progress in Human Geography, 38*(5), 703–711.
Ruan, X., Sun, H., Shou, W., & Wang, J. (2024). The impact of climate change and urbanization on compound flood risks in coastal areas: A comprehensive review of methods. *Applied Sciences, 14*(21), 10019.
Sáez-Martín, A., Haro-de-Rosario, A., & Caba-Perez, C. (2014). A vision of social media in the Spanish smartest cities. *Transforming Government: People, Process and Policy, 8*(4), 521–544. Emerald Group Holdings Ltd.
Smart Parking Ltd. (2024). *We lead the way in car park management* [online]. Retrieved November 19, 2024 from https://www.smartparking.com/uk
Snyder, H. (2019). Literature review as a research methodology: An overview and guidelines. *Journal of Business Research, 104*, 333–339.
Stratigea, A., Papadopoulou, C. A., & Panagiotopoulou, M. (2015). Tools and technologies for planning the development of smart cities [online]. *Journal of Urban Technology, 22*(2), 43–62. Routledge. https://doi.org/10.1080/10630732.2015.1018725
Townsend, A. (2014). *Smart cities: Big data, civic hackers, and the quest for a New Utopia.* W.W. Norton & Company.
Urdain, L. O., Romero, C. P., Doggen, J., et al. (2012). Wireless sensor network protocol for smart parking application experimental study on the Arduino platform. In *AMBIENT 2012: The Second International Conference on Ambient Computing, Applications, Services and Technologies, 2012.* IARIA.
Vrabie, C. (2016). *E-government elements.* ProUniversitaria Publishing House.
Winter, S., & Goel, S. (2021). *Smart parking in fast-growing cities: Challenges and solutions.* TU Wien Academic Press.
Xu, H., Wu, Y., & Hamari, J. (2022). What determines the successfulness of a crowdsourcing campaign: A study on the relationships between indicators of trustworthiness, popularity, and success. *Journal of Business Research, 139*, 484–495.

Lasse Berntzen is a Professor at the University of South-Eastern Norway (USN) specializing in smart cities, digital governance, and emerging technologies. His research focuses on the intersection of technology and urban development, emphasizing digital solutions that enhance efficiency and sustainability in municipal and governmental services. With extensive experience in e-governance, data-driven decision-making, and applied artificial intelligence, Berntzen has contributed to numerous international research projects and publications. He actively collaborates with academia, industry, and government to explore innovative solutions to urban challenges. His research on smart cities has focused on the conceptual level as well as on the areas where technology has led to innovations, such as smart energy, smart transport, smart parking, and smart waste management. Based on his research, he developed a laboratory course on the Internet of Things (IoT) where students conduct practical experiments in the smart city domain. In 2009, he was appointed as a fellow of the International Academy, Research, and Industry Association

(IARIA) and has chaired many IARIA conferences. He is also a regular contributor to the Smart Cities International Conference, which is organized annually by the National University of Political Studies and Public Administration (SNSPA) in Bucharest, Romania. Berntzen is member of AIS, ACM, and IEEE.

Marius Rohde Johannessen is an Associate Professor of Information Systems in the School of Business at the University of South-Eastern Norway (USN) and head of the USN research group on digitalization management. He received his Ph.D. in Information Systems from the University of Agder in 2013 on the topic of digital political communication. His research and teaching focus on digitalization in the public sector (e-participation, smart cities, open government) and sustainability (smart grids, digitalization for sustainable forest management). He has participated in Horizon-funded projects on smart grids, sustainable digital forest management and eHealth. Moreover, Dr. Johannessen has published more than 50 papers in various journals and conferences in the Information Systems and eGovernment fields. He has served as a track/mini-track chair at conferences such as AMCIS, ECIS and EGOV/CEDEM/EPART. He is currently on the editorial board of the Norwegian journal of media studies and on the program committee of the Norwegian practitioner conference on public sector digitalization (NOKIOS).

Jarle Løwe Sørensen is an Associate Professor of Crisis Management in the School of Business at University of South-Eastern Norway (USN). He serves as the Director of the USN Research Center for Security, Crisis Management, and Emergency Preparedness. Dr. Sørensen earned his Doctor of Business Administration in Homeland Security Leadership and Policy from the United States. His primary research areas include emergency and crisis management, with particular emphasis on policy development, interagency collaboration, and crisis exercises. In addition to his research activities, Dr. Sørensen lectures and supervises students at the undergraduate, graduate, and Ph.D. levels. Dr. Sørensen brings over a decade of professional experience in law enforcement and information security to his academic role, enhancing his practical and theoretical contributions to the field.

From Traffic Jams to Climate Solutions: The Micromobility Revolution in Urban Areas Based on Experience from Slovakia

Roland Németh and Marcel Horňák

Abstract Urban areas face increasing challenges due to climate change impacts, prompting the need for innovative transportation solutions that mitigate emissions and enhance resilience. This chapter explores the role of micromobility in urban climate mitigation, with a focus on the unique experience from Slovakia. Micromobility, with bicycles, shared bicycles, and scooters, emerges as a promising solution for curbing greenhouse gas emissions and promoting sustainable urban mobility. These micromobility options can navigate city streets and congested traffic more efficiently than traditional vehicles, presenting efficient alternatives for short-distance travel. However, their potential to help achieve environmental goals is often obstructed by the insufficient knowledge of travel patterns. Drawing on a case study, this chapter employs a data-driven approach to understand the environmental advantages of micromobility, underscoring its role in mitigating climate change impacts within the Slovak urban context. It also examines factors influencing the adoption and success of micromobility programs, including infrastructure development and policy frameworks. The findings underscore micromobility's potential in bolstering urban climate resilience and guide the development of strategies to encourage the mode shift from car travel to more space-efficient and healthier transport options, contributing to a more sustainable and livable urban environment.

Keywords Climate change · Urban environment · Transport · Micromobility · Bicycle leasing · Slovakia · Trnava

R. Németh · M. Horňák (✉)
Department of Economic and Social Geography, Demography and Territorial Development, Comenius University Bratislava, Bratislava, Slovakia
e-mail: marcel.hornak@uniba.sk

R. Németh
e-mail: nemeth79@uniba.sk

M. A. Dulebenets et al. (eds.), *Transport and Climate Change*,
https://doi.org/10.1007/978-3-032-00563-2_6

1 Introduction

Urban environment in developed societies has been facing numerous challenges, among which climate change resilience belongs to the most important ones (Brand et al., 2021; Ceder, 2021). According to the recently published UN report on urban development challenges (World Cities Report, 2024), cities are responsible for a considerable portion of greenhouse emissions worldwide. On the other hand, numerous actions developed by cities to mitigate climate change impacts make them leaders in climate change mitigation (Kumar, 2021; Mi et al., 2019; Sancino et al., 2022). As World Cities Report (2024) indicates, cities should be and in many ways are at the forefront of addressing the challenges of climate change. They usually concentrate relevant political institutions, universities, and financial resources to introduce innovations in climate change agenda (McCowan et al., 2021; Sancino et al., 2022). As a result, in some aspects, urban emissions are reportedly below national averages (Cai et al., 2022; World Cities Report, 2024).

Many studies refer to the fact that the transport sector is one of the major contributors to global greenhouse gas emissions and climate change (Brand et al., 2021; Tang, 2022; Zhang & Fujimori, 2020). Specifically, road transport mode appears to be a substantial source of greenhouse emissions (Zhang & Fujimori, 2020), but it seems that the urban environment offers better conditions to bring solutions to reduce road transport intensity and impacts by applying alternative mobility solutions (Ho & Tirachini, 2024). Not surprisingly, World Cities Report (2024) ranks transport and mobility solutions for climate actions in the core of urban policies.

Principally, there are several basic approaches to urban transport and mobility actions leading to lower greenhouse gas emissions and adaptation of urban environment to climate change. Urban planning and design of a city layout are key elements affecting mobility demands in the urban environment. This fact has led to numerous urban-structure adaptation strategies, with the 15-min city concept being one of the most frequent ones recently (Allam et al., 2022; Khavarian-Garmsir et al., 2023). Based on the 15-min city concept, residents of communities are able to access the majority of their daily needs with a 15-min walk, or bike, or public transit options from their place of residence. Apart from that, Palermo et al. (2020) make an important point that transport upgrading and new mobility solutions remain in the core of urban policies aimed at sustainable urban development. According to numerous studies (Alka et al., 2024; Bebber et al., 2021; Ceder, 2021; Ho & Tirachini, 2024; Kołoś & Taczanowski, 2016; Rišová, 2020), public transport upgrading and enlargement, support of electromobility, micromobility and walkability, and application of new IT solutions into urban mobility management remain the most important sustainable transport and mobility solutions applied in urban policies worldwide.

The position of micromobility in urban mobility in Slovakia is marginal with insufficient representation and rather of marginal importance in the transport performance. According to the most recent population census, only around 4% of urban residents commute to work or to school by bicycle or scooter (SODB, 2021a). The situation is well described by the statement of Pucher and Buehler (2008), who claim that bicycle is marginal in the transport system of many countries and inhabitants primarily perceive it as a means of leisure-time mobility. This also applies to Slovakia, which is clearly the result of the state's transport policy, where bicycle transport has been neglected since the end of the twentieth century. In the post-socialist countries of Central and Eastern Europe, including Slovakia, the period after 1989 was significantly influenced by the rapid increase in passenger-car use (Michniak, 2019), which occurred later compared to Western countries (Pucher & Buehler, 2005). The expected negative impact on the micromobility share is also the increase in the intensity of traffic in and around cities and their surroundings as a result of suburbanization with the preference of passenger cars as the main mean of transport of the residents.

Considering the micromobility development, large Slovak cities can be described as "starter cities". This term refers to cities that have a low share of bicycle transport in their modal split and are without a developed infrastructure, sufficient political support, and with underdeveloped bicycle culture (Dufour, 2010). In such cities, people often do not perceive bicycle's potential and are skeptical of changes. Political and technical skepticism leads to interventions that have only a symbolic character without real effects on the development of this mode of transport (Bicalho et al., 2019). On the other hand, it is positive that in recent years the public interest in the support of micromobility has increased in Slovakia and in some cities the situation has improved significantly. Increasingly, we are witnessing the implementation of shared-bicycle systems throughout the country, although with a delay compared to Western and Northern Europe (Meddin et al., 2024).

According to the authors' knowledge, since the introduction of the first system in 2016, bikesharing services have already appeared in more than 30 cities and towns in Slovakia. Currently, however, we know very little about the behavior patterns of micromobility users in Slovakia, which is a barrier to its effective planning. The cities where the operation of bikesharing has already been analyzed are Žilina (Kubaľák et al., 2021a) and Košice (Kubaľák et al., 2021b). Also, the regime of urban bike routes was evaluated in the territory of the City of Trnava (Németh, 2024). At the level of the country's districts, the spatial distribution of the use of bicycles and scooters was investigated by sociodemographic characteristics of their users (Chromeček, 2024). Based on the limited amount of scientific literature describing the situation in Slovakia, we can conclude that knowledge of the behavior of micromobility users and their position in Slovak cities is insufficient.

In this chapter, our ambition is to focus on a mid-size city experience with micromobility concept support. With about 63,000 residents (as of 2023), the city of Trnava is Slovakia's 7th largest city. The city has recently become the country's leader in biker-friendly solutions and investments, enhancing also new platforms for shared micromobility services. The city authorities in Trnava have also been systematically working on mobility data collection, which allows us to look deeper at the effects of a long-term policy toward sustainable mobility in the city. Our intention is to investigate micromobility usage in the City of Trnava, guided by two main research objectives. First, we examine the socio-demographic profiles and usage patterns of the country's only bike-leasing program users. Specifically, we try to understand whether this unique initiative within urban mobility can reduce reliance on private car use and bring environmental benefits to the city's traffic system, addressing a gap in research on long-term bicycle leasing and user behaviors related to modal shift, particularly in car-dominated urban settings. Second, we aim to understand the overall bicycle usage trends over time within the city environment, focusing on peak-hours car traffic, in response to the claim by Hudde and Wessel (2024) that cycling can only achieve its full potential if individuals cycle consistently across different seasons and weather conditions.

2 Micromobility as a Solution

The importance of micromobility in cities is indisputable (Abduljabbar et al., 2021). There is currently no standard definition of this term. According to ITF (2020), micromobility can be defined as a rapidly developing range of light vehicles. As stated by ITF (2024), there are several ways to classify micromobility, and its definition is focused on safety with an emphasis on the two main factors that influence the severity of accidents—speed and weight. According to this classification, the main feature of individual micromobility vehicles is that they do not exceed a weight of 350 kg and a speed of 45 km/h. This classification includes vehicles powered by both human power and electric energy, including bicycles, e-bikes, scooters, e-scooters, etc.

According to Shaheen and Cohen (2019), micromobility also includes shared schemes such as shared-bike systems (docked or dockless). Micromobility also includes long-term bicycle-lease systems, which have become attractive for their environmental and consumer benefits (ITF, 2024). The most well-known example, Swapfiets, was launched in the Netherlands in 2014 as the world's first bikers' membership model primarily designed for students, providing members with their own bike for a fixed monthly fee (Ma et al., 2020). This leasing model achieves low greenhouse gas emissions per rider-kilometer and offers additional advantages to the users, such as fewer maintenance needs and less frequent servicing requirements since bikes remain with individual users for extended periods (ITF, 2024).

2.1 *Environmental Benefits of Micromobility*

When investigating the environmental benefits of transport, one of the methods used in the literature for assessing the environmental impacts of micromobility is a complex life cycle analysis of vehicles—LCA (de Bortoli & Christoforou, 2020; ITF, 2020, 2024; Severengiz et al., 2020). The assessments take into account emissions that arise from four main components (vehicle production, vehicle use, operating costs, and infrastructure) and are often expressed in g CO_2/pkm for individual vehicles. According to the cited studies, the environmental benefits of private micromobility are higher compared to shared micromobility. These works also show that the most ecological type of micromobility is the private bicycle, whose life cycle emissions are 10.5 times lower for each passenger kilometer compared to a passenger car with a combustion engine and 6 times lower compared to an electric car (ITF, 2024).

The most recent model indicates that the life cycle emissions of the selected vehicles are as follows (ITF, 2024): private bicycle—15 g CO_2/pkm, leased bicycle—17 g CO_2/pkm, shared bicycle—26 g CO_2/pkm, private e-bike—30 g CO_2 /pkm, private e-scooter—42 g CO_2/pkm, shared e-bike—46 g CO_2/pkm, shared e-scooter—61 g CO_2/pkm, passenger car with 1.5 passengers—160 g CO_2/pkm, electric car with 1.5 passengers—91 g CO_2/pkm, bus with 15.3 passengers—90 g CO_2/pkm, taxi ride—275 g CO_2/pkm. Based on these facts, we can conclude that the environmental benefits of using micromobility means differs based on the type of vehicle and the extent to which people use them to switch from other less sustainable modes of transport. These means are a suitable alternative for short to medium-long journeys, most of which are made in Slovak cities by passenger cars. In Trnava, journeys up to 5 km make up approximately 44% of all journeys made by car (Kyp et al., 2020), while in Nitra (another Slovakia's city of a similar size), this value is at the level of 45% (Kyp et al., 2019). In addition, short journeys by car generate a disproportionate amount of emissions due to the so-called "cold starts" (De Nazelle et al., 2010), which can only further strengthen the arguments about the importance of switching to more sustainable modes of transport in urban environments.

Data on the substituted mode of transport are most often obtained in two ways: it is either the mode of transport that the respondents indicated before starting to use the monitored means of transport ("previous mode"), or a hypothetical choice of transport in case the monitored means of transport were not available ("alternative way") (Bigazzi & Wong, 2020). Many micromobility studies report mixed results in terms of commuting substitution. Research on shared bikes shows that sharing services often replace walking and public transport instead of cars. A London study shows that only a small proportion (2%) of bikesharing users rely on this means as a substitute for driving, with the majority of trips replacing walking or other sustainable modes of transport (Fishman et al., 2014). On the other hand, in Dublin, this value of private cars has increased to approximately 20% (Murphy & Usher, 2015). In Lisbon, the research of Teixeira et al. (2023) compared this phenomenon before and during the COVID-19 pandemic. Their results show that in both time periods, public transport was most often replaced, followed by walking. The share of replacement of private cars in the city was at the level of 13.9 and 16.4%.

Conversely, in Barcelona, Roig-Costa et al. (2024) found that only 3.4% of bike-sharing system users switched to the service from private cars. Bieliński et al. (2021) focused on the example of Tricity (Gdańsk-Sopot-Gdynia agglomeration) in Poland and concluded that users of private vehicles do not show a significantly higher probability of renting an e-bike after the introduction of similar service. Greater modal shift at the expense of private cars was achieved in cities outside Europe, e.g. Sacramento, San Francisco, and Melbourne (Barnes, 2019; Fishman et al., 2014; Fukushige et al., 2021). Summarizing the 19 reviewed studies, Teixeira et al. (2021) found that bike-sharing mostly replaces public transport and walking (70%) and partly passenger cars and motorcycles (10%). Fishman et al. (2014) pointed out that the most significant modal shift at the expense of passenger cars would be experienced in cities with a high dependence on passenger car transport.

Research on shared e-scooters also suggests that this alternative often replaces greener modes of transport such as walking or public transport. A study by Fearnley et al. (2020) in Oslo reports that up to 83% of respondents switched to e-scooters from walking and public transport, while only 8% switched from taxi and passenger cars. In Paris, before the use of shared e-scooters, up to 72% of passengers used walking or public transport, and only 10% used private cars, taxis or ride-hailing (Christoforou et al., 2021). In Vienna, Laa and Leth (2020) compared the modal shift of shared e-scooter users with e-scooter owners. The authors found differences between these groups, but walking and public transport were the most affected.

Some studies have calculated the impact of micromobility on greenhouse gas emissions, taking into account emission factors and modes of transport substitution patterns. Reck et al. (2022) reports on using a mode choice model that shared e-bikes and scooters in Zurich, which caused more CO_2 emissions than the means of transport they replaced. On the contrary, there are privately owned versions of these vehicles that generated less CO_2 emissions. These findings about the impacts of shared e-scooters on the environment are also confirmed by de Bortoli and Christoforou (2020) and Moreau et al. (2020), which implies that shared micromobility may not be a low-carbon alternative to urban mobility in every situation.

The outcomes show that the literature dealing with the impact of different types of micromobility on the use of individual modes of transport is richly represented, but there is only limited evidence on leasing bicycles. The only work that partially dealt with the modal shift from leasing bicycles and the factors that influence it is Ma et al. (2020). The authors investigated the dynamics of changes in the use of individual modes of transport caused by shared bike systems, such as OV-fiets (docked bikesharing), Mobike (dockless bikesharing), and Swapfiets (bicycle-lease), and the modal shift of commuting in the City of Delft. Their results show that 30.5% of Swapfiets users have reduced the use of their own bicycle, which was followed by walking (8.6%) and Mobike (3.8%). It should be noted, however, that these data are related to the dynamics of changes in the use of modes of transport and not to their replacement. In this context, most Swapfiets users have switched to this alternative for commuting to work from their private bike. The modal shift from a passenger car to a leased bicycle did not occur at all, which is not so surprising, since 85% of users are students (Ma et al., 2020).

2.2 *Increasing Micromobility at the Expense of Car Transport*

The reviewed literature clearly shows that micromobility has the potential to be one of the strategies that support the transition to sustainable transport in urban environment. However, to maximize its impact on reducing the use of cars, it is necessary to promote measures that encourage the shift to this mode of transport (Fishman et al., 2014). In Slovakia, due to late adoption of the first pro-cycling measures and the absence of historical data on the modal split, it is difficult to assess their impact on the modal shift. How can cities in Slovakia achieve an increase in bicycle traffic at the expense of individual car traffic? Is this change even possible?

According to Gössling (2020), changing the mobility culture is impossible when there are no transport alternatives. A fundamental prerequisite for increasing the share of bicycle traffic in cities with a low level of this mode of transport is the development of cycling infrastructure, which improves the real and perceived safety of cycling (Adam et al., 2020). The importance of cycling infrastructure is also pointed out by Pucher and Buehler (2017), who claim that a significant and rapid increase in the level of bicycle use was recorded in cities that made great progress in building segregated bike lanes in recent years. Moreover, Buehler et al. (2017) show that cities such as Munich, Berlin, Hamburg, Vienna, and Zurich have fundamentally reduced the share of car traffic and increased the level of bike traffic and public transport despite high motorization. According to the authors, the increase in bike traffic was induced by expansion and improvement of cycling infrastructure and public parking for bicycles, better integration of bicycles with public transport, training for children, a wide range of promotional programs, but also the traffic calming.

All the aforementioned cities have more than half of their road network calmed down to 30 km/h or less, which brings safe and comfortable cycling without any need to build additional infrastructure (Buehler et al., 2017; Pucher & Buehler, 2017). A similar conclusion was reached by Caulfield (2014) in Dublin, where the increase in bicycle traffic in the central part of the city can be attributed to significant investments in cycling infrastructure, the reduction of speed limits, and the functioning of the shared bicycle system in the area. This demonstrates that the transition to low-carbon modes of transport is possible. However, this transition not only involves building cycling infrastructure but also much more complex measures. These can also be related to changes in land use, urban development, other solutions to support the use of bicycles or restrictions on car traffic, the success of which is mainly the result of their coordinated implementation and interconnection, which allow to fully using their joint effect (Forsyth & Krizek, 2010). Ballo et al. (2023) state that although pro-cycling policies are becoming more and more attractive in the world, it is still not clear to what extent cycling could replace driving by car and what consequences this would bring.

It is also important to mention the temporal patterns of micromobility use, as O'Brien et al. (2014) reports that shared bicycles in some cities are most intensively used on working days during morning and afternoon commutes. Similar conclusions are shown by Reck et al. (2021) in Zurich on shared e-bikes, indicating the potential of this alternative to play an important role in reducing car traffic during the main commute wave. Wang and Zhou (2017) point to 96 US urban areas where micromobility, specifically shared bicycles, has a direct impact on reducing rush hour traffic congestion with a greater effect in larger cities than smaller ones. Jensen et al. (2010) claim that bicycles also have the ability to compete with cars in terms of travel speed, especially during the morning rush hour. They demonstrate this fact in the central city districts of Lyon (France) using the example of shared bicycles. The fact that shared bicycles can compete with cars is also confirmed by Faghih-Imani et al. (2017) in New York. The results of their work showed that in densely populated urban areas, during traffic jams or on shorter distances of up to 3 km, bicycles often reach or exceed the speed of taxis.

3 Trnava Case Study

3.1 Description of the Study Area

Trnava is a regional centre in western Slovakia, located approximately 50 km northeast of Bratislava. According to the Statistical Office of the Slovak Republic, in 2021, Trnava had 63,803 inhabitants and a population density of 891.88 inhabitants/km^2 (SODB, 2021b). The flat terrain and urban compactness of the city provide ideal conditions for the development of micromobility on its territory, but despite this, its share in the transport modals split is only 5%, while passenger cars make up more than 50% (see Fig. 1). Up to 20% of trips shorter than 1 km and almost 45% of trips up to 5 km are made by car, which is an alarming statistic at first glance. On the other hand, there is a huge potential for replacing these trips with alternative modes of transport. This potential is further enhanced by the fact that approximately 70% of the population has the opportunity to use a classic bicycle at any time, and approximately 2.5% of the population has an electric bicycle (Kyp et al., 2020). Even if the latest statistics do not reflect this sufficiently yet, the city is currently considered one of the leaders in the field of bicycle transport in Slovakia (Cycling Coalition, 2023). How is that possible?

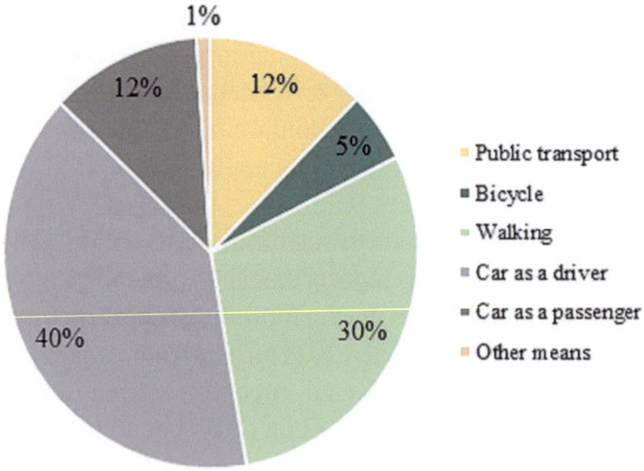

Fig. 1 Modal split of passenger transport in Trnava (as of 2019). *Source* Own elaboration based on Kyp et al. (2020)

3.2 Policy Interventions to Improve Cycling Conditions

Currently, the municipality is focusing on the development of bicycle transport and improving mobility in the city through several measures. These measures are related to the Economic and Social Development Programme of the City of Trnava for the years 2014–2025 with a view promoting the 2030. Within this strategic document, sustainable mobility is one of the four priority areas, including measures to increase the share of environmentally friendly transportation, such as electromobility and cycling, as well as the introduction of new transportation services (City of Trnava, 2024). Support for cycling is also included in another approved strategic document, the "Sustainable Mobility Plan of the Regional City of Trnava and its Functional Area", which aims to enhance the attractiveness and competitiveness of public passenger transport, particularly non-motorized transport. Around 26 km of cycle paths have already been built within the city. As part of the infrastructure, 20 bicycle racks are located at the main commuting points. 560 bike parking lots for bicycles at primary schools were established with a grant from the Ministry of Transport and Construction of the Slovak Republic. There is a Biketower at the railway station, proving to be an effective solution for safe bicycle parking.

One of the very recent steps to support bicycle transport is the "Žltý bajk" (Yellow Bike) project, which started in Trnava in October 2022 as a service primarily intended for high school students. This project, which was inspired by the successful Swapfiets model, represents a long-term rental of city bicycles for a symbolic €30 for the whole year. However, within the first three months, the project failed to gain students' attention (only 25 were rented by students in total). In December 2022, the service was offered to the general public, who showed great interest in Žltý bajk. Students and employees of secondary schools in Trnava and residents of the city can apply for bicycles at any time. The project has gradually expanded since its inception and has 500 active users as of September 2024.

There is also a system of shared e-bikes operating in the city, the first season of which already took place in 2019. Parking of electric bikes works on the basis of virtual stations that users can search for using the Freebike mobile application (Arboriabike, 2024). The seasonal fee for using the service is €20, returning outside the station costs €1, starting fee (up to 10 min) is €0.50, from the 10th minute €0.10 for each started minute, and the maximum rental duration is 120 min (City of Trnava, 2024). There are currently around 100 electric bicycles in operation. Cargo bikes are also an initiative of the municipality available for rent to users of shared bikes or Yellow Bikes.

Traffic calming is another important policy in the city. The goal is to have all local collector roads with a maximum permitted speed of 50 km/h and local streets with a speed of 30 km/h. Zones of regulated parking are automatically transformed into zones 30, which increases the safety of pedestrians and micromobility users. The city also regularly participates in various campaigns to support sustainable mobility (e.g., Bike to Work and European Mobility Week) and even organizes some of them. An example, there was the first Kidical Mass bike ride in 2024 and the distribution of bike shopping bags to the public. The goal of all these initiatives is to create an inclusive city where residents have the opportunity to choose the mode of transport according to their preferences and needs. Thus, Trnava is trying to transform from "car-centric" planning to more sustainable mobility.

3.3 Survey Design and Data

As the principal data collection method, the questionnaire was used in this case study. It focused on the traffic behavior of users of the Yellow Bike and was implemented by the City of Trnava at the beginning of October 2024 in order to better plan the future of the project and improve conditions for cyclists in the city. A set of questions was discussed between the experts from local authorities and the authors of this chapter. The target group consisted of 500 registered users of the Yellow Bike project, to whom an online questionnaire was sent via e-mail service. The first part of the questions in the questionnaire focused on general socio-demographic and mobility-related information of the users, such as gender, age, highest level of education, economic status, access to a car, and the influence of the weather on their decision to use a bicycle. Another set of questions was related to the use of Yellow Bikes, specifically: the frequency of trips, changes in the frequency of bicycle use since the beginning of using this service, the motivation for its use and the combination with public transport. Several questions were related to the most frequently realized ride on the Yellow Bike, requesting its purpose, duration, distance, including a hypothetical question: If there was no Yellow Bike available, which other mode of transport would you use to make this trip? We interpreted the answer to this question (which is crucial for the purposes of this research) as the mode of transport (bicycle, car, public transport, walking, scooter, taxi, other) that was replaced due to the use of this new alternative. A total of 117 respondents filled out the questionnaire.

To analyze the temporal patterns of bicycle use in Trnava, we used quantitative data from automatic bicycle counters, provided by the city hall and covering the entire year of 2023 on daily and hourly levels. The research included 4 selected counters located on segregated bicycle routes running parallel to the main city roads (see Fig. 2), which are typically congested during the working day rush hours. Among other sources of data, we included a survey of the traffic behavior of city residents conducted on a sample of 1,168 households on selected weekdays between October and November 2019 (Kyp et al., 2020). We used these data and the data from the last national population census to describe the basic traffic behavior of the city's residents.

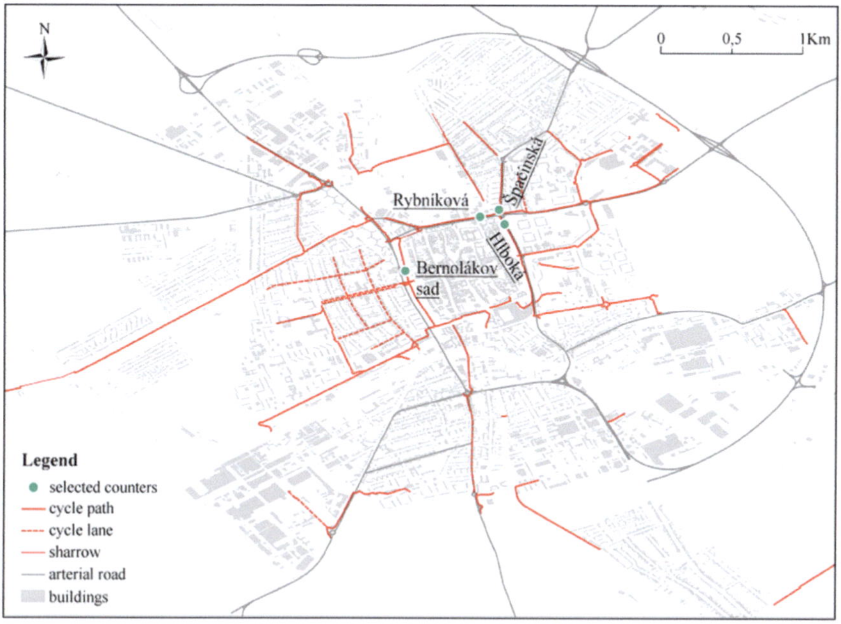

Fig. 2 Cycle network and the location of the bicycle counters used for this study in Trnava. *Source* Own elaboration

3.4 Exploring Socio-demographic Profiles of Yellow Bike Users

The results of the conducted survey show a clear demographic and user profile of the Yellow Bike users (see Table 1). More than half of the respondents (52.99%) are women, while men represent 44.44%. In terms of age characteristics, the largest group consists of persons aged 35–44 years (39.32%), 25–34 years (24.79%), and 45–54 years (20.51%). Younger users aged 15–24 and people in the older age categories (55–64 and over 65) have a combined representation of only 15.39%. This age structure suggests that the Yellow Bike is the most popular especially among middle-aged adults. Other identified characteristics include the respondents' education, which is remarkably high, as up to two-thirds of them have completed university education and approximately one-third have secondary education. A significant part (67.52%) of users are employed and 13.68% are self-employed or entrepreneurs. Students make up only 10.26% of respondents, which is not a high proportion, especially considering the fact that this project was originally intended for them. Pensioners, persons on maternity or parental leave, and the unemployed have the smallest representation (see Table 1).

The questionnaire data also reveal how often the Yellow Bike users have access to a car for daily use. It is interesting that despite the high availability of a passenger car (63.25% of respondents always have access to a personal car), people decide

Table 1 Socio-demographic, mobility-related characteristics, and selected behavior responses of Yellow Bike users in Trnava

Description	Responses	Distribution (%)
Gender	Female	52.99
	Male	44.44
	Prefer not to say	2.56
Age	15–24 years	6.84
	25–34 years	24.79
	35–44 years	39.32
	45–54 years	20.51
	55–64 years	5.13
	Over 65 years	3.42
Highest education	Basic education	0.85
	Secondary education	32.48
	Higher education	66.67
Current work status	Employed	67.52
	Self-employed/entrepreneur	13.68
	Student	10.26
	Senior	2.56
	Maternity/parental leave	4.27
	Unemployed	0.85
	Other	0.85
Access to passenger car	Always	63.25
	Occasionally	23.08
	Never	13.68
Frequency of using Yellow Bike	Daily	26.50
	Several times a week	51.28
	Once a week	5.13
	Several times a month	12.82
	Less than several times a month	4.27
Change in cycling frequency since using Yellow Bike	Ride notably more than before	32.48
	Ride more than before	42.74
	Ride the same amount as before	19.66
	Ride less than before	0.85
	Ride notably less than before	4.27

Source Own elaboration based on survey data

to use Yellow Bike. The majority of respondents are regular users of this service, as 51.28% use it several times a week and 26.50% use it daily. Less frequent use, which includes driving once a week, several times a month, or less than a few times a month, was indicated by 22.22% of respondents. In addition to the regularity of using the Yellow Bike, more than 75% of respondents declare that the overall frequency of using the bicycle has changed positively since they started using this service. About 5% of respondents experienced a decrease, and a fifth did not feel any change in the frequency of using a bicycle (Table 1). These data indicate that the use of micromobility is a routine part of the life of many users, and that leasing bicycles motivates them to use the bicycle more often, thereby effectively contributing to active urban mobility.

3.5 *Usage and Modal Shift Patterns*

In order to understand the impact of the use of Yellow Bike on individual modes of transport, which points to the potential environmental benefits resulting from the use of this alternative, we calculated substitution patterns at the level of individual paths based on the respondents' answers according to the methodology described by Reck et al. (2022). To achieve this goal, we divide the number of trips (answers) made by the substituted mode of transport (e.g., by car) by the total number of trips made by Yellow Bike:

$$subrate_{trip-level}(mode_{chosen}, mode_{substituted}) = \frac{\sum trips(mode_{substituted})}{\sum trips(mode_{chosen})} \quad (1)$$

where $\sum trips(mode)$ indicates the number of trips made by a specific mode of transport. We examined substitution patterns not only at the overall level, but also according to distance categories. The categories up to 1 km and 1–2 km were merged into one group, since only two respondents stated that their most frequent ride is shorter than 1 km.

The origin of the demand for Yellow Bikes is shown in Fig. 3, where we observe that most respondents would use other bicycle (33.33%), walking (32.48%), and a passenger car (24.79%) instead of Yellow Bike. A total of 6.84% of respondents would choose public transport, and only 1.71% would choose scooters. Taxi is the least preferred alternative with a share of 0.85%. However, these substitution rates do not apply equally to all distance categories. For trips shorter than 2 km, the largest part of respondents would choose walking (39.13%), other bicycle (34.78%), and a car (23.91%) as alternatives. For 2–3 km journeys, preferences change slightly, but most respondents would again use sustainable alternatives, with other bicycle (40.91%) and walking (31.82%) being the most preferable options, followed by passenger cars (15.91%) and public transport (6.82%). For journeys longer than 3 km, the preference of a passenger car increases significantly (up to 40.74%), public transport is another popular choice (14.81%), while using other bicycle and walking together make up

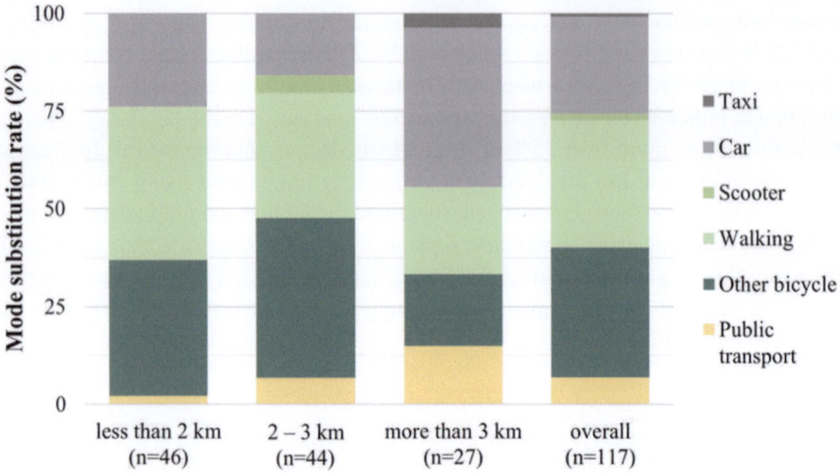

Fig. 3 Substitution rates for Yellow Bike users in Trnava by distance. *Source* Own elaboration based on survey data

40.74%, with an almost equal representation of these two modes. These data show that for shorter distances up to 3 km, walking and own bicycle are the most popular alternatives, while their importance decreases as the distance increases, and more respondents prefer a passenger car or public transport.

We also examined substitution patterns by gender, which allows us to better understand the differences in the preferences of individual groups of users. These results are shown in Fig. 4 for men, women, and respondents who did not specify their gender. According to them, women would most often opt for walking (37.10%), another bicycle (29.03%), and a car (27.42%) if the Yellow Bike was not available. Men's choice of mode of transport is slightly different compared to women, as their first choice is another bicycle (36.54%). On the other hand, fewer of them prefer walking and private cars (26.92% vs. 23.08%). Public transport is approximately at the same level for both genders (6.45% of women and 7.69% of men), but no woman chose a scooter or a taxi as alternative options. According to these findings, we conclude that the most significant difference in the transport preferences of individual genders is that women primarily prefer walking and men prefer bicycles, which may indicate a more sensitive perception of the conditions related to the use of their own bicycles.

Figures 3 and 4 indicate that in case of unavailability of Yellow Bikes, most people would opt for walking, cycling, and private cars. For this reason, we decided to focus on other driving characteristics of users of these three modes. We found interesting differences in the motivations for using Yellow Bikes, which can be observed in Fig. 5. Respondents who would use a passenger car are most motivated by health reasons and exercise need (65.52%), which indicates their interest in maintaining a healthy lifestyle. The speed of transport (37.93%) and the environmental aspect (34.48%) are also important for these respondents. On the other hand, respondents who would travel on foot found the main motivation in the speed of transport (57.89%), as

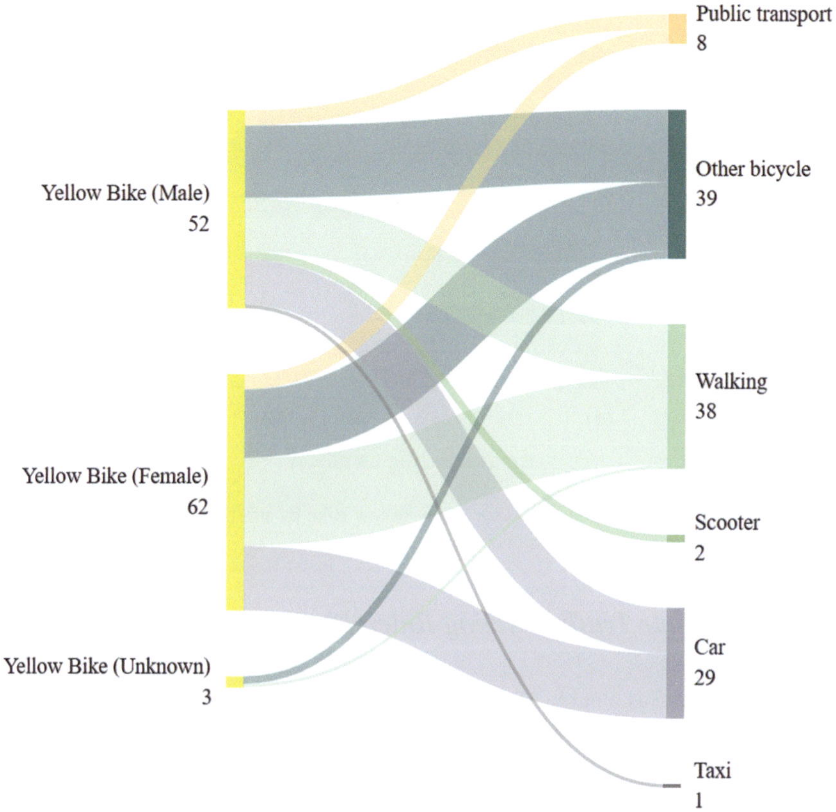

Fig. 4 Modal shift of Yellow Bike users in Trnava by gender. *Source* Own elaboration based on survey data

switching to a bicycle probably reduces the time they spend on roads. Health and exercise are also important for them (50%) but less than for the previous group. Finally, those who would prefer other bicycle are also motivated by the speed of transport (58.97%), but health and exercise affect them less (41.03%) than the first two groups. Compared to the others, this group of the respondents is more motivated by the feeling of freedom and independence (23.08%) that the bicycle provides them. Overall, we can observe that speed of transport is the main motivation across all categories of users. It should be noted that although the price and saving money are not among the most frequent reasons in either group, they are considered important by a total of around 30% of users.

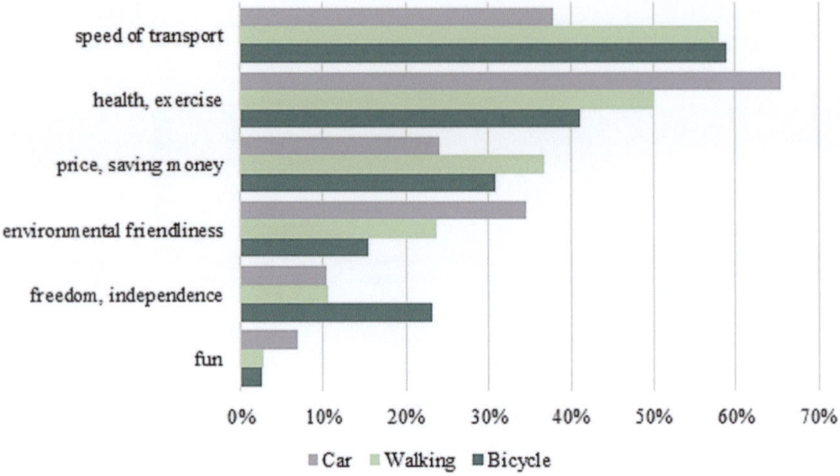

Fig. 5 Comparison of the motivations for using Yellow Bike by substituted mode of transport in Trnava. *Source* Own elaboration based on survey data

3.6 Peak Hour Traffic Cycling Ridership Patterns

According to the Sustainable Mobility Plan of the City of Trnava and its functional region, the peak traffic activity of the city's residents occurs in the morning between 7:00 and 8:00 a.m. and in the afternoon between 3:00 and 6:00 p.m. During these hours, individual car transport accounts for up to 55% of all trips by city residents, while bicycles make up around 5% (Kyp et al., 2020). However, these data do not include the trips of residents commuting to Trnava from the surrounding region, which most likely increase the share of motorized traffic in the total number of trips. Google Maps (2024) indicates that the emergence of traffic jams is a typical phenomenon on several main city roads precisely at the time of the morning and afternoon traffic rush hours.

According to Kyp et al. (2020), the purpose of the trips of residents during rush hours is primarily related to commuting (work, school, shopping, services, accompaniment), while leisure activities make up less than a fifth of all trips. For this reason and following the works of Miranda-Moreno et al. (2013), Nosal and Miranda-Moreno (2014), and Németh (2024), in this research we will consider the journeys of cyclists during rush hour on weekdays also as commuting. We assume that the trips made at this time are among the most regularly made trips by residents, and changes in the intensity of bicycle use during the year can reveal whether the potential impact of bicycle traffic on reducing traffic jams is the same across all seasons.

Seasonal and monthly differences in the use of bicycles during rush hours are investigated on 4 segregated bicycle routes (*Bernolákov sad*, *Rybníková, Špačinská,* and *Žarnova*) using the method of standardized indices inspired by Miranda-Moreno et al. (2013), which expresses the average intensity of bicycle traffic in a specific period (e.g., hours, days, and months) in relation to the average daily intensity during the entire observed period. Thanks to these indices, we can better understand the temporal patterns of bicycle traffic and identify the periods with the highest or lowest intensity of use of the cycling infrastructure during the year. We will apply the following two indices based on the average number of cyclists during peak hours (7:00–7:59, 15:00–17:59) on weekdays:

$$I_S = \frac{\overline{v}_{Seasonal\ peak-hour\ traffic}}{\overline{V}_{Daily\ peak-hour\ traffic}} \tag{2}$$

$$I_M = \frac{\overline{v}_{Monthly\ peak-hour\ traffic}}{\overline{V}_{Daily\ peak-hour\ traffic}} \tag{3}$$

where

I_S, I_M	seasonal or monthly standardized indices,
$\overline{v}_{Seasonal\ peak-hour\ traffic}$	average rush-hour intensity of cyclists on a working day in the given season (winter, spring, summer, autumn),
$\overline{v}_{Monthly\ peak-hour\ traffic}$	average rush-hour intensity of cyclists on a working day in the given month (January to December),
$\overline{V}_{Daily\ peak-hour\ traffic}$	average rush-hour intensity of cyclists on a working day in 2023 (247 days).

By creating seasonal and monthly standardized indices, we tried to identify differences in the intensity of cycle traffic during peak hours on the working day. Table 2 presents the values of these indices in sections of *Bernolákov sad, Rybníková, Špačinská*, and *Žarnova* cycle routes. In winter, the values are the lowest (0.39–0.42), which means that bicycle traffic during peak hours reaches 2.5 times lower values than the annual average. In spring, the intensity increases significantly, with indices reaching the values of around 0.92–0.99. The summer months represent the peak of the number of rides, with the index values ranging between 1.34 and 1.43. These values point to the high popularity of bicycle traffic in the summer. In autumn, the number of journeys decreases slightly, but the values remain high (1.23–1.25), and the cycle routes are still used above average compared to the annual average.

On the other hand, monthly indices provide a more detailed structure of cycle traffic during peak hours in individual months. Not surprisingly, the winter months, especially December and January, show the lowest values of the index (0.33–0.43), which emphasizes the low interest of residents to use bicycles in the cold months. In spring, this value gradually increases from 0.67–0.80 in March to 1.28–1.31 in May. The most popular of the summer months is June, when the average number of trips during rush hour is 1.55–1.62 times higher than the average annual values. July and August also show high values (1.19–1.42) but lower compared to June. All cycling routes show their maximum values in September (1.73–1.80), which indicates the peak of cycling traffic before the start of the colder months. In October, the values are similar to those in May, while in November the average number of bikers drops by almost half (0.67–0.71). The minor differences between cycle routes are probably

Table 2 Standardized monthly and seasonal indices for 4 selected cycle paths in Trnava

Time	Cycle path			
	Bernolákov sad	Rybníková	Špačinská	Žarnova
Winter	0.39	0.41	0.42	0.40
Spring	0.97	0.99	0.98	0.92
Summer	1.36	1.34	1.34	1.43
Autumn	1.25	1.23	1.24	1.23
January	0.38	0.41	0.43	0.38
February	0.45	0.48	0.48	0.42
March	0.75	0.79	0.80	0.67
April	0.86	0.88	0.86	0.81
May	1.31	1.30	1.28	1.28
June	1.62	1.55	1.56	1.60
July	1.27	1.27	1.27	1.42
August	1.19	1.19	1.20	1.26
September	1.80	1.77	1.80	1.73
October	1.30	1.25	1.23	1.28
November	0.67	0.70	0.71	0.70
December	0.33	0.34	0.34	0.38

Source Own elaboration based on survey data

related to their different regimes, which is described in more detail by Németh (2024). The highest values of the indices in the early summer and autumn months indicate that during these months bicycles have the highest potential to contribute to the reduction of traffic jams in Trnava.

A more detailed picture of the seasonality of bicycle traffic is in Fig. 6, which shows the absolute number of bike journeys on *Bernolákov sad* cycle route. This bike path is located next to Hospodárska street, which is congested daily during rush hours. Undoubtedly, the highest use of the cycle route is during summer, when the afternoon peak hours reach up to 4 times higher values than in winter, which means an average of around 70 rides per day in individual hours. It is also interesting to note that the autumn months lag behind the summer months in every peak hour, but the morning, the difference is merely about 120 trips, equating to an average of 2 trips per day. The reason for this insignificant difference may be the holiday season and school holidays in the summer months. Overall, the lowest numbers of journeys are registered in the winter months, when during 60 working days not a single monitored hour has more than 1,400 journeys, which means around 23 rides on average.

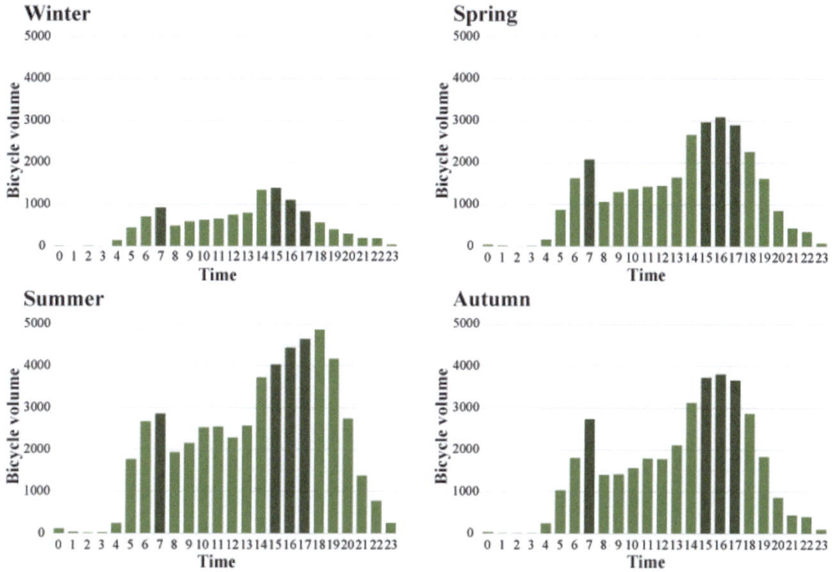

Fig. 6 Cycling levels on the cycle path "Bernolákov sad" during the seasons of 2023. *Source* Own elaboration based on survey data

4 Discussion and Conclusions

In this study, we analyzed the use of bicycles in Trnava, Slovakia, responding to the fact that although many Slovak cities have introduced new alternatives in the field of micromobility, especially bikesharing, it is still not clear whether these initiatives really contribute to solving problems in urban transport and reducing dependence on passenger cars. We paid special attention to the behavior of users of the only bicycle leasing program in Slovakia, with an emphasis on motivational factors and differences related to gender characteristics in the context of the substituted mode of transport. The revealed sociodemographic characteristics of users of this service are partially in line with the findings of Laa and Leth (2020), Ma et al. (2020), Roig-Costa et al. (2024), and Teixeira et al. (2023), since university-educated people represent the main group of users in Trnava as well.

On the other hand, we found differences in the gender and age structure of users compared to users of leasing bicycles in the City of Delft, where men under the age of 25 predominate (Ma et al., 2020). The predominance of younger and male users was also revealed among users of shared bicycles (Teixeira et al., 2023) and scooters (Laa & Leth, 2020). Nikiforiadis et al. (2021) explain that micro-mobility tends to increase gender inequality in urban mobility instead of reducing it. In our study, on the contrary, we see that in Trnava, among users of leasing bicycles, there is a higher representation of women and users aged 35–44, which is contrary to the aforementioned results. An international comparison of the behavior of cyclists shows that in regions with a low share of bicycle traffic (below 7%), women have a significantly lower representation than men, but above this threshold their share equals that of men and even exceeds them in some cases (Goel et al., 2022). Various soft and hard measures that the City of Trnava has been implementing recently to increase the share of bicycle traffic can explain the representation of female users of leased bicycles.

At the same time, these findings may indicate the success of the municipality in improving conditions for cyclists, if we assume that there is a connection between the level of bicycle traffic and the proportion of female users. This hypothesis is also supported by the finding that more than 75% of respondents reported a positive change in the overall frequency of bicycle use since becoming a user of this service, declaring that they ride more or significantly more compared to the previous period. We can conclude that the Yellow Bike contributes to increasing the number of bicycle paths in the city and supports the normalization of bicycle traffic, which is in line with the findings of Teixeira et al. (2021) on shared bicycles.

The revealed substitution patterns according to the number of journeys in Trnava show that around 25% of respondents choose a leased bicycle instead of a passenger car. These values are higher compared to shared bikes or leased bikes in other European cities (Fishman et al., 2014; Ma et al., 2020; Reck et al., 2022; Roig-Costa et al., 2024; Teixeira et al., 2023). We can also observe that public transport in Trnava was only marginally substituted (7%) in contrast to its prominent substitution rates reported in the above-mentioned studies. This dissimilarity, however, can be explained by the relatively low share of public transport use in Trnava, which accounts for approximately 12% of motorized trips in the population (see Fig. 1). Another explanation may be the compactness of the city and the short distances between individual neighborhoods, which are ideal for moving around by bike, but also on foot. Our findings show that two-thirds of users switched to this active transport alternative (bicycle, walking), but the proportions varied depending on the driving distance.

Ma et al. (2020) and Teixeira et al. (2023) found that people most often chose leased bicycles or shared bicycles due to bicycle theft, which may be the answer to the frequent substitution of this means of transport in our sample as well. The potential environmental benefits of using micromobility are manifested especially on journeys longer than 3 km, where Yellow Bikes most often substitute car driving, which is in line with the findings of Reck et al. (2022). In the transport preferences by gender, the only major difference was that women most often substitute walking by Yellow Bikes, while men substitute for other bicycles. Furthermore, the most frequent motivation for more than half of the respondents was the speed of transport, which may signal that bicycle transport in Trnava is becoming competitive even in comparison with cars. The speed of transport as the most common reason for using shared bicycles was also pointed out by Roig-Costa et al. (2024). For the group of respondents who substituted car driving, the main motivation was health and exercise, followed by traffic speed. Environmental concerns were more important for users who substituted car journeys compared to respondents who chose active modes of transport, which is similar to the results presented by Teixeira et al. (2023).

Trends in the bike use in Trnava during the greatest car traffic jams pointed to the existence of seasonality. According to the seasonal indices, the summer months recorded up to 3 times higher value than the winter months. On the other hand, the monthly indices revealed the greatest popularity of cycling in June and September, which may be influenced by the work schedule of the residents and the summer holidays. Based on time and substitution patterns, we assume a positive effect of bicycle traffic on the reduction of car traffic, especially in these months. When comparing our results with Hudde (2023), we found that moderate seasonality in bicycle traffic exists in Germany as well as in the Netherlands, but it is significantly stronger in Trnava. This fact indicates that the bicycle culture in our region is not yet sufficiently developed compared to Western countries, and the seasonal nature prevents the maximum potential of micromobility in solving climate challenges. However, as noted by Hudde (2023), the difference in cycling seasonality between Germany and the Netherlands is not due to differences in climatic conditions but rather to the ways in which people respond to those conditions.

The introduction of measures that encourage a change in the traffic behavior of residents and support the transition from car transport to more sustainable and spatially efficient alternatives still remains a big challenge in Slovakia. Our findings suggest that long-term bicycle rental has the potential to be an effective urban mobility strategy to support this shift. However, practice shows that in order to achieve significant changes, it is necessary to promote a comprehensive approach that combines various measures. In Slovakia, Trnava is exceedingly successful in implementing projects and various initiatives to improve the conditions for micromobility, and strong political support plays a huge role in this. Our study is intended to serve as a starting point for municipalities, operators of bike sharing and leasing systems, and researchers to further investigate the impact of micromobility on reducing dependence on passenger cars.

Based on our findings from Trnava, further recommendations could be formulated. More educational campaigns could contribute to raise awareness among citizens and distribute knowledge about the benefits of cycling and micromobility, as well as the importance of effective infrastructure planning among local governments (for example, more professional educational programs such as those organized by the Trnava Self-Governing Region in the form of the Spring School of Cycling and the Autumn School of Cycling). We also recommend considering the allocation of more generous space for cycling infrastructure and expanding traffic-calmed areas, along with promoting micromobility-friendly solutions, such as dedicated school zones, car-free zones, and other areas designed to prioritize sustainable transport. Pilot projects could also be launched in select neighborhoods or streets to test the effectiveness of these micromobility-friendly approaches. Additionally, municipalities and businesses could enhance bike parking facilities and integrate bikesharing programs into the public transport network.

This study does have certain limitations that should be noted. The first limitation is that a relatively small sample of respondents was reached by using an online survey method distributed via email. In this study we tried to explore whether Yellow Bikes have the potential to replace transportation options, which we considered less sustainable according to the lifecycle assessment of the International Transport Forum. We focused on analyzing trip-based substitution patterns rather than calculating the precise environmental advantages of micromobility, because historical distance-based trip data from Yellow Bikes was not available. Mode substitution patterns were assessed by asking bike-leasing program users to recall their most frequent trip, which is based on participants' memories and may lead to biases. Respondents provided the information on the distance category and the frequency of using this alternative, from which we can infer that three-quarters of them are regular users and that motorized transport is most often replaced on longer trips, pointing to potential environmental benefits. However, trip data from GPS tracking and travel diaries could be another approach to link detailed trip information with mode substitution. The online survey format also required respondents to have access to the Internet and to a smartphone or computer, which may indicate why older adults were underrepresented in the sample.

The second part of our study, which investigated the seasonality of cycling traffic volumes on selected cycle routes during peak car traffic times, also revealed some limitations. Although we noted the potential for bicycles to alleviate traffic congestion, our conclusions would have been more reliable if they had been verified by additional data on motor traffic volumes on the parallel roadway segments.

Future research should prioritize obtaining a more representative sample. A potential solution could involve further collaboration with the City of Trnava to enhance data collection. Additionally, conducting field studies during events like the European Mobility Week, when the traditional bicycle ride through the city's streets attracts a large number of bike users, or Kidical Mass events, could provide valuable insights and help increase the sample size. These approaches would enable a more comprehensive understanding of cycling behavior and its broader implications for urban mobility. Additionally, exploring the factors influencing mode substitution patterns could provide valuable insights, especially if it includes a comparative analysis of shared and leased bikes currently available in Trnava. Such an approach would help to understand the benefits of these micromobility options and clarify their impact on modal shifts. These findings could support Slovak cities in implementing more effective, competitive, and sustainable urban mobility strategies.

Acknowledgements We are grateful for the support of Mgr. Peter Rozsár, who provided valuable data and up-to-date information on cycling transportation in Trnava, greatly contributing to the quality of this work. This work is also supported by Grant UK/1064/2024.

References

Abduljabbar, R. L., Liyanage, S., & Dia, H. (2021). The role of micro-mobility in shaping sustainable cities: A systematic literature review. *Transportation Research Part D: Transport and Environment, 92*, Article 102734.

Adam, L., Jones, T., & Te Brömmelstroet, M. (2020). Planning for cycling in the dispersed city: Establishing a hierarchy of effectiveness of municipal cycling policies. *Transportation, 47*(2), 503–527.

Alka, T. A., Sreenivasan, A., & Suresh, M. (2024). Wheel of change: A systematic literature review on innovation and entrepreneurship in micro mobility solutions. *Transport Economics and Management, 2*, 154–168.

Allam, Z., Nieuwenhuijsen, M., Chabaud, D., & Moreno, C. (2022). The 15-minute city offers a new framework for sustainability, liveability, and health. *The Lancet Planetary Health, 6*(3), e181–e183.

Arboriabike. (2024). *Ako to funguje*. Retrieved September 18, 2024 from https://arboriabike.sk/jak-to-funguje.html

Ballo, L., de Freitas, L. M., Meister, A., & Axhausen, K. W. (2023). The E-Bike City as a radical shift toward zero-emission transport: Sustainable? Equitable? Desirable? *Journal of Transport Geography, 111*, Article 103663.

Barnes, F. (2019). *A scoot, skip, and a JUMP away: Learning from shared micromobility systems in San Francisco*. Retrieved September 11, 2024 from https://doi.org/10.17610/T6QP40

Bebber, S., Libardi, B., Moschen, S. D. A., da Silva, M. B. C., Fachinelli, A. C., & Nogueira, M. L. (2021). Sustainable mobility scale: A contribution for sustainability assessment systems in urban mobility. *Cleaner Engineering and Technology, 5*, Article 100271.

Bicalho, T., Silva, C., Cunha, I., Teixeira, J., & Proença, A. (2019). Planners' attitudes towards the cycling potential of their cities—Creating awareness for attitude change. *Travel Behaviour and Society, 17*, 96–103.

Bieliński, T., Kwapisz, A., & Ważna, A. (2021). Electric bike-sharing services mode substitution for driving, public transit, and cycling. *Transportation Research Part D: Transport and Environment, 96*, Article 102883.

Bigazzi, A., & Wong, K. (2020). Electric bicycle mode substitution for driving, public transit, conventional cycling, and walking. *Transportation Research Part D: Transport and Environment, 85*, Article 102412.

Brand, C., Götschi, T., Dons, E., Gerike, R., Anaya-Boig, E., Avila-Palencia, I., & Nieuwenhuijsen, M. J. (2021). The climate change mitigation impacts of active travel: Evidence from a longitudinal panel study in seven European cities. *Global Environmental Change, 67*, Article 102224.

Buehler, R., Pucher, J., Gerike, R., & Götschi, T. (2017). Reducing car dependence in the heart of Europe: Lessons from Germany, Austria, and Switzerland. *Transport Reviews, 37*(1), 4–28.

Cai, B., Liu, H., Zhang, X., Pan, H., Zhao, M., Zheng, T., & Dhakal, S. (2022). High-resolution accounting of urban emissions in China. *Applied Energy, 325*, Article 119896.

Caulfield, B. (2014). Re-cycling a city—Examining the growth of cycling in Dublin. *Transportation Research Part A: Policy and Practice, 61*, 216–226.

Ceder, A. (2021). Urban mobility and public transport: Future perspectives and review. *International Journal of Urban Sciences, 25*(4), 455–479.

Christoforou, Z., de Bortoli, A., Gioldasis, C., & Seidowsky, R. (2021). Who is using e-scooters and how? Evidence from Paris. *Transportation Research Part D: Transport and Environment, 92*, Article 102708.

Chromeček, A. (2024). Bicykel a kolobežka ako prostriedky dochádzky do zamestnania podľa SODB 2021. *Slovenská štatistika a demografia, 34*(3), 3–17.

City of Trnava. (2024). *Zdieľané elektrobicykle sa vrátia prvý marcový pondelok*. Retrieved October 19, 2024 from https://www.trnava.sk/aktualita/13227/zdielane-elektrobicykle-sa-vratia-prvy-marcovy-pondelok

Cycling Coalition. (2023). *Cyklomestá 2022—bulletin*. Retrieved October 15, 2024 from https://cyklokoalicia.sk/wp-content/uploads/2023/01/Cyklomesta_2022.pdf

De Bortoli, A., & Christoforou, Z. (2020). Consequential LCA for territorial and multimodal transportation policies: Method and application to the free-floating e-scooter disruption in Paris. *Journal of Cleaner Production, 273*, Article 122898.

de Nazelle, A., Morton, B. J., Jerrett, M., & Crawford-Brown, D. (2010). Short trips: An opportunity for reducing mobile-source emissions? *Transportation Research Part D: Transport and Environment, 15*(8), 451–457.

Dufour, D. (2010). *PRESTO cycling policy guide: General framework, PRESTO project: Promoting cycling for everyone as a daily transport mode*. Retrieved September 14, 2024 from http://www.polisnetwork.eu/uploads/Modules/PublicDocuments/presto_cycling-policy-guide-general-framework_english.pdf

Faghih-Imani, A., Anowar, S., Miller, E. J., & Eluru, N. (2017). Hail a cab or ride a bike? A travel time comparison of taxi and bicycle-sharing systems in New York City. *Transportation Research Part A: Policy and Practice, 101*, 11–21.

Fearnley, N., Johnsson, E., & Berge, S. H. (2020). *Patterns of E-scooter use in combination with public transport*. Transport Findings. Retrieved September 10, 2024 from https://doi.org/10.32866/001c.13707

Fishman, E., Washington, S., & Haworth, N. (2014). Bike share's impact on car use: Evidence from the United States, Great Britain, and Australia. *Transportation Research Part D: Transport and Environment, 31*, 13–20.

Forsyth, A., & Krizek, K. J. (2010). Promoting walking and bicycling: Assessing the evidence to assist planners. *Built Environment, 36*(4), 429–446.

Fukushige, T., Fitch, D. T., & Handy, S. (2021). Factors influencing dock-less E-bike-share mode substitution: Evidence from Sacramento, California. *Transportation Research Part D: Transport and Environment, 99*, Article 102990.

Goel, R., Goodman, A., Aldred, R., Nakamura, R., Tatah, L., Garcia, L. M. T., & Woodcock, J. (2022). Cycling behaviour in 17 countries across 6 continents: Levels of cycling, who cycles, for what purpose, and how far? *Transport Reviews, 42*(1), 58–81.

Google Maps. (2024). *Typical traffic in Trnava on Tuesdays*. Retrieved October 8, 2024 from https://www.google.com/maps/@48.3760644,17.5985885,4670m/data=!3m1!1e3!5m1!1e1?entry=ttu&g_ep=EgoyMDI0MTEwNS4wIKXMDSoASAFQAw%3D%3D

Gössling, S. (2020). Why cities need to take road space from cars-and how this could be done. *Journal of Urban Design, 25*(4), 443–448.

Ho, C. Q., & Tirachini, A. (2024). Mobility-as-a-service and the role of multimodality in the sustainability of urban mobility in developing and developed countries. *Transport Policy, 145*, 161–176.

Hudde, A. (2023). It's the mobility culture, stupid! Winter conditions strongly reduce bicycle usage in German cities, but not in Dutch ones. *Journal of Transport Geography, 106*, Article 103503.

Hudde, A., & Wessel, J. (2024). More afraid of the virus than of bad weather? Exploring the link between weather conditions and cycling volume in German cities before and during the COVID-19 pandemic. *Transportation Research Part F: Traffic Psychology and Behaviour, 101*, 267–278.

ITF. (2020). *Safe micromobility*. International Transport Forum Policy Papers, No. 85. OECD Publishing. Retrieved September 7, 2024 from https://doi.org/10.1787/0b98fac1-en

ITF. (2024). *Greener micromobility*. International Transport Forum Policy Papers, No. 131, OECD Publishing. Retrieved September 7, 2024 from https://doi.org/10.1787/9aa6e85a-en

Jensen, P., Rouquier, J. B., Ovtracht, N., & Robardet, C. (2010). Characterizing the speed and paths of shared bicycle use in Lyon. *Transportation Research Part D: Transport and Environment, 15*(8), 522–524.

Khavarian-Garmsir, A. R., Sharifi, A., & Sadeghi, A. (2023). The 15-minute city: Urban planning and design efforts toward creating sustainable neighborhoods. *Cities, 132*, Article 104101.

Kołoś, A., & Taczanowski, J. (2016). The feasibility of introducing light rail systems in medium-sized towns in Central Europe. *Journal of Transport Geography, 54*, 400–413.

Kubal'ák, S., Gogola, M., & Černý, M. (2021a). Options for assessing the impact of the bike-sharing system on mobility in the city Žilina. *Transportation Research Procedia, 55*, 378–386.

Kubal'ák, S., Kalašová, A., & Hájnik, A. (2021b). The bike-sharing system in Slovakia and the impact of COVID-19 on this shared mobility service in a selected city. *Sustainability, 13*(12), 6544.

Kumar, P. (2021). Climate change and cities: Challenges ahead. *Frontiers in Sustainable Cities, 3*, Article 645613.

Kyp, O., Göpfertová, E., Volfová, Z., Želechovská, L., Petrouš, M., Suntych, P., Hlava, V., Buzák, J., Kaňka, P., & Vaňková, Z. (2019). *Plán udržateľnej mobility funkčného územia krajského mesta Nitra. Zoznam príloh: Prieskum dopravného správania obyvateľov*. Retrieved September 14, 2024 from https://nitra.sk/wp-content/uploads/2022/09/Tabulkove_Prilohy_NITRA_A_Mobilita.pdf

Kyp, O., Göpfertová, E., Volfová, Z., Želechovská, L., Petrouš, M., Suntych, P., Hlava, V., & Buzák, J. (2020). *Plán udržateľnej mobility krajského mesta Trnava a jeho funkčného územia. Zoznam príloh: Prieskum dopravného správania obyvateľov*. Retrieved September 14, 2024 from https://www.trnava.sk/stranka/pum#dopravne-prieskumy

Laa, B., & Leth, U. (2020). Survey of E-scooter users in Vienna: Who they are and how they ride. *Journal of Transport Geography, 89*, Article 102874.

Ma, X., Yuan, Y., Van Oort, N., & Hoogendoorn, S. (2020). Bike-sharing systems' impact on modal shift: A case study in Delft, the Netherlands. *Journal of Cleaner Production, 259*, Article 120846.

McCowan, T., Leal Filho, W., & Brandli, L. (2021). *Universities facing climate change and sustainability*. Körber-Stiftung.

Meddin, R., DeMaio, P., O'Brien, O., Rabello, R., Yu, C., Seamon, J., & Mason, J. (2024). *The Meddin bike-sharing world map*. https://bikesharingworldmap.com/#/all/2.3/0/51.5/

Mi, Z., Guan, D., Liu, Z., Liu, J., Viguié, V., Fromer, N., & Wang, Y. (2019). Cities: The core of climate change mitigation. *Journal of Cleaner Production, 207*, 582–589.

Michniak, D. (2019). Ako suburbanizácia mení dopravu v zázemí Bratislavy? In M. Šveda & P. Šuška (Eds.), *Suburbanizácia: Ako sa mení zázemie Bratislavy?* (pp. 253–272). Bratislava (Geografický ústav SAV).

Miranda-Moreno, L. F., Nosal, T., Schneider, R. J., & Proulx, F. (2013). Classification of bicycle traffic patterns in five North American cities. *Transportation Research Record, 2339*(1), 68–79.

Moreau, H., de Jamblinne de Meux, L., Zeller, V., D'Ans, P., Ruwet, C., & Achten, W. M. J. (2020). Dockless e-scooter: A green solution for mobility? Comparative case study between dockless e-scooters, displaced transport, and personal e-scooters. *Sustainability, 12*(5), 1803.

Murphy, E., & Usher, J. (2015). The role of bicycle-sharing in the city: Analysis of the Irish experience. *International Journal of Sustainable Transportation, 9*(2), 116–125.

Németh, R. (2024). Režim mestských cyklotrás: Cyklistická doprava alebo rekreačné bicyklovanie? *Študentská vedecká konferencia Prif UK 2024 [elektronický dokument]. Zborník recenzovaných príspevkov* (pp. 915–920). Univerzita Komenského.

Nikiforiadis, A, Paschalidis, E, Stamatiadis, N, Raptopoulou, A, Kostareli, A, Basbas, S. (2021). Analysis of attitudes and engagement of shared e-scooter users. Transportation research part D: transport and environment, 94, 102790

Nosal, T., & Miranda-Moreno, L. F. (2014). The effect of weather on the use of North American bicycle facilities: A multi-city analysis using automatic counts. *Transportation Research Part A: Policy and Practice, 66*, 213–225.

O'Brien, O., Cheshire, J., & Batty, M. (2014). Mining bicycle sharing data for generating insights into sustainable transport systems. *Journal of Transport Geography, 34*, 262–273.

Palermo, V., Bertoldi, P., Apostolou, M., Kona, A., & Rivas, S. (2020). Assessment of climate change mitigation policies in 315 cities in the Covenant of Mayors initiative. *Sustainable Cities and Society, 60*, Article 102258.

Pucher, J., & Buehler, R. (2005). Transport policies in Central and Eastern Europe. *Transport Strategy, Policy, and Institutions,* 725–744.

Pucher, J., & Buehler, R. (2008). Making cycling irresistible: Lessons from the Netherlands, Denmark and Germany. *Transport Reviews, 28*(4), 495–528.

Pucher, J., & Buehler, R. (2017). Cycling towards a more sustainable transport future. *Transport Reviews, 37*(6), 689–694.

Reck, D. J., Haitao, H., Guidon, S., & Axhausen, K. W. (2021). Explaining shared micromobility usage, competition and mode choice by modelling empirical data from Zurich, Switzerland. *Transportation Research Part C: Emerging Technologies, 124*, Article 102947.

Reck, D. J., Martin, H., & Axhausen, K. W. (2022). Mode choice, substitution patterns and environmental impacts of shared and personal micro-mobility. *Transportation Research Part D: Transport and Environment, 102*, Article 103134.

Rišová, K. (2020). Walkability research: Concept, methods and a critical review of post-socialist studies. *Geografický Časopis, 72*, 219–242.

Roig-Costa, O., Miralles-Guasch, C., & Marquet, O. (2024). Shared bikes vs. private e-scooters. Understanding patterns of use and demand in a policy-constrained micromobility environment. *Transport Policy, 146*, 116–125.

Sancino, A., Stafford, M., Braga, A., & Budd, L. (2022). What can city leaders do for climate change? Insights from the C40 Cities Climate Leadership Group network. *Regional Studies, 56*(7), 1224–1233.

Severengiz, S., Finke, S., Schelte, N., & Wendt, N. (2020). Life cycle assessment on the mobility service E-scooter sharing. In *2020 IEEE European Technology and Engineering Management Summit (E-TEMS)* (pp. 1–6). IEEE.

Shaheen, S., & Cohen, A. (2019). *Shared micromoblity policy toolkit: Docked and dockless bike and scooter sharing*. UC Berkeley Transportation Sustainability Research Center. Retrieved September 8, 2024 from https://escholarship.org/uc/item/00k897b5

SODB. (2021a). Dochádzka (obyvateľstvo odchádzajúce) do zamestnania podľa periodicity dochádzky, spôsobu dopravy a vzdelania. Retrieved October 10, 2024 from https://www.scitanie.sk/obyvatelia/rozsirene-vysledky

SODB. (2021b). *Moja obec*. Retrieved October 18, 2024 from https://www.scitanie.sk/moja-obec/SK0217506745

Tang, K. H. D. (2022). Climate change policies of the four largest global emitters of greenhouse gases: Their similarities, differences and way forward. *Journal of Energy Research and Reviews, 10*(2), 19–35.

Teixeira, J. F., Silva, C., & Sá, F. M. (2021). Empirical evidence on the impacts of bikesharing: A literature review. *Transport Reviews, 41*(3), 329–351.

Teixeira, J. F., Silva, C., & Sá, F. M. (2023). Factors influencing modal shift to bike sharing: Evidence from a travel survey conducted during COVID-19. *Journal of Transport Geography, 111,* 103651.

Wang, M., & Zhou, X. (2017). Bike-sharing systems and congestion: Evidence from US cities. *Journal of Transport Geography, 65*, 147–154.

World Cities Report. (2024). *Cities and climate action*. United Nations Human Settlements Programme (UN-Habitat).

Zhang, R., & Fujimori, S. (2020). The role of transport electrification in global climate change mitigation scenarios. *Environmental Research Letters, 15*(3), Article 034019.

Roland Németh is a Ph.D. student in the Department of Economic and Social Geography, Demography, and Territorial Development at the Faculty of Natural Sciences, Comenius University in Bratislava. His research concentrates on transport geography, with a primary emphasis on urban mobility, particularly micromobility and cycling transportation. In his dissertation, he examines the role of micromobility in Slovakia's urban environments and the potential of emerging data sources for analyzing cycling transportation. In the past, he collaborated with the Cycling

Coalition, a cycling advocacy expert group in Slovakia, and worked in the cycling coordinator's office for the Trnava Self-Governing Region. Currently, he is actively collaborating with the City of Trnava, where he conducts his scientific research.

Marcel Horňák is an Associate Professor in the Department of Economic and Social Geography, Demography, and Territorial Development at the Faculty of Natural Sciences, Comenius University, Bratislava. His lifelong scientific research has focused primarily on transportation geography, the spatial aspects of transport infrastructure, and public transport quality. In recent years, his research has increasingly emphasized the social dimensions of transportation, accessibility, and mobility. This includes addressing issues such as the transport disadvantages faced by various communities in society. He has participated in several international research projects and has been a long-time advocate for cooperation between academia and public authorities on transport and mobility issues. Recently, he has been collaborating with the Metropolitan Institute of Bratislava (Slovakia) on urban mobility initiatives.

Greening Last-Mile Deliveries and Urban Transport: Evaluation of the Potential of E-Cargo Bikes for Sustainable Mobility

Reza Shahin and Maxim A. Dulebenets

Abstract This study addresses the need for a comprehensive systematic literature review on electric cargo bikes (ECBs), focusing on research published since 2020 to account for recent advancements. Given ECBs' potential to mitigate greenhouse gas emissions and alleviate urban congestion, their integration in both commercial and personal transportation holds significant promise. ECBs serve as an efficient intermediary between traditional cargo bikes and motor vehicles, offering increased load capacity and extended travel ranges. Our methodology involves systematically searching major scientific databases, including Google Scholar and Scopus, with keywords "Electric Cargo Bikes" and "ECBs". The present survey study explores the potential role of ECBs, highlighting their practical advantages. Findings from this review contribute to understanding ECBs' potential and pave the way for future research in sustainable transport solutions. Critical research gaps that require more attention from communities in the following years are discussed in detail as well.

Keywords Electric cargo bikes · Sustainable urban logistics · Greenhouse gas emissions · Optimization strategies · Last-mile delivery

1 Introduction

Progress in low-carbon transportation technologies holds considerable potential to mitigate greenhouse gas (GHG) emissions. Local authorities are especially well-positioned to support these sustainable transit solutions and reduce dependence on

R. Shahin (✉)
Department of Mathematics and Industrial Engineering, Polytechnique Montréal, Montréal, QC, Canada
e-mail: Rezaa.shahin.1992@gmail.com

M. A. Dulebenets
Department of Civil and Environmental Engineering, FAMU-FSU College of Engineering, Tallahassee, FL, USA
e-mail: mdulebenets@eng.famu.fsu.edu

M. A. Dulebenets et al. (eds.), *Transport and Climate Change*,
https://doi.org/10.1007/978-3-032-00563-2_7

motor vehicles (Wrighton & Reiter, 2016). Urban areas, characterized by high population densities and diverse land-use patterns, naturally lend themselves to alternative, slower modes of transport, enabled by the quick and easy access to various destinations (Schwanen et al., 2004). However, it is notable that in Swiss urban centers, as an example, 40–50% of car trips are for distances under 5 km, underscoring a continued reliance on cars for even brief trips. For these short-distance journeys, cargo bikes (CBs) represent a practical substitute. Known also as transport bikes, box bikes, long johns, or bakfiets, these bikes can transport more than 100 kg, making them ideal for transporting goods or children (Becker & Rudolf, 2018). A U.S. survey found that 66% of individuals who had previously relied on cars as their primary transport mode shifted to using CBs after acquiring one.

The development of CBs enhances their appeal as an alternative to traditional cars (Lovejoy & Handy, 2012). Originating in the late nineteenth century, CBs were primarily used by merchants and artisans (Kirkels, 2016). Today, they are seeing renewed popularity, particularly among urban families with young children, especially in regions where biking is common, such as the Netherlands and Denmark. For example, in Copenhagen, 17% of families with children own a CB, and in Norway, CB sales more than doubled from 2015 to 2016, reaching 566 units sold in the latter year. Several municipalities in Austria and Germany, alongside Oslo, have introduced permanent or temporary incentives to subsidize electric CB purchases, encouraging adoption of low-carbon transport options (Carvelo2go, 2024). Nevertheless, CBs currently occupy a niche market, and empirical research on their use remains sparse. This research gap is compounded by the lack of a separate classification for CBs within bicycle-related statistical data, complicating efforts to obtain precise usage figures (Siegenthaler, 2017).

Within the European Union (EU), transportation contributes approximately 25% of GHG emissions, with urban vehicle use responsible for around 23% of this total. Additionally, urban freight transport accounts for one-quarter of CO_2 emissions and over half of particulate emissions (Lebeau et al., 2015). To meet the European Green Deal's target of climate neutrality by 2050 (European Commission, 2024), it is essential to reduce transportation emissions by approximately 90%. This ambitious goal may be achievable through the widespread electrification of transportation infrastructure. In urban settings, last-mile logistics is gaining importance, primarily driven by the rapid growth of e-commerce (Bosona, 2020). Sustainable approaches to last-mile logistics can not only substantially reduce emissions that contribute to climate change but also alleviate other pressing urban issues, such as air pollution and traffic congestion. These advancements would consequently enhance the overall quality of life for city residents. Thus, it is crucial to explore and implement innovative technological and strategic solutions to steadily improve the sustainability of last-mile logistics practices.

Historically, CBs have been utilized for a broad spectrum of purposes, including newspaper distribution, postal deliveries, food transportation, and street vending (Cox & Rzewnicki, 2015). However, the rise of motorized vehicles and a decline in door-to-door delivery services led to a reduction in their use. Recently, however, there has been a resurgence in CB adoption, largely driven by their potential as a sustainable transportation option and their environmental benefits. Their reintroduction is increasingly recognized as a viable method for goods transport, aligning with evolving consumer expectations and a growing focus on sustainable logistics practices (Cox & Rzewnicki, 2015). This shift is explicitly supported by the "New EU Urban Mobility Framework" (European Commission, 2022), which highlights CBs' potential role in promoting a more environmentally friendly urban goods transport. The framework also emphasizes the increasing demand for last-mile delivery services, spurred by the expansion of e-commerce throughout the European Union. Notably, several European countries, including Germany, Italy, Austria, Spain, the United Kingdom, and Sweden, have launched pilot programs integrating cargo cycles to support more sustainable logistics operations (European Commission, 2022). According to the European Commission report (European Commission, 2024), urban transportation contributes significantly to GHG emissions within the EU. Beyond the environmental consequences, urban planners are increasingly concerned about vehicular congestion and its substantial economic repercussions. In particular, road congestion across Europe imposes an annual economic burden estimated at approximately 1% of the GDP (Christidis & Rivas, 2012).

A promising approach to mitigating the adverse impacts of traditional transport systems is the integration of electric cargo bikes (ECBs) in both commercial and personal transportation. ECBs occupy an intermediary position between traditional cargo bikes and motor vehicles in terms of cost, load capacity, and operational range (Gruber et al., 2014). Despite the substantial number of research efforts focusing on the urban transport problem with ECBs, an up-to-date review of the literature on this topic is lacking. This survey aims to bridge this research gap by evaluating the relevant literature published since 2020 to identify recent developments in this area. By analyzing the environmental effects, operational features, and policy impacts of ECB deployment, the present research contributes to the holistic understanding of sustainable urban logistics aligning with the EU strategies and policies, including transport, mobility, and climate. Specifically, it provides a structured framework to evaluate ECBs' potential as a low-carbon alternative to traditional urban freight solutions. Additionally, the present research determines unresolved issues, emerging tendencies, and innovative opportunities for ECB adoption, offering a theoretical basis for further developments on sustainable transport alternatives and their integration into wider urban mobility applications. Therefore, we define the research questions for this study as follows:

1. What are the most recent developments in the operational deployment and adoption of ECBs in urban mobility systems since 2020?
2. How do ECBs address the issues surrounding last-mile logistics in terms of environmental effects, efficiency, and economic feasibility?
3. What research gaps appear in the empirical studies on ECBs, and how can future research efforts facilitate their integration into sustainable urban transport applications?

In the remainder of the paper, we start with the description of how the literature search was conducted in Sect. 2. Then, the key characteristics and considerations of ECB deployment for last-mile deliveries are discussed in Sect. 3, followed by the review of the collected and selected studies in Sect. 4. We discuss the major lessons learnt in Sect. 5. The future research directions and the main conclusions are presented in Sects. 6 and 7, respectively.

2 Literature Search

In order to collect the relevant studies on the use of electric cargo bikes, we searched the major databases, such as Google Scholar, Web of Science, Scopus, IEEE Xplore, and ScienceDirect, using the keywords "Electric Cargo Bikes" and "ECB" for locating relevant sources. We then applied a filter to limit results to those published from 2020 onwards. The decision to focus on papers published since 2020 stems from the rapid advancements in electric cargo bike technology and the increasing adoption of ECB in urban logistics during recent years. This timeframe ensures that the reviewed studies reflect the latest developments, innovations, and capturing the most relevant and up-to-date insights in this evolving field. Initially, this search yielded 64 papers. However, through detailed screening, we excluded papers that were not directly relevant to the ECB deployment problem, as well as those that were not in English, lacked sufficient methodological rigor, or did not provide empirical or analytical results. Consequently, we identified 18 highly relevant papers for an in-depth review in this study.

The distribution of the studies selected for review is illustrated in Fig. 1 and Table 1. Figure 1 presents the temporal distribution of the selected studies, categorized by the year of their publication. The results indicate a noticeable increase in research activity in 2022, which is the peak year, reflecting the growing interest in this field during recent years. Table 1 showcases the distribution of the studies across different publication sources, including journals and conference proceedings. This figure highlights that the conferences were the primary venue for publications, followed by journals such as Energies and Transport Policy.

3 Key Characteristics of Electric Cargo Bike Deployment for Last-Mile Deliveries

In this section, we define the nature of ECB deployment for various transportation purposes and outline their key characteristics to provide a comprehensive understanding of this academic inquiry. ECBs in commercial transportation fulfill diverse roles and can be systematically analyzed across three primary dimensions: trip type, fleet configuration, and network structure.

Based on journey characteristics, trips conducted by ECBs are divided into two main categories: service trips and delivery trips. Service trips refer to journeys made

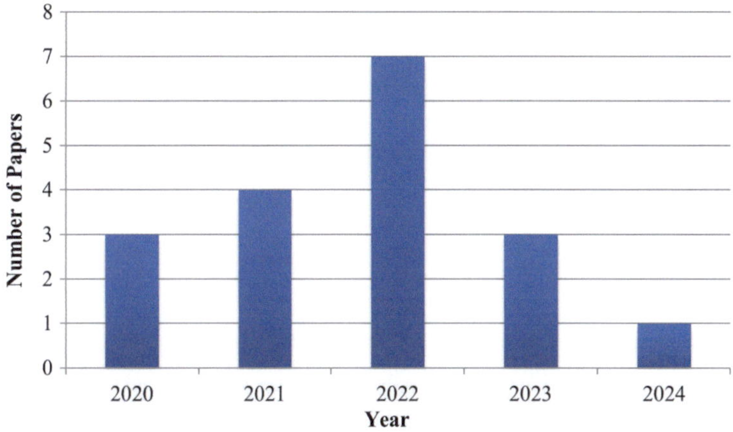

Fig. 1 Distribution of the studies selected for review by year of their publication

Table 1 Distribution of the studies selected for review by publication source

Publication source	Number of publications
Conference Proceedings	8
Energies	3
Transport Policy	2
Future Transportation	1
International Journal of Sustainable Transportation	1
Preprint	1
Renewable and Sustainable Energy Reviews	1
Transportation Research Interdisciplinary Perspectives	1

by skilled professionals or tradespeople who need to transport specialized tools and materials to perform on-site services for clients. Delivery trips encompass five main commercial applications, including postal, courier and parcel services, home deliveries (for restaurants and small retailers), and internal logistics within large facilities. Regarding fleet composition, ECBs are organized in two main configurations. In a dedicated e-cargo fleet, businesses exclusively use ECBs for their operations. In contrast, a mixed fleet integrates ECBs with other vehicle types, such as cars and vans, to support varying transportation needs.

When designing logistical networks, two predominant models are typically used: the direct scheme and the intermediate transshipment model. In the intermediate transshipment model, goods are transferred from other vehicles to ECBs at designated points, enabling efficient last-mile delivery. The direct scheme, by contrast, involves transporting goods directly without intermediary transfers. For example, courier services that require time-sensitive, high-value deliveries often employ the direct scheme, while parcel delivery may utilize transshipment points to facilitate distribution. Service-oriented trips may also adopt a direct scheme for straightforward A-to-B travel.

The choice of ECB integration factors will vary depending on specific contexts and industry requirements. Decision-makers are thus tasked with identifying the most relevant factors for defining problem parameters in each case.

4 Literature Review

In this section, we conduct a thorough review of the academic literature addressing the integration and optimization of ECBs within commercial transportation systems and urban logistics. This review examines studies focused on the effectiveness of ECBs in urban settings. By analyzing the existing research, we aim to identify the key approaches, methodologies, and findings that contribute to a deeper understanding of ECBs' role in last-mile delivery, service-related transportation, and sustainable logistics. Through this examination, we highlight critical insights and gaps in the literature, providing a foundation for the modeling and analysis undertaken in subsequent sections. All the collected studies were classified into different categories, including the following: (1) logistics and mobility; (2) transportation and health; (3) transportation and life cycle assessment; (4) travel behavior; and (5) relevant survey studies.

4.1 Logistics and Mobility

Caggiani et al. (2020) introduced an eco-friendly Decision Support System (DSS) designed to optimize last-mile delivery using ECBs. In this context, "eco-friendly" refers to the system's focus on reducing the environmental impact by incorporating logistics performance and minimizing the driver's exposure to emissions. The DSS enables route optimization based on two objectives, minimum travel time or minimum emission exposure, aligning with sustainable urban logistics goals. The system employed a Stochastic User Equilibrium (SUE) model with the Method of Successive Averages—Flow Averaging (MSA-FA) for traffic flow estimation, utilizing real-time traffic and emissions data. Additionally, a Fuzzy Inference System (FIS) was integrated, allowing drivers to choose between the fastest route or the healthiest route based on their preferences. By facilitating these environmentally-conscious choices, the DSS contributes to reducing pollution exposure for drivers without significant increases in travel time, thereby promoting a more sustainable approach to last-mile delivery operations.

Ewert and Deniz (2020) investigated the specific vehicle requirements for ECBs in commercial transport, providing a detailed analysis of different use cases, such as last-mile delivery, courier services, and municipal services. The study employed a combination of competitive analysis, stakeholder interviews, and a user survey. These methods helped identify the operational and vehicle requirements for different use cases, focusing on parameters, such as payload capacity, driving profiles, operating times, and vehicle downtime. The paper concluded that ECBs have substantial potential to reduce emissions and traffic congestion in urban areas. Based on the study findings, payload capacity requirements for electric cargo bikes vary by use case: municipal services need up to 100 kg, delivery services range from 25 to 200 kg, and CEP (Courier, Express, Parcel) services require up to 178 kg for parcels. Adaptable designs, such as removable boxes and cooling systems, are essential to meet these diverse demands.

Milenković et al. (2020) presented a comprehensive framework for designing and implementing an ECB system for urban logistics, focusing specifically on postal logistics in Belgrade (Serbia). The study added value by addressing the key dimensions, such as organizational, software, hardware, and policy frameworks, providing an integrated methodology for the successful deployment of ECBs in urban areas. This approach could guide cities and logistics operators in reducing traffic congestion and improving sustainability in last-mile deliveries.

Aiello et al. (2021) presented a comprehensive approach to optimizing urban delivery systems by using ECBs with modular battery configurations. It addressed the trade-off between battery size, vehicle endurance, and payload capacity, which was crucial for efficient last-mile logistics. The modular design approach allowed for customization based on specific operational contexts, enabling better adaptability to various urban delivery scenarios. The methodology involved a multi-criteria decision-making approach (TOPSIS) to optimize the configuration of ECBs in relation to battery size, payload capacity, and delivery requirements. The paper modeled energy consumption, travel distance, and service time while analyzing different vehicle configurations. The study also included a cost model for evaluating the economic feasibility of each configuration based on delivery cost, service area, and vehicle capacity. The primary objective was to find the optimal configuration of ECBs in urban delivery systems by balancing battery size (affecting endurance) and payload capacity (affecting the number of serviceable customers) to minimize delivery time and costs while maximizing system efficiency. The study demonstrated that modular ECBs with optimized configurations can significantly reduce delivery costs and time. Cost efficiency improved by up to 50% when comparing optimized configurations with non-optimized ones. The paper concluded that a modular approach to ECB design, with a focus on customizing battery size and payload capacity, can optimize urban delivery systems.

Caggiani et al. (2021) presented a novel approach to last-mile logistics by introducing a Two-Echelon Electric Vehicle Routing Problem with Time Windows and Partial Recharging (2E-EVRPTW-PR). This model incorporated the use of both e-vans and ECBs, providing a sustainable and cost-effective logistics solution for urban and restricted traffic zones. The study highlighted the environmental benefits of zero-emission vehicles, demonstrating how combining e-vans and ECBs can optimize urban freight deliveries. The authors formulated a Mixed-Integer Linear Programming (MILP) model for the 2E-EVRPTW-PR, which involved two echelons: e-vans in urban zones and ECBs in restricted traffic zones. The study concluded that combining e-vans and ECBs for last-mile deliveries is an effective solution for minimizing costs and reducing emissions in urban logistics.

Naumov (2021) presented a novel simulation-based methodology to choose the optimal location of loading hubs for ECB delivery systems. The method incorporated the stochastic nature of transport demand and detailed routing procedures, offering a more dynamic and practical approach than the traditional facility location problem (FLP) models. It addressed urban logistics challenges, particularly in areas with restricted traffic, and was applied to the real-world case of the Old Town district in Krakow, Poland. The study used a simulation framework to model the transport system for goods deliveries by electric cargo bikes. It simulated the demand for transport services and the routing process, taking into account factors like vehicle capacity, road network, and customer locations. Monte Carlo simulations and the Clarke–Wright heuristic algorithm were employed for route optimization. The objective was to determine the optimal loading hub location in urban areas with restricted traffic to minimize the total transport work. The simulation results showed that for the Old Town district in Krakow, two locations out of all the available locations offered the lowest total transport work, making them the best options for hub placement.

Carracedo and Mostofi (2022) offered a comprehensive review of the role of ECBs in urban transportation, specifically focusing on private transportation. It filled a research gap by analyzing ECBs from a personal mobility perspective rather than focusing solely on logistics, which has been the primary focus of past studies. The objective was to highlight the role of ECBs as a sustainable alternative to private car usage. The study found that ECBs have the potential to significantly reduce car dependency in urban areas, particularly for short trips related to shopping, errands, and child transportation. It was found that the main barriers to adoption are safety concerns, high costs, and inadequate cycling infrastructure. The paper concluded that ECBs are a promising sustainable mobility solution for urban areas, especially for private transportation.

Philips et al. (2022) provided a comprehensive analysis of the potential for ECBs to reduce CO_2 emissions by replacing private car travel in England. The main objective of the study was to estimate the maximum possible reduction in car CO_2 emissions if ECBs use was maximized across different geographic areas and demographic profiles. The study highlighted the specific areas where ECB adoption could have the greatest impact on emissions reduction, offering valuable insights for policymakers targeting place-based decarbonization. Spatial microsimulation, a population-synthesis method, was deployed to simulate the travel behavior of the adult population across England. It estimated the distance individuals could realistically travel by ECBs, based on various factors, such as fitness, geographic location, and typical car trip lengths. The results were aggregated by small geographic areas (Lower Super Output Areas) to assess the CO_2 reduction potential of replacing car trips with ECBs. The authors highlighted that median car travel distances of 10 km were observed in rural settings and 5 km in urban settings. These distances were both shorter than the average maximum ECB distances of up to 20 km.

Gil et al. (2022) presented a data-driven analysis of the ECB network of a Lisbon-based last-mile logistics startup, Yoob. It offered new insights into the spatiotemporal patterns of ECB operations and used machine learning techniques to identify optimal locations for the expansion of micro-hubs and nano-hubs. The study applied the Cross-Industry Standard Process for Data Mining methodology, which included business understanding, data understanding, data preparation, modeling, and evaluation. Data from Yoob's ECB fleet was collected over four months. The analysis employed K-means clustering to identify operational patterns and to propose new hub locations. The study used traffic and fleet data to cluster route characteristics and created a gravity center model for hub placement. The primary goal was to analyze and optimize the spatiotemporal behavior of Yoob's ECB fleet and to propose optimal locations for expanding the network of logistics hubs. The study identified five distinct clusters characterizing ECB operations in Lisbon based on various factors, such as distance traveled, speed, and delivery volume. It proposed three new optimal locations for hub expansion in three neighborhoods of Lisbon.

In another study, Malik et al. (2023) explored the critical factors influencing the adoption and use of ECBs by small businesses. It adds value by applying an intelligent modeling framework to real-world data from a small business ECB rental pilot scheme in Dublin, Ireland. The study provided insights into how weather conditions, trip length, and other variables impact the choice to use ECBs for deliveries. The study found that weather conditions, particularly temperature and rainfall, are the most critical factors influencing trip length and the likelihood of choosing ECBs for deliveries. Higher temperatures and dry weather increased the attractiveness of using ECBs for longer trips, while cold and wet conditions significantly reduce their appeal. Additionally, trip time was identified as a more significant factor than trip length when considering rainy conditions.

Papaioannou et al. (2023) presented an optimization model for last-mile logistics using ECBs, focusing on energy efficiency and workload balancing among bike couriers. It was one of the few studies that took into account road grade (affecting energy consumption) and workload balancing in last-mile delivery. The problem was formulated as a Capacitated Vehicle Routing Problem (CVRP) with constraints related to road grade and workload balancing. The model minimized the energy consumed by ECBs and ensured a fair distribution of delivery points and distance among couriers. A Genetic Algorithm (GA) was employed to solve the problem, and the results were validated against an Ant Colony Optimization (ACO) algorithm. The developed GA was able to achieve better solution quality compared to the ACO algorithm. The results suggested that increasing the capacity of ECBs could reduce the number of routes and lower total energy consumption, although at the cost of longer individual routes for couriers.

4.2 Transportation and Health

Cudok et al. (2022) contributed to understanding how to increase the acceptance of refurbished products, focusing on ECBs. It drew from marketing, psychology, and circular economy principles to derive strategies for promoting refurbished ECBs as part of sustainable mobility solutions. This added value by offering insights on how to encourage consumer adoption of environmentally friendly alternatives through targeted marketing strategies. The primary goal was to identify and propose measures that can increase the acceptance of refurbished ECBs. The paper aimed to support the wider adoption of such products by addressing customer concerns and promoting sustainable mobility. The paper identified several key factors that influence customer acceptance of refurbished ECBs, including price, functionality, transparency in the refurbishment process, and marketing strategies. It showed that consumers were willing to accept refurbished ECBs if they perceive high functionality and appropriate cost savings compared to new models.

Philips et al. (2024) aimed to investigate the potential of electric micromobility (e-micromobility) technologies, including ECBs, in domestic use. The study focused on both environmental and health benefits, specifically examining how ECBs can improve physical activity levels while reducing carbon emissions in suburban and peri-urban areas. This research addressed a gap in understanding how ECBs can be integrated into everyday suburban mobility. The study employed a multi-center, mixed-methods approach combining quantitative and qualitative data collection. Methods included nationally representative and city-specific surveys, household trials with GPS tracking, smartwatches to monitor physical activity, and travel diaries. Additionally, semi-structured interviews were conducted with stakeholders, and a survey was used to assess public perceptions and potential barriers to adopting e-micromobility. The paper outlined the planned methods and data collection techniques that could be used to assess the impacts of ECBs on health, carbon emissions, and transportation habits.

4.3 Transportation and Life Cycle Assessment

Schünemann et al. (2022) contributed to understanding the environmental impact of using ECBs for urban freight transportation in a developing country context, specifically Ghana. Through a life cycle assessment (LCA), the study provided insights into how ECBs compare to traditional diesel vans in terms of GHG emissions, highlighting the potential environmental benefits of ECBs when charged with renewable energy sources like solar power. The results showed that ECBs have significantly lower GHG emissions when charged with solar energy.

Temporelli et al. (2022) contributed by conducting a LCA to compare the environmental impacts of different vehicle types (ECB, electric van, plug-in hybrid van, and diesel van) in the context of urban last-mile logistics. It provided a comprehensive assessment of the environmental performance of these vehicles, focusing on their entire life cycle, from production to end-of-life. The study used primary data, making it more accurate and reliable, especially in the use phase. The authors applied the ISO 14040-compliant LCA methodology, using primary data for vehicle usage and emissions. The assessment covered all life cycle phases, including raw material extraction, manufacturing, use phase, maintenance, and end-of-life processes. The environmental impacts were measured using distinct categories, such as climate change, photochemical ozone formation, and resource use. Data were collected through on-road tests and laboratory simulations. The paper concluded that ECBs are the most sustainable option for last-mile delivery, with significantly lower environmental impacts than vans and other transport alternatives considered.

4.4 Travel Behavior

Keler et al. (2021) presented early insights into the movement patterns of ECB commuters through field experiments and trajectory analyses. The study contributed by analyzing the spatial accuracy of Global Navigation Satellite System (GNSS) data to map specific bike maneuvers in urban environments. The study evaluated how urban infrastructure influences ECB usage and provides valuable data on the potential for ECBs to replace motorized transport, offering practical findings for urban mobility and infrastructure planning. The methodology involved field experiments where test subjects use different ECB models in daily commuting scenarios. These experiments were supplemented with GNSS tracking data to analyze travel trajectories. The study also used questionnaires to gather qualitative feedback on user experiences. The recorded trajectories were analyzed to map and evaluate the performance of ECBs under different urban conditions, focusing on infrastructure limitations, maneuverability, and travel efficiency. The paper concluded that while ECBs have potential as an alternative to motorized transport, there are substantial challenges related to urban infrastructure.

Thomas (2022) explored how ECBs are used by parents in auto-centric communities to continue biking despite the challenges posed by car-dominant environments. This research filled a gap in the literature by focusing on family use of ECBs for child transportation. The study also examined how ECBs can help reduce car dependence while offering physical and mental health benefits. The study employed qualitative methods, specifically in-depth interviews with 20 parents or caretakers in the San Francisco Bay Area (USA) who used ECBs for child transportation. Participants were recruited through local bike organizations and online networks. The interviews were transcribed and analyzed using the “Atlas.ti” software, focusing on themes, such as motivations, benefits, and challenges of ECBs use. The paper aimed to understand how families in car-oriented urban areas integrate ECBs into their daily routines, focusing on the motivations behind using ECBs and the challenges they face. It also sought to explore the potential for ECBs to replace car trips, particularly for parents with young children. The findings showed that ECBs help parents overcome physical barriers (e.g., hills, heavy loads) and reduced the stress associated with car trips, such as parking and traffic. Participants cited the ability to avoid car drop-off lines at schools as a key advantage. ECBs also allowed families to engage in physical activity, spend more quality time together, and experience greater freedom and accessibility within their communities.

4.5 Relevant Survey Studies

Narayanan and Antoniou (2022) provided a comprehensive review of the existing literature on ECBs, particularly focusing on their applications in commercial transport. It systematically categorized the existing research based on various aspects, such as typology, penetration, impacts, operational frameworks, and policy requirements. The review included journal papers, conference papers, policy papers, and technical reports published between 1950 and 2021. This paper provided a comprehensive synthesis of the collected studies on ECBs, offering valuable insights into their typologies, potential impacts on different areas, and policy implications. It helped policymakers and businesses better understand the potential of ECBs, addressing research gaps and guiding effective deployment strategies. The present survey study is supplementary to the survey study conducted by Narayanan and Antoniou (2022) and focuses mainly on the recent studies, whereas Narayanan and Antoniou (2022) mainly concentrated on the studies published before 2021.

5 Major Lessons Learnt

As a result of the conducted research, it was found that integration of ECBs into urban logistics provides valuable advantages across environmental, operational, and social dimensions (see Table 2 and Fig. 2). One of the key findings discovered during the present literature survey is the necessity of aligning ECB usage with sustainable energy policies. As an example, Schünemann et al. (2022) demonstrated that the environmental benefits of ECBs, especially in reducing GHG, are significantly enhanced when powered by renewable energy sources like solar energy. This suggests that cities aiming to improve air quality and cut emissions should support ECB infrastructure with sustainable energy initiatives. In regions where non-renewable energy is predominant, the environmental benefits could be substantially limited without this alignment.

From an operational standpoint, studies underscore the strategic value of incorporating ECBs within a multi-modal logistics model. The research conducted by Caggiani et al. (2021) illustrated that combining ECBs with other zero-emission vehicles, such as electric vans, optimize last-mile delivery while enhancing logistical flexibility. This approach allows ECBs to efficiently cover densely populated urban

Table 2 Summary of the reviewed studies

References	Methodology	Objective	Domain
Caggiani et al. (2020)	Decision Support System (DSS), Fuzzy Inference System (FIS)	Minimize travel time and minimize emission exposure	Logistics and mobility
Ewert and Deniz (2020)	Stakeholder interviews and user survey	Identify the operational and vehicle requirements	Logistics and mobility
Milenković et al. (2020)	Value case methodology	Reducing traffic congestion and improving sustainability in last-mile deliveries	Logistics and mobility
Aiello et al. (2021)	Multi-Criteria Decision-Making Approach (TOPSIS)	Optimizing urban delivery systems by using ECBs with modular battery configurations	Logistics and mobility
Caggiani et al. (2021)	Mixed-Integer Linear Programming (MILP)	Cost-effective logistics solution for urban and restricted traffic zones	Logistics and mobility
Keler et al. (2021)	Trajectory analysis combined with user studies and Global Navigation Satellite System (GNSS) tracking experiments	Evaluate how urban infrastructure influences ECB usage and provides valuable data on the potential for ECBs to replace motorized transport, offering practical findings for urban mobility and infrastructure planning	Travel behavior
Naumov (2021)	Monte Carlo simulations and the Clarke–Wright heuristic algorithm	Determine the optimal loading hub location in urban areas with restricted traffic to minimize the total transport work	Logistics and mobility
Carracedo and Mostofi (2022)	Comprehensive review of the role of ECBs in urban transportation, specifically focusing on private transportation	The objective was to highlight the role of ECBs as a sustainable alternative to private car usage	Logistics and mobility
Cudok et al. (2022)	Persona method, complemented by an online survey, to characterize typical user groups and derive measures to increase the acceptance of refurbished e-cargo bikes	Identify and propose measures that can increase the acceptance of refurbished ECBs	Transportation and health

(continued)

Table 2 (continued)

References	Methodology	Objective	Domain
Narayanan and Antoniou (2022)	Systematic literature review	Comprehensive review of the existing literature on ECBs, particularly focusing on their applications in commercial transport	Relevant survey studies
Philips et al. (2022)	Spatial microsimulation	Estimate the maximum possible reduction in car CO_2 emissions if ECBs use was maximized across different geographic areas and demographic profiles	Logistics and mobility
Schünemann et al. (2022)	Life cycle assessment	To understand the environmental impact of using ECBs for urban freight transportation in a developing country context	Transportation and life cycle assessment
Temporelli et al. (2022)	Life cycle assessment	To compare the environmental impacts of different vehicle types (ECB, electric van, plug-in hybrid van, and diesel van) in the context of urban last-mile logistics	Transportation and life cycle assessment
Thomas (2022)	The study employed qualitative methods, specifically in-depth interviews with 20 parents or caretakers in the San Francisco Bay Area (USA) who used ECBs for child transportation	Examine how ECBs can help reduce car dependence while offering physical and mental health benefits	Travel behavior
Gil et al. (2022)	Cross-industry standard process for data mining	Analyze and optimize the spatiotemporal behavior of ECB fleet and to propose optimal locations for expanding the network of logistics hubs	Logistics and mobility

(continued)

Table 2 (continued)

References	Methodology	Objective	Domain
Malik et al. (2023)	An intelligent modelling framework consisting of three components: a data acquisition system, an intelligent learning unit, and an output unit. This framework integrates artificial intelligence, statistical, and mathematical approaches to analyze and model e-cargo bike usage	Identify the critical factors influencing the adoption and use of ECBs by small businesses	Logistics and mobility
Papaioannou et al. (2023)	A Genetic Algorithm (GA) was employed to solve the problem, and the results were validated against an Ant Colony Optimization (ACO) algorithm	Minimize the energy consumed by ECBs and ensure a fair distribution of delivery points and distance among couriers	Logistics and mobility
Philips et al. (2024)	Methods included nationally representative and city-specific surveys, household trials with GPS tracking, smartwatches to monitor physical activity, and travel diaries. Additionally, semi-structured interviews were conducted with stakeholders, and a survey was used to assess public perceptions and potential barriers to adopting e-micromobility	The study focused on both environmental and health benefits, specifically examining how ECBs can improve physical activity levels while reducing carbon emissions in suburban and peri-urban areas	Transportation and health

centers, while electric vans manage suburban deliveries, providing a comprehensive solution that meets both emissions reduction and operational challenges. These insights indicate that a hybrid model could serve as a cost-effective and environmentally beneficial strategy for logistics firms, bolstering resilience in urban logistics systems.

The complexity of broader ECB adoption is further highlighted by consumer perception and behavioral factors. The study by Cudok et al. (2022) emphasized the need for targeted marketing and education to foster acceptance of ECBs, particularly refurbished models. Findings suggest that when consumers are informed of the economic and functional advantages of refurbished ECBs, they are more likely to adopt them as a sustainable alternative. Policymakers and logistics providers could use these insights to develop campaigns that clarify the practical benefits of ECBs, positioning them as feasible replacements for conventional vehicles, and addressing both psychological and financial barriers to adoption.

The literature also underscores the importance of technological advancements, such as data analytics and decision support systems, in maximizing ECB efficiency.

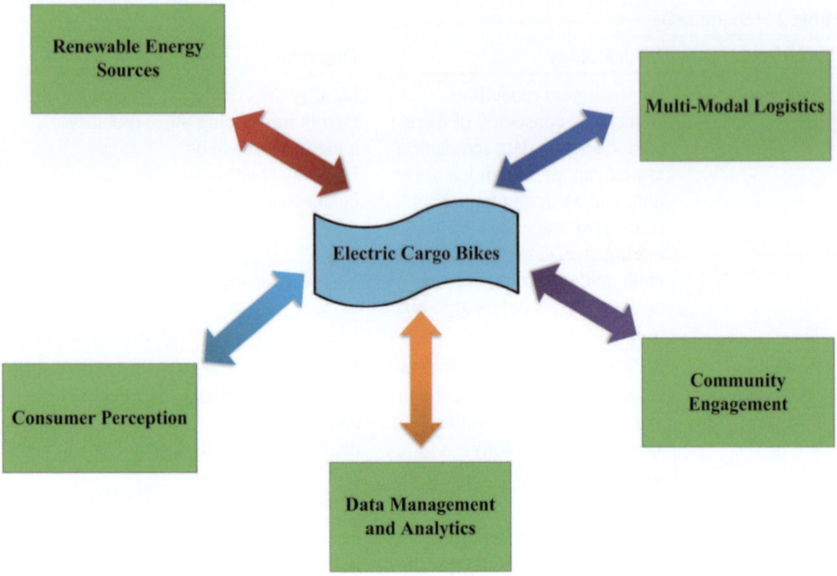

Fig. 2 Important insights discovered during the literature survey

For instance, Caggiani et al. (2020) introduced an eco-friendly decision support system that balances route efficiency with driver health by minimizing travel time and exposure to emissions. This technology not only enhanced logistical efficiency but also improved delivery personnel's well-being. The future logistics solutions should focus on various sustainability dimensions (e.g., prioritizing economic sustainability at the expense of environmental and social sustainability is not viewed as an effective alternative anymore).

Lastly, the societal benefits of ECBs extend beyond logistics, fostering public health and enhancing community engagement. The research by Philips et al. (2022) revealed that by reducing car dependency, ECBs encouraged physical activity, cleaner air, and reduced traffic noise, which all contribute to higher urban living standards. These findings supported the integration of ECB policies into a comprehensive urban planning approach that values public health and livability alongside transportation efficiency. Policies promoting bike-friendly infrastructure and family-oriented urban designs could increase the accessibility of ECBs for diverse populations, as illustrated in Thomas (2022). Therefore, while ECBs present significant potential within logistics, their broader social impact positions them as integral components of sustainable, community-centered urban environments.

6 Research Gaps and Future Research Directions

In this section, we provide an exploration of potential future research directions aimed at advancing the deployment and sustainability of ECBs in urban logistics. The outlined avenues focus on several critical dimensions. Together, these directions offer a roadmap for addressing the technological, environmental, economic, and policy-related challenges associated with integrating ECBs into modern logistics and delivery systems, providing actionable insights for researchers, policymakers, and industry stakeholders alike.

6.1 Simulation-Based and Scenario Analysis Approaches

Future research could significantly enhance simulation-based models by applying them at city and regional levels to evaluate a variety of scenarios of ECB deployment. In particular, simulations could be used to understand system behavior under different demand configurations, such as distinguishing between weekday and weekend dynamics, rush hours versus off-peak times, and using uniform or other probabilistic demand distributions. These differentiated scenarios would provide valuable insights for policymakers, allowing them to anticipate the effects of varied demand densities, fleet compositions, and network configurations on critical metrics, such as vehicle-kilometers traveled and delivery times. Additionally, dynamic simulations could assess emissions from car fleets under different evolution pathways, focusing on the environmental benefits of integrating ECBs within multimodal travel scenarios. Such research could capture modified trip distributions and could offer opportunities to replace short car trips with ECB deliveries.

6.2 Environmental and Life Cycle Impact Assessments

Future studies should prioritize a comprehensive analysis of emissions across the entire life cycle of ECBs, encompassing production, operational use, and end-of-life recycling stages. LCAs can offer insights into the total GHG emissions associated with various stages of ECB usage, from the extraction of raw materials to manufacturing processes and disposal. Such assessments could be tailored to regional differences, as countries with distinct industrial practices, energy mixes, and recycling capabilities may show significant variance in life cycle emissions. For instance, regions reliant on coal-based energy for manufacturing may yield higher embedded emissions in the production phase compared to regions with greener energy sources. A crucial aspect of this analysis would be performing sensitivity tests (Shahin et al., 2024) on the lifespan and replacement cycles of key components, particularly batteries. Battery production is known to be one of the most GHG-intensive stages of ECB manufacturing. Future research could explore the environmental benefits of alternative battery types, such as solid-state batteries, which may offer longer lifespans and higher energy densities, potentially reducing the frequency of battery replacements and, consequently, the environmental footprint. In regions with high ambient temperatures, which can accelerate battery degradation, such studies would be particularly valuable.

6.3 Policy and Incentive Mechanisms for ECB Adoption

Future research could explore the broader implications of supportive policies for the deployment of ECBs, emphasizing their potential to influence adoption rates and urban sustainability. For example, financial incentives, such as tax rebates, targeted subsidies, and reduced registration fees, could be evaluated to determine their effectiveness in overcoming the high upfront costs associated with ECBs compared to conventional bicycles. A comparative analysis could highlight which financial measures are most impactful for businesses and individual consumers, particularly in congested urban areas.

Additionally, regulatory measures could play a pivotal role in accelerating ECB adoption. Policies, such as "low emission zones" that restrict high-emission vehicles in central areas, coupled with designated ECB lanes and parking spaces, could enhance their practicality and appeal. Incentives for zero-emission delivery vehicles might further support this transition. Future studies could model the combined effects of financial and regulatory policies on ECB adoption, air quality improvements, and traffic decongestion. By providing actionable insights, such research would be instrumental in guiding policymakers and practitioners toward designing effective strategies for sustainable urban mobility.

6.4 Cost, Efficiency, and Design Optimization

Future research on the cost-reduction potential of ECBs could focus on identifying specific strategies to reduce both initial investment and ongoing operational expenses, making ECBs a more financially viable option for various commercial applications. For instance, research could assess the comparative costs of different materials and manufacturing processes, such as the use of lighter, more durable materials that reduce both weight and maintenance needs over time. By simulating the lifecycle costs of different configurations (e.g., comparing aluminum, steel, and composite frames), researchers could offer insights into which designs maximize longevity and reduce repair costs, especially under high-frequency, last-mile delivery conditions.

Optimizing ECB designs tailored for different commercial needs would be another valuable research avenue. For last-mile delivery in densely populated urban areas, compact designs with high maneuverability and tight turning radii may be preferred.

These bikes could include modular storage systems that allow for rapid loading and unloading, thereby reducing turnaround times. In contrast, for suburban or rural applications where delivery distances are longer, designs might prioritize battery life, load capacity, and rider comfort over compactness. Research could also test prototypes in real-world delivery scenarios, collecting data on the key performance metrics like speed, cargo stability, and battery efficiency under different conditions.

Last but not least, the potential for ECBs to alleviate traffic congestion is another valuable research avenue. As smaller, more maneuverable vehicles, ECBs can help reduce traffic density by shifting certain deliveries away from larger vehicles that contribute significantly to congestion in busy areas. Future studies could develop models to quantify these impacts, perhaps comparing traffic flow in scenarios with varying proportions of ECBs versus traditional delivery trucks.

7 Conclusion

Electric cargo bikes (ECBs) represent a promising solution to the pressing issues of urban transportation, specifically in addressing GHG, congestion, and last-mile delivery challenges. Given the rapid growth of e-commerce and urbanization, cities face increased demands for efficient, low-impact transportation alternatives. ECBs offer a viable option due to their extended range, larger load capacity, and accessibility in dense urban environments compared to traditional bicycles. Their integration into urban logistics aligns with the objectives of the European Green Deal and other global climate commitments, which aim to reduce the transportation sector's carbon footprint. Consequently, ECBs not only present an environmentally friendly alternative but also a means to enhance urban quality of life through reduced congestion and pollution.

Despite these advantages, ECBs still face challenges that limit their widespread adoption, such as infrastructure constraints, policy gaps, and insufficient empirical data on their operational efficiency in diverse urban settings. The lack of standardized data on ECBs usage makes it difficult to assess their full impact and hinders their incorporation into urban transportation planning. Additionally, there remains a need for optimization in various aspects, from route planning and fleet management, in order to effectively integrate ECBs within existing transportation networks. The present literature review highlighted these areas as crucial for future research to support the strategic deployment of ECBs in urban logistics.

By systematically reviewing recent studies since 2020, this paper aimed to consolidate existing knowledge on ECBs and identify gaps in research that, if addressed, could enhance the viability and effectiveness of ECBs in urban transportation. As a result, we proposed several avenues for future research, including developing optimization models that consider dynamic urban environments, evaluating the long-term cost-effectiveness of ECB adoption, and exploring the impact of supportive policies on ECB deployment. These directions offer significant potential for advancing

ECB integration in urban logistics, contributing to sustainable urban mobility and supporting climate action objectives.

References

Aiello, G., Quaranta, S., Certa, A., & Inguanta, R. (2021). Optimization of urban delivery systems based on electric assisted cargo bikes with modular battery size, taking into account the service requirements and the specific operational context. *Energies, 14*(15), 4672.

Becker, S., & Rudolf, C. (2018). Exploring the potential of free cargo-bikesharing for sustainable mobility. *Gaia-Ecological Perspectives for Science and Society, 27*(1), 156–164.

Bosona, T. (2020). Urban freight last mile logistics—Challenges and opportunities to improve sustainability: A literature review. *Sustainability, 12*(21), 8769.

Caggiani, L., Colovic, A., Prencipe, L. P., & Ottomanelli, M. (2021). A green logistics solution for last-mile deliveries considering e-vans and ECBs. *Transportation Research Procedia, 52*, 75–82.

Caggiani, L., Prencipe, L. P., Colovic, A., & Dell'Orco, M. (2020, June). An eco-friendly decision support system for last-mile delivery using ECBs. In *2020 IEEE International Conference on Environment and Electrical Engineering and 2020 IEEE Industrial and Commercial Power Systems Europe (EEEIC/I&CPS Europe)* (pp. 1–6). IEEE.

Carracedo, D., & Mostofi, H. (2022). Electric cargo bikes in urban areas: A new mobility option for private transportation. *Transportation Research Interdisciplinary Perspectives, 16*, Article 100705.

Carvelo2go. (2024). *Jahresbericht 2023—eCargo-Bike-Sharing in der Schweiz*. Retrieved January 25, 2025 from https://www.carvelo.ch/wp-content/uploads/2024/02/Jahresbericht_carvelo_2023_DE.pdf

Christidis, P., & Rivas, J. N. I. (2012). *Measuring road congestion*. https://publications.jrc.ec.europa.eu/repository/bitstream/JRC69961/congestion%20report%20final.pdf

Cox, P., & Rzewnicki, R. (2015). Cargo bikes: Distributing consumer goods. *Cycling Cultures, 1*, 130–151.

Cudok, A., Neugebauer, L., & Vietor, T. (2022). Increasing acceptance for refurbished products at the example of ECBs. *Procedia CIRP, 105*, 571–576.

European Commission. (2022). *Communication from the commission to the European parliament, the Council, the European economic and social committee and the committee of the regions*. The New EU Urban Mobility Framework. Retrieved June 17, 2022 from https://transport.ec.europa.eu/system/files/2021-12/com_2021_811_the-new-eu-urban-mobility.pdf

European Commission. (2024). *Mid-term evaluation of the 8th Environment Action Programme*. Retrieved January 25, 2025 from https://www.umweltministerkonferenz.de/documents/bericht-zu-top-8_2_3_1734638901.pdf

Ewert, A., & Deniz, Ö. (2020). Vehicle requirements for electric cargo bikes in commercial transport. In *Proceedings of 8th Transport Research Arena TRA 2020*, April 27–30, 2020, Helsinki, Finland.

Gil, B., Albuquerque, V., Dias, M. S., Abranches, R., & Ogando, M. (2022, December). Data driven spatiotemporal analysis of ECB network in Lisbon and its expansion: The Yoob case study. In *International Conference on Intelligent Transport Systems* (pp. 23–39). Springer Nature Switzerland.

Gruber, J., Kihm, A., & Lenz, B. (2014). A new vehicle for urban freight? An ex-ante evaluation of electric cargo bikes in courier services. *Research in Transportation Business & Management, 11*, 53–62.

Keler, A., Kessler, L., Fehn, F., & Bogenberger, K. (2021, December). Movement patterns of electric cargo bike commuters—First insights from field experiments and trajectory analyses. In *Proceedings of the ICA* (Vol. 4, pp. 1–4). Copernicus GmbH.

Kirkels, M. (2016). *Short history of the cargo bike*. Retrieved February 16, 2016 from http://www.cargobikefestival.com/news/short-history-of-the-cargo-bike/

Lebeau, P., De Cauwer, C., Van Mierlo, J., Macharis, C., Verbeke, W., & Coosemans, T. (2015). Conventional, hybrid, or electric vehicles: Which technology for an urban distribution centre? *The Scientific World Journal, 2015*(1), Article 302867.

Lovejoy, K., & Handy, S. (2012). Developments in bicycle equipment and its role in promoting cycling as a travel mode. *City Cycling,* 75–104.

Malik, F. A., Egan, R., Dowling, C. M., & Caulfield, B. (2023). Factors influencing e-cargo bike mode choice for small businesses. *Renewable and Sustainable Energy Reviews, 178*, Article 113253.

Milenković, M., Knežević, N., de Yuso, A. M., & Bojović, N. (2020). *Methodological framework for the development of urban electric cargo bike system in shipment distribution*. Zbornik Radova Trideset Osmog Simpozijuma O Novim Tehnologijama U PoŠtanskom I Telekomunikacionom SaobraĆaju–Postel.

Narayanan, S., & Antoniou, C. (2022). Electric cargo cycles—A comprehensive review. *Transport Policy, 116*, 278–303.

Naumov, V. (2021). Substantiation of loading hub location for electric cargo bikes servicing city areas with restricted traffic. *Energies, 14*(4), 839.

Papaioannou, E., Iliopoulou, C., & Kepaptsoglou, K. (2023). Last-mile logistics network design under ECBs. *Future Transportation, 3*(2), 403–416.

Philips, I., Anable, J., & Chatterton, T. (2022). E-bikes and their capability to reduce car CO_2 emissions. *Transport Policy, 116*, 11–23.

Philips, I., Azzouz, L., De Sejournet, A., Anable, J., Behrendt, F., Cairns, S., Cass, N., Darking, M., Glachant, C., Heinen, E., & Marks, N. (2024). *Domestic use of ECBs and other E-micromobility: Protocol for a multi-centre, mixed methods study.* https://www.preprints.org/manuscript/202409.2049/v1

Schünemann, J., Finke, S., Severengiz, S., Schelte, N., & Gandhi, S. (2022). Life cycle assessment on electric cargo bikes for the use-case of urban freight transportation in Ghana. *Procedia CIRP, 105*, 721–726.

Schwanen, T., Dijst, M., & Dieleman, F. M. (2004). Policies for urban form and their impact on travel: The Netherlands experience. *Urban Studies, 41*(3), 579–603.

Shahin, R., Beaulieu, M., Bélanger, V., & Cousineau, M. (2024). Optimizing strategic personal protective equipment stockpiling: An assessment using Morris sensitivity analysis. *Engineering Proceedings, 76*(1), 29.

Siegenthaler, P. (2017). *Cargo bikes: Cheaper, faster, greener*. SWI swissinfo.ch. Retrieved May 9, 2018 from https://www.swissinfo.ch/eng/society/sustainable-mobility_cargo-bikes-cheaper-faster-greener/43395840

Temporelli, A., Brambilla, P. C., Brivio, E., & Girardi, P. (2022). Last mile logistics life cycle assessment: A comparative analysis from diesel van to ECB. *Energies, 15*(20), 7817.

Thomas, A. (2022). Electric bicycles and cargo bikes—Tools for parents to keep on biking in auto-centric communities? Findings from a US metropolitan area. *International Journal of Sustainable Transportation, 16*(7), 637–646.

Wrighton, S., & Reiter, K. (2016). CycleLogistics—Moving Europe forward! *Transportation Research Procedia, 12*, 950–958.

Reza Shahin has a Ph.D. in Computer Science with a focus on Operations Research and specializes in developing and implementing advanced optimization models, including Mixed-Integer Linear Programming (MILP) and machine learning techniques, for complex industrial and supply chain systems. He completed his Ph.D. program at University of Gustave Eiffel in France on the Flex-Route Transit problem. His Ph.D. focus was on integrating ad hoc stops into public transport in which the flexibility of demand responsive transit was merged with the low cost of public transportation, providing a door-to-door pick-up and drop-off to customers. Currently, he is a postdoctoral researcher at University of Polytechnique Montreal, where he works on scheduling tasks in a Canadian underground mine. His expertise includes applying machine learning methods, such as reinforcement learning and data-driven decision-making, to enhance operations in logistics, scheduling, and intelligent manufacturing.

Maxim A. Dulebenets is an Associate Professor and Graduate Program Director in the Department of Civil and Environmental Engineering at Florida A&M University-Florida State University (FAMU-FSU) College of Engineering. Dr. Dulebenets holds B.Sc. and M.Sc. degrees in Railway Construction from the Moscow State University of Railroad Engineering (Moscow, Russia) and M.Sc. and Ph.D. degrees from the University of Memphis (Memphis, TN—USA) in Civil Engineering with concentration in Transportation. His research interests include, but are not limited to, operations research, optimization, simulation modeling, NP-hard problems, mathematical programming, metaheuristics, hybrid algorithms, evolutionary computation, transportation engineering, freight transportation, intermodal freight facilities, railroads, liner shipping scheduling, and GPS data processing. Dr. Dulebenets has been involved in a variety of research projects with the overall value ~$19.5 million (~$2.9 million as a PI), sponsored by the United States Department of Defense, United States Department of Transportation, National Science Foundation, Center for Accessibility and Safety for an Aging Population, Florida Department of Transportation, Tennessee Department of Transportation, Center for Freight and Infrastructure Research and Education, Federal Highway Administration, Canadian National Railway, Federal Express, Intermodal Freight Transportation Institute, North Caucasus Railroad Authority, and Moscow Railroad Authority.

He serves as a referee for 250+ international journals. According to the 2024 Scopus citation report, he is ranked as #14 in the world out of 28,891 scholars in Logistics and Transportation based on the composite c-score c(ns) and last year performance and #148 based on the lifetime performance. His work has been published in the leading international journals, including the International Journal of Production Economics, Information Sciences, Reliability Engineering and System Safety, and IEEE Transactions on Intelligent Transportation Systems. Dr. Dulebenets is actively involved in activities of more than 10 Standing Committees and Subcommittees of the Transportation Research Board of the National Academies of Sciences, Engineering, and Medicine. He is an Affiliate Member of the INFORMS Optimization Society and a Senior Member of the IEEE. Dr. Dulebenets is a licensed Professional Engineer registered in the State of Michigan (License № 6201063266).

The Impact of Decarbonization Instruments on Shaping Innovative and Zero-Emission Last-Mile Delivery Strategies

Urszula Motowidlak

Abstract This chapter focuses on the changes in the last-mile delivery market in terms of its adaptation to ongoing processes and challenges related to both climate policy requirements and growing expectations of authorities and society. The study fits into the broader context of the transformation of the entire transport sector towards achieving climate neutrality. The aim of the study is to identify promising transport and logistics decarbonization tools and assess their importance from the perspective of shaping innovative and zero-emission last-mile delivery strategies. The techniques and research methods used included: content analysis, cause-effect analysis, and applicable quantitative and qualitative analysis methods. These methods were employed to discover and elaborate on the results of the study conducted among representatives of entities carrying out last-mile deliveries. The evaluation of instruments for decarbonizing last-mile deliveries refers to the theory of targeted technological progress in the environmental context. The low-emission transformation of urban transport is one of the priorities of the European Union. The challenges related to last-mile deliveries are causing growing interest in concepts for implementing them in an increasingly sustainable manner. Their implementation is to be supported by actions to reduce the transport intensity of the economy, as well as actions to better organize last-mile logistics.

Keywords Zero-emission transport · Decarbonization of last-mile deliveries · Decarbonization instruments · Decarbonization of deliveries

U. Motowidlak (✉)
Faculty of Economics and Sociology, Department of Logistics and Innovation, University of Lodz, Lodz, Poland
e-mail: urszula.motowidlak@uni.lodz.pl

M. A. Dulebenets et al. (eds.), *Transport and Climate Change*,
https://doi.org/10.1007/978-3-032-00563-2_8

1 Introduction

The development of a low-carbon economy is a manifestation of the urgent need to cope with global warming and achieve sustainable economic development. In theoretical terms, it is defined as an economy pursuing long-term development based on endogenous factors, while striving to minimize the negative impact on the natural environment, including through the reduction of greenhouse gas (GHG) emissions and efficient use of resources (Xin et al., 2011). Wang et al. (2019) emphasize that it is a completely new transformation of the pattern of energy consumption, economic development, and human lifestyle, constituting a huge progress from industrial civilization to ecological economic civilization. Its goal is to limit anthropogenic climate change, rationally manage resources, improve the quality of life of society, increase competitiveness and ensure sustainable economic development. In general, it includes activities "generating products or services that provide low-carbon results" (Sengupta et al., 2020).

The active involvement of the European Union (EU) in actions supporting the global fight against climate change is a political priority of the European Commission (EC). In 2023, the total EU net greenhouse gas emissions fell to 37% below 1990 levels (European Environment Agency, 2024). This is a significant achievement for the EU towards climate neutrality, considering that in 2022 it accounted for 6.7% of global CO_2 emissions, behind China and the United States (Crippa et al., 2023). In the EU strategy aimed at developing a competitive and low-emission economy, the basic factors that are to determine growth and development are continuous investment in human capital and technological progress (European Commission, 2018). Investments related to reducing the emission and resource intensity of economies are to lead to achieving net zero greenhouse gas emissions by 2050 (Carteni, 2018).

The demand for new technologies, as well as investment and consumer goods contributing to the decarbonization of the economy and increasing energy efficiency, remains a function of many variables (Geels, 2022; Sovacool et al., 2024). Government policies have been recognized as key instruments for decarbonizing the economy. Johnstone and Newell (2018) confirm that the key role of the state has been indicated in the lively scientific and political discussion on the challenges related to climate change. Radical and rapid transformation of the economy requires political and economic interventions of the state, starting from research and technological innovation to industrial policy and regulations. In turn, Ehnert et al. (2018) and Pegels and Lütkenhorst (2014), recognizing the efficiency of the self-regulating market, point to the indirect role of the state in managing industrial policy instruments to solving market problems and coordinating actions within the strategy of supporting innovative and low-emission solutions.

Legal regulations that tighten environmental standards and impose sanctions on economic entities and individuals who do not meet the designated environmental criteria force innovation. At the same time, using technological potential as a factor of economic growth requires ensuring an appropriate level of diffusion of innovative solutions. This involves taking action to increase social acceptance of new solutions and create effective investment and consumption demand (Argyriou & Barry, 2021). In addition, it requires state support for activities that stimulate the supply of modern low-emission technologies and resource-efficient goods and services that initially do not generate profits but are necessary for the public good. Actions have already been taken in the EU to align private investment with climate and resource-efficient management goals using political strategies and strategic public investments (Furmankiewicz et al., 2021). Financial instruments also make a significant contribution to climate protection activities (Motowidlak & Motowidlak, 2023). From the perspective of striving for global decarbonization, investments in the development of "clean" technologies are an important element of the strategies of the functioning of the economies of many countries in the world. In the years 2020–2024, the governments of these countries allocated a total of USD 2 trillion to financial support for actions under many new decarbonization policies (International Energy Agency, 2024).

Recognizing the need to decarbonize the economy, the EU has embarked on an ambitious European Green Deal, adopting a package of initiatives in 2021, 'Fit for 55' (European Commission, 2021a). The Communication, published in December 2019, committed to achieving climate neutrality by 2050, including a 'green transition' in key cross-cutting programs and sectorial policies (European Commission, 2019). Nevertheless, a net-zero economy and society will not be feasible without having a more innovative and sustainable transportation system. In 2022, transport activities in the EU reached 31% of final energy consumption, making it the largest consumer of the available energy, ahead of households (27%) and industry (25%). The largest demand was reported by road transport. In 2022, it was responsible for 74% of total transport fuel needs (Eurostat, 2024).

In the period 1990–2022, due to the increasing demand for freight and passenger transport, GHG emissions from transport increased by over 33%. In 2022, they accounted for over 26% of GHG emissions in the EU. At the same time, ¾ of these emissions come from road transport. Forecasts show that transport activity will continue to grow, especially in the area of road freight transport. The EU Reference Scenario 2020 (European Commission, 2021b) predicts an increase in transport performance, expressed in tonne-kilometres (tkm), by 31% by 2030 and by 55% by 2050 compared to 2015. In order to achieve climate neutrality and reduce GHG emissions from transport by 90% by 2050 compared to 1990, there is an urgent need to focus decarbonization instruments on organizing deliveries and road freight transport.

Despite the gradual implementation of decarbonization strategies in recent years, most vehicles worldwide still rely on gasoline and diesel, which emit GHG emissions contributing to climate change. Achieving global climate neutrality requires, above all, a fundamental shift towards sustainable transport development in the world's largest economies. In 2023, China was the world's largest GHG emitter, accounting for about 30% of global emissions. The United States (US) came second, emitting about 11% of global GHG emissions (UNEP, 2024). In the US, the largest share of GHG emissions is the transport sector. In 2023, transport accounted for 28% of national emissions (EPA, 2024). At the same time, data indicate that transport emissions in the US have been increasing continuously since 2020, despite the growing popularity of electric vehicles and other zero-emission modes of transport. In China, the transportation industry is also one of the main GHG emitters. Emissions generated by transportation account for about 10% of China's total CO_2 emissions. As part of the adopted strategies, China has taken concrete actions to green the transportation sector (Zhang et al., 2024).

Cities have been identified as one of the strategic areas where significant benefits are expected from implementing a strategy to decarbonize transport and logistics. With the growing number of people migrating to urban areas, the issue of sustainable development of transport in cities is becoming increasingly important (McKinnon, 2023). The mobility system is being completely transformed, and the delivery of goods, which is mainly based on car transport, poses both enormous challenges and great opportunities for city authorities and businesses. The exiting demand for passenger mobility in urban locations is expected to double by 2050. Even greater growth is anticipated in the area of freight transport, especially in densely populated urban areas, as a result of the increasing importance of e-commerce and the associated large increase in demand for last-mile deliveries (Janinhoff et al., 2024).

At the same time, research confirms that last-mile deliveries will be influenced by three clear trends in the coming years: sustainability, technology, and personalization of deliveries (Escudero-Santana et al., 2022). The introduction of new and more efficient technologies is expected to play a key role in the decarbonization of these deliveries. While underlining the importance of decarbonizing last-mile deliveries in cities, the possibilities for its implementation are limited and difficult to implement. Moreover, in the transport policies of the EU and national policies of the member states and the strategies of the Transport-Freight-Logistics (TFL) companies, this topic has not been the subject of particular attention for a long time (Hammond et al., 2020).

This chapter focuses on identifying solutions/actions in last-mile supply chains in the context of companies striving to achieve the EU climate policy goals and responding to changing social and economic needs, treating innovative and low/zero emission solutions as a fundamental pillar of strengthening competitiveness. In conditions of increasing variability and complexity of the environment and consumer needs, the current and future growth in added value will be achieved by those supply chains that are not only able to efficiently establish but also actively manage last-mile delivery logistics. The aim of this study is to identify promising transport and logistics decarbonization tools and assess their importance from the perspective of shaping innovative and zero-emission last-mile delivery strategies. In order to achieve the aim, three research questions were formulated as follows:

- What problems/challenges in last-mile supply chains do companies responsible for their operation face?
- Does the pursuit of decarbonization of the EU economy create new opportunities for the organization/operation of last-mile deliveries?

- What actions/strategies/business models are key to the process of decarbonizing last-mile deliveries and maintaining the competitiveness of TFL sector companies?

The present study focuses on theoretical and pragmatic aspects of shaping competitive and low-emission last-mile deliveries and the main instruments/measures determining their development.

2 Literature Review

The development of competitive and low-emission freight transport is one of the currently discussed topics in the context of local, national and international economic policies, climate and environmental protection (Magalhães & Santos, 2022; Paddeu et al., 2024). By connecting international economies, it directly and indirectly affects the implementation of all sustainable development goals (Cascetta & Henke, 2023; Silvestri et al., 2021). At the same time, the functioning of EU transport in the field of freight transport is subject to strong influences of exogenous conditions that result from its role in the economy. Transport significantly supports international cooperation and plays a fundamental role in creating a global network of consumer goods trade and in the transfer of investment goods between countries (McKinnon, 2023; Santos, 2017). Modern supply chains, often belonging to dispersed logistics networks, determine a wider scope of transport activities, which is directly related to the increase in the transshipment coefficient.

The increasingly common division of the production process into stages carried out in different countries results in multiple transfers of products at different stages of their production process, which affects the rate of growth of demand for last-mile deliveries (Kiba-Janiak et al., 2024). This demand is additionally strengthened by the progressive implementation of the lean management concept. This concept, focused on the maximum reduction of inventory levels and the organization of supplies in a just-in-time system, forces significant changes in the functioning of transport. As a result, the demand for the movement of bulk cargo decreases, while the demand for the transport of general cargo increases. The departure from an economy based on inventories and the emergence of an economy based on flows requires fast, regular, flexible, and timely cargo flows. As a result, there is a dynamic increase in demand for so-called last-mile logistics services, implemented mainly using the least energy-efficient means of road transport.

The production process and the goods sales market connected by e-commerce is another factor generating demand for last-mile deliveries, which is characterized by very high growth dynamics. In 2020, the value of the global e-commerce logistics services market was EUR 368.1 billion, recording an increase of 27.3% compared to 2019 (Statista, 2024a). In Europe, in 2020, the value of e-commerce logistics services increased by 26.5% compared to 2019, reaching EUR 10.8 million (Statista, 2024b). In Poland, purchases via e-commerce channels in 2020 increased by 31.4% compared to 2019, reaching PLN 15.4 billion (Statista, 2024c). The increase in trust in online distribution channels, consolidated during the lockdowns related to the COVID-19 pandemic, has redefined the approach to purchasing processes, which results in a continued increase in online sales volumes. In 2023, the global e-commerce services market was valued at EUR 487 billion, which meant an increase of 32.4% compared to 2020 (ECDB, 2024). A similar trend characterizes the sales market in Europe, including Poland.

The dynamic development of last-mile freight transport is also a significant source of inconvenience and problems. Direct and indirect external costs of transport services are increasingly analyzed through the prism of threats not only of a local and national nature, but also global (Sovacool et al., 2024). From the point of view of the development of competitive and low-emission last-mile deliveries, they are primarily related to the use of fossil fuels, mainly crude oil, and the generation of GHG emissions. Other challenges concern the low rate of cargo space utilization, high fuel prices, sales peaks, and the appropriate level of customer service in the context of the growing number of shipments and returns. "The last mile" is a term referring to the final stage of the supply chain, which includes the transport of goods from the distribution center to the destination—the customer, store, or other collection point. Last-mile delivery, also called last-mile logistics or last-mile distribution, is a key element of the competitive advantage of the supply chain (Van Duin et al., 2024).

At the same time, it is considered one of the most expensive and generally inefficient supply chain processes. As estimated by logistics market experts, last-mile service often accounts for 40–55% of the logistics costs of the chain. These costs result from the need to deliver to many, often geographically dispersed points, instead of one central distribution point. Customers want to receive their parcels at home, at a parcel locker, at collection points (Pick-Up and Drop-Off, PUDO) or directly in the store (Click and Collect model). Therefore, it is necessary to ensure efficient service of several delivery channels. The challenge is to combine these deliveries into often complex routes in order to optimize the total costs. The growth of online sales also means that shipments are becoming more diverse in terms of size, content, and the time in which the goods must be delivered. Therefore, proper planning and management of routes becomes extremely important (Dell'Amico et al., 2022). Modern technologies, such as route management systems, can come to the rescue, which will help avoid delays and increase efficiency.

The management of last-mile processes varies depending on the industry, the company's business model, and the type of supply chain. Last-mile delivery in the B2B (business-to-business) supply chain involves delivering raw materials or finished products to other companies' warehouses. On the other hand, the B2C (business-to-consumer) supply chain involves more complex processes related to delivering products to individual customers, e.g. to a collection point indicated by them. In the B2C segment, organizing deliveries becomes a challenge mainly due to the need to balance high delivery costs with on-time delivery (Kiba-Janiak et al., 2024).

At the same time, the last-mile mainly involves delivering small quantities of goods to many different locations. Customers expect fast, reliable deliveries that are tailored to their lifestyle. The final customer satisfaction depends on whether the goods are delivered on time and without problems. Meeting these expectations can be a challenge, especially in urban areas, where issues such as traffic congestion can affect delivery times. Companies must therefore deliver within the time windows specified by customers, demonstrate flexibility (e.g., by enabling direct contact with the courier), and allow for real-time tracking of the shipment. These actions reduce the risk of the recipient being absent from home and having to redeliver the shipment, having a positive impact not only on meeting customer expectations, but also on the economic and environmental dimensions of last-mile deliveries.

The transport of parcels is also becoming a challenge, especially in the city, and it is not only related to traffic congestion. Problems include not only random events on the roads or parking difficulties, but also various types of regulations limiting the possibility of unloading (Paddeu & Aditjandra, 2020). In addition, there is an increase in the number of delivery addresses, the number of priority deliveries, the transport of perishable goods with short shelf lives (food and medicines), seasonality of deliveries (sales, holidays, etc.), overflowing parcel lockers, lack of space for parcels of certain dimensions, even multiple delivery attempts for direct pickups, etc. All these factors can cause delays, lead to errors, and generate additional costs (Lin et al., 2022). As a result, they can have a negative impact on the climate and the environment due to increased fuel consumption, exhaust emissions, and environmental pollution. Ecological and energy-saving solutions are therefore becoming an increasingly important criterion when choosing a courier company.

Last-mile delivery is experiencing a number of positive effects of technological progress and innovation driven by the fourth industrial revolution. At the same time, it is under strong pressure from customers' expectations for fast, reliable and personalized deliveries, which are environmentally and climate-friendly (Van Audenhove et al., 2018). Logistics and transport companies must be more diligent in modifying the last-mile supply chains in which they operate, taking into account both top-down and bottom-up requirements for low-emission and competitive deliveries (Maxner et al., 2022). To this end, they are using available solutions and designing new ones that may lead to completely new last-mile delivery systems in the future (Table 1).

The development of low-emission and competitive last-mile freight delivery requires a combination of technological innovations, social and behavioral changes, and organizational changes, both within the organization and in cooperation with last-mile supply chain actors. Possible transition paths may be (re)evolutionary. Examples of changes that evolve gradually are technological changes, such as electric vehicles, courier bikes, or eco-friendly packaging. Decarbonizing last-mile deliveries also includes social and behavioral changes. Depending on the timing and interaction between social and technological events, these changes can be more radical transformations, expressed in changes in behavior, norms, or culture.

In the process of decarbonizing last-mile deliveries, the development of electromobility and new fuels (especially hydrogen) has been determined as the most promising alternatives for substituting petroleum fuels. However, the potential economic benefits and reduction of GHG emissions have not brought the expected level of saturation for freight transport fleets (Bosona, 2020). Despite the high sales growth

Table 1 Possible actions from the perspective of developing competitive and low-emission last-mile deliveries

Main activities	Detailed activities	Effects of actions
Legal regulation and public policy (F_1)	Green zone and low emission zone CO_2 emission standards Subsidies and tax breaks Emission charges Financing green investments for last-mile logistics Extending the EU ETS to road transport	R&D spending Development of start-ups creating climate- and economy-friendly last-mile delivery models Development of alternative fuels and energy Improvement of the environment Climate protection
Transport organization and last-mile delivery management (F_2)	Route management systems Time slot delivery planning Matching loads to vehicle types Delivery flexibility and consolidation Real-time shipment tracking Pick-up points and parcel machines IT applications and infrastructure Real-time virtual management centers	Route optimization and volume consolidation Increasing delivery efficiency and avoiding delivery delays Reducing the number of vehicles on the route Minimizing the risk of re-delivery Improving the level of service and the company's image Reducing costs and GHG emissions Improving fuel/energy efficiency
Logistics in last-mile deliveries (F_3)	Systems supporting comprehensive management of warehouse processes in last-mile deliveries Automation and robotization in the processes of unloading, picking, and packing orders Intelligent systems supporting sales and operations planning based on advanced data analytics Logistics centers in strategic locations	Saving space, reducing the risk of errors Increasing the accuracy of demand forecasting Shortening the time of last-mile logistics Improving the efficiency of managing internal transport, distribution, and personnel Reducing costs Reducing GHG emissions Ecological packaging

(continued)

Table 1 (continued)

Main activities	Detailed activities	Effects of actions
Infrastructure and modern technologies in transport (F_4)	Increased share of electric vehicles and hydrogen technology in the last-mile Development of charging infrastructure Increased share of renewable energy in electricity and fuel production Autonomous vehicles Drones in deliveries Integration of used forms of delivery with road/air and communication infrastructure	Development of an innovative, integrated last-mile delivery network Increased technological risk Large investment outlays A threat to the finances of transport companies, especially small and medium-sized ones Reduction of GHG emissions Reduced competitiveness, especially in the initial investment period Improved trust and company image
New business and financial models (F_5)	Fleet sharing Business collaboration (shipping and receiving network) Green last-mile supply chains Use of cargo bikes Co-financing green transport projects Clusters for low-emission last-mile deliveries	Higher fleet utilization rates Increased throughput Efficient handling of peaks in last-mile deliveries Reduced costs for climate and environment Increased trust and transparency Improved competitiveness Increased added value in the chain A holistic last-mile freight network
Social and behavioral changes (F_6)	Implementing new communication strategies Sharing knowledge and educating consumers about their decisions Engaging stakeholders in collaboration	Increased customer satisfaction Reduced number of returns Reduced GHG emissions Increased competitiveness of "green" companies/ deliveries Increased awareness and choice of ecological last-mile deliveries

Source Own elaboration based on literature review

dynamics, their share in the EU light commercial vehicle market in 2023 was about 4%. Meanwhile, with courier deliveries in cities, their use shows great potential for reducing GHG emissions (Bjørge et al., 2022; Figenbaum, 2018). The development of charging infrastructure, financial support, and acceptance of new technologies remain a challenge (Bjørgen & Ryghaug, 2022). An alternative to vehicles, especially for deliveries made in compact city centers and zero-emission zones, is electric bicycles. More future-oriented technologies include hydrogen-powered vehicles, drones, and autonomous vehicles, which could be the next solutions for efficient and low-emission last-mile deliveries. Research confirms that their implementation in the delivery market still requires overcoming significant barriers, including technological, regulatory, urban planning, and social acceptance.

Another important path to decarbonizing last-mile deliveries is changes in the organization of delivery logistics and transport management. Fleet and logistics management systems allow for optimizing the delivery route and shortening the delivery time of parcels, which has a positive impact on climate, environment, and company efficiency. Location plays a key role in effective management of last-mile logistics (Bjørgen et al., 2021). Establishing logistics centres in strategic locations contributes to greater efficiency of the delivery process. This allows for reducing the number of vehicles on the route, which translates into cost reduction, lower GHG emissions, and better traffic management. In e-commerce deliveries, a system of parcel machines and collection/dispatch points is beneficial from the perspective of decarbonizing deliveries, which allow for reducing GHG emissions by up to approximately 80% compared to deliveries made to the customer's home (Dong et al., 2023).

For collection points to become a viable way to reduce CO_2 emissions, courier companies should design last-mile delivery networks with a high density of collection points, so that potential customers can reach them by foot or bicycle. Case studies in urban, suburban, and rural areas in the Netherlands show that the most positive effects can be accomplished in urban areas. In rural areas, the CO_2 reduction benefits resulting from enhanced delivery route effectiveness are diminished by the carbon emissions produced during the customer journey (Niemeijer & Buijs, 2023). Research conducted in the UK, the US, France, Germany, Spain, the Netherlands, and Italy has shown that convenient location of collection points increases their value to customers (Vakulenko et al., 2018). From a customer service and potential environmental benefits perspective, collection points should be located in nearby frequently visited locations, such as shopping areas, supermarkets, or public transport (Kedia et al., 2020).

Micro-centres have great potential for last-mile grocery deliveries, as confirmed by studies conducted in Germany and France. To optimize routes, some retailers surveyed have begun to create and use micro-hubs between their customers and online warehouses (Seidel, 2021). This model is often chosen when the online warehouse serves a significant catchment area. An efficient last-mile process is often considered crucial to ensuring overall customer satisfaction (Vakulenko et al., 2019). An example of innovation in this area is the use of lockers and parcel machines at airports (Gowans, 2023). They provide a safe and convenient option for storing or shipping items challenged during security screening, which is a positive from an environmental and operational cost perspective.

Behavioral change is increasingly being explored in research on the decarbonization of last-mile delivery. Interpreted as changes in established behavioral rules and collaborative practices, they are an essential component of a holistic approach to shaping low-carbon and competitive last-mile supply chains (Ma et al., 2022). Collaboration between organizations in the last-mile supply chain can result in the optimization of operational activities and thus greater system efficiency and lower GHG emissions. The main barriers to operationalizing these concepts in the urban last-mile delivery system relate to concerns about sharing information and resources with other entities, including sharing costs and benefits (Hölscher et al., 2018). Despite the economic and environmental benefits, including GHG reduction, collaboration and greater transparency in the last-mile supply chain are often perceived as conflicting with the interests of achieving success in a competitive market.

3 Methodology

The research process for shaping competitive and low-emission last-mile supply chains is presented in Fig. 1. In order to accomplish the main objectives of this and address the key research questions, diverse primary and secondary sources of data were considered. All the data were collected by employing the applicable research techniques and methods. The overall research process relied on two separate stages.

The first stage (theoretical approach) was used to identify thematic content regarding the main trends and policies for shaping innovative and zero-emission last-mile delivery strategies. The desk research method proved useful for this purpose. The literature review was carried out following a classic methodology: identification of sources, keyword search, selection and evaluation of articles, thorough reading of chosen literature directly related to the theme of the present research. The time range covered by the search covers the period 2015–2024. The number of publications found after the combination of keywords was 1,106. Focusing on low-emission and competitive solutions, the number of available publications was 443. In the next step, the search was limited to instruments for decarbonizing last-mile deliveries. At the same time, it was assumed that the period since the announcement of the European Green Deal strategy presents the latest scientific achievements, on the basis of

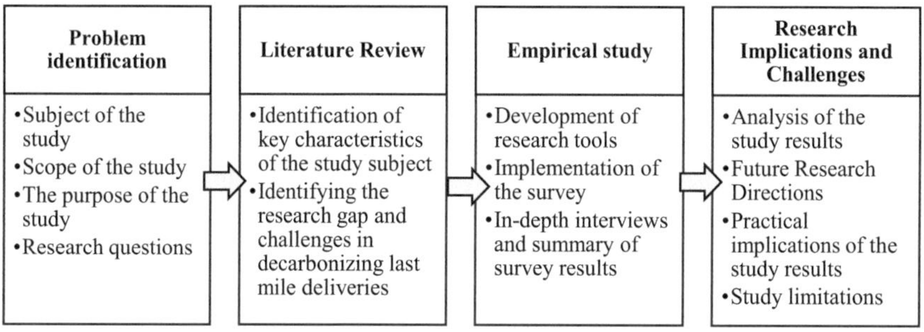

Fig. 1 Diagram of the research process for the development of innovative and zero-emission and last-mile deliveries. *Source* Own elaboration

which it is possible to formulate directions for the development of low-emission and competitive last-mile deliveries. According to this criterion, the number of identified articles was narrowed down to 157.

After analyzing the titles of publications and abstracts, adopting the criteria of place and type of publication, number of citations, and preferring the latest research results, the final number of publications used in the research process was reduced to 50. The selected scientific articles appear in the popular scientific databases: Science Direct, Scopus, MDPI, along with Google Scholar. Publications, expert reports, and relevant statistical information used in the study were published by reputable publishers and domestic and foreign institutions. Based on these sources, key characteristics describing the functioning of last-mile deliveries in the conditions of constant technological, social, and environmental changes were identified. A review of the subject literature allowed identifying a general gap in knowledge on the holistic and inclusive approach to supporting decisions in the development of low-emission and competitive solutions in last-mile deliveries through appropriate targeting of selected decarbonization instruments. Actions aimed at strengthening the transformation of last-mile logistics towards low-emission and competitive solutions were evaluated as well.

In the second stage of the research procedure (empirical approach), the focus was on assessing the effectiveness of the development of low-emission and competitive last-mile deliveries. Assuming that the developed set of actions can contribute to the decarbonization of freight transport in cities, a qualitative model of the development of low-emission and competitive last-mile deliveries was developed. It was then subjected to a declarative assessment by experts in three dimensions: economic (Ec), environmental (En), and social (So). In order to use the qualitative model to assess the effects of the development of low-emission and competitive last-mile deliveries as a result of the use of specific instruments/actions, it was given quantitative features. These features were expressed in assigning point values to the main actions (F_1, ..., F_6) and external factors (F_7), determining the values of partial action indicators (Ec, En, and So), and the values of the indicators of the direction of their effect.

It was assumed that the total number of points evaluating actions within the six action categories is 100. In accordance with the fuzzy set theory, the sum of percentage values for each of the actions is 100%. These values create a four-level scale of the effectiveness of partial actions. The highest values correspond to the most effective partial action, while the lowest indicates the least effective. The indicators of the direction of the effect of partial actions are input variables of the model. They allow for the expression of the nature of the impact of partial actions on the implementation of partial goals. The nature of this impact can be "favorable" (in which case the indicator of the direction of impact takes the value of "1"), "neutral" (the indicator of the direction of impact takes the value of "0"), or "unfavorable" (the indicator of the direction of impact takes the value of "-1"). The indicator of the implementation of the main goal and the indicators of the implementation of partial goals act as the output variables of the model. The values of the indicators of the implementation of partial goals are determined for each of the main actions and for each partial goal.

The study used selected methods, techniques and tools, including the following: cause-effect analysis, qualitative analysis, survey, and interviews. The results of the survey were collected using the PAPI (Paper and Pen Personal Interview) method. Participation in the study involved answering 42 closed questions, listed as part of the main activities (F_1, …, F_6). These activities aim to strengthen and coordinate efforts to develop competitive and low-emission last-mile deliveries. At the same time, on the one hand, these activities support this development, but on the other hand, they are often associated with high costs. For each activity, the respondents assessed the extent to which it would translate into the implementation of the main goal ("O") and partial goals (O_{Ec}, O_{En}, O_{So}). For example, they were asked to assess activities in the field of electrification of the last-mile fleet, new delivery models using cargo bikes, development in the area of warehouse processes or new solutions in customer service. At the same time, the isolation of external factors (F_7) allows us to take into account the potential impact of events that remain in the sphere of conjecture, but cannot be ruled out. Examples of these events include the emergence of new, highly efficient propulsion technologies or the occurrence of geopolitical chaos. The former could accelerate the decarbonization of last-mile deliveries, while the latter could slow it down or abandon it.

The results of the study were detailed during in-depth interviews and 5 study visits to logistics operators responsible for last-mile delivery operations. The logistics operators included global operators conducting local operations in the relevant cities and smaller logistics companies operating locally mainly in Europe. The visits were carried out in 4 locations in the Łódź region and one branch in the vicinity of Warsaw. The study was conducted from January to May 2024. The selection of the research sample was purposeful. The invitation was sent to experts representing international TFL companies with many years of professional experience in managerial and management positions in the field of transport and logistics services. A total of 26 experts took part in the study. The companies represented by the experts were logistics operators dealing mainly with transport, but also storage (15 experts), entities focused on warehouse management, and e-commerce services carrying out last-mile deliveries in cooperation with subcontractors (11 experts). The study was a pilot study. It was carried out as part of statutory research.

4 Research Results

A qualitative model of last-mile supply chain decarbonization was built with particular emphasis on the dependencies resulting from the implementation of innovative and sustainable solutions (Fig. 2). The model was also built taking into account the interactions between transport and supply logistics and the elements of associated with the economy, environment, and society. The development of competitive and low-emission last-mile deliveries is one of the available solutions with great potential to reduce GHG emissions from transport.

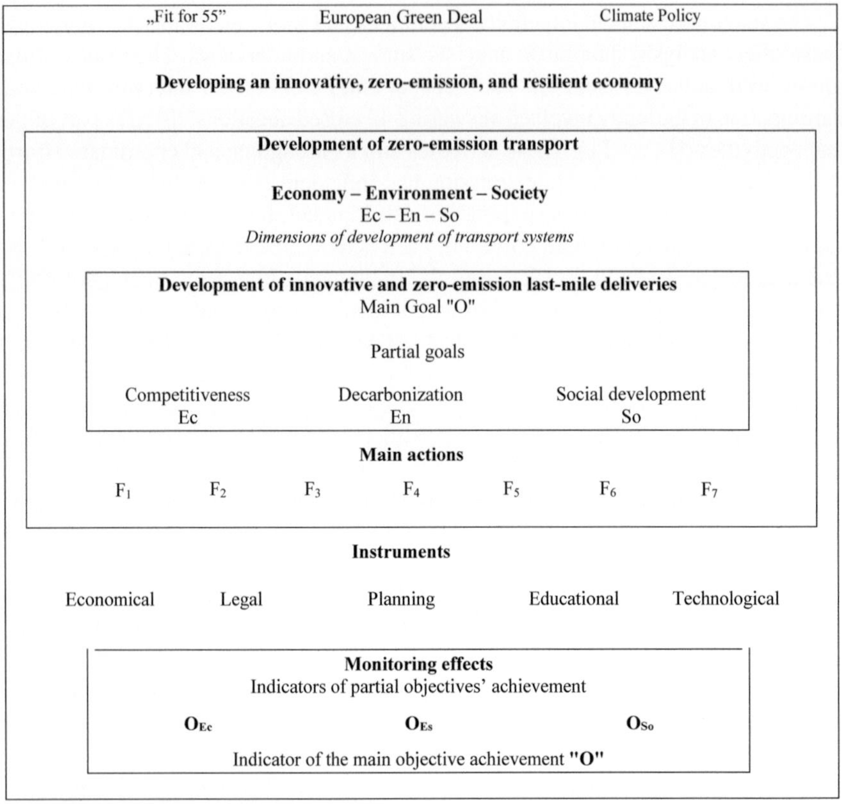

Fig. 2 Simplified qualitative model for the development of innovative and zero-emission last-mile supply chains. *Source* Own elaboration

Decarbonization of last-mile deliveries is associated with the implementation of six fundamental groups of actions that aim to strengthen and coordinate efforts to develop low-emission transport. In addition to actions resulting directly from legal regulations and public policy (F_1), strategies for the development of competitive and low-emission last-mile deliveries also include actions assuming: increased efficiency in the organization of transport and management of deliveries (F_2) and last-mile logistics processes (F_3), development of infrastructure and modern technologies in transport (F_4), introduction of new business and financial models (F_5) ensuring efficiency and continuity of deliveries, and promoting changes in consumer behavior (F_6). The implementation of the specified actions is to be enabled by a set of instruments, including economic, legal, financial, technological, planning, and educational instruments. The final effect of these actions will also depend on the impact of external factors (F_7), including autonomous and random factors.

The effects of last-mile decarbonization are therefore the result of actions taken as part of the strategy for the development of competitive and low-emission last-mile supply chains implemented using available instruments and the impact of external factors. The multidimensionality of the impact of the actions taken means that, in addition to the main objective (objective "O"), partial objectives are also implemented. The partial objectives assume improving the competitiveness of last-mile transport and logistics (sub-objective O_{Ec}), reducing GHG emissions (sub-objective O_{En}), and meeting social needs (sub-objective O_{So}). The implementation of the main objective and partial objectives is subject to monitoring. If a deviation between the achieved and expected effects is detected, the strategy and/or instruments for implementing actions for the development of low-emission and innovative last-mile deliveries are corrected.

The average assessment of the impact of the actions and/or instruments on the development of innovative and zero-emission last-mile supply chains carried out by the experts participating in the study is presented in Table 2.

The findings clearly demonstrate the high efficiency of actions for the development of innovative and low/zero emission last-mile deliveries, in line with the assumptions of the European Green Deal, while ensuring the achievement of objectives in the scope of improving competitiveness, reducing GHG emissions, and increasing the level of quality of life. The value of the main objective implementation indicator, presenting the degree of effectiveness, is 88.00 points. Based on the adopted description of the quantitative model features to the individual categories of actions, it can be stated that by assigning 24, 21, and 17 points to the actions "Transport organization and management of last-mile deliveries" (F_2), "Logistics in last-mile deliveries" (F_3), and "Infrastructure and modern technologies in transport" (F_4), the participants of the study considered them to be the "levers" of the development. The partial action indicators for actions F_2, F_3, and F_4 signal that they will be, to the

Table 2 Quantitative model of innovative and zero-emission last-mile supply chains

Main actions (symbol)	Value	Indicators of partial actions (%)			Indicators of the direction of partial actions impact[a]			Indicators of partial objectives' achievement		
		Ec	En	So	Ec	En	So	O_{Ec}	O_{En}	O_{So}
F_1	10	33	47	20	1	1	0	3.30	4.70	0.00
F_2	24	48	32	20	1	1	1	11.52	7.68	4.80
F_3	21	45	25	30	1	1	1	9.45	5.25	6.30
F_4	17	48	32	20	1	1	1	8.16	5.44	3.40
F_5	12	47	18	35	1	1	1	5.64	2.16	4.20
F_6	6	38	22	40	1	1	1	2.28	1.32	2.40
F_7	10	25	25	25	0	0	0	0.00	0.00	0.00
Total	100	x	x	x	x	x	x	**40.35**	**26.55**	**21.10**
Indicator of the main objective achievement "O"								**88.00**		

Source Own elaboration
[a] 1 (favorable), 0 (neutral), −1 (unfavorable)

greatest extent, involved in the implementation of economic goals (Ec). At the same time, it was assumed that 32% of the value of actions F_2 and F_4 and 25% of action F_3 will directly translate into achieving the goals in the scope of decarbonization of last-mile deliveries. The experts also considered action F_5 "New business and financial models" as relatively effective, assigning it 12.00 points. The distribution of indications concerning the next two actions, i.e. "Legal regulations and public policy" (F_1) and "Social and behavioral changes" (F_6) shows their lower level of effectiveness, in accordance with the results of the study. At the same time, experts considered action F_1 to have the greatest impact (47%) on achieving the goal of reducing GHG emissions.

The determined values of the O_{Ec}, O_{En}, O_{So} target implementation indicators show that actions aimed at developing innovative and low/zero emission last-mile deliveries will allow for achieving the highest level of effectiveness in terms of strengthening competitiveness, reaching 40.35 points. The level of effectiveness of actions in the context of decarbonization of deliveries is 26.55 points, as indicated by the value of the O_{En} target implementation indicator. The lowest level of effectiveness can be achieved for social aspects. The value of the O_{So} target implementation indicator is 21.10.

5 Summary

The issue of shaping competitive and zero-emission last-mile supply chains is currently taking on a new dimension in the context of the dynamic development of e-commerce and increasing reverse flows, as well as changing social expectations regarding order handling and the need to protect the environment and reduce GHG emissions (Pilati & Tronconi, 2022). At the same time, their functioning is to ensure an appropriate level of competitiveness by maintaining operational continuity at the desired level of connections and control over the structure and function of the chain. Available research confirms the gradual increase in interest in the transformation of last-mile deliveries and provides guidance on the direction in which it should go, taking into account the reduction of GHG emissions from transport and improvement of efficiency (Wu et al., 2022). To date, last-mile supply decarbonization strategies have been based mainly on two pillars.

The first pillar, considered crucial, concerned the implementation of technological innovations and digital solutions. Electrification and new fuels, especially hydrogen, were identified as the most likely and effective low-emission technologies (Cascetta & Henke, 2023; Tsoi et al., 2022). Substitution of conventionally powered delivery vehicles requires not only new low/zero emission vehicles, but also the necessary charging or hydrogen refueling infrastructure (Cartenì, et al., 2020; Fehling & Saraceni, 2023). As a result, the path of GHG emission reduction through the implementation of new technologies is associated with relatively high investment outlays (Arias et al., 2020). On the other hand, these outlays can be a source of new impulses for economic development and employment growth. The rate of diffusion of new technologies is primarily influenced by financial resources allocated within the framework of earmarked funds (Chang & Han, 2020). Access to public funding is an impulse to engage private funds in the development of innovative and zero-emission last-mile deliveries. Local government units, especially city authorities, have a special role in undertaking comprehensive actions to decarbonize deliveries in cities. It is the responsibility of city authorities to properly direct financial and regulatory instruments that can accelerate the decarbonization of freight transport in cities. In turn, rural communes are the most competent in assessing local renewable energy resources of logistics operators and the possibilities of using them in their delivery activities.

The second pillar of the transformation of last-mile deliveries towards low and zero-emission solutions results from ongoing organizational changes, both within the organization and in the supply chain. These activities demonstrate a higher level of maturity of companies creating last-mile supply chains. Cooperation, sharing knowledge, information, and resources show great potential in improving the efficiency of supply operations and reducing GHG emissions, especially in cities. This supports the creation of new business models, the development of innovations, improving the quality of deliveries and, as a consequence, decarbonizing the last mile. Against this background, the essence of the third pillar emerges, encompassing the ability to cooperate, behavioral and social changes, which can be a key factor in accelerating the decarbonization of last-mile deliveries. However, its potential is still underestimated. On the other hand, recognizing decarbonization as a collective responsibility often requires a deep transformation of the organization or a redefinition of social attitudes and behaviors.

Concluding that the effectiveness of last-mile decarbonization is a result of actions included in the three pillars, which are characterized by a high level of complementarity, this chapter provides several findings. They contribute to the development of research on the issue of shaping competitive and zero-emission last-mile deliveries from the perspective of targeting promising decarbonization instruments and tools. This is one of the available paths to transport decarbonization with a high potential to implement effective low- and zero-emission solutions. An important role in these actions is played by properly targeted environmental, technological policy tools and social programs.

The proposed qualitative model for shaping innovative and zero-emission last-mile deliveries shows the necessity for a holistic method from the perspective of the value chain and aiming to foster the decarbonization of supplies. This model allows for the identification of interactions between activities and the environment. The effects of these activities have economic, environmental, and social dimensions. The analyses and conclusions from previous studies show that last-mile decarbonization is a complex process. It requires the implementation of new technologies, organizational solutions, and social and behavioral changes. This process involves identifying weak points that make last-mile deliveries ineffective and determining attributes that strengthen the decarbonization and competitiveness of supplies. This is confirmed by the research of Paddeu and Aditjandra (2020). The researchers argue that the decarbonization of last-mile supply chains can be partially accomplished by the increase in the ability of supply chain stakeholders to cooperate. Co-creating knowledge and developing acceptable solutions is essential for effective decarbonization of supplies. At the same time, Cleophas et al. (2019) emphasize the need to apply a holistic approach in the last-mile transformation process.

In a study using a quantitative model, experts confirmed that the foundation of the transformation is the belief that creating an innovative and zero-emission last-mile supply chain is a way to get ahead of the competition. At the same time, the pressure to decrease the overall costs is still high, because customers expect competitive prices and company boards expect specific margins. The challenge for last-mile supply leaders and their organizations is therefore to develop compromises between the costs of implementing technological innovations, increasing the ability to reduce GHG emissions, and increasingly high customer expectations. They considered technological and digital innovations to be key actions to strengthen competition and reduce GHG emissions. At the same time, social and behavioral changes received a lower level of effectiveness.

In a world that is constantly changing, last-mile supply chains must be able to adapt and improve their current operations to remain competitive while taking care of the environment and climate. The transition to innovative and zero-emission solutions in last-mile freight delivery is therefore becoming a priority, not only for political strategies but also for business strategies. This study strengthens the understanding of the importance of last-mile delivery in the transformation of transport towards zero-emission. It points to the need for a holistic approach to last-mile delivery transformation, which is consistent with previous research results.

Meanwhile, the experts participating in this study, referring to the conditions of the last-mile delivery market, highlighted the importance of prioritization of technological changes, taking into account the connections within global or international supply chains. Strengthening the perception of the importance of social and behavioral changes may prove crucial for accelerating the decarbonization of last-mile deliveries. Strategies for the development of innovative and zero-emission last-mile deliveries therefore require a balanced focus of decarbonization instruments on the implementation of activities within the three pillars. The third pillar should be an area of particular interest. Acceptance of new technologies, practices, or business models will not be achieved only with the implementation of financial instruments. In addition to limiting the environmental externalities and the consequences of innovative and zero-emission technologies, it requires cooperation between stakeholders and building trust. Meanwhile, social and behavioral changes will not happen spontaneously. Often having the nature of transformational changes, they require active support within broader and local government policies and programs, as well as business initiatives.

The obtained results of the present research facilitate the further advancement of scientific studies and show various practical insights. The outcomes of this research also indicate the need for expanding empirical research on the social and behavioral changes that determine cooperation and knowledge sharing for the decarbonization of urban freight transport.

Moreover, the present research has several practical implications for various stakeholders. Given the importance of the topic, decision-makers, when designing specific policies and incentives, should encourage companies to take actions that contribute to the decarbonization of last-mile deliveries. Public–private partnerships are an opportunity to effectively implement green solutions in last-mile logistics. Solutions developed in foresight projects have great potential to mitigate the impact of last-mile deliveries on the climate. As the digital era progresses, companies are not only leaning towards innovative delivery locations, but also trying to ensure that their operations are climate- and environmentally friendly. Although the challenges related to last-mile deliveries are focused on current operations, they require a strategic approach to developing urban infrastructure, improving data security, and removing regulatory barriers.

Coordinating the cooperation of decision-makers, companies, and technology providers is crucial to accelerating the development of sustainable, low-emission last-mile deliveries. A collaborative approach with consumers can help build trust and acceptance for new delivery options, eco-friendly packaging, and the propensity to adopt reusable packaging in e-commerce deliveries. Consumers can also learn about the nature and level of their impact (negative or positive) on the implementation of various actions reducing GHG emissions from freight transport. The implications for business representatives are broad as well. Firstly, showing the impact of the implemented actions on competitiveness, decarbonization of transport, and the level of social development would play an essential role. Secondly, awareness of the potential benefits and challenges related to the transformation of social and behavioral changes can contribute to strengthening the decarbonization of last-mile deliveries. Today's digital consumers expect precision, transparency, and personalization. At the same time, the development of green initiatives, such as electric vehicles or efficient route planning, are in line with modern consumer values and expectations. Dynamic pricing strategies for off-peak deliveries have the potential to reduce environmental impact and ease urban congestion. Developing collaboration, building trust, and ensuring smooth coordination between last-mile supply chain participants are essential actions for greater sustainability, efficiency, and customer satisfaction in last-mile logistics.

Despite the practical and theoretical implications, this study has a number of limitations. The chapter does not assess the impact of instruments on the transformation process of last-mile deliveries from the perspective of the size of enterprises. The requirement for the successful achievement of such a task is the availability of a detailed set of data for small, medium, and large enterprises. The outcomes of these analyses would encourage further comprehensive studies in this area.

References

Argyriou, I., & Barry, J. (2021). The political economy of socio-technical transitions: A relational view of the state and bus system decarbonization in the United Kingdom. *Energy Research & Social Science, 79*, Article 102174. https://doi.org/10.1016/j.erss.2021.102174

Arias, N. B., Hashemi, S., Andersen, P. B., Træholt, C., & Romero, R. (2020). Assessment of economic benefits for EV owners participating in the primary frequency regulation markets. *International Journal of Electrical Power & Energy Systems, 120*, Article 105985. https://doi.org/10.1016/j.ijepes.2020.105985

Bjørge, N. M., Hjelkrem, O. A., & Babri, S. (2022). Characterisation of Norwegian battery electric vehicle owners by level of adoption. *World Electric Vehicle Journal, 13*(8), 150. https://doi.org/10.3390/wevj13080150

Bjørgen, A., Bjerkan, K. Y., & Hjelkrem, O. A. (2021). E-groceries: Sustainable last mile distribution in city planning. *Research in Transportation Economics, 87,* 100805. https://doi.org/10.1016/j.retrec.2019.100805

Bjørgen, A., & Ryghaug, M. (2022). Integration of urban freight transport in city planning: Lesson learned. *Transportation Research Part D: Transport and Environment, 107,* Article 103310. https://doi.org/10.1016/j.trd.2022.103310

Bosona, T. (2020). Urban freight last mile logistics—Challenges and opportunities to improve sustainability: Literature Review. *Sustainability, 12*(21), 8769. https://doi.org/10.3390/su12218769

Carteni, A. (2018). A cost-benefit analysis based on the carbon footprint derived from plug-in hybrid electric buses for urban public transport services. *WSEAS Transactions on Environment and Development, 14*, 125–135.

Carteni, A., Henke, I., Molitierno, C., & Errico, A. (2020). Towards E-mobility: Strengths and weaknesses of electric vehicles. In L. Barolli, F. Amato, F. Moscato, T. Enokido, & M. Takizawa (Eds.), *Web, artificial intelligence and network applications. WAINA 2020*. Advances in intelligent systems and computing (Vol. 1150). Springer. https://doi.org/10.1007/978-3-030-44038-1_126

Cascetta, E., & Henke, I. (2023). The seventh transport revolution and the new challenges for sustainable mobility. *Journal of Urban Mobility, 4*, Article 100059. https://doi.org/10.1016/j.urbmob.2023.100059

Chang, N., & Han, C. H. (2020). Cost-push impact of taxing carbon in China: A price transmission perspective. *Journal of Cleaner Production, 248*, Article 119194. https://doi.org/10.1016/j.jclepro.2019.119194

Cleophas, C., Cottrill, C., Ehmke, J. F., & Tierney, K. (2019). Collaborative urban transportation: Recent advances in theory and practice. *European Journal of Operational Research, 273*(3), 801–816. https://doi.org/10.1016/j.ejor.2018.04.037

Crippa, M., et al. (2023). *GHG emissions of all world countries*. Publications Office of the European Union, Luxembourg, JRC134504. https://doi.org/10.2760/953322

Dell'Amico, M., Montemanni, R., & Novellani, S. (2022). Pickup and delivery with lockers. *Transportation Research Part C: Emerging Technologies, 148*, Article 104022. https://doi.org/10.1016/j.trc.2023.104022

Dong, B., Hovi, I. B., & Pinchasik, D. R. (2023). Analysis of service efficiency of parcel locker in last-mile delivery: A case study in Norway. *Transportation Research Procedia, 69*, 918–925. https://doi.org/10.1016/j.trpro.2023.02.253

ECDB. (2024). *State of digital retail 2024*. Global eCommerce Market.

Ehnert, F., Kern, F., Borgström, S., Gorissen, L., Maschmeyer, S., & Egermann, M. (2018). Urban sustainability transitions in a context of multi-level governance: A comparison of four European states. *Environmental Innovation and Societal Transitions, 26*, 101–116. https://doi.org/10.1016/j.eist.2017.05.002

EPA. (2024). *Transportation and climate change*. Retrieved May 14, 2024 from https://www.epa.gov/transportation-air-pollution-and-climate-change/carbon-pollution-transportation

Escudero-Santana, A., Munuzuri, J., Lorenzo-Espejo, A., & Munuz-Díaz, M.-L. (2022). Improving e-commerce distribution through last-mile logistics with multiple possibilities of deliveries based on time and location. *Journal of Theoretical and Applied Electronic Commerce Research, 17*(2), 507–521. https://doi.org/10.3390/jtaer17020027

European Commission. (2018). *A clean planet for all: A European strategic long-term vision for a prosperous, modern, competitive and climate neutral economy*. Brussels, 28.11.2018 COM(2018) 773 final.

European Commission. (2019). *European Green Deal*. Communication from the Commission to the European Parliament, the European Council, the Council, the Economic and Social Committee and the Committee of the Regions. COM(2019) 640.

European Commission. (2021a). *European Green Deal: The commission proposes transformation of EU economy and society to meet climate ambitions*, 14 July 2021, online. Retrieved September 20, 2024 from https://ec.europa.eu/commission/presscorner/detail/en/IP_21_3541

European Commission. (2021b). *EU reference scenario 2020. Energy, transport and GHG emissions: Trends to 2050*. Publications Office of the European Union.

European Environment Agency. (2024). *Trends and projections in Europe*. Report, 11, EEA, Copenhagen.

Eurostat. Final energy consumption in transport—Detailed statistics. Retrieved September 25, 2024 from https://ec.europa.eu/eurostat/statistics-explained/index.php?title=Final_energy_consumption_in_transport_-_detailed_statistics

Fehling, C., & Saraceni, A. (2023). Technical and legal critical success factors: Feasibility of drones & AGV in the last-mile-delivery. *Research in Transportation Business & Management, 50*, Article 101029.

Figenbaum, E. (2018). Can battery electric light commercial vehicles work for craftsmen and service enterprises? *Energy Policy, 120*(May), 58–72. https://doi.org/10.1016/j.enpol.2018.04.076

Furmankiewicz, M., Hewitt, R. J., & Kazak, J. K. (2021). Can rural stakeholders drive the low-carbon transition? Analysis of climate-related activities planned in local development strategies in Poland. *Renewable and Sustainable Energy Reviews, 150*, Article 111419. https://doi.org/10.1016/j.rser.2021.111419

Geels, F. W. (2022). Conflicts between economic and low-carbon reorientation processes: Insights from a contextual analysis of evolving company strategies in the United Kingdom petrochemical industry (1970–2021). *Energy Research & Social Science, 91*, Article 102729.

Gowans, G. (2023). *InPost CEO unveils parcel locker with a difference at Warsaw airport*. Journalist Trans.INFO. Retrieved August 24, 2023 from https://trans.info/en/warsaw-airport-359688

Hammond, W., Axsen, J., & Kjeang, E. (2020). How to slash greenhouse gas emissions in the freight sector: Policy insights from a technology-adoption model of Canada. *Energy, 137*, Article 111093. https://doi.org/10.1016/j.enpol.2019.111093

Hölscher, K., Wittmayer, J. M., & Loorbach, D. (2018). Transition versus transformation: What's the difference? *Environmental Innovation and Societal Transitions, 27*, 1–3. https://doi.org/10.1016/j.eist.2017.10.007

International Energy Agency. (2024). *World energy outlook*. IEA Publications.

Janinhoff, L., Klein, R., Sailer, D., & Schoppa, J. M. (2024). Out-of-home delivery in last-mile logistics: A review. *Computers & Operations Research, 168*, Article 106686. https://doi.org/10.1016/j.cor.2024.106686

Johnstone, P., & Newell, P. (2018). Sustainability transitions and the state. *Environmental Innovation and Societal Transitions, 27*, 72–82. https://doi.org/10.1016/j.eist.2017.10.006

Kedia, A., Kusumastuti, D., & Nicholson, A. (2020). Locating collection and delivery points for goods' last-mile travel: A case study in New Zealand. *Transportation Research Procedia, 46*, 85–92. https://doi.org/10.1016/j.trpro.2020.03.167

Kiba-Janiak, M., Cheba, K., Mucowska, M., de Oliveira, L. K., Piecyk, M., Evangelista, P., Prockl, G., & Rześny-Cieplińska, J. (2024). How to design a sustainable last-mile delivery and returns business model from E-Customers' expectations perspective? *Research in Transportation Business & Management, 56*, Article 101194. https://doi.org/10.1016/j.rtbm.2024.101194

Lin, Y. H., Wang, Y., Lee, L. H., & Chew, E. P. (2022). Profit-maximizing parcel locker location problem under threshold Luce model. *Transportation. Research, Part E: Logistics and Transportation Review, 157,* 102541. https://doi.org/10.1016/j.tre.2021.102541

Ma, B., Wong, Y. D., & Teo, C. C. (2022). Parcel self-collection for urban last-mile deliveries: A review and research agenda with a dual operations-consumer perspective. *Transportation Research Interdisciplinary Perspectives, 16,* 100719. https://doi.org/10.1016/j.trip.2022.100719

Magalhães, I., & Santos, E. (2022). Evaluating the potential of mobility plans for achieving sustainable urban development. *Research in Business & Management, 43.* https://doi.org/10.1016/j.rtbm.2021.100743

Maxner, T., Dalla Chiara, G., & Goodchild, A. (2022). Identifying the challenges to sustainable urban last-mile deliveries: Perspectives from public and private stakeholders. *Sustainability, 14*(8). https://doi.org/10.3390/su14084701

McKinnon, A. (2023). Environmentally sustainable city logistics: Minimising urban freight emissions. In E. Marcucci, V. Gatta, & M. Le Pira (Eds.), *Handbook on city logistics and urban freight* (pp. 463–482).

Motowidlak, U., & Motowidlak, T. (2023). Development and financing of infrastructure investment to shape sustainable and resilient mobility. *Scientific Journal Series Silesian University of Technology—Organization and Management, 170,* 349–370. https://doi.org/10.29119/1641-3466.2023.170.22

Niemeijer, R., & Buijs, P. (2023). A greener last mile: Analyzing the carbon emission impact of pickup points in last-mile parcel delivery. *Renewable and Sustainable Energy Reviews, 186*, Article 113630. https://doi.org/10.1016/j.rser.2023.113630

Paddeu, D., & Aditjandra, P. (2020). Shaping urban freight systems via a participatory approach to inform policy-making. *Sustainability, 12*(1), 441. https://doi.org/10.3390/su12010441

Paddeu, D., Parkhurst, G., Rosenberg, G., Carhart, N., & Taylor, C. (2024). Promoting sustainable urban freight through stakeholder engagement to co-create decarbonisation pathways in the UK. *Research in Transportation Economics, 104*, Article 101424. https://doi.org/10.1016/j.retrec.2024.101424

Pegels, A., & Lütkenhorst, W. (2014). Is Germany's energy transition a case of successful green industrial policy? Contrasting wind and solar PV. *Energy Policy, 74*, 522–534. https://doi.org/10.1016/j.enpol.2014.06.031

Pilati, F., & Tronconi, R. (2022). Social-economic sustainable vehicle routing problem for local e-commerce platforms. *IFAC PapersOnLine, 55–10*, 470–475.

Santos, G. (2017). Road transport and CO_2 emissions: What are the challenges? *Transport Policy, 59*, 71–74.

Seidel, S. (2021). One goal, one approach? A comparative analysis of online grocery strategies in France and Germany. *Case Studies on Transport Policy, 9*(4), 1922–1932. https://doi.org/10.1016/j.cstp.2021.10.013

Sengupta, P., Choudhury, B., KMitra, S., & Agrawal, K. M. (2020). Low carbon economy for sustainable development. In S. Hashmi & I. A. Choudhury (Eds.), *Encyclopedia of renewable and sustainable materials* (pp. 551–560). Elsevier. https://doi.org/10.1016/B978-0-12-803581-8.11217-2

Silvestri, L., Forcina, A., Silvestri, C., & Traverso, M. (2021). Circularity potential of rare earths for sustainable mobility: Recent developments, challenges and future prospects. *Journal of Cleaner Production, 72*, Article 101885. https://doi.org/10.1016/j.erss.2020.101885

Sovacool, B. K., Herman, K. S., Iskandarova, M., & Hall, J. K. (2024). "Oh Yes! Net-Zero": Sociotechnical capabilities and regional innovation systems for British industrial decarbonization. *Environmental Innovation and Societal Transitions, 51*, Article 100852. https://doi.org/10.1016/j.eist.2024.100852

Statista. (2024a). E-commerce worldwide. *Statistics & Facts.* Retrieved February 05, 2024 from https://www.statista.com/topics/871/online-shopping/#topicOverview

Statista. (2024b). E-commerce in the European Union. *Statistics & Facts.* Retrieved January 10, 2024 from https://www.statista.com/topics/3792/e-commerce-in-europe/#topicOverview

Statista. (2024c). E-commerce in Poland. *Statistics & Facts.* Retrieved May 29, 2024 from https://www.statista.com/topics/5128/e-commerce-in-poland/#editorsPicks

Tsoi, K. H., Loo, B. P., Tal, G., & Sperling, D. (2022). Pioneers of electric mobility: Lessons about transport decarbonisation from two bay areas. *Journal of Cleaner Production, 330*, Article 129866.

UNEP. (2024). *Emissions gap report*. https://www.unep.org/resources/emissions-gap-report-2024

Vakulenko, Y., Hellström, D., & Hjort, K. (2018). What's in the parcel locker? Exploring customer value in e-commerce last mile delivery. *Journal of Business Research, 88*, 421–427. https://doi.org/10.1016/j.jbusres.2017.11.033

Vakulenko, Y., Shams, P., Hellström, D., & Hjort, K. (2019). Online retail experience and customer satisfaction: The mediating role of last mile delivery. *The International Review of Retail, Distribution and Consumer Research, 29*(3), 306–320. https://doi.org/10.1080/09593969.2019.1598466

Van Audenhove, F., Korn, A., Steylemans, N., Smith, A., Rominger, G., Bettati, A., Zintel, M., & Haon, S. (2018). *The future of Mobility 3.0—Reinventing mobility in the era of disruption and creativity*. Arthur D. Little.

Van Duin, J. H. R., Munir, A., & Wiegmans, B. (2024). Using E-fulfilment centres for direct last-mile parcel delivery: An activity cost based analysis. *Transportation Research Procedia, 79*, 265–272. https://doi.org/10.1016/j.trpro.2024.03.036

Wang, Q., Wang, Y., Hang, Y., & Zhou, P. (2019). An improved production-theoretical approach to decomposing carbon dioxide emissions. *Journal of Environmental Management, 252*, Article 109577. https://doi.org/10.1016/j.jenvman.2019.109577

Wu, H., Sun, Y., Su, Y., Chen, M., Zhao, H., & Li, Q. (2022). Which is the best supply chain policy: Carbon tax, or a low-carbon subsidy? *Sustainability, 14*(10), 6312. https://doi.org/10.3390/su14106312

Xin, X., Yuding, W., & Jianzhong, W. (2011). The problems and strategies of the low carbon economy development. *Energy Procedia, 5*, 1831–1836. https://doi.org/10.1016/j.egypro.2011.03.312

Zhang, R., Hanaoka, T., Liu, I., Li, Z., & Sun, L. (2024). Air pollution reduction co-benefits associated with low-carbon transport initiatives for carbon neutrality in China by 2060. *Energy, 313*, Article 134090. https://doi.org/10.1016/j.energy.2024.134090

Urszula Motowidlak is a Professor at the University of Lodz, employee of the Department of Logistics and Innovation at the Faculty of Economics and Sociology. She is also a Member of the Forecasts Committee of the Polish Academy of Sciences in Warsaw and the Polish Association of Environmental and Natural Resource Economists, conducting research on the green transformation of transport and logistics (climate section). Dr. Motowidlak is a graduate of the Faculty of Biology and Earth Sciences, specializing in Environmental Biology and the Faculty of Economics and Sociology of the University of Lodz. She obtained her doctoral and postdoctoral degrees in the field of economic sciences in the discipline of economics. In her scientific research, she uses an interdisciplinary approach to the analysis of economic and environmental issues. The scope of her scientific interests concerns the theoretical and applied aspects of sustainable development of transport and logistics. Dr. Motowidlak is a specialist in the field of low- and zero-emission transport and mobility, including electromobility, transport policy, as well as the concept of sustainable and resilient supply chains. She is an author and co-author of numerous publications and expert opinions. Moreover, Dr. Motowidlak is a member of teams in over 40 research, R&D, and application projects related to sustainable transport, mobility and logistics. She is also a speaker at many conferences organized by foreign and domestic research centers. As an expert, she cooperates with transport and logistics companies and public sector entities.

Redesigning a Closed-Loop Supply Chain Network in the Tire Industry

Amir M. Fathollahi-Fard

Abstract Network design is a critical topic in supply chain management, integrating forward and reverse flows to create a closed-loop supply chain (CLSC) network. In the tire industry, CLSC networks must strategically address facility location and allocation across multiple levels, including suppliers, manufacturers, retailers, and markets for forward flows, and collectors, recyclers, and secondary markets for reverse flows. With the rapid growth in demand for tire manufacturing, existing CLSC networks often fail to meet the escalating of demand satisfaction, necessitating a comprehensive redesign to enhance facility location and capacity planning performance. This paper proposes a novel framework for redesigning CLSC networks in the tire industry by formulating a mixed-integer programming (MIP) model that incorporates unique challenges specific to network redesign. Most importantly, the network design for the CLSC differs significantly from the CLSC network redesign. In the latter, facilities are divided into two subsets: a set of open facilities and a set of candidate facilities. The decisions involved not only pertain to the opening of candidate facilities but also to the closure of existing open facilities, while balancing the varying fixed and variable costs associated with these transitions. The developed MIP model for the network redesign problem explicitly accounts for these decisions, providing a practical tool for addressing real-world complexities within the CLSC for the tire industry. Extensive sensitivity analyses are conducted to evaluate the impact of key parameters, such as network redesign costs and recycling efficiency. The results help the CLSC networks to optimize their CLSC to achieve higher efficiency and better demand satisfaction. This research not only highlights the importance of network

Electronic Supplementary Material The online version of this chapter (https://doi.org/10.1007/978-3-032-00563-2_9) contains supplementary material, which is available to authorized users.

A. M. Fathollahi-Fard (✉)
School of Business Administration, Laurentian University, 935 Ramsey Lake Road, Sudbury, ON P3E 2C6, Canada
e-mail: afathollahifard@laurentian.ca; amirfard.ie@gmail.com

Department of Engineering Science, Faculty of Innovation Engineering, Macau University of Science and Technology, Taipa, Macau, China

M. A. Dulebenets et al. (eds.), *Transport and Climate Change*,
https://doi.org/10.1007/978-3-032-00563-2_9

adaptability but also provides an efficient decision-support tool for redesigning CLSC systems in high-demand industries.

Keywords Network design · Network redesign · Closed-loop supply chain · Mixed-integer program

1 Introduction

The tire industry is a vital sector contributing significantly to economic growth, particularly in developed countries. For instance, the tire manufacturing industry in Canada accounted for 10% of the manufacturing sector's gross domestic product (GDP).[1] Additionally, Canada has demonstrated a strong commitment to sustainability and resource efficiency by achieving an impressive average scrap tire recycling rate of approximately 98% over the past decade. This achievement underscores the country's dedication to promoting environmental responsibility within the tire industry.[2]

The design and maintenance of effective closed-loop supply chain (CLSC) networks are critical for addressing the environmental impacts of tire production and disposal. CLSCs seamlessly integrate tire manufacturing, distribution, and comprehensive scrap tire recycling processes (Fathollahi-Fard et al., 2021). These networks not only enhance resource usage efficiency but also play a pivotal role in aligning industrial operations with sustainability criteria and promoting circular economy principles in the tire industry (Banguera et al., 2018).

Closed-loop supply chains are inherently linked to environmental concerns, as they aim to minimize waste and maximize resource recovery, thereby reducing the overall environmental footprint of industrial activities (Jian et al., 2021). In the context of climate change, the adoption of CLSCs in the tire industry helps mitigate greenhouse gas emissions associated with the disposal of scrap tires and the extraction of raw materials (Mohammed et al., 2017). For example, the recycling of scrap tires significantly reduces the need for virgin rubber and fossil fuel-derived materials, which are major contributors to carbon emissions. Moreover, efficient CLSCs facilitate the transition to a circular economy, where resources are continuously reused, contributing to the reduction of landfill waste and environmental degradation (MahmoumGonbadi et al., 2021).

The implementation of CLSCs in the tire industry is further supported by global efforts to address environmental and climate challenges. International frameworks, such as the Paris Agreement and the United Nations Sustainable Development Goals (SDGs), emphasize the need for sustainable production and consumption practices. CLSCs align with these objectives by fostering the responsible management of resources and reducing the environmental impact of industrial processes (Lozano-Oviedo et al., 2024). For instance, Goal 12 of the SDGs explicitly focuses on ensuring

[1] https://ised-isde.canada.ca/app/ixb/cis/manufacturing-fabrication/32621.

[2] https://www.catraonline.ca/.

sustainable consumption and production patterns, which are central to the concept of CLSCs.

However, the global landscape is evolving rapidly, driven by population growth, urbanization, and the increasing proliferation of automobiles. Modernization has fueled a significant rise in vehicle ownership, which has, in turn, directly intensified the demand for tires (Manupati et al., 2024). For instance, the Canadian tire market, valued at $5.23 billion in 2023, is projected to reach $7.84 billion by 2032, reflecting the substantial growth in demand within the sector.[3] This surge in demand presents significant challenges to the existing CLSC networks, which are often unable to accommodate the rising volumes of both new tire production and end-of-life tire recycling efficiently (Sahebjamnia et al., 2018). Consequently, tire manufacturers face growing pressures to adapt their supply chain networks to meet these new demands while maintaining their environmental and operational commitments.

The necessity of redesigning CLSC networks in the tire industry arises from the aforementioned challenges. Network redesign involves addressing a distinct set of assumptions and requirements that differ significantly from network design models. In a typical CLSC network design, all facilities are considered candidate facilities, and the focus is on creating the network from scratch. However, in the network redesign concept, the problem is more complex as it involves both existing facilities that are already open and candidate facilities that may extend this network. Key differences between network design and redesign problems include the necessity to evaluate the closure of existing facilities, where the associated costs differ significantly from those incurred when opening new facilities.

This paper aims to tackle these unique challenges by proposing a comprehensive framework specifically for redesigning CLSC networks in the tire industry. By developing a mixed-integer programming (MIP) model, this research addresses the intricate decisions involved in network redesign and provides actionable insights for improving efficiency, sustainability, and demand satisfaction. Unlike basic CLSC design models, this framework integrates the complexities of maintaining or modifying existing facilities alongside the potential for new facility investments. Through this study, tire industry stakeholders can be equipped with practical tools and strategies to navigate the evolving landscape and ensure the resilience and sustainability of their supply chain networks.

The remainder of this paper is organized as follows. Section 2 presents a comprehensive review of the existing literature on supply chain management in the tire industry, highlighting relevant studies and identifying gaps addressed by this research. Section 3 outlines the problem settings for the proposed CLSC network, detailing the scope and structure of the redesign process. Section 4 formulates the CLSC network redesign problem using an MIP model, incorporating key considerations unique to the tire industry. Section 5 discusses the results of the model, including detailed analyses and sensitivity evaluations to assess its efficiency and applicability. Finally, Sect. 6 is a discussion for the conclusions and perspectives from the results of this research.

---

[3] https://tracanada.ca/industry-news/canada-tire-market-forecast-2018-2028/.

2 Literature Review

Over the past decade, more than a hundred supply chain network design models have been developed and analyzed (Battini et al., 2017; Lozano-Oviedo et al., 2024). However, research specifically focused on supply chains within the tire industry remains relatively nascent. This study provides a comprehensive review of all MIP models and optimization algorithms applied to supply chain management in the tire industry in recent years. Through this review, we highlight the key contributions from the existing literature and present a critical assessment of their strengths and limitations. Furthermore, we identify significant research gaps and outline the primary contributions of this work in addressing these gaps to advance the field.

One of the pioneering studies in this field is by Ferrer (1997), who demonstrated that scrap tires have the potential to be recycled multiple times, with the possibility of undergoing more than two cycles. Building on this, Chopra and Meindl (2007) further analyzed that scrap tires can be recycled up to 10 times, significantly extending their usability. They also highlighted that the recycled products could serve various industrial applications beyond remanufacturing, such as being repurposed in amusement parks. These findings emphasize the importance of designing and implementing a CLSC specifically tailored for the tire industry to maximize recycling and remanufacturing efficiency, while also exploring secondary uses for recycled materials.

Sasikumar et al. (2010) made a significant contribution to the design of reverse supply chains for the tire industry. They proposed a mixed-integer nonlinear programming (MINLP) model aimed at optimizing the location of recycling centers and planning transportation within the network. The primary objective of their model was to maximize the total profit generated from the reverse supply chain operations. Subulan et al. (2015) introduced a fuzzy goal programming approach to design a tire CLSC with recycling and remanufacturing options for scrap tires. Their model accounted for a multi-period framework considering different types of tires, aiming to optimize two primary objectives: maximizing total profit and minimizing environmental emissions.

Sarkar et al. (2017) developed a CLSC network for the tire industry, focusing on a single vendor and multiple retailers to optimize inventory decisions while incorporating the possibilities of recycling and remanufacturing scrap tires. In a separate study, Pedram et al. (2017) formulated a tire CLSC problem using a two-stage stochastic programming approach aimed at minimizing total costs, addressing uncertainties in demand and supply. Additionally, Amin et al. (2017) examined a multi-period tire CLSC problem for the Canadian tire industry, designing their model to maximize total profit.

Banguera et al. (2018) developed an MIP model to tackle the reverse logistics challenges in tire remanufacturing within the context of Chile. Their work highlighted the critical role of manufacturers in making strategic location and allocation decisions. Sahebjamnia et al. (2018) extended the scope of tire CLSC network design by incorporating three key sustainability objectives: economic, environmental, and social. Their study proposed a multi-product framework that accounted for diverse manufacturing and recycling technologies, utilizing hybrid metaheuristic algorithms inspired by the red deer algorithm (RDA) and water wave optimization (WWO).

Fathollahi-Fard et al. (2018) introduced a tri-level programming model for a multi-product CLSC network in the tire industry. Their approach structured decision-making across manufacturing centers, distribution centers, and collection centers, emphasizing economic considerations. To solve the model efficiently, they implemented hybrid metaheuristics combining the Keshtel algorithm (KA) and stochastic fractal search (SFS). In a separate study, Yadollahinia et al. (2018) focused on a multi-period, multi-product reverse logistics network, incorporating customer classification and relationship constraints. They employed robust optimization integrated with goal programming to strike a balance between profitability and customer satisfaction. Ebrahimi (2018) proposed a stochastic programming model for designing CLSC networks in the tire industry. The model addressed quantity discounts and pursued dual objectives: minimizing costs and environmental impact while enhancing network responsiveness, a vital social dimension.

Abdolazimi et al. (2020) explored the profitability of a CLSC network by optimizing three key objectives: maximizing total profit, total product sales, and minimizing delivery time. Environmental emissions were incorporated as a constraint with specified allowable limits. To address uncertainties in demand and return rates for scrap tires, they employed a robust optimization approach. A notable innovation in their study was the inclusion of both internal and external suppliers in the network. Fathollahi-Fard et al. (2021) examined a dual-channel CLSC network for the tire industry, integrating online and offline retail channels. Utilizing a fuzzy programming methodology, they introduced two hybrid metaheuristic algorithms based on the RDA and the whale optimization algorithm (WOA).

Mehrjerdi and Shafiee (2021) investigated the role of information-sharing strategies within a CLSC network for the tire industry, applying a stochastic programming model. Their approach sought to minimize total costs and environmental impact while simultaneously maximizing job creation. In a complementary study, Ghasemzadeh et al. (2021) formulated a bi-objective stochastic CLSC network design model aimed at maximizing profit and minimizing environmental pollution. Their research uniquely incorporated customs duties and exchange rate variations, reflecting the complexities faced by international suppliers.

Amirian et al. (2022) focused on optimizing the total profit and carbon emissions in a tire CLSC network, accounting for uncertainty in the pricing of sold and recycled products. To manage this uncertainty, they adopted a fuzzy programming approach, ensuring flexibility and robustness in their model. Soleimani et al. (2022) proposed a sustainable CLSC model where there are three recovery options based on the quality of the scrap tires. They separated the secondary markets from other industrial applications for the recycled scrap tires. For solving their model, they applied a Lagrangian-based heuristic method. Similarly, Ahmed et al. (2023) proposed a multi-period, multi-product tire CLSC network model aimed at optimizing multiple objectives: total profit, carbon emissions, supplier performance scores, and surface water withdrawal. Their formulation also relied on a fuzzy programming methodology to effectively handle uncertainties, showcasing the increasing reliance on advanced computational approaches to enhance sustainability and efficiency in CLSC networks.

Recently, Hoeke et al. (2024) explored the significant role of tires in contributing to microplastic pollution and aimed to lay the groundwork for an effective mitigation strategy. They mapped the tire supply chain and quantified microplastic emissions using a material system analysis for the Netherlands in 2021, incorporating stakeholder involvement for data collection and fostering collaboration. In a related study, Manupati et al. (2024) employed an evolutionary algorithm-based approach to optimize facility locations, material flows, and demand distribution and minimize the total cost. Their results demonstrated that the Modified Genetic Algorithm (MGA) outperformed other algorithms in identifying cost-effective solutions, offering valuable insights for enhancing end-of-life strategies.

While numerous studies have explored various aspects of CLSC in the tire industry, including tire recycling, remanufacturing, and environmental impact minimization, none have focused on redesigning a CLSC network to address the possibility of closing the existing facilities and the opening of new ones to accommodate increased demand. Most research has concentrated on optimizing facility locations, allocations, inventory decisions, and sustainability goals, without considering the dynamic nature of facility openings and closings with differing cost structures in response to demand shifts. For instance, several studies have proposed models for optimizing the tire reverse supply chain and recycling operations, but they do not account for the complexities of changing facility configurations or the costs associated with opening and closing facilities based on evolving demand (Banguera et al., 2018; Sasikumar et al., 2010; Subulan et al., 2015).

Additionally, studies like those by Fathollahi-Fard et al. (2018), Sahebjamnia et al. (2018), Ghasemzadeh et al. (2021), and Manupati et al. (2024) have integrated advanced modeling techniques, such as hybrid metaheuristics and advanced programming methods, but they focus on static or predefined network structures rather than redesigning the network dynamically to better respond to market fluctuations and demand increases. This gap suggests an opportunity to redesign a CLSC network and address the strategic and tactical planning levels. Therefore, this study proposes an MIP to redesign a CLSC network for the tire industry.

3 Problem Settings

The problem settings section highlights the key characteristics of the network redesign, including the sets and parameters specific to the CLSC in the tire industry. This section provides a detailed overview of the network structure, outlining the various elements and variables that define the tire industry's closed-loop supply chain and the factors influencing the redesign process.

The rest of this section is as follows. Section 3.1 outlines the physical network for the CLSC in the tire industry. Section 3.2 defines the parameters associated with suppliers, focusing on their contribution to the forward flow of the network. Section 3.3 examines the role of manufacturers in this network redesign problem. Section 3.4 explores the role of retailers in distributing tires. Section 3.5 defines the demands of the markets, providing a clear characterization of how demand is quantified and addressed. Section 3.6 discusses the role of collectors in the reverse flow. Section 3.7 studies the role of recyclers in processing collected scrap tires to ensure sustainability. This comprehensive definition of the network elements establishes the foundation for developing the proposed CLSC network redesign framework.

3.1 Physical Network

Figure 1 illustrates a conceptual physical network for the proposed tire CLSC network, encompassing both the production and recycling processes. The forward supply chain starts with suppliers, $s \in \mathcal{S}$, who provide raw materials to manufacturers, $m \in \mathcal{M}$. These manufacturers produce the final tire products, which are then distributed to retailers, $d \in \mathcal{D}$. The retailers serve as intermediaries, distributing the tires to satisfy the demands of markets, $k \in \mathcal{K}$. This sequence of movements, from suppliers to markets, constitutes the forward flow within the CLSC network.

In the reverse supply chain, the process starts with collectors, $c \in \mathcal{C}$, who retrieve scrap tires and wasted materials from the markets. These collectors sort and classify these wasted materials before transferring them to recyclers, $r \in \mathcal{R}$. The recyclers process scrap tires, ensuring their efficient reuse. A portion of the recycled materials, particularly those of higher quality, is sent back to manufacturers for remanufacturing the recycled tires, thus closing the loop of this tire supply chain. The remaining

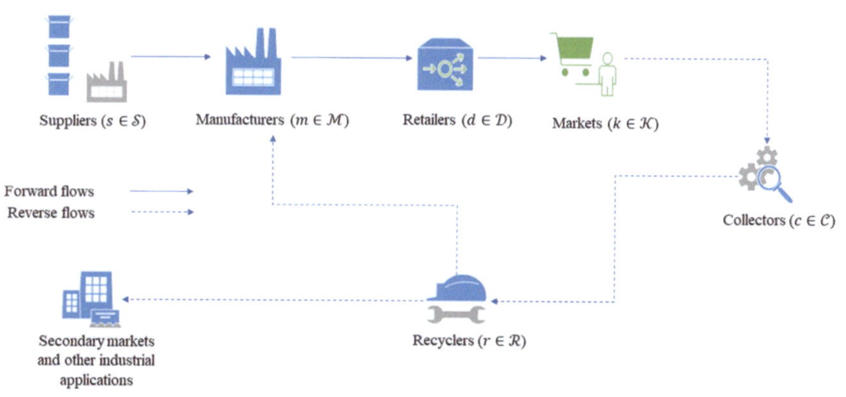

Fig. 1 Conceptual physical network

recycled materials are directed to secondary markets and other industrial applications, where they are utilized for diverse purposes, contributing to sustainability and resource efficiency.

This study focuses on redesigning the CLSC network with an emphasis on strategic and tactical planning levels. The primary decision determines the optimal locations of facilities and their allocations through different levels shown in Fig. 1. This study addresses both strategic and tactical planning levels. Specifically, the decisions regarding the opening and closing of facilities are part of the strategic planning, while the allocation of facilities across different levels of the CLSC is integral to the tactical planning. These combined efforts ensure that the network redesign is optimized for both long-term and medium-term objectives. However, the short-term objectives from the operational planning level of the CLSC, such as pickup and delivery scheduling, routing, and detailed operations management, are beyond the scope of this paper. These elements, while crucial for day-to-day operations, are considered separate from the higher-level planning decisions addressed in this study.

3.2 Suppliers

The raw materials are provided by the suppliers. Based on the network redesign framework for the tire CLSC problem, suppliers operate independently and are not owned by the network (Bhatia et al., 2024). Consequently, the investigated problem does not involve making location decisions for suppliers. However, their geographical locations play a critical role, as they significantly influence transportation costs for the network redesign framework.

We define $\vartheta^{\mathcal{S}}_{sm}$ as the distance between suppliers $s \in \mathcal{S}$ and manufacturers $m \in \mathcal{M}$, and $tc^{\mathcal{S}}_{sm}$ is the unit transportation cost. Additionally, the raw materials are priced at $pc^{\mathcal{S}}_{s}$ per unit, reflecting the cost incurred by manufacturers for procurement. Last but not least, the capacity of suppliers is limited to $u^{\mathcal{S}}_{s}$. This capacity limitation for the suppliers necessitates careful planning and allocation to ensure that the CLSC network operates efficiently without exceeding this threshold.

3.3 Manufacturers

Manufacturers play a pivotal role in the CLSC network design for the tire industry, ensuring that market demands for tires are adequately met. For the redesign of the CLSC network, the set of manufacturers is categorized into two subsets: a set of facilities which are already located, $\mathcal{M}^{\mathcal{L}}$, and a set of potential facilities which have not been located yet, $\mathcal{M}^{\mathcal{NL}}$, and can be added to network as an extension. The union of these two subsets defines the set of all the manufacturers, $\mathcal{M} = \{\mathcal{M}^{\mathcal{L}} \cup \mathcal{M}^{\mathcal{NL}}\}$. The strategic planning for the manufacturers contributes to both the opening and closing fixed costs. An open manufacturer $m \in \mathcal{M}^{\mathcal{L}}$ has the fixed closing cost of

$f_m^{\mathcal{ML}}$. A candidate manufacturer $m \in \mathcal{M}^{\mathcal{NL}}$ has the fixed opening cost of $f_m^{\mathcal{MNL}}$. It is assumed that the average cost of opening a new manufacturer is greater than the average cost of closing an existing manufacturer.

The tactical planning for the manufacturers contributes to their limited capacity $u_m^{\mathcal{M}}$ where $m \in \mathcal{M}$. They also have a capacity, $\overline{u}_m^{\mathcal{M}}$, where $m \in \mathcal{M}$ for remanufacturing the recycled tires from the reverse supply chain. Another important part of the tactical planning refers to the unit cost of manufacturing each tire, $vc_m^{\mathcal{M}}$ for $m \in \mathcal{M}$. The transportation of finished products from manufacturers to retailers is another key component of the CLSC network. We assume that $\vartheta_{md}^{\mathcal{M}}$ is the geographical distance between a manufacturer $m \in \mathcal{M}$ and a retailer $d \in \mathcal{D}$, and $tc_{md}^{\mathcal{M}}$ is the unit transportation cost per tire.

3.4 Retailers

Retailers play a crucial role in the CLSC network, acting as intermediaries that distribute tires from manufacturers to markets. In the proposed redesign, retailers are divided into two groups: those already open, $\mathcal{D}^{\mathcal{L}}$, and those that are potential additions to the network, $\mathcal{D}^{\mathcal{NL}}$. The combined set of all retailers is represented as $\mathcal{D} = \{\mathcal{D}^{\mathcal{L}} \cup \mathcal{D}^{\mathcal{NL}}\}$. This categorization reflects the need to balance the existing infrastructure with opportunities for expansion to improve the network's efficiency and capacity.

Strategic planning for retailers involves making decisions about facility location. For existing retailers, where $d \in \mathcal{D}^{\mathcal{L}}$, a fixed cost is incurred for closing these retailers, $f_d^{\mathcal{DL}}$. On the other hand, opening a new retailer, where $d \in \mathcal{D}^{\mathcal{NL}}$, requires a fixed cost, $f_d^{\mathcal{DNL}}$. These costs, where the cost of opening a new retailer is greater than the cost of closing a retailer, are significant considerations in determining the feasibility of network adjustments and must align with the overall objectives of minimizing costs while ensuring market demands are met. Tactical considerations focus on the tactical aspects of retailer activities. Each retailer, $d \in \mathcal{D}$, incurs a unit distribution cost, $vc_d^{\mathcal{D}}$, which represents the expense of handling and delivering each unit of tires. Additionally, retailers operate within capacity limits, $u_d^{\mathcal{D}}$. The distance, $\vartheta_{dk}^{\mathcal{D}}$, between a retailer $d \in \mathcal{D}$ and the markets $k \in \mathcal{K}$, and the unit transportation cost per tire, $tc_{dk}^{\mathcal{D}}$, compute the transportation costs.

3.5 Markets

Each market, represented by $k \in \mathcal{K}$, serves as a demand origin, encompassing a set of customers distributed within the region of that market. These demands consist of a specified volume of tires, q_k, which can be satisfied from one or more retailers. According to this demand from the forward supply chain, the demands for the reverse supply chain are defined. For returned products and scrap tires, it is assumed that only a maximum of α_k from the total demand volume can be collected by collectors. Last but not least, the transportation cost for the reverse supply chain is based on the distance, $\vartheta_{kc}^{\mathcal{K}}$, between a market, $k \in \mathcal{K}$, and a collector, $c \in \mathcal{C}$, as well as the unit transportation cost per scarp tire, $tc_{kc}^{\mathcal{K}}$.

3.6 Collectors

The collection centers, $c \in \mathcal{C}$, play a pivotal role in the CLSC network for the tire industry, functioning as facilities for gathering and classifying waste and returned tires from markets. In the redesign of this network, these centers are categorized into two groups. The first group, $\mathcal{C}^{\mathcal{L}}$, consists of collectors that are already open and active within the network. The second group, $\mathcal{C}^{\mathcal{NL}}$, includes potential collection centers that could be added to enhance the network's capacity and resiliency. The comprehensive set of all collectors in the CLSC network is thus defined as $\mathcal{C} = \{\mathcal{C}^{\mathcal{L}} \cup \mathcal{C}^{\mathcal{NL}}\}$.

The strategic planning for the collectors involves fixed costs associated with opening and closing these facilities. Each open collector, $c \in \mathcal{C}^{\mathcal{L}}$, has a fixed closing cost $f_c^{\mathcal{C}\mathcal{L}}$. As such, each candidate collector $c \in \mathcal{C}^{\mathcal{NL}}$, has a fixed opening cost, $f_c^{\mathcal{C}\mathcal{NL}}$. Striking the right balance between these costs where the closing cost of a collector is lower than the opening cost of a new one, ensuring the network's cost-effectiveness while meeting the requirements of the reverse supply chain for scrap tires. These strategic decisions significantly influence the network's capacity to handle the flow of returned tires efficiently.

On the tactical planning level, each collector has a variable collection cost, $vc_c^{\mathcal{C}}$. Furthermore, collectors have a capacity constraint, $u_c^{\mathcal{C}}$. Transportation costs also play a significant role in determining the overall efficiency of the reverse supply chain. These costs depend on the distance, $\vartheta_{cr}^{\mathcal{C}}$, between a collector, $c \in \mathcal{C}$, and a recycler, $r \in \mathcal{R}$, and the unit transportation cost per scrap tire, $tc_{cr}^{\mathcal{C}}$. In conclusion, collectors are a cornerstone of the CLSC network, bridging the gap between waste generation at markets and recycling processes.

3.7 Recyclers

Recyclers serve as the backbone of the reverse supply chain in the tire industry, playing a crucial role in transforming scrap tires into valuable products and enabling sustainable operations (Sahebjamnia et al., 2018). Through recycling, scrap tires are processed and classified for remanufacturing, secondary markets, and diverse industrial applications. For example, in the construction industry, recycled rubber is used as crumb rubber in asphalt, enhancing road durability and noise reduction, or incorporated into concrete and lightweight materials for embankments and retaining walls (Medina et al., 2018). In sports and recreation, shredded rubber finds applications in playground surfaces, artificial turf fields, and running tracks (Benazzouk et al.,

2006). In the automotive sector, recycled materials contribute to insulation components and even the production of new tires, showcasing the wide-ranging potential of recycled tires across industries (Lozano-Oviedo et al., 2024).

In the CLSC network for the tire industry, recyclers are categorized into two groups. A subset, $\mathcal{R}^{\mathcal{L}}$, comprises recycling centers that are already open and operational, while the candidate sites, $\mathcal{R}^{\mathcal{NL}}$, represent potential locations for future expansion. The complete set of recyclers is defined as $\mathcal{R} = \{\mathcal{R}^{\mathcal{L}} \cup \mathcal{R}^{\mathcal{NL}}\}$. From a strategic perspective, open recycling centers, $r \in \mathcal{R}^{\mathcal{L}}$, incur a fixed cost, $f_r^{\mathcal{RL}}$, if closed, whereas establishing new recycling centers, $r \in \mathcal{R}^{\mathcal{NL}}$, involves a fixed opening cost, $f_r^{\mathcal{RNL}}$. These strategic costs, where the cost of opening a new facility is larger than the cost of closing an existing one, influence the structure and efficiency of the reverse supply chain.

On the tactical level, each recycler, $r \in \mathcal{R}$, incurs a unit recycling cost associated with processing each scrap tire, $vc_r^{\mathcal{R}}$, with capacity constraints, $u_r^{\mathcal{R}}$, limiting the maximum volume they can manage. After recycling, it is estimated that a proportion, β_r, of high-quality recycled tires at each recycling centers is sent to manufacturers for remanufacturing, while the remainder is allocated to secondary markets and other industrial applications, which operate outside the CLSC network's ownership. The process generates added value by creating revenue streams: recycling centers earn $pc_{rm}^{\mathcal{R}}$, per unit of recycled tires sold to manufacturers, while tires sold to secondary markets or other industries yield a comparatively lower revenue, $pc_r^{\mathcal{R}O}$, where $pc_r^{\mathcal{R}O} < pc_{rm}^{\mathcal{R}}$. Consequently, recycling centers prioritize supplying manufacturers over other outlets to maximize profitability.

Transportation further influences the operational costs, as the recycled products must be moved from recycling centers to manufacturers. The distance between a recycler, $r \in \mathcal{R}$, and a manufacturer, $m \in \mathcal{M}$, is represented by $\vartheta_{rm}^{\mathcal{R}}$, while the unit transportation cost per recycled tire is $tc_{rm}^{\mathcal{R}}$. Together, these parameters determine the total transportation cost, significantly impacting the profitability and feasibility of remanufacturing processes. The recycling centers, by integrating strategic location planning, capacity management, and transportation planning optimization, thus play a pivotal role in enhancing the economic and environmental sustainability of the CLSC network for the tire industry.

4 Proposed Model

Here, we first outline the assumptions underlying our optimization model (Sect. 4.1). Subsequently, we formulate and present the proposed MIP model (Sect. 4.2) for redesigning a CLSC network, as previously described in Sect. 3. Since the sets and parameters are already defined in Sect. 3, they are not repeated here.

4.1 Assumptions

This study incorporates a set of new assumptions specific to this study as follows:

- Facilities are divided into two groups: already open and candidate facilities.
- The existing facilities are already open, and there is a fixed cost associated with closing them.
- The candidate facilities are potential new facilities that can be added to the network, with a fixed cost for opening them.
- It is assumed that the cost of closing a facility is comparatively lower than the cost of opening a new one.

There are also a set of assumptions based on the literature review for the CLSC network design for the tire industry as follows (Fathollahi-Fard et al., 2021; Sahebjamnia et al., 2018):

- There is no flow of materials between facilities at the same hierarchical level. For example, suppliers do not exchange materials among themselves. As illustrated in Fig. 1, flows only occur between adjacent levels of the CLSC network, such as between suppliers and manufacturers.
- Facilities at each level have a predefined capacity limit. For example, manufacturers have capacity constraints for both tire production and remanufacturing of recycled scrap tires.
- Transportation costs are determined by the geographical distances between facilities and the per-unit cost of transferring tires.
- The demands at market locations are known and deterministic.
- All the demands must be met and there is no shortage.
- It is estimated that only a fraction of scrap tires at each market is collected by the collectors.
- Recycled tires are classified based on quality, where higher-quality recycled tires are sent to manufacturers for remanufacturing. Lower-quality recycled tires are directed to secondary markets or utilized for other industrial applications.
- Recycling centers generate revenue by selling to secondary markets, although this revenue is lower compared to remanufactured products sold by manufacturers.
- Suppliers and secondary markets operate independently and are not owned by the CLSC network.

4.2 Mixed-Integer Program (MIP)

To define the proposed MIP, the following binary and non-binary variables must be defined:

$y_m^{\mathcal{M}}$ 1, if manufacturer, $m \in \mathcal{M}$, is open; otherwise, 0
$y_d^{\mathcal{D}}$ 1, if retailer, $d \in \mathcal{D}$, is open; otherwise, 0
$y_c^{\mathcal{C}}$ 1, if collector, $c \in \mathcal{C}$, is open; otherwise, 0
$y_r^{\mathcal{R}}$ 1, if recycler, $r \in \mathcal{R}$, is open; otherwise, 0
$x_{sm}^{\mathcal{S}}$ Amount of raw materials transferred from supplier, $s \in \mathcal{S}$, to manufacturer, $m \in \mathcal{M}$
$x_{md}^{\mathcal{M}}$ Amount of manufactured tires transferred from manufacturer, $m \in \mathcal{M}$, to retailer, $d \in \mathcal{D}$
$x_{dk}^{\mathcal{D}}$ Amount of distributed tires transferred from retailer, $d \in \mathcal{D}$, to market, $k \in \mathcal{K}$
$x_{kc}^{\mathcal{K}}$ Amount of scrap tires transferred from market, $k \in \mathcal{K}$, to collector, $c \in \mathcal{C}$
$x_{cr}^{\mathcal{C}}$ Amount of scrap tires transferred from collector, $c \in \mathcal{C}$, to recycler, $r \in \mathcal{R}$
$x_{rm}^{\mathcal{R}}$ Amount of recycled tires transferred from recycler, $r \in \mathcal{R}$, to manufacturer, $m \in \mathcal{M}$
$x_r^{\mathcal{O}}$ Amount of recycled tires transferred from recycler, $r \in \mathcal{R}$, to secondary markets and other industrial applications

To calculate the objective function, it is necessary to determine the total cost, Z, which comprises the following components: the fixed cost of closing facilities, $Z_{\mathcal{CF}}$, the fixed cost of opening facilities, $Z_{\mathcal{OF}}$, the variable costs, $Z_{\mathcal{VC}}$, and the transportation costs, $Z_{\mathcal{TC}}$, as defined below.

$$min(Z = Z_{\mathcal{CF}} + Z_{\mathcal{OF}} + Z_{\mathcal{VC}} + Z_{\mathcal{TC}}) \tag{1}$$

$$\begin{aligned} Z_{\mathcal{CF}} = & \sum_{m \in \mathcal{M}^{\mathcal{L}}} f_m^{\mathcal{ML}}\left(1 - y_m^{\mathcal{M}}\right) + \sum_{d \in \mathcal{D}^{\mathcal{L}}} f_d^{\mathcal{DL}}\left(1 - y_d^{\mathcal{D}}\right) + \sum_{c \in \mathcal{C}^{\mathcal{L}}} f_c^{\mathcal{CL}}\left(1 - y_c^{\mathcal{C}}\right) \\ & + \sum_{r \in \mathcal{R}^{\mathcal{L}}} f_r^{\mathcal{RL}}\left(1 - y_r^{\mathcal{R}}\right) \end{aligned} \tag{2}$$

$$Z_{\mathcal{OF}} = \sum_{m \in \mathcal{M}^{\mathcal{NL}}} f_m^{\mathcal{MNL}} y_m^{\mathcal{M}} + \sum_{d \in \mathcal{D}^{\mathcal{NL}}} f_d^{\mathcal{DNL}} y_d^{\mathcal{D}} + \sum_{c \in \mathcal{C}^{\mathcal{NL}}} f_c^{\mathcal{CNL}} y_c^{\mathcal{C}} + \sum_{r \in \mathcal{R}^{\mathcal{NL}}} f_r^{\mathcal{RNL}} y_r^{\mathcal{R}} \tag{3}$$

$$\begin{aligned} Z_{\mathcal{VC}} = & \sum_{s \in \mathcal{S}} \sum_{m \in \mathcal{M}} pc_s^{\mathcal{S}} x_{sm}^{\mathcal{S}} + \sum_{m \in \mathcal{M}} \sum_{d \in \mathcal{D}} vc_m^{\mathcal{M}} x_{md}^{\mathcal{M}} + \sum_{d \in \mathcal{D}} \sum_{k \in \mathcal{K}} vc_d^{\mathcal{D}} x_{dk}^{\mathcal{D}} + \sum_{c \in \mathcal{C}} \sum_{r \in \mathcal{R}} vc_c^{\mathcal{C}} x_{cr}^{\mathcal{C}} \\ & + \sum_{r \in \mathcal{R}} \sum_{m \in \mathcal{M}} vc_r^{\mathcal{R}}\left(x_{rm}^{\mathcal{R}} + x_r^{\mathcal{O}}\right) - \sum_{r \in \mathcal{R}} \sum_{m \in \mathcal{M}} pc_{rm}^{\mathcal{R}} x_{rm}^{\mathcal{R}} - \sum_{r \in \mathcal{R}} pc_r^{\mathcal{RO}} x_r^{\mathcal{O}} \end{aligned} \tag{4}$$

$$\begin{aligned} Z_{\mathcal{TC}} = & \sum_{s \in \mathcal{S}} \sum_{m \in \mathcal{M}} \vartheta_{sm}^{\mathcal{S}} tc_{sm}^{\mathcal{S}} x_{sm}^{\mathcal{S}} + \sum_{m \in \mathcal{M}} \sum_{d \in \mathcal{D}} \vartheta_{md}^{\mathcal{M}} tc_{md}^{\mathcal{M}} x_{md}^{\mathcal{M}} + \sum_{d \in \mathcal{D}} \sum_{k \in \mathcal{K}} \vartheta_{dk}^{\mathcal{D}} tc_{dk}^{\mathcal{D}} x_{dk}^{\mathcal{D}} \\ & + \sum_{k \in \mathcal{K}} \sum_{c \in \mathcal{C}} \vartheta_{kc}^{\mathcal{K}} tc_{kc}^{\mathcal{K}} x_{kc}^{\mathcal{K}} + \sum_{c \in \mathcal{C}} \sum_{r \in \mathcal{R}} \vartheta_{cr}^{\mathcal{C}} tc_{cr}^{\mathcal{C}} x_{cr}^{\mathcal{C}} + \sum_{r \in \mathcal{R}} \sum_{m \in \mathcal{M}} \vartheta_{rm}^{\mathcal{R}} tc_{rm}^{\mathcal{R}} x_{rm}^{\mathcal{R}} \end{aligned} \tag{5}$$

For defining the objective, Eq. (1) serves as the objective function, aiming to minimize the total cost, which includes the fixed costs of closing and opening facilities, variable costs, and transportation costs. Equation (2) calculates the fixed costs associated with closing existing facilities, including manufacturers, retailers, collectors, and recyclers. Similarly, Eq. (3) computes the fixed costs of opening new facilities in the same categories. These equations capture the trade-offs involved in redesigning the CLSC network. Equation (4) evaluates the total variable costs incurred within the CLSC network. These costs encompass purchasing raw materials, manufacturing tires, distributing products to markets, collecting scrap tires, and recycling operations. Additionally, the model accounts for the revenue generated from recycled tires, both through remanufacturing and sales to secondary markets. Finally, Eq. (5) determines the transportation costs at various stages of the CLSC network. This includes costs for transferring raw materials to manufacturers, finished tires to retailers and markets, scrap tires to collectors and recyclers, and recycled tires back to manufacturers for remanufacturing. Together, these equations provide a comprehensive framework for minimizing the total cost of the CLSC network redesign.

The objective function components in Eqs. (2)–(5) are subject to several constraints to ensure the feasibility of the CLSC network. Network balance constraints given in Eqs. (6)–(9) maintain the flow consistency between facilities, while demand satisfaction constraints given in Eqs. (10) and (11) ensure that tire demands are met. Recycling constraints given in Eq. (12) govern the allocation of recycled scrap tires for remanufacturing, whereas capacity constraints given in Eqs. (13)–(18) limit the facility capabilities. Finally, boundary constraints given in Eqs. (19)–(29) define the permissible ranges for decision variables.

- **Network balance constraints**

The network balance constraints ensure that the inputs and outputs at each level of the CLSC network remain consistent and feasible. Equation (6) enforces balance for manufacturers, ensuring that the raw materials supplied to them are transformed into tires, which are then distributed to retailers. Equation (7) ensures balance for retailers, verifying that the tires received are appropriately distributed to meet market demands. Equation (8) ensures that the amount of scrap tires collected from markets by collectors matches the amount transferred to recyclers. Finally, Eq. (9) confirms that the total recycled scrap tires are correctly allocated between manufacturers for remanufacturing and secondary markets or other industrial applications.

$$\sum_{s\in\mathcal{S}} x_{sm}^{\mathcal{S}} = \sum_{d\in\mathcal{D}} x_{md}^{\mathcal{M}}, \quad \forall m \in \mathcal{M} \tag{6}$$

$$\sum_{m\in\mathcal{M}} x_{md}^{\mathcal{M}} = \sum_{k\in\mathcal{K}} x_{dk}^{\mathcal{D}}, \quad \forall d \in \mathcal{D} \tag{7}$$

$$\sum_{k\in\mathcal{K}} x_{kc}^{\mathcal{K}} = \sum_{r\in\mathcal{R}} x_{cr}^{\mathcal{C}}, \quad \forall c \in \mathcal{C} \tag{8}$$

$$\sum_{c\in\mathcal{C}} x_{cr}^{\mathcal{C}} = \sum_{m\in\mathcal{M}} x_{rm}^{\mathcal{R}} + x_{r}^{\mathcal{O}}, \quad \forall r \in \mathcal{R} \tag{9}$$

- **Demands satisfaction constraints**

Equation (10) ensures that the retailers fulfill all market demands, guaranteeing that the required quantity of tires is supplied to meet customer needs. Equation (11) accounts for the portion of the original demand that becomes waste, returned, or scrap tires, specifying that these must be collected by the collectors for further processing within the CLSC network.

$$\sum_{d \in \mathcal{D}} x_{dk}^{\mathcal{D}} = q_k, \quad \forall k \in \mathcal{K} \tag{10}$$

$$\sum_{c \in \mathcal{C}} x_{kc}^{\mathcal{K}} = \alpha_k q_k, \quad \forall k \in \mathcal{K} \tag{11}$$

- **Recycling constraints**

Equation (12) imposes a constraint set that limits the proportion of recycled tires sent to manufacturers for remanufacturing, ensuring that this quantity does not exceed a predefined maximum percentage of the total recycled tires.

$$\sum_{m \in \mathcal{M}} x_{rm}^{\mathcal{R}} \leq \beta_r \sum_{c \in \mathcal{C}} x_{cr}^{\mathcal{C}}, \quad \forall r \in \mathcal{R} \tag{12}$$

- **Capacity constraints**

The capacity constraints in the CLSC network design represent essential constraints in MIPs, ensuring that the allocation of materials and products aligns with the capabilities of various facilities (Aykin, 1994). Equation (13) limits the amount of raw materials purchased by manufacturers from each supplier based on the suppliers' capacity. Equation (14) ensures that the number of tires produced by each manufacturer and sent to retailers does not exceed the manufacturers' production capacity. Equation (15) defines the remanufacturing capacity of manufacturers for recycled tires in the reverse supply chain. Equation (16) restricts the distribution of tires from retailers to markets according to the retailers' capacity. Equation (17) limits the quantity of waste, returned, and scrap tires collected from markets based on the collectors' capacity. Finally, Eq. (18) ensures that recyclers do not exceed their capacity to process and recycle scrap tires.

$$\sum_{m \in \mathcal{M}} x_{sm}^{\mathcal{S}} \leq u_s^{\mathcal{S}}, \quad \forall s \in \mathcal{S} \tag{13}$$

$$\sum_{d \in \mathcal{D}} x_{md}^{\mathcal{M}} \leq u_m^{\mathcal{M}} y_m^{\mathcal{M}}, \quad \forall m \in \mathcal{M} \tag{14}$$

$$\sum_{r \in \mathcal{R}} x_{rm}^{\mathcal{R}} \leq \overline{u}_m^{\mathcal{M}} y_m^{\mathcal{M}}, \quad \forall m \in \mathcal{M} \tag{15}$$

$$\sum_{k \in \mathcal{K}} x_{dk}^{\mathcal{D}} \leq u_d^{\mathcal{D}} y_d^{\mathcal{D}}, \quad \forall d \in \mathcal{D} \tag{16}$$

$$\sum_{k \in \mathcal{K}} x_{kc}^{\mathcal{K}} \leq u_c^{\mathcal{C}} y_c^{\mathcal{C}}, \quad \forall c \in \mathcal{C} \tag{17}$$

$$\sum_{c \in \mathcal{C}} x_{cr}^{\mathcal{C}} \leq u_r^{\mathcal{R}} y_r^{\mathcal{R}}, \quad \forall r \in \mathcal{R} \tag{18}$$

- **Boundary constraints**

Equations (19)–(22) define the binary variables used in the model, which represent decisions, such as whether to open or close facilities within the CLSC network. On the other hand, Eqs. (23)–(29) define the non-binary variables, capturing continuous values for the allocation of materials flows, production quantities, and tires and scrap tires between the facilities at different levels.

$$y_m^{\mathcal{M}} \in \{0,1\}, \quad \forall m \in \mathcal{M} = \{\mathcal{M}^{\mathcal{L}} \cup \mathcal{M}^{\mathcal{NL}}\} \tag{19}$$

$$y_d^{\mathcal{D}} \in \{0,1\}, \quad \forall d \in \mathcal{D} = \{\mathcal{D}^{\mathcal{L}} \cup \mathcal{D}^{\mathcal{NL}}\} \tag{20}$$

$$y_c^{\mathcal{C}} \in \{0,1\}, \quad \forall c \in \mathcal{C} = \{\mathcal{C}^{\mathcal{L}} \cup \mathcal{C}^{\mathcal{NL}}\} \tag{21}$$

$$y_r^{\mathcal{R}} \in \{0,1\}, \quad \forall r \in \mathcal{R} = \{\mathcal{R}^{\mathcal{L}} \cup \mathcal{R}^{\mathcal{NL}}\} \tag{22}$$

$$x_{sm}^{\mathcal{S}} \geq 0, \quad \forall s \in \mathcal{S}, m \in \mathcal{M} = \{\mathcal{M}^{\mathcal{L}} \cup \mathcal{M}^{\mathcal{NL}}\} \tag{23}$$

$$x_{md}^{\mathcal{M}} \geq 0, \quad \forall m \in \mathcal{M} = \left\{\mathcal{M}^{\mathcal{L}} \cup \mathcal{M}^{\mathcal{NL}}\right\}, d \in \mathcal{D} = \{\mathcal{D}^{\mathcal{L}} \cup \mathcal{D}^{\mathcal{NL}}\} \tag{24}$$

$$x_{dk}^{\mathcal{D}} \geq 0, \quad \forall d \in \mathcal{D} = \left\{\mathcal{D}^{\mathcal{L}} \cup \mathcal{D}^{\mathcal{NL}}\right\}, k \in \mathcal{K} \tag{25}$$

$$x_{kc}^{\mathcal{K}} \geq 0, \quad \forall k \in \mathcal{K}, c \in \mathcal{C} = \{\mathcal{C}^{\mathcal{L}} \cup \mathcal{C}^{\mathcal{NL}}\} \tag{26}$$

$$x_{cr}^{\mathcal{C}} \geq 0, \quad \forall c \in \mathcal{C} = \left\{\mathcal{C}^{\mathcal{L}} \cup \mathcal{C}^{\mathcal{NL}}\right\}, r \in \mathcal{R} = \{\mathcal{R}^{\mathcal{L}} \cup \mathcal{R}^{\mathcal{NL}}\} \tag{27}$$

$$x_{rm}^{\mathcal{R}} \geq 0, \quad \forall r \in \mathcal{R} = \left\{\mathcal{R}^{\mathcal{L}} \cup \mathcal{R}^{\mathcal{NL}}\right\}, m \in \mathcal{M} = \{\mathcal{M}^{\mathcal{L}} \cup \mathcal{M}^{\mathcal{NL}}\} \tag{28}$$

$$x_r^{\mathcal{O}} \geq 0, \quad \forall r \in \mathcal{R} = \{\mathcal{R}^{\mathcal{L}} \cup \mathcal{R}^{\mathcal{NL}}\} \tag{29}$$

5 Computational Experiments

In this section, we present the results of a series of computational experiments conducted to evaluate the proposed model's performance in terms of computing time and obtained solutions. Additionally, we analyze the impact of various settings that are directly related to the problem for strategic and tactical planning in redesigning a CLSC network for the tire industry. This analysis focuses on exploring the overall performance and structural features of the solutions generated by the proposed model.

Section 5.1 introduces the main attributes of the instances generated for the computational experiments. These instances are carefully designed based on realistic case scenarios to ensure the validity and relevance of the results. By leveraging real-world data, we aim to provide robust insights into the model's practical applications. Section 5.2 examines the impact of adding new facilities to the network during the redesign process. This analysis evaluates how the inclusion of new candidate facilities influences the overall network structure, costs, and efficiency, providing critical insights for strategic decision-making in CLSC network planning. Then, Sect. 5.3 presents sensitivity analyses focusing on the reverse supply chain. The effects of key parameters on the performance of recyclers are assessed, offering a deeper understanding of how these parameters influence the network's performance and efficiency. Finally, Sect. 5.4 offers a discussion on the managerial insights and recommendations for policymakers regarding CLSC networks.

It should be noted that the proposed MIP model was solved using GAMS 24.7.4 software based on the CPLEX solver. The computational experiments were conducted on a system equipped with an Intel(R) Core(TM) i7–10850H CPU and 32 GB of RAM.

5.1 Test Instances

To develop various test instances for a comprehensive analysis of the presented MIP model, we adopted a systematic approach to parameter selection. This approach was informed by the logic and principles established in prior studies, particularly those of Sahebjamnia et al. (2018) and Fathollahi-Fard et al. (2021), ensuring a consistent and robust framework for data generation. Since this study does not contribute to a heuristic or metaheuristic algorithm for solving the proposed network redesign problem in the tire CLSC management, only small-scale instances were considered, as described in Table 1. We identified four real-world problem sizes and generated four test instances for each size. For problem size P1, the instances T1 through T4 were created. Similarly, instances T5 through T8 correspond to size P2, while instances T9 through T12 were developed for size P3. Finally, for problem size P4, T13 to T16 were created. All these instances are available in Appendix.

Table 1 Test instances

Sizes	Test instances	$\vert\mathcal{S}\vert$	$\vert\mathcal{M}^{\mathcal{L}}\vert$	$\vert\mathcal{M}^{\mathcal{NL}}\vert$	$\vert\mathcal{D}^{\mathcal{L}}\vert$	$\vert\mathcal{D}^{\mathcal{NL}}\vert$	$\vert\mathcal{C}^{\mathcal{L}}\vert$	$\vert\mathcal{C}^{\mathcal{NL}}\vert$	$\vert\mathcal{R}^{\mathcal{L}}\vert$	$\vert\mathcal{R}^{\mathcal{NL}}\vert$	$\vert\mathcal{K}\vert$
P1	T1 to T4	4	2	2	2	2	2	2	2	2	10
P2	T5 to T8	5	3	4	3	4	3	4	3	4	20
P3	T9 to T12	5	3	6	3	6	3	6	3	6	40
P4	T13 to T16	8	4	12	4	12	4	12	4	12	50

The instances summarized in Table 1 were generated using random numbers based on the following structured approach benchmarked from Sahebjamnia et al. (2018) and Fathollahi-Fard et al. (2021):

- We began by randomly assigning geographical locations to each facility within a two-dimensional coordinate system (x, y), where both x and y are uniformly distributed between 0 and 1,000. The distances between facilities were then calculated based on these locations.
- Transportation costs were randomly valued between 3 and 8.
- The demand volume was ranging from 50 to 200 units.
- The fixed closing cost of a facility was determined by calculating the transportation expenses for all potential connections to that facility, considering half of the total demand. For instance, for a manufacturer, this includes the transportation costs from all associated suppliers, retailers, and recyclers. In contrast, the fixed opening cost of a facility was set to be twice the fixed closing cost of that facility.
- Suppliers' purchasing costs were randomly valued from 3 to 5 units.
- The demand's percentage which was collected from each market was estimated to fall between 50 and 100%.
- Facility capacities were estimated to be between 20 and 50% of the total demand.
- Regarding recycled tires, the proportion allocated for remanufacturing was set between 50 and 100% of the recycled tires at each recycler.
- Variable costs for operations at each facility were randomly estimated within a range of 3–5 units.
- The price of recycled tires sent for remanufacturing was generated between 5 and 8 units, while the price for recycled tires designated for secondary markets was between 3 and 5 units.

5.2 *Analyzing the Network Redesign Performance*

In this study, we assess the performance of the proposed CLSC network redesign by solving 16 test instances of varying sizes, as summarized in Table 2. These instances provide a comprehensive analysis of the model's behavior under different network

Table 2 Computational results

Test instances	Total cost	CPU time (s)	Number of open facilities before network redesign	Number of open facilities after network redesign
T1	178,032,872	18.81	8	12
T2	208,388,373	20.99	8	12
T3	257,646,344	18.98	8	13
T4	218,294,466	19.12	8	13
T5	23,732,852	18.4	12	12
T6	354,771,616	18.32	12	14
T7	162,551,349	18.854	12	13
T8	165,015,992	19.13	12	13
T9	318,940,232	28.71	12	13
T10	921,351,963	29.9	12	14
T11	1,126,203,330	29.9	12	14
T12	284,679,337	29.81	12	13
T13	1,995,220,840	40.43	16	18
T14	1,047,352,278	39.82	16	17
T15	51,501,532	38.91	16	16
T16	911,102,762	40.71	16	17

configurations, offering valuable insights into the dynamics of CLSC redesign. The results highlight the significant role of various factors, including the number of facilities, particularly candidate facilities, in influencing the computational complexity of the proposed MIP model.

Figure 2 illustrates the differences across instances in terms of the number of new facilities added to the network. Among all the instances, the minimum-cost solutions are observed in T5 and T15. These two instances exhibit remarkable stability in their network configurations: no new facilities are added, and the number of open facilities remains unchanged before and after the redesign. This lack of structural adjustment reflects a highly stable network design, resulting in both reduced computational complexity and lower total costs. This outcome demonstrates the efficiency of the model in scenarios where minimal changes are required to achieve cost-effective network configurations. Such stability suggests that the model performs exceptionally well in situations with a limited need for reconfiguration, further reinforcing its applicability in real-world scenarios where structural changes are often costly or undesirable.

On the other hand, the instances with the most extensive network changes are T3 and T4, where five new facilities are added to the network. These instances represent more substantial network reconfigurations, reflecting the model's ability to accommodate significant structural adjustments. The addition of multiple new facilities increases the overall network complexity, leading to higher computational efforts

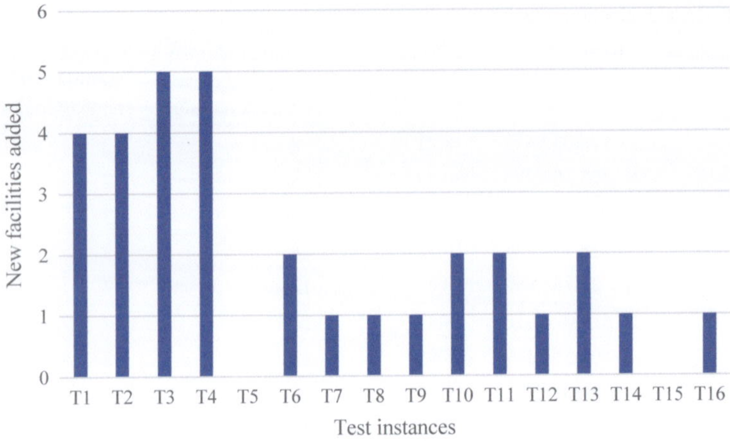

Fig. 2 Number of new facilities added during the solution of each instance

and increased total costs. While these changes provide opportunities for enhanced network performance, they also highlight a key trade-off between network improvement and the associated complexities of redesign. Specifically, the results from T3 and T4 emphasize the balance between optimizing network performance and managing the computational and cost-related challenges that come with more extensive redesign steps.

Furthermore, the findings suggest that the computational complexity of the proposed MIP model increases with the number of candidate facilities, underscoring the importance of carefully selecting which facilities to include in the redesign process. As the number of potential facilities rises, the model faces a more complex decision-making process, requiring additional computational resources to identify optimal solutions. This is particularly evident in larger instances, where the scope of potential facility configurations exponentially increases, further challenging the scalability of the model.

These results provide important insights into the interplay between redesign complexity, cost, and network performance. They emphasize the proposed model's scalability and robustness in addressing CLSC network redesign problems across a range of scenarios. The ability to handle both minimal and substantial changes in the network configuration makes the model versatile and adaptable to different industrial contexts. Moreover, the findings underline the practical significance of the model in both strategic and tactical decision-making, particularly for industries like tire manufacturing, where optimizing the CLSC network design is crucial for improving cost efficiency and operational performance.

In conclusion, this study highlights the dual aspects of the model's performance: its efficiency in scenarios requiring minimal changes and its flexibility in accommodating larger-scale redesign steps. By capturing the trade-offs between complexity and cost, the proposed model provides valuable support for decision-makers seeking to optimize CLSC networks in dynamic and complex environments.

5.3 *Analyzing the Effects of the Recycling Process on the Performance of the CLSC Network*

To further analyze the proposed model, we selected instance T4 as a representative and insightful test case for conducting a series of sensitivity analyses on the key parameters. Since T4 is one of the instances where the network redesign process is significantly impacted, as shown in Fig. 2, we chose it as a challenging case for the sensitivity analyses. Specifically, we focused on the percentage of recycled tires used for remanufacturing, a critical factor influencing the performance of the CLSC network. This parameter was systematically varied across different scenarios to evaluate its impact on solution quality and computational time.

The results, presented in Table 3, clearly illustrate the significant impact of the percentage of recycled tires on the revenue generated from the reverse supply chain. Specifically, the analysis reveals that remanufacturing yields substantially higher revenue than sales to secondary markets, emphasizing the value of recycling in the CLSC network. As we gradually reduced the percentage of recycled tires in our analyses, we observed no significant change in computational time, suggesting that the model's performance remains relatively stable despite variations in recycling levels. However, a clear trend emerged in the total cost, which increased progressively as the percentage of recycled tires decreased. This cost escalation can be primarily attributed to the reduction in the revenue from remanufacturing activities, as the system increasingly relied on less profitable sales to secondary markets.

The cost increase associated with reduced recycling underscores the pivotal role of remanufacturing in the CLSC network design. As recycling efforts decrease, the reverse supply chain becomes less profitable, highlighting the necessity of maintaining or even enhancing recycling capabilities. The observed increase in total costs, while notable, was relatively minor in magnitude. This can be explained by other fixed and variable costs, such as facility operation and transportation, as these costs have a more dominant influence on the overall cost structure. However, the results make it clear that even modest reductions in recycling can lead to increased reliance on alternative, less profitable revenue streams, affecting the overall network efficiency.

Table 3 Sensitivity analyses on the recycling percentage for remanufacturing

β_r	Total cost	CPU time (s)
{0.55, 0.63, 0.88, 0.91}	218,294,466	19.12
{0.5, 0.5, 0.5, 0.5}	218,294,537	18.30
{0.25, 0.25, 0.25, 0.25}	218,294,700	18.24
{0, 0, 0, 0}	218,295,000	18.85

These findings provide important insights into the financial dynamics of the CLSC network. They emphasize that optimizing recycling processes is crucial not only for environmental sustainability but also for maximizing the economic benefits of remanufacturing. For CLSC managers, these results highlight the strategic importance of investing in and enhancing recycling capabilities, as this can serve as a key lever for improving cost efficiency and overall network performance. By prioritizing recycling, managers can achieve better revenue generation from remanufacturing, thereby reducing the reliance on secondary market sales and mitigating cost increases.

Moreover, these insights have practical implications for the tire industry and other sectors that rely on CLSC networks. The findings suggest that organizations should focus on developing robust recycling infrastructure and technologies to increase the proportion of recycled materials in their supply chain. Doing so will not only improve the economic viability of reverse supply chain operations but also contribute to more sustainable and circular business practices.

5.4 Managerial Insights

The results from our analyses offer valuable managerial insights for optimizing CLSC networks. First, the significant impact of recycling efforts on revenue generation, particularly through remanufacturing, underscores the importance of maintaining or even enhancing recycling capabilities. Managers should prioritize investments in recycling infrastructure to maximize the economic benefits derived from remanufacturing, which yields higher revenue compared to sales to secondary markets. By maintaining a high percentage of recycled materials, CLSC managers can reduce reliance on less profitable market channels and mitigate rising costs.

Furthermore, the trade-off between network complexity and network redesign cost, which was observed in both facility redesign and recycling strategies, highlights the need for strategic decision-making. Managers must balance the benefits of substantial network improvements with the associated costs and computational challenges, ensuring that redesign efforts align with the broader strategic goals of the CLSC networks to satisfy the markets' demands. The findings suggest that cost efficiency and network performance can be significantly enhanced by focusing on recycling optimization, which not only drives revenue but also supports more sustainable practices within the supply chain.

Ultimately, these insights emphasize the strategic role of recycling in reducing total costs and enhancing network performance. By investing in recycling capabilities and optimizing reverse supply chain processes, managers can improve both the profitability and sustainability of their operations.

6 Conclusions and Perspectives

This paper introduces a novel framework for redesigning CLSC networks specifically tailored to the tire industry. Through the formulation of an MIP model, the study addresses the key challenges in network redesign, including facility openings and closures, cost-balancing, and capacity adjustments. The proposed framework captures the unique dynamics of forward and reverse flows, serving as a practical decision-support tool for optimizing facility location, allocation, and capacity planning.

For analyzing the model, 16 test instances in different small-scale sizes were solved by the CPLEX solver. The computational results reveal several important findings. Instances with minimal structural changes, such as T5 and T15, resulted in the lowest total costs, highlighting the efficiency of maintaining network stability. On the other hand, instances involving significant reconfiguration, such as T3 and T4, showcased the model's adaptability but incurred higher costs and computational complexity. Sensitivity analyses further demonstrated that reductions in recycling efficiency directly increase total costs, emphasizing the economic importance of maintaining effective remanufacturing practices.

This study offers actionable insights for supply chain managers in the tire industry. Managers are encouraged to carefully assess trade-offs between redesign complexity and cost, prioritizing minimal disruptions whenever possible. Additionally, investments in recycling processes and remanufacturing capabilities can help reduce costs while improving network sustainability. Finally, the model's adaptability to significant structural changes provides a valuable tool for strategic decision-making, enabling businesses to respond effectively to evolving market demands and regulatory requirements.

This study has several limitations. First, the MIP model primarily focuses on the financial and operational aspects of network redesign, without incorporating broader environmental or social sustainability factors that may significantly influence decision-making in the tire industry. Additionally, the sensitivity analysis is based on a limited set of parameters, and future research could expand this by including factors such as demand fluctuations, transportation costs, and the utilization rates of open facilities. Furthermore, this study does not account for potential disruptions in the supply chain, such as natural disasters or global supply chain shocks, which could considerably affect network performance.

This research also opens several promising avenues for future work. For example, developing MIP models that account for dynamic demand fluctuations and incorporate stochastic elements could improve the model's applicability to real-world scenarios. Further investigation into the integration of emerging technologies, such as digital twins, blockchain, and the Internet of Things (IoT), could enhance CLSC network transparency and operational efficiency. Lastly, to address the complexity of large-scale networks with numerous candidate facilities and demand points, the development of efficient metaheuristic algorithms would be a valuable extension, providing a scalable solution to more complex instances.

Acknowledgements The author acknowledges the seminal publications of Prof. Teodor G. Crainic which served as a significant source of inspiration for the network redesign problem for the tire closed-loop supply chain.

References

Abdolazimi, O., Esfandarani, M. S., Salehi, M., & Shishebori, D. (2020). Robust design of a multi-objective closed-loop supply chain by integrating on-time delivery, cost, and environmental aspects, case study of a tire factory. *Journal of Cleaner Production, 264*, Article 121566.

Ahmed, J., Amin, S. H., & Fang, L. (2023). A multi-objective approach for designing a tire closed-loop supply chain network considering producer responsibility. *Applied Mathematical Modelling, 115*, 616–644.

Amin, S. H., Zhang, G., & Akhtar, P. (2017). Effects of uncertainty on a tire closed-loop supply chain network. *Expert Systems with Applications, 73*, 82–91.

Amirian, J., Amoozad Khalili, H., & Mehrabian, A. (2022). Designing an optimization model for green closed-loop supply chain network of heavy tire by considering economic pricing under uncertainty. *Environmental Science and Pollution Research,* 1–14.

Aykin, T. (1994). Lagrangian relaxation based approaches to capacitated hub-and-spoke network design problem. *European Journal of Operational Research, 79*(3), 501–523.

Banguera, L. A., Sepúlveda, J. M., Ternero, R., Vargas, M., & Vásquez, Ó. C. (2018). Reverse logistics network design under extended producer responsibility: The case of out-of-use tires in the Gran Santiago city of Chile. *International Journal of Production Economics, 205*, 193–200.

Battini, D., Bogataj, M., & Choudhary, A. (2017). Closed loop supply chain (CLSC): Economics, modelling, management and control. *International Journal of Production Economics, 183*, 319–321.

Benazzouk, A., Douzane, O., Mezreb, K., & Quéneudec, M. (2006). Physico-mechanical properties of aerated cement composites containing shredded rubber waste. *Cement and Concrete Composites, 28*(7), 650–657.

Bhatia, M. S., Kumar, S., Gangwani, K. K., & Kaur, B. (2024). Key capabilities for closed-loop supply chain: Empirical evidence from manufacturing firms. *Technological Forecasting and Social Change, 208*, Article 123717.

Chopra, S., & Meindl, P. (2007). Supply chain management. Strategy, planning & operation. In *Das summa summarum des management* (pp. 265–275). Gabler.

Ebrahimi, S. B. (2018). A stochastic multi-objective location-allocation-routing problem for tire supply chain considering sustainability aspects and quantity discounts. *Journal of Cleaner Production, 198*, 704–720.

Fathollahi-Fard, A. M., Dulebenets, M. A., Hajiaghaei-Keshteli, M., Tavakkoli-Moghaddam, R., Safaeian, M., & Mirzahosseinian, H. (2021). Two hybrid meta-heuristic algorithms for a dual-channel closed-loop supply chain network design problem in the tire industry under uncertainty. *Advanced Engineering Informatics, 50*, Article 101418.

Fathollahi-Fard, A. M., Hajiaghaei-Keshteli, M., & Mirjalili, S. (2018). Hybrid optimizers to solve a tri-level programming model for a tire closed-loop supply chain network design problem. *Applied Soft Computing, 70*, 701–722.

Ferrer, G. (1997). The economics of tire remanufacturing. *Resources, Conservation and Recycling, 19*(4), 221–255.

Ghasemzadeh, Z., Sadeghieh, A., & Shishebori, D. (2021). A stochastic multi-objective closed-loop global supply chain concerning waste management: A case study of the tire industry. *Environment, Development and Sustainability, 23*(4), 5794–5821.

Hoeke, S., van Wijnen, J., Krikke, H., Löhr, A., & Ragas, A. M. (2024). Mapping the tire supply chain and its microplastics emissions using a multi-stakeholder approach. *Resources, Conservation and Recycling, 203*, Article 107389.

Jian, J., Li, B., Zhang, N., & Su, J. (2021). Decision-making and coordination of green closed-loop supply chain with fairness concern. *Journal of Cleaner Production, 298*, Article 126779.

Lozano-Oviedo, J., Cortés, C. E., & Rey, P. A. (2024). Sustainable closed-loop supply chains and their optimization models: A review of the literature. *Clean Technologies and Environmental Policy, 26*(4), 999–1023.

MahmoumGonbadi, A., Genovese, A., & Sgalambro, A. (2021). Closed-loop supply chain design for the transition towards a circular economy: A systematic literature review of methods, applications and current gaps. *Journal of Cleaner Production, 323*, Article 129101.

Manupati, V. K., Kulatunga, A. K., Ray, P., & Rofin, T. M. (2024). Sustainable management of EOL tyres: A remanufacturing supply chain network design. *Journal of Cleaner Production, 434*, Article 140115.

Medina, N. F., Garcia, R., Hajirasouliha, I., Pilakoutas, K., Guadagnini, M., & Raffoul, S. (2018). Composites with recycled rubber aggregates: Properties and opportunities in construction. *Construction and Building Materials, 188*, 884–897.

Mehrjerdi, Y. Z., & Shafiee, M. (2021). A resilient and sustainable closed-loop supply chain using multiple sourcing and information sharing strategies. *Journal of Cleaner Production, 289*, Article 125141.

Mohammed, F., Selim, S. Z., Hassan, A., & Syed, M. N. (2017). Multi-period planning of closed-loop supply chain with carbon policies under uncertainty. *Transportation Research Part D: Transport and Environment, 51*, 146–172.

Pedram, A., Yusoff, N. B., Udoncy, O. E., Mahat, A. B., Pedram, P., & Babalola, A. (2017). Integrated forward and reverse supply chain: A tire case study. *Waste Management, 60*, 460–470.

Sahebjamnia, N., Fathollahi-Fard, A. M., & Hajiaghaei-Keshteli, M. (2018). Sustainable tire closed-loop supply chain network design: Hybrid metaheuristic algorithms for large-scale networks. *Journal of Cleaner Production, 196*, 273–296.

Sarkar, B., Ullah, M., & Kim, N. (2017). Environmental and economic assessment of closed-loop supply chain with remanufacturing and returnable transport items. *Computers & Industrial Engineering, 111*, 148–163.

Sasikumar, P., Kannan, G., & Haq, A. N. (2010). A multi-echelon reverse logistics network design for product recovery—A case of truck tire remanufacturing. *The International Journal of Advanced Manufacturing Technology, 49*(9), 1223–1234.

Soleimani, H., Chhetri, P., Fathollahi-Fard, A. M., Mirzapour Al-e-Hashem, S. M. J., & Shahparvari, S. (2022). Sustainable closed-loop supply chain with energy efficiency: Lagrangian relaxation, reformulations and heuristics. *Annals of Operations Research, 318*(1), 531–556.

Subulan, K., Taşan, A. S., & Baykasoğlu, A. (2015). Designing an environmentally conscious tire closed-loop supply chain network with multiple recovery options using interactive fuzzy goal programming. *Applied Mathematical Modelling, 39*(9), 2661–2702.

Yadollahinia, M., Teimoury, E., & Paydar, M. M. (2018). Tire forward and reverse supply chain design considering customer relationship management. *Resources, Conservation and Recycling, 138*, 215–228.

Amir M. Fathollahi-Fard is an Assistant Professor at the School of Business Administration at Laurentian University, Sudbury, Canada. He is also an Adjunct Professor at Macau University of Science and Technology, Taipa, Macau. Dr. Fathollahi-Fard earned his PhD from École de Technologie Supérieure (ÉTS), Montreal, Canada, and holds a B.Sc. and M.Sc. from the University of Science and Technology of Mazandaran, Behshahr, Mazandaran, Iran. His research interests span supply chain management, healthcare systems, sustainable logistics and production management, optimisation algorithms, heuristics, and metaheuristics. He has authored over 100 papers in high-ranked journals such as the Journal of Cleaner Production, Annals of Operations

Research, Computers & Industrial Engineering, Information Sciences, Applied Soft Computing, Expert Systems with Applications, and Engineering Applications of Artificial Intelligence. In recognition of his scholarly influence, he was named among the Top 1% Highly Cited Researchers worldwide by Clarivate (Web of Science) in 2024.

Alternative Forms of Transport and Possibilities of Sustainable Transport in the European Union

Csongor Máthé, János Pénzes, Gábor Horváth, Máté Zakar, and Tamás Tóth

Abstract Climate change and the energy crisis have become pressing problems nowadays like never before. As a result, the European Union (EU) is increasingly promoting energy-efficient modes in all aspects of life, and the transport sector is no different. Initiatives to reform the transport sector in the EU started as early as the 2000s. At that time, these objectives focused on various strategies, including the promotion of biofuels. The global events of the following decades (i.e., economic recessions, war conflicts, the increasingly tangible consequences of climate change, etc.) required a change of attitude and a new approach to transport from the EU. These processes and their subsequent effects have contributed towards a target of 42.5% renewable share by 2030, mainly driven by facilitating the increasing utilization of electric cars. At the moment, it seems that thanks to the new directive, the use of biofuels will be pushed into the background—at least no further increase in the blending rate is expected—and the green transport reform is expected from electric vehicles, while the promotion of hydrogen-powered vehicles is also imminent. The development of electric cars is certainly remarkable, but their limitations should not be overlooked. Electric cars can only be considered truly zero-emission vehicles if the electricity used as "fuel" is generated from renewable energy. An important

C. Máthé
Department of Geography in Hungarian, Faculty of Geography, Babeş-Bolyai University, Cluj-Napoca, Romania
e-mail: csongor.mathe@ubbcluj.ro

J. Pénzes (✉)
Department of Social Geography and Regional Development, Faculty of Science and Technology, Institute of Earth Sciences, University of Debrecen, Debrecen, Hungary
e-mail: penzes.janos@science.unideb.hu

G. Horváth · M. Zakar · T. Tóth
Department of Meteorology, Faculty of Science and Technology, Institute of Earth Sciences, University of Debrecen, Debrecen, Hungary
e-mail: horvath.gabor@science.unideb.hu

M. Zakar
e-mail: zakar.mate4@mailbox.unideb.hu

T. Tóth
e-mail: toth.tamas@science.unideb.hu

M. A. Dulebenets et al. (eds.), *Transport and Climate Change*,
https://doi.org/10.1007/978-3-032-00563-2_10

question is that whether electric cars can really change the entire transportation sector and lead to a more sustainable and cleaner life, or they are just a transition towards further innovations. The present chapter explores these issues and provides interesting insights.

Keywords Sustainability · EU2030 · Biofuel · Electromobility · Electric network · Hydrogen

1 Introduction

According to Nordin et al. (2024), today, the transport sector accounts for 20% of global greenhouse gas (GHG) emissions. The European Union (EU) explored numerous measures to reduce the emissions of the transport sector in recent decades. Initially, the development and application of biofuels was a priority, but since the late 2010s, the EU has increasingly seen the green reform of transport through the expansion of electric cars.

The indisputable advantage of electric cars is that the journey does not directly produce CO_2 emissions (Luo et al., 2022). At the same time, they can still become polluters indirectly, since the actual emissions depend on how the electricity used as fuel is produced. The energy mix therefore has a major impact on global environmental and air pollution caused by electric cars. Boudmen et al. (2024) found that if electricity is generated from renewable energy sources, the carbon footprint of electric vehicles can be significantly reduced. At the same time, the power network varies at the national level, and weather-dependent energy sources can cause significant problems, as it is difficult to regulate their electricity production. Several researchers pointed out that government incentives and subsidies are of great importance in the spread of electromobility and the increase in the number of EVs (Bridi et al., 2024; Jaman et al., 2023; Rego et al., 2023).

The aim of this study is to analyze the sustainable transport potential of the EU member states based on the current situation and prospects with a primary focus on the directives applied in the transport policy of the EU in recent decades. Since the EU is committed to the increasing deployment of EVs based on the existing directives, the current situation and the potential of transport with sustainable electric cars in the EU member states are evaluated as well.

2 Efforts in the European Union to Make Transport More Sustainable

The European Union has been concerned with sustainability and sustainable transport for decades. In recent years, many attempts have been made to make the transport sector more sustainable and environmentally friendly, but none of them have been able to completely replace fossil fuels. The main innovations of the past few decades have been plant-based biofuels, electric cars, and hydrogen. The EU Directive 2003/30/EC laid down the legal basis for the use of biofuels and other renewable energy sources in the transport sector (European Union, 2003). This EU mechanism set the direction for biofuel production and targets for its use till 2010. The member states have committed themselves to achieving a biofuel blending target of 2% by 31st December 2005 and 5.75% by 31st December 2010 on the basis of energy content. According to the 2013 progress report, 22 member states ultimately failed to meet this blending target in 2010.

In 2006, the Road Map for Renewable Energy was published (Commission of the European Communities, 2006). The Renewable Energy Road Map discusses the use of renewable energy resources in the transport sector. At that time, the Commission set a minimum target of 10% of biofuel share by 2020. The EU clearly considered biofuels the key to making transport more sustainable, aiming to specifically increase the share of biofuels. This later also appeared in Directive 2009/28/EC of 23 April 2009 (Renewable Energy Directive, "RED"), which also required a 10% share of biofuels for all member states. Directive 2009/28/EC on renewable energy resources established a European framework for the support of renewable energy resources (European Union, 2009). It also prescribed a mandatory overall share of 20% of renewable energy resources in the total energy consumption and at least 10% in the transport sector. These efforts are strategic because they contribute to reducing emissions, improving energy security, and reducing dependence on energy imports.

The RED directive was amended in 2015 to include the Indirect Land Use Change (ILUC) Directive EU 2015/1513 (European Union, 2015). This includes a 7% cap on the production of biofuels from food and fodder crops. In the transport sector, therefore, the target for renewable energy resources is 10%, but a maximum of 7% can be covered by biofuels produced from food and fodder crops. At the same time, the RED directive prescribed that biofuel use must reduce greenhouse gas emissions by at least 50% compared to fossil fuels by 2017. From 2018, greenhouse gas reduction of 60% is expected for newly established plants. The greenhouse gas reduction should be calculated over the total life cycle (well-to-wheel) of the biofuel, i.e. from planting to using it as biofuel, including all processes.

In February 2017, a proposal to revise the RED directive was published as the RED II directive. Based on the 20% renewable energy target for 2020, a new target for the EU for 2030 was set by amending the directive EU 2018/2001 (European Union, 2018). The RED II directive requires member states to cover at least 32% of gross final consumption of energy in the EU from renewable energy by 2030. According to the directive, a target of 14% of renewables has been set for the transport sector. RED

II states that after 2030 the blending rate of biofuels produced from food and fodder crops should be limited to 3.8%. The RED II directive further marginalized the first-generation biofuels. With this new directive, the EU wants to limit the inclusion of food crops in biofuel production in response to land-use changes caused by biofuel production. At the same time, 3.5% of the 14% transport target must be achieved for advanced biofuels (i.e., the second- and third-generation biofuels). Their energy content is assumed to double. Ligno-cellulosic bioethanol shall have a target of at least 1.75% (3.5% calculated energy content) and biodiesel produced from waste (used edible oil and animal fats) shall account for at least 1.7% (calculated energy content of 3.4%). Thus, the actual share of advanced biofuels will be 3.5%, but in the calculation, it will be 7%. RED II also requires biofuel plants deployed from 2021 to reduce GHG emissions by at least 65% compared to fossil fuels.

On 14 July 2021, the Commission presented the climate targets of the EU for 2030, including a proposal to amend the RED II directive. The aim was to increase the target of 32% renewable energy to 40% in the overall energy mix of the EU by 2030. The amended RED II directive (EU/2023/2413) sets a binding target of 42.5% renewable energy by 2030, from the previous target of 32%, while aiming for reaching up to 45% (European Union, 2023). The document stresses the need to accelerate the clean energy transition while phasing out energy imports from Russia. To achieve this, it is essential that member states increase the share of renewable energy resources in energy production, industry, buildings, and transport to 45% by 2030.

Overall, it can be concluded that in order to achieve the transport targets set for 2030, the EU no longer wants to increase the share of biofuels but rather sees the reform of transport in the spread of electric cars. At the same time, in the case of internal combustion cars, it can be observed that biocomponents are still mandatory in fuels. The EU therefore does not reject biofuels entirely but calls for restrictions on the first-generation (1G) biofuels in order to reduce the impact on farmland and thus ensure sustainability. Based on this, EU decision-makers realized that biofuels alone cannot replace fossil fuels. One of the goals of the EU is to phase out the registration of internal combustion cars by 2035, which again highlights the current trend of promoting sustainable transport mainly in the spread of electric cars.

The EU's long-term goals include achieving zero emissions by 2050. To achieve this objective, it is essential to exploit renewable energy resources on an even wider scale, assuming further development of technology. Most automakers have committed to producing electric cars. Therefore, the spreading of electric cars is likely to be the fastest in the coming decades. At the same time, the EU's plans for 2030 include a complex hydrogen strategy, based on which hydrogen-powered cars are expected to play a role in the EU transport. Hydrogen cars are also being developed globally.

3 Biofuel Generations and Cars Powered by Biofuel

Biofuels are any *solid*, *liquid*, or *gaseous* fuel produced from plant or animal biomass. The main advantages of biofuels are that they are renewable energy resources, and their production does not require extraction. The base material can even be grown "locally", and the use of the new generation biofuels does not result in additional greenhouse gas (GHG) emissions.

3.1 Biofuel Generations

In transport, mainly liquid biofuels, such as biodiesel and bioethanol, have become widespread. However, the use of biogas is considered somewhat limited. Based on the raw materials and technological process used in production, four generations of biofuels (Deora et al., 2022) can be distinguished:

- **The first-generation biofuels** are produced from food and fodder crops. The first-generation biofuels typically have a negative impact on food production and indirect land-use change (ILUC).
- **The second-generation biofuels** are not made from food and fodder crops but from plant by-products with a high lignocellulose content. Their clear advantage is that food production is not hampered, as it is mainly produced from plant by-products (e.g., corn stalks, grain straw, forest by-products, wood industry waste) (Carpio & De Souza, 2017). Economically, the production of the second-generation biofuels is currently not competitive with that of first-generation biofuels.
- **The third-generation biofuels** are produced using raw materials grown exclusively for biofuel production, such as algae and seaweed. Bearing in mind the limitations of other types of biofuels (Kumar et al., 2021), the production of the raw materials is highly efficient from an environmental and sustainability point of view, as it utilizes wastewater, waste heat, and carbon dioxide.
- **The fourth-generation biofuels** are produced using non-arable crops, organic waste, artificial photosynthesis using atmospheric CO_2, and thus even carbon–neutral fuel can be produced. The development of 4G biofuels is currently very rudimentary, and the technology is still in the R&D phase.

As production progresses between biofuel generations, their environmental performance continues to increase, with some challenges in terms of the economic performance. The third- and fourth-generation biofuels are generally viewed as more advantageous than the first- and second-generation biofuels.

3.2 Global Trends of Biofuels

According to the analysis by the Agrarmarkt-Informations-Gesellschaft (mbH) (UFOP, 2024), various grains, protein, sugar and fibre crops, fruits, vegetables and more were grown on about 1.2 billion hectares in 2022. Of this, around 6% of the area was cultivated for biofuel production (see Fig. 1). In most cases, biofuels are produced in countries where there is already a surplus of suitable raw material (mainly maize, soy, palm oil). Biofuel production can address the problem of surplus of certain materials and effectively reduce the need to import fossil fuels. High-quality protein fodder can be used for biofuel production as well.

According to Statista (2023), the U.S. is currently the first in biofuel production (biodiesel and bioethanol combined), followed by Brazil. These two countries accounted for more than 60% of total biofuel production in 2023 (see Fig. 2). It is important to mention that the feedstock for biofuel production differs between the countries shown in Fig. 2, as Brazil uses sugar cane, China and the United States use maize, India mainly molasses, and EU member states mainly cereals and sugar beet. It is clear that Germany, which plays a prominent role within the EU, accounts for only 3.5% of global production. Moreover, the four largest producers in the EU account for only 8.3% of global biofuel production, while accounting for almost two-thirds of the total EU production. They therefore account for just over 10% of the total production as a whole. The easiest way to increase production would be to bring in additional farmland for biofuel production. At the same time, this would contradict to the current objectives of the EU, since according to the RED II Directive, the aim would be to reduce the use of land for biofuel production. It is possible

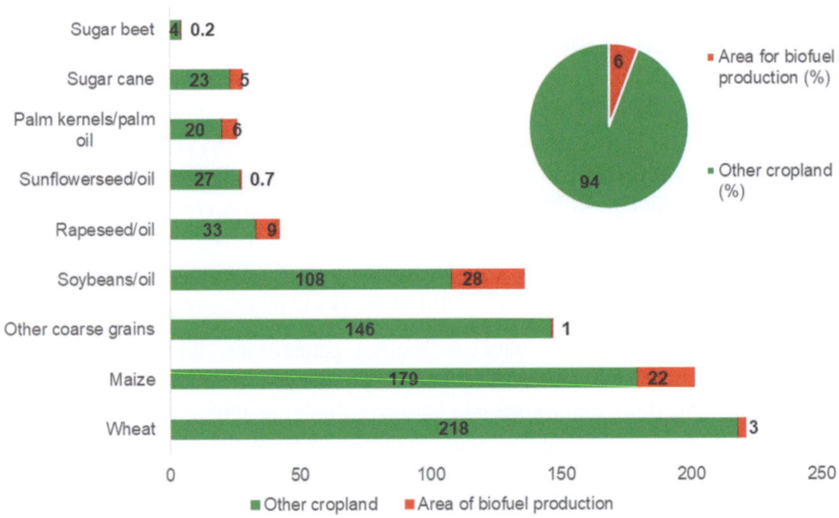

Fig. 1 Shares of global cropland planted with selected crops for biofuel production in 2022 (arable + permanent crops). *Source* Own editing based on UFOP (2024)

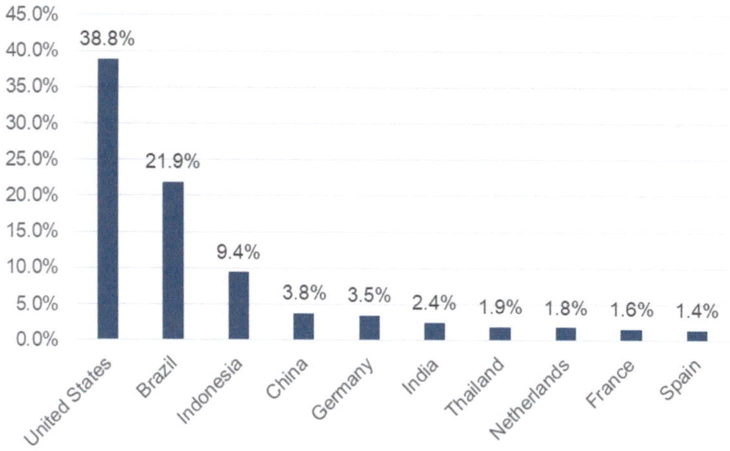

Fig. 2 Leading biofuel producing countries worldwide in 2023 by share of global production. *Source* Own editing based on Statista (2023)

to increase yields without involving additional production areas, but a number of questions may arise in this regard. Raw materials with a low ILUC risk, in particular forest or agricultural waste, as well as crops specifically intended for energy production, seem to be promising options.

The U.S. is able to achieve such significant results in biofuel production because high-yield GMO crops (mainly maize and soybean plantations) are used for biofuel production. Due to health risks, food safety and environmental impacts, the cultivation of GMO crops is a controversial topic, even if they were grown exclusively on land intended for biofuel production. At the same time, along with the U.S., Brazil and Argentina use large quantities of GMO crops for biofuels. Currently, their production in the EU is strictly regulated and varies from one member state to another.

The EU has recognized that achieving the emissions targets of the 2015 Paris Agreement (UNFCCC, 2015) is not possible without decarbonising the transport sector. Transport is responsible for more than 25% of greenhouse gas emissions in the EU. According to the objectives of European Commission (2020a), reducing emissions from the transport sector is therefore key to achieving climate neutrality in the EU by 2050. As electromobility takes longer to take off, a widespread use of biofuels is inevitable.

3.3 *Biofuel Production Growth and Food Production Security*

Biofuels can be considered a transitional reservoir of CO_2 while they are used or can even be regarded as carbon–neutral fuels. It is then could be asked why biofuels do not play a greater role in the energy market. The main reason is that the total area of land that can be used for agriculture is finite. The expansion of this land may result in significant environmental damage (e.g., burning of virgin forests). Another problem may be that newly obtained agricultural land is quickly depleted and its productivity deteriorates. According to the study of Ripple et al. (2024), the world's population is currently growing by about 200,000 people per day, a fact that seriously affects food security, among other issues. According to the report by UN-FAO (2024), around 733 million people suffered from hunger in 2023, affecting one in eleven people worldwide and one in five in Africa. According to the report of the World Resource Institute (Searchinger & Heimlich, 2015), the world faces a "food shortage" of around 70%, taking into account the amount of plant calories available in 2006 and calories needed by 2050, provided that the market penetration of biofuels globally would not significantly exceed the current share of 2.5%. The ambition for bioenergy to cover 20% of the world's total energy demand by 2050 would mean that humanity would have to at least double the world's annual production of plant raw materials in all its forms. This is both unrealistic and unsustainable. Against this background, large-scale production of biofuels that could completely replace current fuels is not a viable option from an economic, technical, environmental, or moral points of view.

4 Methodology

The aim of the procedure is to develop a uniform and easily reproducible methodology for assessing the situation of electromobility in the member states of the European Union, with a particular focus on the key indicators related to sustainable transport. The methodology focuses on three key dimensions: spreading of electric vehicles (EV penetration), the development of charging infrastructure, and the share of renewables in the energy mix. These indicators provide a comprehensive picture of the level of development of sustainable transport and its contribution to decarbonization goals, as supported by analyses by Moro and Lonza (2018) and IEA (2022).

Studying the three fundamental dimensions of sustainable transport—EV penetration, charging infrastructure, and the share of renewable energy—together is also essential because they represent different but complementary aspects. EV penetration focuses on modernizing the vehicle fleet, charging infrastructure on accessibility and usability, and renewable energy on reducing greenhouse gas emissions. The complementary role of these three dimensions is supported by several studies, such as those of IEA (2022) and Hall and Lutsey (2020). It is important to note, however, that while the proposed methodology provides a comprehensive picture of sustainability indicators for individual transport means, it does not cover a detailed analysis of public transport, freight transport, or micromobility.

4.1 Data Sources

The analysis was based on the database published by the European Alternative Fuels Observatory (EAFO), which contains the key information on the deployment of electric vehicles, the development of the charging infrastructure, and the share of renewable energy sources in the energy mix, thus providing an accurate picture of the situation in each country (EAFO, 2023). The analysis took into account the dynamic development rate of the sustainable transport market, which can vary significantly from country to country even in a short period of time. Therefore, the results can be interpreted rather as temporary snapshots reflecting the current state of affairs, which serve as a reliable basis for subsequent comparisons and trend studies, as suggested by the work of Chen et al. (2016) and IEA (2022).

4.2 Selection of Indicators

EV penetration (number of electric vehicles per a thousand people)

The penetration of electric vehicles, battery electric vehicles (BEVs) and plug-in hybrid electric vehicles (PHEVs), is one of the most important indicators of electromobility transition, as it directly refers to the extent of the transition away from internal combustion engines. The use of an indicator per 1,000 inhabitants eliminates biases due to population size. Therefore, the results of countries with different populations can be more accurately compared, as demonstrated by Sierzchula (2014), Hardman (2018), and Wappelhorst et al. (2020).

Development of charging infrastructure (number of chargers per a thousand people)

A well-developed and accessible charging infrastructure is an essential condition for the spreading of electric vehicles. Although charging points per 1,000 EVs were to be tested initially, this provided apparently distorted results in the case of smaller EV markets. Therefore, a population-proportional approach was adopted, as it provides a more accurate picture of accessibility and usability of infrastructure (Chen et al., 2016; Hall & Lutsey, 2020).

Share of renewable energy resources in the energy mix (%)

The environmental benefits of electric vehicles will only be fully realized if all or at least a significant part of the electricity they use comes from renewable resources. A low share of renewable energy can lead to significant indirect emissions due to electricity from fossil resources (IEA, 2022; Moro & Lonza, 2018).

Equal weighting of indicators used in the assessment of sustainable transport allows results to remain dimension-independent, i.e. no indicator dominates the assessment. This is particularly important as different types of data, such as vehicle penetration, infrastructure density, and energy mix, are analyzed during the assessment. The balanced approach thus ensures objective comparison, as emphasized in the studies by Sierzchula (2014) and Moro and Lonza (2018).

During the ranking, a scale of 1–27 was created based on each indicator, where 1 indicates the best result and 27 indicates the weakest result. The sum of the separate rankings by the three indicators formed the final score of the countries. This final score reflects each country's sustainability performance (see Table 1), making it easier to identify the development directions and priorities, as shown by the analyses of Sierzchula (2014) and IEA (2022).

This methodology provides an objective basis for assessing the sustainability performance of each country. The ranking can be used to identify countries with the greatest potential for development, as well as those that require further development in order to achieve sustainable transport goals, or even to identify which of the three dimensions should be given more focus.

4.3 Creating Clusters

Based on the final scores, the countries were classified into four groups:

(1) **Developing Markets**: Technologically advanced countries with high EV penetration and charging infrastructure where renewable energy is also significant.
(2) **Fast-Growing Markets**: Rapidly developing countries where infrastructure expansion and the share of renewable energy resources have not yet reached the desired level. These countries can be clearly separated into growing markets and unsustainable fast-growing markets.
(3) **High Potential Markets**: EV penetration is low; infrastructure and renewable energy use are still in their infancy.
(4) **Emerging Markets**: Less developed markets where EV technology and infrastructure are still emerging.

4.4 Methodological Assessment and Recommendations

The methodology presented provides a fundamentally simplified but informative framework for assessing sustainable transport in the EU member states. Due to the equal weighting, the method can be flexibly applied to compare countries with different levels of development. For example, for a country with a developed infrastructure, EV penetration may be the biggest barrier, while in other countries increasing the share of renewable energy is a priority. Thanks to the equal weighting, the method provides an opportunity to identify the development needs in different sub-fields in a balanced way, as proposed by Mohammed and Horváth (2023). This

Table 1 Sustainable transport indicators for the EU member states in 2023. Own editing based on EAFO (2023)

Category	Country	EV (BEV+PHEV combined) per 1,000 inhabitants	Chargers (Slow+Fast combined) per 1,000 inhabitants	Renewable Energy (%)	Final Points
Developing Markets	Denmark	56.8	3.95	59.0	6
	Sweden	56.4	3.58	65.1	9
	Finland	34.8	2.03	49.5	16
	Austria	25.2	2.09	38.0	20
Fast-Growing Markets	Portugal	20.0	0.71	38.4	26
	Netherlands	41.8	8.27	16.4	27
	Luxembourg	69.5	3.66	13.9	31
	Belgium	36.9	3.84	15.5	32
	Germany	29.7	1.45	21.7	32
	France	23.6	1.77	21.4	34
High-Potential Markets	Slovenia	6.8	0.76	25.6	38
	Lithuania	7.0	0.47	29.7	39
	Italy	8.6	0.69	22.3	40
	Spain	7.8	0.64	24.5	40
	Estonia	4.3	0.45	32.8	41
	Ireland	19.3	0.56	18.8	44
	Latvia	3.2	0.28	43.6	46
Emerging Markets	Greece	3.2	0.30	24.0	56
	Slovakia	2.8	0.44	19.5	57
	Croatia	2.1	0.27	31.1	57
	Czech Republic	3.3	0.44	17.0	58
	Hungary	6.8	0.34	14.5	61
	Romania	2.6	0.14	26.3	62
	Bulgaria	1.1	0.24	25.3	63
	Malta	14.7	0.20	12.6	64
	Cyprus	2.8	0.37	15.7	65
	Poland	2.7	0.16	18.2	70

Notes In the table, the colors indicate the level of development with respect to each indicator. Green means that the given indicator plays a significant role for the member country. Red is intended to show the need for urgent intervention. The intermediate colors, along with their various shades, indicate whether intervention is necessary and to what extent—the closer they are to red, the more urgent the intervention is needed

approach therefore guarantees objective and comparable results that support the development of effective sustainable transport strategies and the achievement of the decarbonization targets of the EU, as pointed out by Hardman (2018), Moro and Lonza (2018), and Hall and Lutsey (2020).

4.4.1 Significance and Application of Temporal Trend Monitoring

The integration of temporal trend monitoring significantly increases the practical value of the methodology. Indicators such as EV penetration, the development of charging infrastructure, and the share of renewable energy are regularly updated to quantitatively measure the rate of development and progress of countries towards sustainable transport. This approach is particularly important for achieving the long-term goals of the EU, including the net-zero emissions by 2050. The analysis of trends over time allows not only to compare the development directions, but also to identify development paths with different rates, as emphasised by Chen et al. (2016), Wappelhorst et al. (2020), and Mohammed and Horváth (2023). In addition, the impact of policy decisions for decarbonization can be assessed through trend monitoring if multi-year values are available, and on this basis, policymakers can even take further measures to change directions.

4.4.2 Evaluation of Differences Between the Markets of Each Country

The speed and direction of electromobility transition vary from country to country. Therefore, one of the main advantages of the proposed methodology is that it allows for the quantitative identification of different development patterns. For example, a slow increase in the EV penetration may point to regulatory or infrastructure development gaps, while a rapid increase in the share of renewables may point to successful energy strategies, as demonstrated by analyses by Wappelhorst et al. (2020) and Hall and Lutsey (2020). Therefore, if the EV penetration in a particular country shows stagnation, it may indicate that additional incentive measures are needed for the wider adoption of EVs. Similarly, slow expansion of charging infrastructure may indicate that the infrastructure development is not given sufficient priority, as highlighted in the study by IEA (2022).

The analysis of trends over time also supports the identification of priority areas as a basis for targeted policy interventions. Furthermore, the methodology is suitable for evaluating policy impacts and reviewing their usefulness. For example, for some countries, the focus may be on the rapid expansion of renewable energy resources, while in others, infrastructure development or vehicle price reductions may be the most effective strategies. This allows for the better tailoring of the transport decarbonization strategies of the EU to the needs of individual member states, as the analyses of Moro and Lonza (2018) and Hall and Lutsey (2020) have shown.

4.4.3 Methodological Recommendations

A number of recommendations can be provided to ensure the continuous effectiveness of the proposed methodology over the next years, including the following:

(1) ***Introduction of regular data updating***: data on the EV penetration, charging infrastructure, and the share of renewable energy resources can be updated regularly in order to make temporal trends accurately measurable.
(2) ***Graphical visualisation of trends over time***: visualisation of data—in the form of temporal graphs and interactive maps or others—may help the work of political decision makers as the development directions and defects can be understood easier.
(3) ***Supporting international comparisons***: apart from the performance of the EU member states, the methodology could be applied in global comparisons as well in the course of, for example, parallel analysis with other developed economies (U.S., Japan, South Korea).

The proposed methodology is informative and flexible. At the same time, making the assessment of the current situation of electromobility and the detection of trends over time are possible. Based on the results, the development differences between countries can be identified objectively, supporting targeted political interventions and the achievement of the long-term sustainability goals of the EU.

5 Alternatively Powered Cars in the European Union

5.1 Electric Car Penetration and Charging Networks in Europe

With the rise of EVs and the dynamic growth of the alternative fuels transport market, the European transport sector is moving in new directions that will facilitate transport transformation and directly support sustainability objectives. In addition to the growing role of environmental awareness in the EU member states, emission reduction standards are becoming stricter, and thanks to government subsidies, electric and alternatively powered vehicles are increasingly becoming the determining factor of future transport solutions. The aim of the following market analysis is to present the current situation and development stages of electromobility in Europe with a primary focus on electric cars based on the latest data and trends, highlighting the market structure of EVs, the level of the development of infrastructure, and the importance of renewable energy resources.

Within the countries of EU27, the electric vehicle market has become a central issue in the transport sector, shaped by environmental goals, transport development strategies, and EU regulatory frameworks. The continuous expansion of charging infrastructure is progressing in parallel with the growth of the electric car fleet, which is also significantly supported by the increasing share of renewable energy resources. According to the database of EAFO (2023), in 2020, there were 156,000 slow and 14,000 fast charging stations across EU27, and by 2023 these numbers have risen to 550,000 and 81,000 respectively, significantly improving the availability of sustainable charging options. The share of renewable energy sources in the EU is constantly increasing, which is essential for the sustainable transformation of the transport sector.

In 2023, 24.5% of energy consumed in the European Union was obtained from renewable resources, an increase from 23.0% in 2022. In the transport sector, the share of energy from renewable sources also increased, reaching 10.8% in 2023 compared to 9.6% in 2022. This increase is an important step towards achieving the EU's sustainability and decarbonization goals, in particular with regard to the transformation of the transport sector (Eurostat, 2024). The TEN-T (Trans-European Transport Network) directive encourages the deployment of electric and alternative fuel charging networks to ensure sustainable transport in Europe in the long term. The Alternative Fuel Infrastructure Directive (AFID) of the European Commission obliges member states to develop national action plans for the deployment of alternative fuel infrastructure, including the servicing of electric, hydrogen, and natural gas vehicles.

5.2 *Market Analysis*

5.2.1 Developing Markets

The European electric vehicle market is developing at different rates from region to region, reflecting the technological and infrastructural development of each country. Developing markets include countries which are outstanding in all aspects and where sustainable transport is well-founded. Mature markets are examples of successful integration of sustainable transport, where the spreading of electric vehicles not only benefits the environment but also supports sustainable development at economic and social levels. These countries have become the key players in sustainable transport, supported not only by public incentives and a favourable regulatory environment, but also by increasing social awareness and active participation. In such markets, electromobility will become part of everyday life, serving as a model for other regions that are also striving for sustainability. Developing markets are excellent examples of a combination of green energy and electric vehicles that work in line with global and regional policies for sustainability, making significant progress in decarbonizing the transport sector. Developing market countries have everything they need for sustainable transport if they continue along this line (i.e., continue to expand the EV fleet at an outstanding pace, while developing the density and number of charging points at a satisfactory pace and increasing the share of renewable energy even further). The foundations of sustainable transport are now developed, and further expansion of renewable energy resources can greatly contribute to increasing sustainability. At the same time, the spreading of EVs will greatly contribute to reducing emissions.

5.2.2 Fast-Growing Markets

Fast-growing markets include European countries where the electric vehicle fleet is growing dynamically. The integration of renewable energy resources is essential for these countries to support the spreading of electric vehicles and the development of a green transport sector in a sustainable way. Changing the energy mix, expanding the charging network and related infrastructure are essential for the sustainable development of electric vehicles. In areas where the share of renewable energy sources is lower, the promotion of green transport requires more serious interventions. In such countries, government incentives, policies to increase energy efficiency, and the integration of renewable energy resources are essential to ensure the long-term sustainability of electric cars. Measures such as EV charging incentives, subsidies for the development of green transport infrastructure, and incentives for renewable energy use are all important ones to ensure that sustainable transport not only supports the growth of the EV fleet but also makes a real contribution to meeting environmental goals.

The countries of fast-growing markets are divided into two sub-types based on their main parameters: growing markets and unsustainable fast-growing markets. Portugal, Germany, and France are included in the growing markets sub-type, as in these three countries, all the three factors are growing together. Thus, countries are moving towards sustainable transport, although not at the fastest pace, and the share of renewable energy resources is also adequate. The countries of unsustainable fast-growing markets are the Benelux countries, where the biggest problem is the low share of renewable energy. In these countries, increasing the share of renewable energies is particularly important for sustainable transport (as well as for the sustainable economy). Therefore, in the Benelux countries, government incentives need to be introduced to significantly increase the use of renewable energies, thereby increasing the share of renewables in the energy mix.

5.2.3 High-Potential Markets

High-potential markets include European countries where the share of renewables mostly creates favourable conditions, and it serves as the basis for sustainable transport. However, the spreading of EVs in these countries is not progressing fast enough. In terms of the level of the development of charging infrastructure, there is space for improvement in these countries as well. The advantage of high-potential markets is that they quickly adapt to policies aimed at sustainability and carbon reduction, thus directly benefiting from the high share of renewable energy resources. By committing to low-carbon transport solutions, these countries can sustain and accelerate growth in the market. Sustainable transport is well established in high-potential markets countries, thanks to the high share of renewables, but government incentives are needed to promote EVs (and thus charging points). The only exception is Ireland, where the population is committed to EV use, but energy policy measures currently have a room for improvements. Therefore, in case of Ireland, it is necessary to introduce appropriate measures as soon as possible in order to ensure clean/green transport.

5.2.4 Emerging Markets

As for the countries in the "emerging markets" category, the EV market is still in its early stages, with low EV penetration, smaller shares of renewables, and less developed charging infrastructure—at least two of the three indicators are particularly disadvantageous in the countries of this cluster. In order to stimulate the development of electric mobility, it is essential to expand charging infrastructure and further use renewable energy sources, which can provide opportunities for sustainable transport and environmentally friendly transport goals. The countries of the emerging markets are equally disadvantaged in the relevant respects. Poland closes the line in the ranking, mainly due to the low number of EVs. At the same time, it should be mentioned that the Polish population is open to the use of alternative fuels, and liquefied petroleum gas (LPG)-powered cars are particularly widespread there. Furthermore, the share of zero-emission buses, including electric buses, is extremely high in the country, thanks mainly to the Polish company Solaris, which is one of the most successful EV bus manufacturers in Europe. In 2024, 1,472 EV buses were registered in Poland, an increase of around 25% compared to the previous year, according to Statista (2025).

Croatia, Romania, and Bulgaria should be highlighted due to their relatively high share of renewables, as well as Malta and Hungary in terms of the number of EVs. At the same time, Malta has a very low share of renewables in the overall energy mix, and the number of charging points cannot fully meet demand, thus energy policy ambitions are certainly justified. The rest of the countries are lagging behind in terms of EVs per 1,000 inhabitants, charging points per 1,000 inhabitants, and the share of renewables. In these countries, significant changes and support are needed to kick-start the spread of electric cars and to implement it in a sustainable way. The deployment of adequate charging infrastructure and the wider use of renewable energy resources are essential for the growth of EV fleets in these countries, contributing to the achievement of sustainable transport goals and the achievement of the EU environmental goals.

5.3 *Relationship Between Electric Cars and the Energy Mix*

The spreading of electric vehicles (EVs) in the European transport system is closely linked to the development of charging infrastructure. The EU member states have taken significant steps in recent years to promote the use of EVs, in particular by expanding charging networks (Cieslik et al., 2021). With the growing EV fleet, energy balance and network load are becoming increasingly important. Cieslik et al. (2021) pointed out that ensuring a balance between electricity generation and consumption is essential for the spreading of EVs, especially with the integration of an increasing share of renewable energy resources. Factors affecting network stability include the key system characteristics, as it must remain constant at 50 Hz in the European system (Di Natale et al., 2021). Renewable energy resources, such as solar and wind energy, often provide fluctuating production levels, which can also affect frequency stability, especially for sensitive electrical equipment. Wang et al. (2024) discussed how blockchain-based models can support distributed energy systems in industrial park microgrids to address these challenges.

The ability to react quickly and incorporate storage solutions is key to stabilizing fluctuations in renewables and the energy balance. Storage systems, such as batteries, allow energy to be stored during periods of low demand and fed back into the network during peak periods. Huang et al. (2020) explored a multi-objective optimization strategy for distribution networks that integrates vehicle-to-grid-enabled EVs to enhance energy system efficiency.

The additional energy demand during peak periods of EV charging poses a challenge to electricity transmission and distribution networks. Energy is often transferred over long distances, which requires infrastructure with sufficient capacity and reliability, especially if the charging needs to be increased significantly. According to the findings of Di Natale et al. (2021), rapid response systems, such as gas power plants, are necessary for network stability, which can supply electricity demand in a short period of time, thereby contributing to maintaining the balance of the network. Overall, the balance between electricity generation and distribution can be ensured if the system is able to dynamically adapt to consumer demand and changes in renewable energy production. El-Hendawi et al. (2022) emphasized the importance of centralized and distributed optimization strategies for vehicle-to-grid applications, particularly in frequency regulation, to achieve a stable and reliable network that supports the widespread use of EVs and sustainable transport goals.

5.4 Vehicle-to-Grid (V2G) Technology and Energy System Stability

Vehicle-to-Grid (V2G) technology acts as a special solution for energy storage that allows electric vehicle (EV) batteries to be used as temporary energy storage. Through this technology, EVs are able to feed stored energy back into the network during peak periods, stabilizing the power system and reducing the network load when energy demand exceeds supply (Bianchi et al., 2023). V2G systems can be utilized to optimize the use of renewable energy resources, such as solar and wind energy. Wan et al. (2024) highlighted that these systems enable dynamic energy flow between EVs and the network, enhancing grid stability and providing the flexibility required to maintain the energy balance.

The development and widespread use of V2G technology is particularly significant in urban and industrial environments, where increasing energy efficiency and ensuring sustainable energy supply are essential. The effectiveness of V2G technology largely depends on the aggregate management of EVs, which allows vehicle fleets to collectively support the reduction of network load, especially during peak periods. According to Amamra and Marco (2019), aggregate systems provide economic benefits for both users and the network, as EV owners can be compensated for the energy regenerated, providing an additional incentive to use the system.

However, the spreading of V2G systems requires further regulatory and infrastructure improvements that allow bidirectional energy flows between electric vehicles and the power network. Escoto et al. (2024) stated that the development of smart energy distribution systems capable of dynamically managing network loads and integrating fluctuating production from renewable energy resources is essential for achieving sustainable transport and low-emission energy supply.

5.5 *Hydrogen Power in the European Union*

In addition to electromobility, another possible alternative to achieving environmentally friendly transport is technological solutions using hydrogen. The current state of technology makes it possible to utilize hydrogen to revolutionize transport. The really difficult question, however, is to what extent large-scale production of "green" hydrogen at a competitive price can be an achievable objective. The European Union adopted its hydrogen strategy at the end of 2020, which, together with REPowerEU (2022), proposed a comprehensive framework to support the spreading of low-carbon and renewable hydrogen to help the decarbonization of the EU. The European hydrogen policy framework was first suggested by the European Commission in July 2021 as part of the "Fit for 55" package and contains restrictive targets for the industrial and transport deployment of renewable hydrogen by 2030. This framework is based on the European Commission's hydrogen strategy (COM/2020/301) building on 20 key hydrogen targets in 5 areas (European Commission, 2020b).

The hydrogen strategy of the EU aims to make hydrogen a key player in the EU's energy transition and contribute to achieving climate goals, in particular, reducing carbon emissions. The objectives of the strategy include the following:

1. **To increase the role of hydrogen in energy production and use**: The aim is to make hydrogen a major player in industry, transport, and energy storage.
2. **To increase the production of renewable hydrogen**: The EU aims to produce 10 million tonnes of renewable hydrogen per year and import the same amount by 2030.
3. **To develop sustainable infrastructure**: The aim is to build optimal hydrogen infrastructure, including hydrogen supply chains and storage solutions.
4. **Market development for hydrogen**: The aim is to create a competitive and functioning hydrogen market that facilitates trade and the use of hydrogen in the EU.

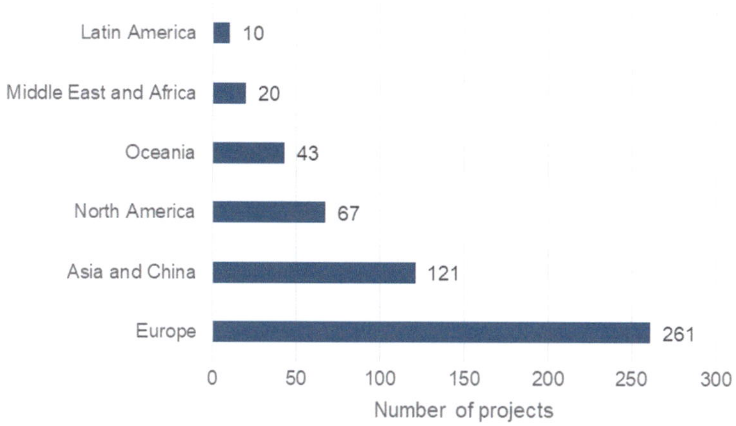

Fig. 3 Number of hydrogen projects announced worldwide by region as of 2021. *Source* Own editing based on Hydrogen Council (2021)

5. **Fostering innovation and research**: Supports hydrogen innovations and research.

These goals are in line with the climate goals set by the EU, particularly the achievement of a carbon-free future by 2050. Europe accounted for roughly half of global investments and projects in 2021 for most of the hydrogen developments and projects announced (see Fig. 3). Therefore, the EU has serious plans for hydrogen, which, assuming the state of technology, could also play an important role for transport in the future.

6 Critics in Relation to the Transport Policy of the European Union

The renewable energy directives of the EU in transport have repeatedly encountered problems. While the targets are constantly increasing, it is important to note that the targets have not been achieved smoothly. In 2007–2008, the food price crisis sparked a heated debate about the use of food as fuel to power the cars of the rich (Tenenbaum, 2008; Thompson, 2012). In addition, evidence was gathered that plant-based fuels are not capable of significant GHG savings (Correa et al., 2019; Fargione et al., 2008; Searchinger et al., 2008). In 2012, recognizing that biofuels may not contribute to the desired GHG reduction due to the ILUC effect and other adverse effects, such as loss of biodiversity, depletion of water resources, water pollution, the European Commission proposed to limit the share of the first-generation biofuels to 5% by 2020 (European Commission, 2012). However, this proposal was not supported by the Council and Parliament, and therefore, the limit was raised to 7% while ignoring the

ILUC effects in the calculation of GHG emissions (Asveld, 2016; Cadillo-Benalcazar et al., 2021).

A statement from the European Biodiesel Board accused the Commission of "purposely causing the death of the entire EU biodiesel industrial sector" (Euroactive, 2013), as the ILUC proposal would close a large number of production sites worth billions of euros and result in the loss of hundreds of thousands of jobs. According to the estimation of Charles et al. (2013a, 2013b), the EU spent roughly €5.5–6.9 billion on biofuels in 2011, while the total value of the EU biofuel market that year was estimated at €13–16 billion. With the amendment of RED and later the adoption of RED II, the first-generation biofuel production has clearly been pushed into the background. The new version supports and encourages advanced cellulosic biofuels. However, the question arises whether the cost of developing advanced biofuels is acceptable, since we are facing persistent difficulties in these respects and only double accounting enables targets to be met. So far, it seems that the use of biofuels cannot really replace fossil fuels.

The EU wants to phase out the registration of cars with internal combustion engines by 2035. It also wants to completely abolish transport using internal combustion engines by 2050 and, according to current prospects, intends to reform transport with electric cars. The EU countries have different levels of development, both in economic terms and in terms of sustainable transport indicators. Many member states have a particularly low share of renewable energies in the energy mix, which is the most important factor in sustainable transport (Abdelsattar et al., 2024; Alam et al., 2023; Sinha et al., 2023). For example, Malta, Hungary, and Luxembourg are lagging behind significantly in the share of renewables. To develop these countries, significant investments are required, and achieving zero emissions requires a significant development and pace of renewable resources and capacities. The EU still obliges all countries to the increasing use of EVs, even though the low share of renewables in many member states is a vital problem, making electric cars not an option for an environmentally friendly alternative. For this reason, it is particularly important to monitor the development of sustainable transport indicators in the EU member states, and in case of slower pace of development, to promote the development of certain factors with various incentives and tax relieves.

The application of the methodology proposed in this study can be particularly useful in developing guidelines that differ from country to country and in monitoring the effectiveness of these measures with the explicit aim of developing sustainable transport in each EU country. Therefore, different development paths can be established for the member states, which could help achieve the 2050 decarbonization targets more efficiently and quickly for each member state.

7 Conclusion

The aim of this study was to analyze the current situation in each EU country based on an overview of the relevant transport policies, examining the most fundamental factors of sustainable transport (i.e., number of EVs per 1,000 people, number of charging points per 1,000 people, share of renewable energy). Biofuels were the subject of the EU transport policy action, which the EU had previously seen as the way for green transport reform. In terms of the use of biofuels, the EU is currently committed to the production and utilization of advanced biofuels, regardless of the fact that this is not an economical and competitive solution (only the first-generation biofuels can compete economically with fossil fuels). The EU decided to do so because the first-generation biofuels have been criticized for their ILUC effects. Therefore, the first-generation biofuels will be phased out by 2030, while advanced biofuels will be accounted for based the existing targets. Biofuels are unable to replace fossil fuels for a number of reasons (i.e., ILUC effects, economic issues, finite farmland). The inclusion of GMO crops may increase the competitiveness of biofuels in the future, but there is currently a lot of controversy surrounding these crops, mainly because of the various health risks.

The EU wants to gradually phase out the registration of vehicles with internal combustion engines by 2035, based on which the "zero emissions" targets set for 2050 are planned to be solved specifically with electric cars. By using electric cars, it is indeed possible to achieve sustainable transport if the electricity—which is used as fuel—is generated from purely renewable energy resources. The further spreading of electric cars could cause significant problems for the stability of the electricity network in some countries, if the charging of electric cars does not develop into an off-peak charging habit. At the same time, with the widespread use of V2G technology, EVs can significantly help in reducing network fluctuations (and thus network peaks).

The aim of the EU is therefore to promote strongly the spreading of EVs, reform the transport sector, and make transport sustainable with the help of electric cars. For this reason, we examined the current situation in the EU member states using the relevant indicators. Based on the three indicators examined (number of EVs per 1,000 people, number of charging points per 1,000 people, share of renewable energy), a total of four clusters were formed for the member states: developing markets, fast-growing markets, high-potential markets, and emerging markets. Member states in each market are on a similar level of development in terms of sustainable transport. The countries of developing markets are particularly advanced in all the three indicators. The fast-growing cluster is further divided into two sub-types according to the main characteristics. In the countries of growing markets, the spreading of EVs and charging points is at a slower pace and needs to catch up; at the same time, the share of renewables is fairly high, thus providing good conditions for sustainability. In contrast, in the countries of unsustainable fast-growing markets, the number of EVs and that of charging points are particularly high, while the share of renewables is low. The renewable share in countries of high potential markets is also very high and therefore has great potential, and the foundations for sustainable transport are available. The countries of emerging markets are lagging behind in all respects and require urgent intervention to achieve sustainable transport.

All in all, it is clear that the EU faces many challenges, as the member states are in very different situations when it comes to sustainable transport, taking into account the three factors studied here. In order to implement the "zero emission" plan set for 2050, an individual plan must be developed for each country. It is usually the countries with weaker economic indicators that have a less developed sustainable transport system; therefore, these require increased attention. With the methodology proposed in this study, it is possible to monitor the development of sustainability goals for each member state year by year.

The EU's ability to adapt to achieving sustainable transport is illustrated by the shift from encouraging the widespread use of biofuels to the "green" production of electric and, more recently, hydrogen-powered vehicles, the infrastructure elements supporting them, and sustainable energy sources (electricity, hydrogen) in a relatively short period of time. It is clear that the EU is open to different technological solutions that can contribute to the sustainability of transport and, if expectations are not met, change its transport development policy. Directives adopted at the EU level, binding member states, and the EU funding provided for the purpose of achieving strategic goals play a key role in achieving transport development goals and establishing sustainable transport.

References

Abdelsattar, M., Ismeil, M., Aly, M. M., & Abu-Elwfa, S. S. (2024). Analysis of renewable energy sources and electrical vehicles integration into microgrid. *IEEE Access, 12*, 66822–66832. https://doi.org/10.1109/ACCESS.2024.3399124

Alam, M. S., Chowdhury, T. A., Dhar, A., Al-Ismail, F. S., Choudhury, M. S. H., Shafiullah, M., Hossain, M. I., Hossain, M. A., Ullah, A., & Rahman, S. M. (2023). Solar and wind energy integrated system frequency control: A critical review on recent developments. *Energies, 16*(2), 812, 1–31. https://doi.org/10.3390/en16020812

Amamra, S.-A., & Marco, J. (2019). Vehicle-to-grid aggregator to support power grid and reduce electric vehicle charging cost. *IEEE Access, 7*, 178528–178538. https://doi.org/10.1109/ACCESS.2019.2958664

Asveld, L. (2016). The need for governance by experimentation: The case of biofuels. *Science and Engineering Ethics, 22*, 815–830. https://doi.org/10.1007/s11948-015-9729-y

Bianchi, F., Falsone, A., & Vignali, R. (2023). Vehicle-to-grid and ancillary services: A profitability analysis under uncertainty. *IFAC-PapersOnLine, 56*(2), 7077–7083. https://doi.org/10.48550/arXiv.2309.11118

Boudmen, K., El ghazi, A., Eddaoudi, Z., Aarab, Z., & Rahmani, M. D. (2024). Electric vehicles, the future of transportation powered by machine learning: A brief review. *Energy Inform, 7*(80), 1–19. https://doi.org/10.1186/s42162-024-00379-3

Bridi, R. M., Ben Jabra, M., Al Hosani, N., & Almurshidi, A. H. (2024). The propensity to adopt electric vehicles in the United Arab Emirates: An analysis of economic and geographic factors. *Sustainability, 16*(2), 770, 1–19. https://doi.org/10.3390/su16020770

Cadillo-Benalcazar, J. J., Bukkens, S. G., Ripa, M., & Giampietro, M. (2021). Why does the European Union produce biofuels? Examining consistency and plausibility in prevailing narratives with quantitative storytelling. *Energy Research & Social Science, 71*(101810), 1–14. https://doi.org/10.1016/j.erss.2020.101810

Carpio, L. G. T., & De Souza, F. S. (2017). Optimal allocation of sugarcane bagasse for producing bioelectricity and second generation ethanol in Brazil: Scenarios of cost reductions. *Renewable Energy, 111*, 771–780. https://doi.org/10.1016/j.renene.2017.05.015

Charles, C., Gerasimchuk, I., Bridle, R., Moerenhout, T., Asmelash, E., & Laan, T. (2013a). *Biofuels: At what cost? A review of costs and benefits of EU biofuel policies, 2013*. Retrieved December 29, 2024 from https://www.iisd.org/gsi/sites/default/files/biofuels_subsidies_eu_review.pdf

Charles, C., Gerasimchuk, I., Bridle, R., Moerenhout, T., Asmelash, E., & Laan, T. (2013b). *Addendum to biofuels—At what cost? A review of costs and benefits of EU biofuel policies, 2013*. Retrieved December 29, 2024 from https://www.iisd.org/gsi/sites/default/files/biofuels_eu_addendum.pdf

Chen, T. D., Kockelman, K. M., & Hanna, J. P. (2016). Operations of a shared, autonomous, electric vehicle fleet: Implications of vehicle & charging infrastructure decisions. *Transportation Research Part A, 94*, 243–254. https://doi.org/10.1016/j.tra.2016.08.020

Cieslik, W., Szwajca, F., Golimowski, W., & Berger, A. (2021). Experimental analysis of residential photovoltaic (PV) and electric vehicle (EV) systems in terms of annual energy utilization. *Energies, 14*(4), 1085, 1–21. https://doi.org/10.3390/en14041085

Commission of the European Communities. (2006). *Communication from the Commission to the Council and the European Parliament. Renewable energy road map: Renewable energies in the 21st century: Building a more sustainable future*. COM (2006) 848 final. Retrieved December 29, 2024 from https://www.europarl.europa.eu/RegData/docs_autres_institutions/commission_europeenne/com/2006/0848/COM_COM(2006)0848_EN.pdf

Correa, D. F., Beyer, H. L., Fargione, J. E., Hill, J. D., Possingham, H. P., Thomas-Hall, S. R., & Schenk, P. M. (2019). Towards the implementation of sustainable biofuel production systems. *Renewable and Sustainable Energy Reviews, 107*, 250–263. https://doi.org/10.1016/j.rser.2019.03.005

Deora, P. S., Verma, Y., Muhal, R. A., Goswami, C., & Singh, T. (2022). Biofuels: An alternative to conventional fuel and energy source. *Materials Today: Proceedings, 48*, 1178–1184. https://doi.org/10.1016/j.matpr.2021.08.227

Di Natale, L., Funk, L., Rüdisüli, M., Svetozarevic, B., Pareschi, G., Heer, P., & Sansavini, G. (2021). The potential of vehicle-to-grid to support the energy transition: A case study on Switzerland. *Energies, 14*(16), 4812, 1–22. https://doi.org/10.3390/en14164812.

EAFO. (2023). *European alternative fuels observatory*. Road. Retrieved December 29, 2024 from https://alternative-fuels-observatory.ec.europa.eu/transport-mode/road

El-Hendawi, M., Wang, Z., & Liu, X. (2022). Centralized and distributed optimization for vehicle-to-grid applications in frequency regulation. *Energies, 15*(12), 4446, 1–22. https://doi.org/10.3390/en15124446

Escoto, M., Guerrero, A., Ghorbani, E., & Juan, A. A. (2024). Optimization challenges in vehicle-to-grid (V2G) systems and artificial intelligence solving methods. *Applied Sciences, 14*(5211), 1–19. https://doi.org/10.3390/app14125211

Euroactive. (2013). *Biofuels industry sent 'three mails an hour' in ILUC lobby offensive*. EURACTIV.com. Retrieved December 29, 2024.

European Commission. (2012). *Proposal for a Directive of the European Parliament and of the Council amending Directive 98/70/EC relating to the quality of petrol and diesel fuels and amending Directive 2009/28/EC on the promotion of the use of energy from renewable sources*. COM (2012) 595 final. Retrieved December 29, 2024 from https://www.europarl.europa.eu/registre/docs_autres_institutions/commission_europeenne/com/2012/0595/COM_COM(2012)0595_FR.pdf

European Commission. (2020a). *Energy, climate change, environment.* European Commission. Retrieved October 30, 2024 from https://commission.europa.eu/energy-climate-change-environment_en
European Commission. (2020b). *Key actions of the EU hydrogen strategy.* European Commission. Retrieved November 08, 2024 from https://energy.ec.europa.eu/topics/energy-systems-integration/hydrogen/key-actions-eu-hydrogen-strategy_en
European Union. (2003). *Directive 2003/30/EC of the European Parliament and of the Council of 8 May 2003 on the promotion of the use of biofuels or other renewable fuels for transport.* Retrieved December 29, 2024 from https://eur-lex.europa.eu/eli/dir/2003/30/oj/eng
European Union. (2009). *Directive 2009/28/EC of the European Parliament and of the Council of 23 April 2009 on the promotion of the use of energy from renewable sources and amending and subsequently repealing Directives 2001/77/EC and 2003/30/EC.* Retrieved December 29, 2024 from https://eur-lex.europa.eu/eli/dir/2009/28/oj/eng
European Union. (2015). *Directive (EU) 2015/1513 of the European Parliament and of the Council of 9 September 2015 amending Directive 98/70/EC relating to the quality of petrol and diesel fuels and amending Directive 2009/28/EC on the promotion of the use of energy from renewable sources.* Retrieved December 29, 2024 from https://eur-lex.europa.eu/eli/dir/2015/1513/oj
European Union. (2018). *Directive (EU) 2018/2001 of the European Parliament and of the Council of 11 December 2018 on the promotion of the use of energy from renewable sources (recast).* Retrieved December 29, 2024 from https://eur-lex.europa.eu/legal-content/EN/TXT/PDF/?uri=CELEX:32018L2001
European Union. (2023). *Directive (EU) 2023/2413 of the European Parliament and of the Council of 18 October 2023 amending Directive (EU) 2018/2001, Regulation (EU) 2018/1999 and Directive 98/70/EC as regards the promotion of energy from renewable sources, and repealing Council Directive (EU) 2015/652.* Retrieved December 29, 2024 from https://eur-lex.europa.eu/legal-content/EN/TXT/PDF/?uri=OJ:L_202302413
Eurostat. (2024). *Renewable energy statistics.* European Commission. Retrieved January 10, 2025 from https://ec.europa.eu/eurostat/statistics-explained/index.php?title=Renewable_energy_statistics
Fargione, J., Hill, J., Tilman, D., Polasky, S., & Hawthorne, P. (2008). Land clearing and the biofuel carbon debt. *Science, 319*(5867), 1235–1238. https://doi.org/10.1126/science.1152747
Hall, D., & Lutsey, N. (2020). *Electric vehicle charging guide for cities.* The International Council on Clean Transportation (ICCT). Retrieved November 08, 2024 from https://theicct.org/publication/electric-vehicle-charging-guide-for-cities/
Hardman, S. (2018). Understanding the impact of reoccurring and non-financial incentives on plug-in electric vehicle adoption—A review. *Transportation Research Part A: Policy and Practice, 119*, 1–14. https://doi.org/10.1016/j.tra.2018.11.002
Huang, Z., Fang, B., & Deng, J. (2020). Multi-objective optimization strategy for distribution network considering V2G-enabled electric vehicles in building integrated energy system. *Protection and Control of Modern Power Systems, 5*(7), 1–8. https://doi.org/10.1186/s41601-020-0154-0
Hydrogen Council. (2021). *Hydrogen for net-zero. A critical cost-competitive energy vector.* Hydrogen Council, McKinsey & Company. Retrieved November 08, 2024 from https://hydrogencouncil.com/wp-content/uploads/2021/11/Hydrogen-for-Net-Zero.pdf
International Energy Agency (IEA). (2022). *Global EV Outlook 2022, Paris.* Retrieved November 08, 2024 from https://www.iea.org/reports/global-ev-outlook-2022
Jaman, S., Abdel-Monem, M., Geury, T., & Hegazy, O. (2023). Development and validation of an integrated EV charging station with grid interfacing inverter for residential application. *IEEE Access, 11*, 115751–115774. https://doi.org/10.1109/ACCESS.2023.3323219
Kumar, B. R., Mathimani, T., Sudhakar, M. P., Rajendran, K., Nizami, A. S., Brindhadevi, K., & Pugazhendhi, A. (2021). A state of the art review on the cultivation of algae for energy and other valuable products: Application, challenges, and opportunities. *Renewable and Sustainable Energy Reviews, 138*, Article 110649. https://doi.org/10.1016/j.rser.2020.110649

Luo, F., Zhao, Z., & Rashidi, T. H. (2022). Mobile energy-to-home integration: An adaption of mobility as a service in urban energy systems. *IET Energy Systems Integration, 4*(2), 220–234. https://doi.org/10.1049/esi2.12056

Mohammed, D., & Horváth, B. (2023). Travel demand increment due to the use of autonomous vehicles. *Sustainability, 15*(8937), 1–20. https://doi.org/10.3390/su15118937

Moro, A., & Lonza, L. (2018). Electricity carbon intensity in European Member States: Impacts on GHG emissions of electric vehicles. *Transportation Research Part D: Transport and Environment, 64*, 5–14. https://doi.org/10.1016/j.trd.2017.07.012

Nordin, I., Elofsson, K., & Jansson, T. (2024). Cost-effective reductions in greenhouse gas emissions: Reducing fuel consumption or replacing fossil fuels with biofuels. *Energy Policy, 190*(114138), 1–16. https://doi.org/10.1016/j.enpol.2024.114138

Rego, N., Castro, R., & Silva, C. S. (2023). Assessment of current smart house solutions: The case of Portugal. *Energies, 16*(22), 7469, 1–23. https://doi.org/10.3390/en16227469

REPowerEU. (2022). *A plan to rapidly reduce dependence on Russian fossil fuels and fast forward the green transition Brussels, 18 May 2022*. Retrieved December 29, 2024 from https://ec.europa.eu/commission/presscorner/api/files/document/print/en/ip_22_3131/IP_22_3131_EN.pdf

Ripple, W. J., Wolf, C., Gregg, J. W., Rockström, J., Mann, M. E., Oreskes, N., Lenton, T. M., Rahmstorf, S., Newsome, T., Xu, C., Svenning, J.-C., Cardoso Pereira, C., Law, B. E., & Crowther, T. W. (2024). The 2024 state of the climate report: Perilous times on planet Earth. *BioScience, 74*(12), 812–824. https://doi.org/10.1093/biosci/biae087

Searchinger, T., Heimlich, R., Houghton, R. A., Dong, F., Elobeid, A., Fabiosa, J., Tokgoz, S., Hayes D., & Yu, T. H. (2008). Use of U.S. Croplands for biofuels increases greenhouse gases through emissions from land-use change. *Science, 319*(5867), 1238–1241. https://doi.org/10.1126/science.1151861

Searchinger, T. D., & Heimlich, R. E. (2015). *Installment 9 of "Creating a sustainable food future"*. Avoiding Bioenergy Competition for Food Crops and Land, World Resource Institute. Working paper.

Sierzchula, W. (2014). The influence of financial incentives and other socio-economic factors on electric vehicle adoption. *Energy Policy, 68*, 183–194. https://doi.org/10.1016/j.enpol.2014.01.043

Sinha, P., Paul, K., Deb, S., & Sachan, S. (2023). Comprehensive review based on the impact of integrating electric vehicle and renewable energy sources to the grid. *Energies, 16*(6), 2924, p.1–39. https://doi.org/10.3390/en16062924

Statista. (2023). *Biofuel production in Europe in 2023, by country (in 1,000 barrels of oil equivalent per day)*. Retrieved December 29, 2024 from https://www.statista.com/statistics/274169/top-biofuel-producing-countries-worldwide/

Statista. (2025). *Total number of electric buses with gross vehicle weight >3.5t in Poland from 2019 to 2024*. Retrieved January 21, 2025 from https://www.statista.com/statistics/1411992/poland-number-of-electric-buses/

Tenenbaum, D. J. (2008). *Food vs. fuel: Diversion of crops could cause more hunger*. *Environmental Health Perspectives, 116*(6), 254–257. https://doi.org/10.1289/ehp.116-a254

Thompson, P. (2012). The agricultural ethics of biofuels: The food vs fuel debate. *Agriculture, 2*(4), 339–358. https://doi.org/10.3390/agriculture2040339

UFOP. (2024). *Report on global market supply 2023/2024*.Retrieved November 28, 2024 from https://www.ufop.de/files/8217/0548/9837/UFOP-2116_Report_Global_Market_Supply_A5_EN_23_24_160124.pdf

UN-FAO. (2024). *The state of food security and nutrition in the world 2024—Financing to end hunger, food insecurity and malnutrition in all its forms*. Rome. https://doi.org/10.4060/cd1254en

UNFCCC. (2015). *Paris Agreement to the United Nations Framework Convention on Climate Change, T.I.A.S. No. 16-1104*. Retrieved December 29, 2024 from https://unfccc.int/sites/default/files/resource/parisagreement_publication.pdf

Wan, M., Yu, H., Huo, Y., Yu, K., Jiang, Q., & Geng, G. (2024). Feasibility and challenges for vehicle-to-grid in electricity market: A review. *Energies, 17*(3), 679, 1–23. https://doi.org/10.3390/en17030679

Wang, L., Zhang, Z., Fan, J., Zeng, S., Pan, S., & Chen, H. (2024). Blockchain-based joint auction model for distributed energy in industrial park microgrids. *Energies, 17*(13), 3140, 1–15. https://doi.org/10.3390/en17133140

Wappelhorst, S., Hall, D., & Lutsey, N. (2020). *Analyzing policies to grow the electric vehicle market in European cities.* The International Council on Clean Transportation (ICCT). Retrieved November 08, 2024 from https://theicct.org/publication/analyzing-policies-to-grow-the-electric-vehicle-market-in-european-cities/

Csongor Máthé graduated as geographer from the Babeş-Bolyai University, Cluj-Napoca (Romania) and received his Ph.D. degree in the field of Geography in 2011 from the same university. He is an Assistant Professor in the Department of Geography in Hungarian Language, Faculty of Geography at the Babeş-Bolyai University. His research areas include the analysis of the spatial aspects and impacts of transport infrastructure, the examination of the social and economic effects of electromobility and accessibility, as well as the study of processes related to sustainable tourism, ecotourism products, and services.

János Pénzes graduated as a geographer at the University of Debrecen, Hungary and received his Ph.D. degree in Earth Sciences in 2010 in Debrecen. He is an Associate Professor in the Department of Social Geography and Regional Development Planning at the University of Debrecen. His fields of research are the spatial inequalities, the social and economic processes of the peripheral areas, quantitative and GIS methods in the regional development besides social issues of energy consumption and renewable energy resources.

Gábor Horváth is a Ph.D. student in Earth Sciences at the University of Debrecen, specializing in Vehicle-to-Everything (V2X) technology and sustainable energy systems. His research focuses on the integration of electric vehicles with intelligent infrastructures and their role in energy supply, transportation efficiency, and grid stability. His work encompasses smart grids, mobile energy storage, renewable energy integration, and the economic, technical, and regulatory aspects of electrified mobility. His goal is to advance V2X-based systems that support energy transition, sustainable transportation solutions, and the development of intelligent energy networks.

Máté Zakar earned his geography degree from the University of Debrecen. He is currently a Ph.D. student in Earth Sciences in the Department of Meteorology at the University of Debrecen. His research mainly focuses on greenhouse gas emissions in the transportation sector. Within this area, he investigates the aspects related to biofuels, electric cars, and hydrogen-powered vehicles. The impacts of green reforms in the transportation sector are subject to complex analyses concerning economic, environmental, and practical applicability. The spread and widespread implementation of car-sharing services play an important role in the reform of transportation, as public transport can significantly contribute to reducing urban noise pollution and air pollution, as well as alleviating traffic jams.

Tamás Tóth graduated from the University of Debrecen, Hungary as a geographer and geography teacher. He received his Ph.D. degree in the field of Earth Sciences in Debrecen in 2013. He works as an Assistant Professor in the Department of Meteorology at the University of Debrecen. His research fields are the energy awareness and energy savings, the natural, social, and economic aspects of the utilization of renewable energy resources. He also deals with the issues of the usage of biomass as fuel, biogas or biofuel, besides the urban climate and low emission city transport.

Low Emission Zones as a Tool for Climate Change Mitigation and Adaptation

Marta Borowska-Stefańska, Szymon Wiśniewski, and Magdalena Magiera

Abstract The purpose of this chapter is to outline the operating principles of low emission zones (LEZs) in European countries and to evaluate their effectiveness. The impact of LEZs on urban sustainable development and climate change mitigation is assessed through a systematic review of the literature and practical applications. The chapter reviews a number of operational principles based on legal frameworks, technology-driven enforcement, and access control systems, which are essential for successful LEZ implementation. It also addresses the challenges faced and the costs of compliance for residents and businesses, emphasising the need for supportive urban policies and alternative transportation options. The study highlights significant achievements in pollutant reduction, with certain LEZ models demonstrating remarkable effectiveness in improving air quality. The key findings indicate that well-implemented LEZs can significantly reduce urban transport emissions, contributing positively to public health and environmental goals. The conclusion synthesises the primary contributions of the research, suggesting that effective LEZs are crucial for advancing health, equity, and urban policy. The chapter underscores the need to integrate LEZs with broader urban development strategies to ensure that future sustainability goals are met. The findings advocate policies that will enhance public acceptance and promote equitable outcomes.

Keywords Low emission zone · Sustainable mobility · Urban transport system · Spatial development · Local economy · Urban environment

M. Borowska-Stefańska (✉) · S. Wiśniewski · M. Magiera
Faculty of Geographical Sciences, University of Lodz, Łódź, Poland
e-mail: marta.borowska@geo.uni.lodz.pl

S. Wiśniewski
e-mail: szymon.wisniewski@geo.uni.lodz.pl

M. Magiera
e-mail: magdalena.magiera@edu.uni.lodz.pl

S. Wiśniewski
Research Center for European Spatial Policy and Local Development, University of Lodz, Łódź, Poland

M. A. Dulebenets et al. (eds.), *Transport and Climate Change*,
https://doi.org/10.1007/978-3-032-00563-2_11

1 Introduction

The World Health Organisation estimates that 99% of the global population is exposed to dangerous levels of pollution (WHO, 2022). Air pollution contributes to a rise in diseases—primarily cardiovascular diseases—and premature mortality (Burnett et al., 2018). Substantial evidence indicates that high levels of air pollution also carry significant social costs, including increased rates of school absenteeism (Chen et al., 2018; Currie et al., 2009), diminished academic performance (Mohai et al., 2011), decreased productivity (Chang et al., 2019; He et al., 2019), and reduced life expectancy (Liu et al., 2024). The transport industry is a significant contributor to air pollution, accounting for over 40% of emissions (Anderson, 2020; Anenberg et al., 2019; Khomenko et al., 2023). These transport-related emissions are particularly pronounced in urban areas (Zhai & Wolff, 2021).

To safeguard public health, the European Union has established directives (as stated in Directive 2008/50/EC) to control ambient air quality. This is done by setting limits for various pollutants, such as particulate matter (PM_{10}) and nitrogen dioxide (NO_2). According to Directive 2008/50/EC, member states are required to develop action plans for areas that exceed these air pollution limits. Within these plans, many European cities have adopted and executed comprehensive traffic management strategies aimed at reducing air pollution, with a particular emphasis on emissions from road transport (Dias et al., 2016). Among the tools to reduce traffic-induced pollution, including direct user charges, congestion pricing, green vehicles, and licence plate-based restrictions (Margaryan, 2021; Simeonova et al., 2021; Wolff, 2014), one promising alternative that can reduce air pollution in cities is the introduction of low emission zones and relevant policy actions (Khomenko et al., 2023; Tassinari, 2024).

Low emission zones (LEZ) are areas where access is restricted for vehicles that do not meet specific emission standards. The aim is to improve air quality by limiting the number of high-polluting vehicles, particularly older diesel and petrol-powered vehicles. LEZs are mostly implemented in urban areas to combat air pollution, which has been linked to various health issues and environmental degradation (Dias et al., 2016; Ferreira et al., 2012). LEZs help reduce emissions of fine particles, nitrogen dioxide, and, indirectly, ozone, which are the three major air pollutants in Europe (Azdad et al., 2022). Introduction of the relevant action plans shows a robust and intensifying effort across Europe to reduce vehicle emissions, enhance public health, and support environmental sustainability through urban traffic management.

LEZ schemes vary widely in their size, the area they cover, the hours during which they are in force, the emission standards which must be met, the types of vehicles subject to the regulations, and the methods used for LEZ implementation and enforcement. Three main criteria are considered when creating LEZs:

1. vehicle—access to the zone mainly depends on fuel type (e.g., diesel, petrol), however, the age of the vehicle may also be a factor;
2. area—the size of the zone; it may cover only selected streets or entire neighbourhoods;

3. time—regulations define when entry into the zone is possible, e.g., cars are permitted to enter a specific zone only within defined hours (Quak & de Koster, 2009).

In 2022, over 300 cities in the EU member states had implemented LEZ schemes, mostly affecting heavy-goods vehicles and remaining in force for 365 days and 24 h per day (Azdad et al., 2022). The purpose of this chapter is to outline the operating principles of LEZs in European countries and to evaluate their effectiveness. By analysing different LEZ models and their outcomes, the present research seeks to provide valuable insights into how well these zones serve their purpose. In order to achieve this objective, a systematic review of the literature on the operating principles of LEZs in European countries is conducted to assess their effectiveness.

2 History of LEZ Implementation and Individual Case Studies

In Europe, the implementation of low emission zones to manage air pollution dates back to 1996, in Sweden, where pilot schemes, known as "Environmental Zones", were initiated in the three largest cities: Stockholm, Gothenburg, and Malmö. These first low emission zones focused solely on older, large diesel-powered vehicles (i.e., heavy-duty vehicles or HDVs), including buses and heavy-goods vehicles. It is only in recent years that LEZs have begun to target passenger cars as well. The first LEZ outside of Sweden was established for the Mont Blanc Tunnel, between France and Italy, in 2002. It restricted access for HDVs that did not meet the Euro 3 or higher standards (Andriolli & Silva, 2024; Holman et al., 2015). In Europe, the categorisation of vehicles for LEZs has been based on the Euro class requirements (Ezeah et al., 2015). The number of active low emission zones in European cities has been growing steadily, increasing from 228 to 320 between 2019 and 2022, which reflects the significant commitment in the region to improve urban air quality. By 2025, this number is expected to rise further, to 507, which represents a 58% increase from mid-2022 (Fig. 1). This expansion includes both the tightening of existing LEZ zones and the introduction of new ones due to regulations introduced in countries like France, Spain, and Poland. Alongside LEZs, some cities are also planning to implement zero-emission zones (ZEZs), which will target all vehicles with internal combustion engines (Azdad et al., 2022).

2.1 *Stockholm*

Sweden has established a national framework for low emission zones, referred to as "miljözon" ("environmental zone"), under traffic regulation SFS 1998:1276. This regulation divides LEZs into three classes, each with increasing levels of stringency:

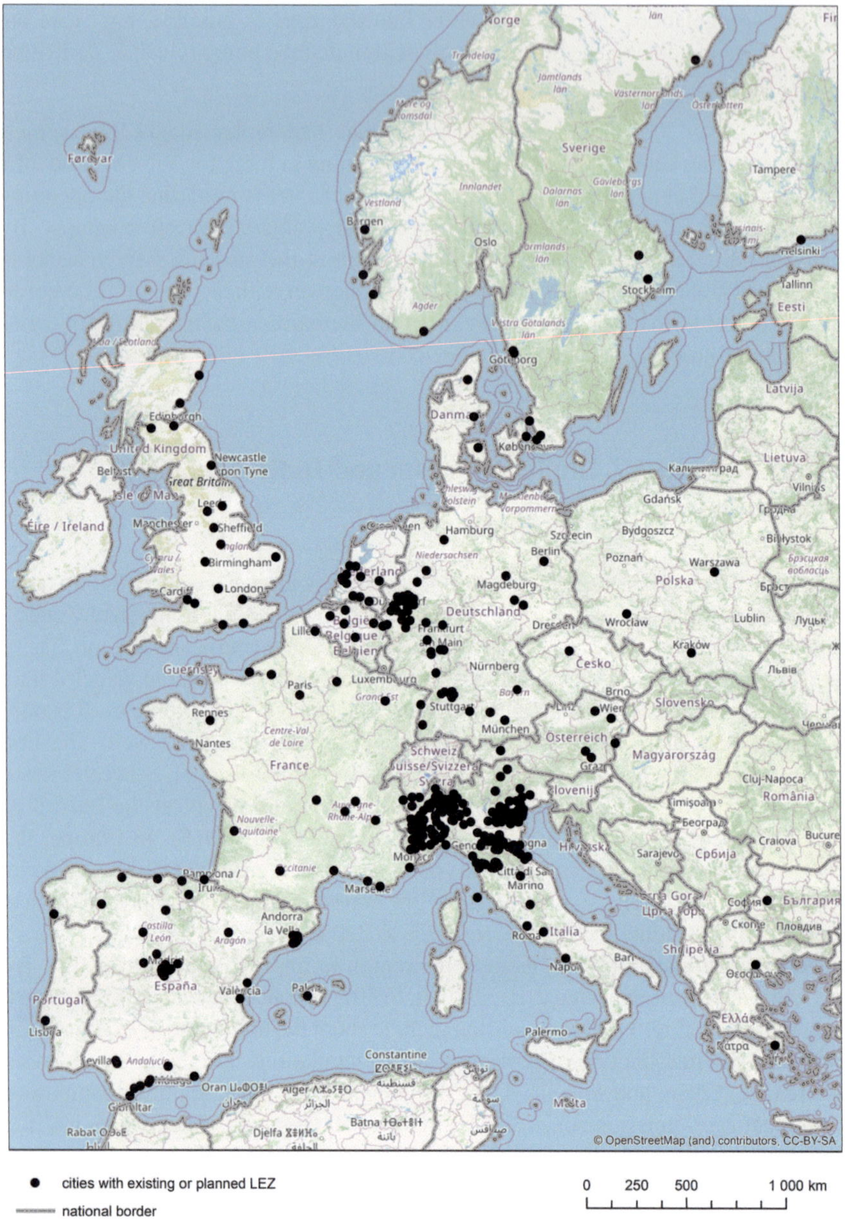

Fig. 1 Existing or planned LEZs in Europe. *Source* Own study based on https://urbanaccessregulations.eu/ (accessed: 10.10.2024) and OpenStreetMap

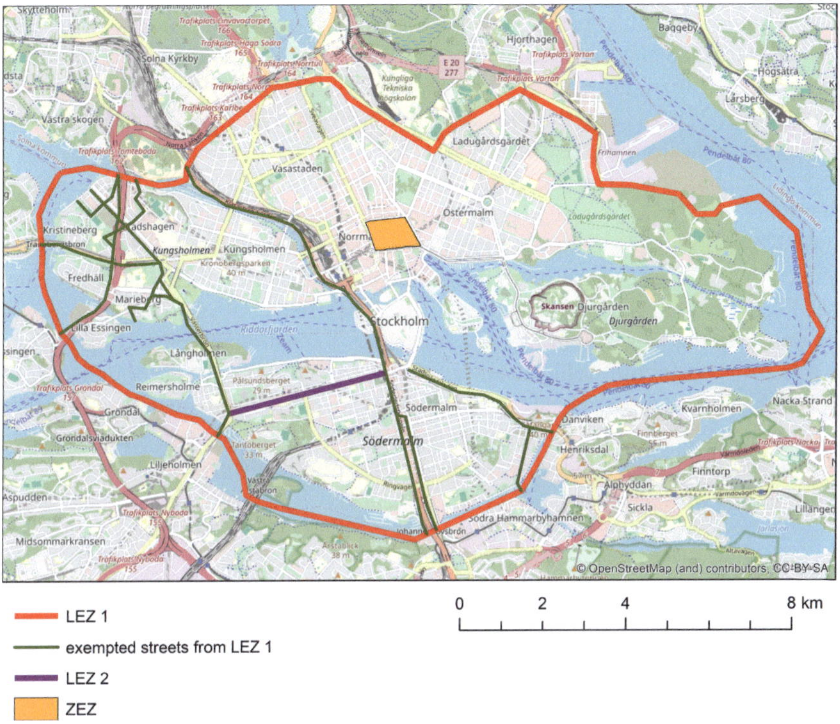

Fig. 2 LEZ categories in Stockholm (Sweden). *Source* Own study based on https://urbanaccessregulations.eu/ (accessed: 10/10/2024) and OpenStreetMap

Class 1, Class 2, and Class 3. Stockholm serves as a comprehensive case study, having implemented—or be planning to implement—one LEZ of each class, which are designated as LEZ 1, LEZ 2, and ZEZ (Zero Emission Zone), respectively, with their locations shown in Fig. 2. Buses and heavy-goods vehicles are affected by LEZ Class 1, which covers the entire city centre. LEZ 2 was launched in 2020 along Hornsgatan affecting different types of vehicles (passenger cars, minibuses, and vans). All vehicles entering the Stockholm area has to pay a congestion charge. The ZEZ is scheduled for introduction by the end of 2024 following the approval from the European Commission in April 2024, which confirmed that its introduction does not impede the free movement of trade in fossil-fuelled vehicles (Bergeling, 2024).

2.2 Berlin

In Germany, LEZs are implemented in three distinct stages: LEZ 1, LEZ 2, and LEZ 3. The decision to establish an LEZ, including its coverage and stringency, is made by a given municipality. LEZ 3 imposes the strictest requirements, only permitting entry to petrol vehicles that meet the Euro 1 emission standard and diesel vehicles that comply with Euro 4, Euro 3 with diesel particulate filters (DPF), or higher standards (Gu et al., 2022). The low emission zone in Berlin came into force at the beginning of 2008 and was introduced primarily because the limits for particulate matter PM_{10} imposed in 2005 had been exceeded. The first environmental zones were defined in air quality plans as a measure to reduce particulate matter PM_{10} (Cyrys et al., 2018). The LEZ in Berlin expands over a large area—88km^2 (10% of the city's entire area defined by the inner S-Bahn ring), thus covering the majority of Berlin's city centre (see Fig. 3) (Aydin & Kürschner Rauck, 2023; Morfeld et al., 2014). Berlin is still the only city that has moved directly from LEZ 1 (restricting vehicles up to Euro 1) to LEZ 3 (restricting vehicles up to Euro 3) (Aydin & Kürschner Rauck, 2023). Only vehicles that meet low-emission standards are permitted to operate in the city's environmental zone as part of the efforts to curb air pollution from diesel soot and nitrogen oxides. Such vehicles are required to display a green sticker.

2.3 *Milan*

Since 2008, Milan has been enforcing a Low Traffic Zone policy, initially through the Ecopass programme from 2008 to 2011. This programme primarily involved a pollution charge within a designated Limited Traffic Zone. Subsequently, this policy was replaced by Area C, a congestion charge scheme introduced in 2012 (Ku et al., 2020). In 2019, Area B was launched, which is currently the largest LEZ in Italy, covering about 72% of the city's municipal territory (Fig. 4) and impacting all those who enter the city by motor vehicle. This area restricts access for the most polluting vehicles, including vans on older diesel and petrol vehicles. The goal of Area B is to significantly reduce particulate matter and nitrogen oxides, aiming for a substantial decrease in traffic emissions by 2026. This zone is part of a broader strategy to enhance urban air quality and reduce pollution-related health risks, showing the committed effort taken towards sustainable urban mobility.

2.4 *Paris*

Paris pioneered the introduction of low emission zones in France. An LEZ zone was introduced, inter alia, to enhance air quality, diminish noise pollution, and alleviate traffic congestion. In 2015, the French State Council established a national policy for low emission zones in urban areas, introducing the *Zone à Faible Émissions* (ZFE) in Paris. The LEZ in Paris was implemented in phases—each targeting a different category of vehicles.

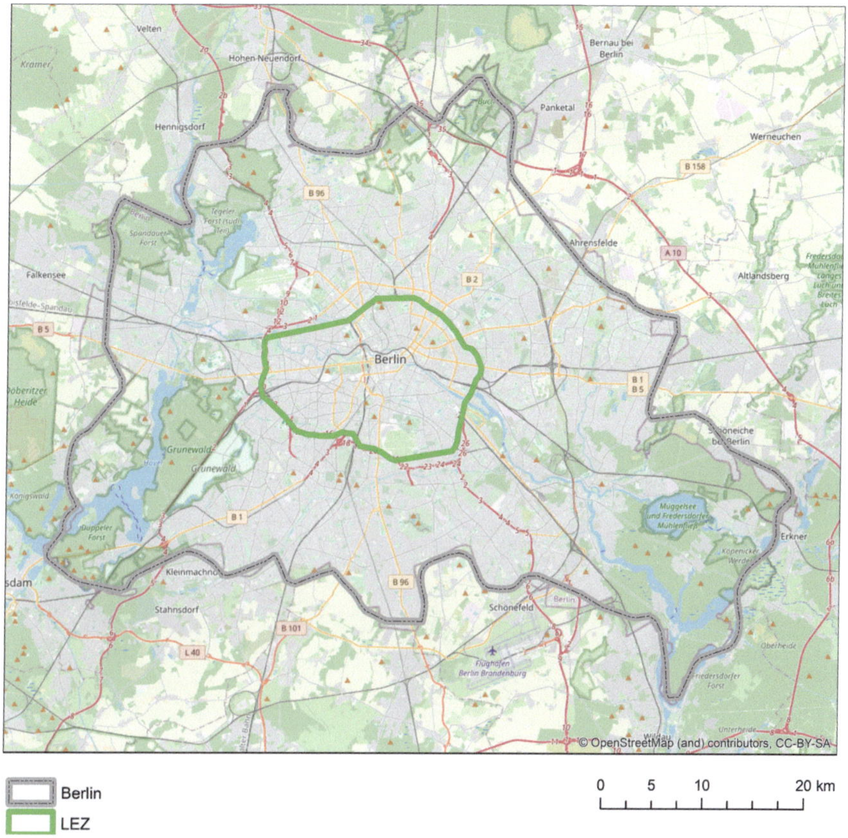

Fig. 3 LEZ in Berlin (Germany). *Source* Own study based on https://urbanaccessregulations.eu/ (accessed: 10/10/2024) and OpenStreetMap

1. **Phase 1 (September 2015)**: This phase prohibited Euro 2 heavy-goods vehicles, coaches, and buses from entering and circulating in the city centre.
2. **Phase 2 (July 2016)**: Restrictions were expanded to include pre-Euro 2 cars and motorcycles, further reducing the number of older, more polluting vehicles. The LEZ curbed access to vehicles with a Crit'Air 5 classification.
3. **Phase 3 (July 2017)**: This phase banned Euro 2 diesel cars and Euro 3 heavy-goods vehicles from entering the zone. The LEZ required vehicles to meet at least Crit'Air 4.
4. **Phase 4 (July 2019)**: The LEZ was progressively expanded to cover larger areas of Paris and other districts beyond the city centre in order to manage air quality more effectively over a wider area. The entry of vehicles only meeting Euro 3 (petrol cars) and Euro 4 (diesel) emission standards was prohibited. The LEZ requires that vehicles have at least a Crit'Air 3 certificate to enter the City of

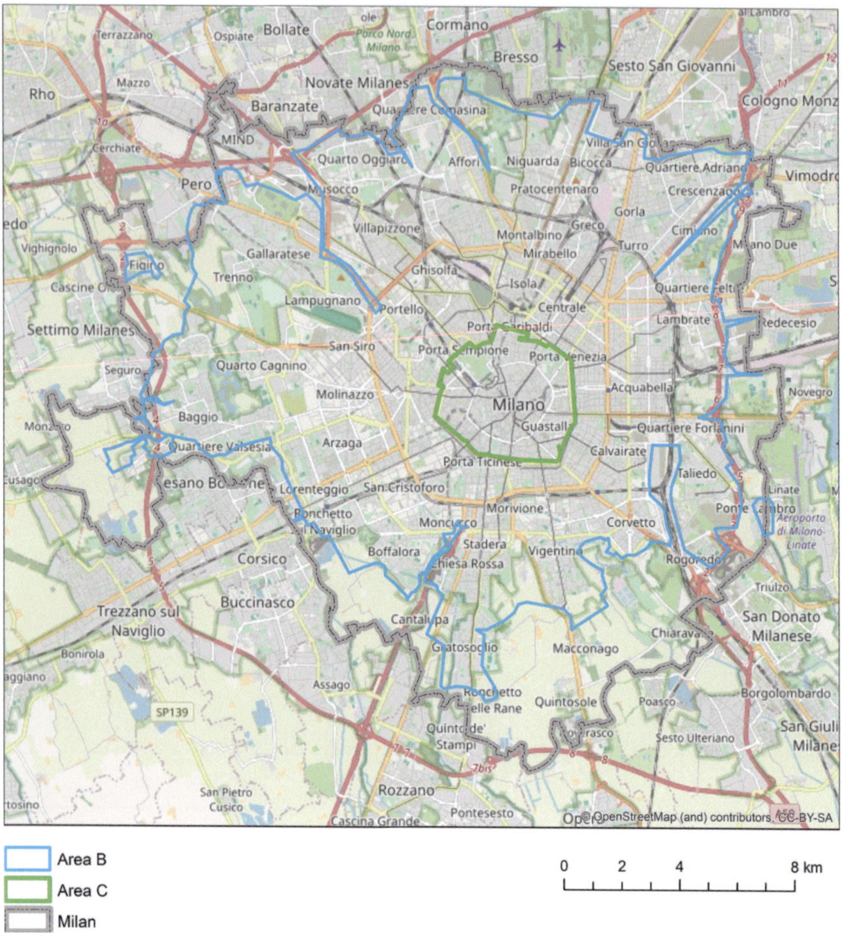

Fig. 4 LEZ in Milan (Italy). *Source* Own study based on https://urbanaccessregulations.eu/ (accessed: 10.10.2024) and OpenStreetMap

Paris. For access to the Greater Paris area, vehicles must meet at least a Crit'Air 4 certificate.

5. **Phases 5–7** for the City of Paris and Phase 7 for Greater Paris involve further tightening of the regulations for vehicle entry. The long-term goal for the LEZ is to grant access only to vehicles with a green Crit'Air certificate by 2030 (Phase 7) (Bernard et al., 2020; Moreno et al., 2022).

Since 2016, the criteria for the Paris LEZ have been progressively tightened based on the national Crit'Air classification. These measures are in effect on weekdays from 8 a.m. to 8 p.m. to significantly decrease air pollution during peak hours. Paris continues to adapt and tighten these regulations to promote cleaner air and encourage the use of environmentally-friendly transport alternatives (Font et al., 2019; Moreno et al., 2022) (see Fig. 5).

The city and its greater area encompass a set of distinct low emission zones, each with specific rules and boundaries.

1. ***Paris Greater Area ZFE***: All municipalities within the second ring road, the A86, are included in the permanent low emission zone, although the A86 itself is not part of the zone. Additionally, several municipalities located outside the A86 are also part of the ZFE. The restrictions are in effect from Monday to Friday,

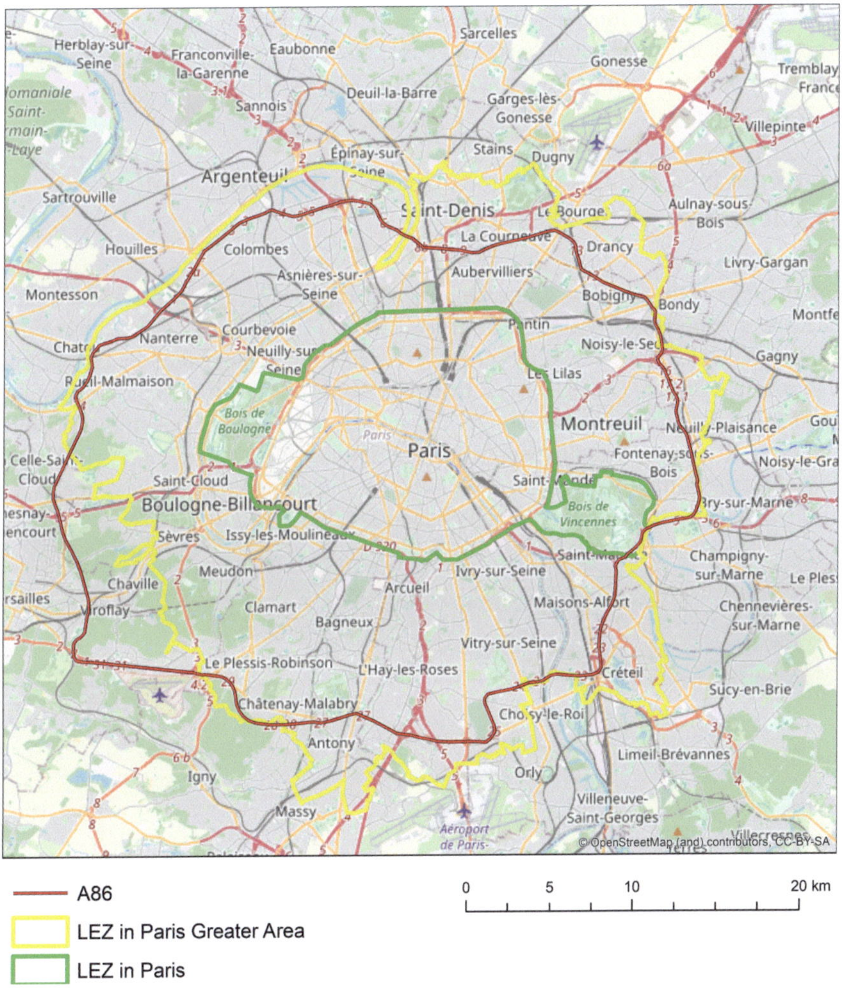

Fig. 5 LEZ in Paris (France). *Source* Own study based on https://urbanaccessregulations.eu/ (accessed: 10.10.2024) and OpenStreetMap

between 8 a.m. and 8 p.m. For lorries and buses, these restrictions extend to weekends and public holidays, also from 8 a.m. to 8 p.m.

2. ***Paris Greater Area ZPA***: All communities within the A86 are included in the temporary zone, though the A86 itself is not. The Paris airports, Roissy Charles-de-Gaulle and Orly, as well as Disneyland Paris, are located outside the A86 and are therefore not subject to any restrictions. Vehicle access is determined by taking precedence over the regulations of the Paris Greater Area ZFE. The A86 motorway is not impacted by these restrictions.
3. ***Paris A86 (Inside and Outside) ZFE***: Initially, the zone was intended to cover all 79 municipalities, those within the A86 motorway. However, since participation is not mandatory for municipalities, the regulations apply in only 47 of the 79 areas. Notably, the busy A86 ring road is exempt from the Grand Paris ZCR/ZFE zone regulations (https://www.lez-france.fr/).

3 Impacts of LEZs on the Natural Environment: Collecting Facts from Various Cities

The implementation of low emission zones in European cities has become a key measure to combat air pollution from road traffic. LEZs restrict access for high-emission vehicles, particularly diesel-powered ones, aiming to reduce concentrations of particulate matter (PM_{10}, $PM_{2.5}$), nitrogen oxides (NO_x), nitrogen dioxide (NO_2), and black carbon (BC), thereby improving public health (Fig. 6). This study critically examines the results of studies conducted in large cities, such as London, Paris, Milan, Warsaw, and Sofia, highlighting both benefits and limitations of these measures within diverse local contexts and complementary policies.

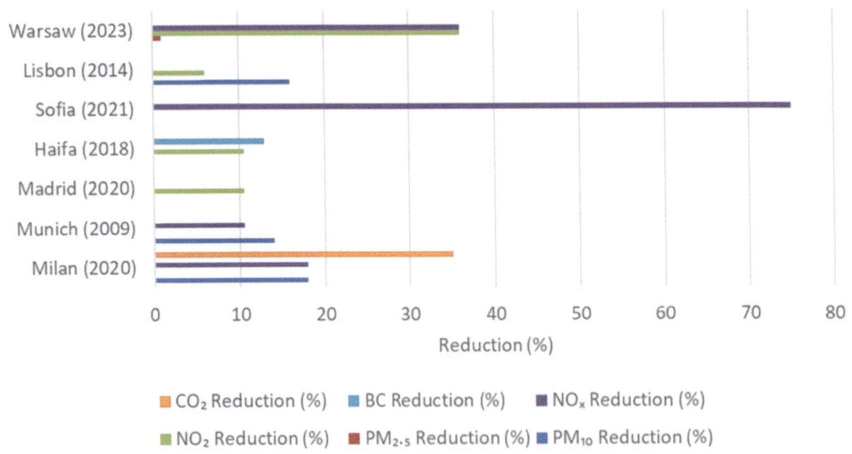

Fig. 6 Levels of reduction of various air pollutants following the introduction of LEZs in different cities

3.1 Western Europe: Effectiveness of LEZs in Cities with Diverse Strategies

London is a leading example of advanced measures to improve air quality. The introduction of the LEZ in 2008 followed by the ultra-low emission zone (ULEZ) in 2019 has significantly reduced emissions. Ku et al. (2020) found that PM_{10} and NO_x emissions decreased after ULEZ implementation. However, its effectiveness was limited due to exemptions for vehicles meeting Euro 6 standards, which can sometimes emit more pollutants than the standards allow. Hajmohammadi and Heydecker (2022) observed NO_x reductions of 19–20% within the ULEZ and smaller reductions (11–15%) across the LEZ in Greater London. These reductions were more pronounced during peak traffic periods, highlighting the pivotal role of the ULEZ in managing transport emissions. Nevertheless, the authors emphasised the necessity of integrating the LEZ with supporting measures, such as enhancing public transport, expanding cycling infrastructure, and introducing mechanisms to curb overall demand for private vehicle travel.

The Crit'Air vignette system in Paris, introduced in 2016, progressively restricts access for the most polluting vehicles. Bernard et al. (2020) showed that the initial phases yielded modest reductions in NO_x emissions, with significant improvements only expected after full implementation in 2024, potentially cutting NO_x emissions by 91%. However, Pasquier and André (2016) highlighted the displacement effect, where traffic is diverted to adjacent areas outside the LEZ, increasing pollution in these zones. This finding underscores the need for comprehensive urban transport planning.

In the Netherlands, Boogaard et al. (2012) evaluated the impacts of LEZs in five cities and found limited reductions in NO_x and "soot" concentrations. However, areas with a significant reduction in traffic density saw marked improvements. For instance, on one street with a 50% traffic reduction, soot levels dropped by 41%, and NO_x by 36%. Treurniet (2018) pointed out that insufficient monitoring stations and external factors, including weather and non-traffic emissions, hindered the ability to detect significant changes in Rotterdam.

3.2 Southern Europe: Significant Changes in Fleet Composition

In Milan, the Area C scheme reduced PM_{10} and NO_x emissions by 18% and CO_2 by 35% (Ku et al., 2020). Rome restricted access to its historic centre, resulting in a 5% decrease in car journeys and a 3.6% increase in public transport use. Despite these achievements, Holman et al. (2015) noted that LEZ effectiveness in reducing PM_{10} and $PM_{2.5}$ in Milan was limited, with significant reductions observed only for black carbon levels. Although Madrid's SPLEZ (Superilla Zona de Bajas Emisiones) significantly reduced NO_2 levels in central areas (Cárdenas-Montes, 2020), Morillas et al.

(2024) highlighted that NO_2 emissions during winter were predominantly driven by the heating of buildings rather than traffic. This finding underscores the need to expand LEZ policies to address emissions from the residential and commercial sectors. Silva et al. (2014) reported that the two-phase implementation of Lisbon's LEZ led to PM_{10} and NO_2 reductions of 16% and 6%, respectively. These improvements were primarily attributed to fleet modernisation, including the replacement of older taxis with newer, low-emission models. The study emphasised the importance of collaboration between local authorities and the national government in order to further modernise fleets and support low-emission public transport initiatives.

3.3 Central and Eastern Europe and the Middle East: Regional Contexts

Warsaw's LEZ, which is to be introduced in stages from 2024 to 2032, is expected to reduce NO_x levels by 36% (Holnicki et al., 2023). However, its impact on $PM_{2.5}$ concentrations will be minimal (1 $\mu g/m^3$) primarily due to the predominance of coal-based heating in Poland. The authors therefore stressed that the full potential of the LEZ can only be realized through parallel implementation of anti-smog programmes and decarbonisation of the municipal heating sector. In Sofia, older diesel vehicles are responsible for 85% of particulate matter (PM) emissions and 56% of NO_x emissions (Lee et al., 2021). Introducing an LEZ with annual tightening of restrictions could accelerate NO_x reduction by 6–8 years and PM reduction by 4–5 years compared to natural fleet turnover. The authors also emphasised the need for financial support mechanisms to mitigate the social impact on less affluent populations. Tartakovsky et al. (2020) analysed the introduction of the LEZ in Haifa, noting reductions of 12.9% in BC and 10.6% in NO_x levels. Night-time and pre-holiday reductions were particularly pronounced, underscoring the effectiveness of restrictions on diesel vehicles in improving air quality.

3.4 Area-Specific Conclusions and Recommendations

Low emission zones effectively contribute to reducing the concentrations of major air pollutants, such as PM_{10}, $PM_{2.5}$, NO_2, NO_x, and black carbon (BC). However, their effectiveness varies depending on the city and local conditions. The best results are achieved in locations where strict restrictions are imposed on older vehicles (e.g., diesel vehicles only meeting Euro 4 standards or lower) and where low-emission transport alternatives are simultaneously developed. Even if this is implemented, the effectiveness of LEZs may be limited during winter, as residential heating based on solid fuels becomes a major source of emissions, as observed in such cities as Munich, Madrid, and Warsaw. The long-term benefits of LEZs, including improved air quality and reduced transport emissions, are particularly evident in cities like London and Paris. However, fully realising their potential requires sustained efforts and the systematic tightening of regulations over time. In some locations (e.g., Rotterdam), the effectiveness of LEZs has been limited due to local factors and insufficient alignment with broader transport policies.

To enhance the efficiency of LEZs in reducing air pollution and to better prepare cities for climate challenges, several actions should be undertaken. First, emission standards for vehicles entering LEZs must be regularly tightened, eventually allowing access only to electric and hybrid vehicles. Secondly, financial support programmes enabling households and businesses to purchase zero-emission vehicles are essential. These supportive measures should extend beyond the boundaries of LEZs to prevent traffic and emissions from being displaced to adjacent areas. Investments in public transport, cycling and pedestrian infrastructure, and charging stations for electric vehicles also need to be intensified. During winter, particular attention should be paid to reducing emissions from the residential heating sector by modernising heating systems and curbing the use of solid fuels.

Regular monitoring of the effectiveness of LEZs alongside adaptive management, such as adjusting LEZ boundaries or implementing more stringent restrictions, would assist in ensuring their success. Incentives for public transport users, e.g., tax reliefs or free rides, should also be introduced to make public transport more attractive than private vehicle use. Public education campaigns highlighting the benefits of LEZs, including improved air quality and public health, would help in gaining community support. Further research into the long-term impacts of LEZs, including their effects on air quality and changes in residents' mobility patterns is necessary. Integrating LEZs into broader climate policies should become a priority. A holistic approach would fully leverage the potential of LEZs and ensure sustainable urban development in the face of environmental challenges.

4 Impacts of LEZs on the Major Elements of the Urban Transport System

Low emission zones are among the most effective urban policy tools to combat air pollution. Their primary aims are: to restrict the circulation of vehicles that do not meet the specified emission standards, thereby improving air quality; to reshape transport systems; and to influence user behaviour. Research conducted in various cities, such as Beijing, Madrid, Medellín, London, Berlin, and Rotterdam, provides evidence of the effectiveness of LEZs while also highlighting the challenges faced in their implementation. This section of the chapter specifically focuses on the following aspects: the impact of LEZs on reduction in emissions, fleet modernisation, changes in transport behaviour, public acceptance, and the development of alternative transport modes (Fig. 7).

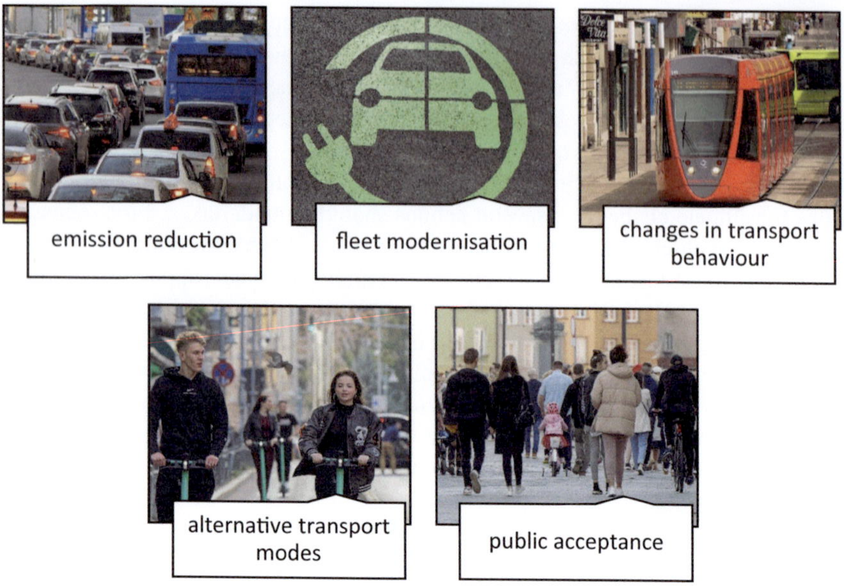

Fig. 7 Areas where the effectiveness of LEZ impacts can be assessed

4.1 Emission Reduction and Fleet Modernisation

The introduction of LEZs yields measurable results in reducing emissions and modernisation of vehicle fleets. Zhang et al. (2018) conducted research in Beijing using an urban vehicle emissions inventory that combined local emission factors with dynamic and static data. They demonstrated that LEZs reduced emissions of carbon monoxide (CO) and hydrocarbons (HC), while increasing the average traffic speed from 30.1 km/h to 34.3 km/h, thus alleviating congestion. However, the authors highlighted that long-term success requires regular updates to LEZ regulations. Settey et al. (2019) found that the number of LEZs in Germany increased from 0 in 2008 to 58 in 2018, which significantly boosted registrations of new vehicles compliant with emission standards, particularly in the heavy-duty transport sector. In Madrid, Balaguer et al. (2023) observed an increase in the number of vehicles being scrapped following the introduction of LEZs, particularly among older diesel models. In the first quarter, the number of vehicles being scrapped rose by 8%, reflecting the adaptation of vehicle owners to the new regulations.

Boriboonsomsin et al. (2024) studied the impact of emission-based fees in Southern California and found that 60% of non-compliant heavy-duty trucks changed their routes, leading to a reduction in nitrogen oxides (NO_x), particulate matter ($PM_{2.5}$), and carbon dioxide (CO_2) emissions within the LEZ by 37%, 38%, and 25%, respectively. However, the impact on the overall emissions when including the study area outside the LEZ underscores the need to consider the wider regional effects of LEZ implementation.

4.2 Changes in Transport Behaviour and Public Acceptance

LEZs influence transport behaviour in support of the goals of urban policies. Peters et al. (2021) observed that the LEZs in Madrid significantly increased registrations of alternative fuel vehicles (AFVs), particularly plug-in hybrid electric vehicles (PHEVs). However, the impact on CO_2 emissions was limited, suggesting that the more stringent zero-emission zones (ZEZs) may be required for effective decarbonisation of urban fleets. In Norway, Seter et al. (2023) analysed driver responses to LEZs and road usage charges (RUC) among PHEV users. The study revealed that individuals who increased their use of electric driving modes within LEZs displayed more positive attitudes toward the system. Public acceptance of LEZs also depended on practical factors, such as the availability of charging infrastructure and other supportive measures. In some cases, unintended consequences of LEZ implementation had been observed. Moral-Carcedo (2024) noted that although traffic intensity decreased significantly in Madrid's central zone, it increased in neighbouring areas. This displacement effect highlights the need for integrated spatial planning to mitigate regional side effects and ensure equitable outcomes.

4.3 Development of Alternative Transport Modes

Dablanc and Montenon (2015) observed that LEZs in London and Berlin reduced the number of logistics operators in cities, as the introduction of the LEZ forced companies to invest in low-emission fleets and optimise delivery routes. While operational efficiency improved, smaller companies often exited the market, which underscores the serious challenges LEZs could pose for the logistics sector. In Rotterdam, Attia (2020) reported an increase in electric vehicle adoption alongside improvements in public transport and cycling infrastructure following LEZ implementation. However, the expanded parking facilities paradoxically led to an increase in car traffic, indicating the need for the better integration of LEZ policies with broader transport planning strategies.

LEZs play a crucial role in promoting more sustainable transport options. Ding et al. (2023) investigated the impact of London's ultra-low emission zone (ULEZ) on the public bicycle scheme (London Cycle Hire, LCH). They found a 27.9% increase in bike use, particularly for short and medium trips, which demonstrates that LEZs can effectively encourage users to shift from car dependency to greener transport alternatives. Saldarriaga and Patiño (2024) analysed traffic patterns within the LEZ, identifying seasonal variations in traffic volumes and vehicle occupancy as critical factors affecting vehicle speeds. Their study revealed that small vehicles (up to 3 m in length) dominated traffic and strongly correlated with overall traffic/density, emphasising the importance of monitoring such variables in urban transport policies.

4.4 Area-Specific Conclusions and Recommendations

Low emission zones represent a critical urban policy instrument aimed at reducing air pollution through restrictions placed on high-emission vehicles. Analysis of the impact of LEZs across a number of cities reveals significant transformations in urban transport systems, with both positive and less advantageous outcomes. LEZs have proven to be effective in reducing air pollutant emissions, including NO_2, NO_x, PM_{10}, and $PM_{2.5}$, but their long-term success depends on regular policy updates and support for vehicle fleet modernisation. While low emission zones encourage the adoption of alternative fuel vehicles (AFVs), their impact on CO_2 emission has been limited, highlighting the necessity to move to the more stringent zero-emission zones. Furthermore, LEZs often influence urban logistics by reducing the number of delivery companies, as observed in London and Berlin. Although the introduction of LEZs has led to improved operational efficiency and greener logistics practices, smaller enterprises face significant challenges in conforming, which results in some market exits or relocation.

LEZs reshape urban mobility patterns, encouraging the use of sustainable transport modes, such as bicycles, public transit, and electric vehicles. In London, the introduction of LEZs has increased the use of the public bicycle hire system by almost 28%, demonstrating their potential to shift users away from car dependency. However, these benefits are sometimes offset by drawbacks, such as the spillover of traffic to adjacent areas or increased road traffic. In Rotterdam, this increased traffic was due to expanded parking availability. Such effects underscore the need for comprehensive urban strategies that align LEZ policies with broader transport planning objectives. Another consideration is the social and economic equity of LEZ policies. In some cases, these zones may disproportionately affect lower-income groups or redistribute environmental burdens to less affluent areas. Addressing these challenges requires integrating principles of environmental justice into LEZ planning and implementation. To maximise the effectiveness of LEZs, a number of recommendations emerge:

(1) Policy frameworks should be regularly updated to incorporate technological advancements and evolving transport needs.
(2) Governments should provide financial incentives and support to facilitate the transition to cleaner vehicles, particularly in regions with older vehicle fleets.
(3) Spillover effects should be mitigated through the creation of buffer zones and coordinated regional transport policies.
(4) The development of sustainable transport infrastructure, including public transit and cycling facilities, should be prioritised to offer viable alternatives to car use.
(5) Public education and clear communication are crucial to building societal support and minimising political resistance.
(6) Continuous monitoring and assessment of the impact of LEZs is essential to determine their effectiveness and refine strategies for achieving long-term urban sustainability goals.

In conclusion, while LEZs have demonstrated considerable potential in reducing urban air pollution and transforming transport systems, their success depends upon adaptive policymaking, comprehensive planning, and a commitment to addressing social and environmental equity. These measures will ensure that LEZs contribute not only to cleaner cities but also to more inclusive and sustainable urban environments.

5 Impacts of LEZs on the Health of City Residents

Low emission zones have emerged as a significant policy tool for mitigating air pollution and its adverse effects on human health in urban areas. By restricting access to vehicles with high emissions in specific zones, LEZs aim to reduce major pollutants, such as nitrogen oxides (NO_2) and particulate matter ($PM_{2.5}$, PM_{10}). A number of studies have documented the harmful impact of these pollutants on respiratory and cardiovascular health/systems, as well as on overall life expectancy. While the primary goal of LEZs is to enhance public health through improved air quality, their implementation has raised questions about their effectiveness, the equitable distribution of benefits, and the need for context-specific approaches. This section of the chapter delves into three major themes/topics: the health benefits of LEZs, their implications for socio-economic equity, and the variability of their effectiveness across different contexts.

5.1 Health Benefits of Low Emission Zones

The core rationale for introducing LEZs is their potential to reduce air pollution and its detrimental effects on health. Research consistently demonstrates that LEZs can bring measurable improvements in air quality, which in turn translate into significant public health benefits. Host et al. (2020) conducted an analysis of four LEZ scenarios in Paris. These varied in size and the degree of restrictions placed on vehicles to assess their potential impact on pollutant concentrations and public health. The study found that the most stringent scenario could lower the maximum NO_2 exposure level from 55 $\mu g/m^3$ to 42 $\mu g/m^3$. This reduction would prevent 340 deaths annually, 170 cases of low birth weight, 130 new cases of coronary heart disease, and 2,930 asthma diagnoses among the city's 9.4 million residents. These benefits would extend beyond the LEZ boundaries, demonstrating the wide-reaching impact of such measures on public health.

Flanagan et al. (2022) modelled the potential impact of a hypothetical LEZ in Malmö (Sweden), estimating that it could reduce NO_2 concentrations by 13.4%. This reduction could prevent between 9 and 26 deaths per year and avoid many cases of respiratory ailments, including childhood asthma and hypertension during pregnancy. Similarly, Kang et al. (2024) found that LEZ implementation in Tokyo led to a 7.17% reduction in suspended particulate matter (PM) and a 9.08% decrease in nitrogen oxides (NO_x). These improvements had a pronounced effect on infant health, with reductions in infant mortality rates comparable to outcomes observed in California after similar interventions. In addition to health benefits, the study noted an average increase of 2.74% in land values, reflecting the broader social and economic gains from cleaner air. Collectively, these studies underline the strong link between air quality improvements and public health gains, particularly for vulnerable groups, such as children and individuals with pre-existing health conditions.

5.2 *Socio-economic Equity and Health Disparities*

While LEZs generally contribute to improvements in public health, their benefits are not always equitably distributed across socio-economic groups, which raises important concerns about environmental justice and the unintended consequences of such policies. Cesaroni et al. (2012) discussed the environmental justice and social equity issues associated with the LEZs in Rome. While the policy successfully reduced NO_2 exposure, the health benefits were largely concentrated among residents of higher socio-economic status. These individuals often lived in areas with lower baseline pollution levels or had greater access to resources that facilitated compliance with LEZ regulations. The study highlighted the need for targeted measures, such as subsidies for cleaner vehicles or investments in public transport to ensure that the benefits of LEZs are more evenly distributed.

In Germany, Pestel and Wozny (2019) analysed hospital data to assess the health impacts of the LEZs introduced since 2007. They found significant reductions in hospital admissions for chronic circulatory and respiratory diseases in areas within or near LEZs. However, economically disadvantaged groups often faced greater challenges in adapting to these regulations, which underscores the importance of policies to address both environmental and social inequities. These findings emphasise that while LEZs can be powerful tools for improving public health, their design and implementation must consider the needs of vulnerable populations to avoid deepening socio-economic disparities. In Paris, Moreno et al. (2022) found that expanding the existing LEZs and enforcing stricter vehicle standards could prevent 810 deaths and 3,200 asthma cases annually. Importantly, the study noted that the health benefits were disproportionately experienced by lower-income groups, as these populations are often more exposed to air pollution. However, the financial burden of adapting to LEZ regulations, e.g., upgrading vehicles or relying on alternative modes of transport, may exacerbate existing socio-economic inequalities.

More details regarding the representative studies assessing the socioeconomic and environmental impacts of LEZs are provided in Table 1. To maximise the benefits of LEZs and address potential negative impacts, comprehensive urban planning and thoughtful policymaking are crucial. It is essential that LEZs do not disproportionately disadvantage lower-income groups and small businesses. Strategies to achieve this balance include:

Table 1 Socioeconomic and environmental impacts of low emission zones: key findings from representative studies

Authors	Title	Conclusions
Cesaroni et al. (2012)	Health benefits of traffic-related air pollution reduction in different socioeconomic groups: the effect of low emission zoning in rome	The study indicates that low emission zones primarily benefitted the health of higher socioeconomic groups
Morton et al. (2018)	A framework for assessing spatial vulnerability to the introduction of low emission zones: a case study of Edinburgh, Scotland	LEZs may disproportionately impact lower-income groups by reducing their accessibility owing to vehicle upgrade costs and limited public transport alternatives
De Vrij and Vanoutrive (2022)	"No-one visits me anymore:" low emission zones and social exclusion via sustainable transport policy	LEZs disproportionately impact low-income groups as these groups are often unable to afford the necessary vehicle upgrades or access public transport alternatives. Additionally, LEZs may displace pollution to adjacent neighborhoods, affecting those outside the LEZ
Blandin et al. (2024)	Assessing the ex-ante impacts of a low emission zone on transport poverty and vulnerability with the Vulmob indicator—application to the low emission zone in Grenoble, France	LEZs disproportionately affect highly vulnerable households that experience poor access to public transport alternatives, which necessitates purchasing a conforming vehicle to comply and may lead to increased transport poverty. In addition, potential pollution displacement to adjacent neighborhoods remains a concern, impacting overall environmental equity

- adopting a phased approach to the introduction of an LEZ and then gradually tightening the restrictions;
- actively incorporating the needs and experiences of those most likely to be affected by LEZ policies, ensuring their voices are heard and addressed in the planning process;
- offering support mechanisms, e.g., subsidies for cleaner vehicle options or alternative means of transport;
- encouraging businesses to adopt sustainable mobility policies for their employees;
- expanding cycling and pedestrian infrastructure.

5.3 *Effectiveness Across Different Contexts*

Whether LEZs successfully achieve the intended goals depends greatly on local conditions, policy design, and enforcement mechanisms. For instance, Wood et al.

(2015) and Griffiths et al. (2016) examined the impact of the LEZs in London and found no significant improvements in air quality or children's respiratory health during the initial years of implementation. These studies suggest that the effectiveness of LEZs in their early stage of implementation may be limited by such factors as insufficient enforcement, inadequate size of zones, and delays in compliance by vehicle owners. Conversely, Tokyo is a good example of an effective LEZ implementation. Kang et al. (2024) reported clear reductions in key pollutants and significant public health benefits following the introduction of stringent LEZ measures. The success in Tokyo was attributed to comprehensive planning, robust enforcement, and complementary policies, including investments in public transportation and incentives for cleaner vehicles.

In Germany, Gehrsitz (2017) examined the impact of LEZs on newborn health and found that while the zones have reduced the number of days that pollution targets are exceed, these reductions have been insufficient to produce measurable improvements in neonatal health indicators, such as birth weight. This suggests that the health impacts of LEZs may be more pronounced for chronic rather than acute conditions, or that greater reductions in pollution levels are needed to achieve broader health benefits. These examples illustrate the importance of tailoring LEZ policies to local contexts. Different factors, such as baseline pollution levels, socio-economic conditions, and the availability of alternative transport options, can significantly influence the outcomes of these interventions. To maximise their effectiveness, LEZs should be integrated into broader urban planning strategies and supported by robust monitoring and evaluation mechanisms.

5.4 Area-Specific Conclusions and Recommendations

The implementation of low emission zones has demonstrated significant health benefits, including reductions in mortality rates, hospitalisations, and cases of diseases linked to air pollution, including asthma and cardiovascular conditions. Studies show that the most substantial health benefits are achieved when LEZs cover large areas and enforce strict vehicle emission standards, thereby ensuring a broader and more equitable distribution of positive outcomes among residents.

Nonetheless, the implementation of an LEZ can exacerbate social inequalities. Populations with lower socio-economic status often face challenges in adapting to the regulations, e.g., the financial burden of replacing non-compliant vehicles. The research conducted in London indicated limited improvements in air quality and no detectable health benefits during the initial years of LEZ operation, emphasising the need for more effective and comprehensive measures. Similarly, while LEZ policies in Germany have reduced the number of days when pollution targets are exceeded, they have not led to measurable improvements in newborn health. Furthermore, the research in Rome highlighted disparities in health benefits, with higher socio-economic groups experiencing more significant gains.

LEZ policies have the potential to yield substantial public health and environmental benefits, but their success depends on effective planning, implementation, and consideration of social equity. Ensuring that these zones are both inclusive and impactful requires ongoing monitoring and adjustment of their scope and regulations to address emerging challenges and disparities. To maximise the effectiveness of LEZ policies, zones should be designed to cover extensive areas and enforce stringent emission standards, as these approaches yield the most substantial health and environmental benefits. It is crucial to address socio-economic disparities by introducing financial support programmes, such as subsidies for low-income households to upgrade to compliant vehicles. By continuous monitoring the long-term effects of LEZs on air quality and public health, strategies to optimise their effectiveness can be implemented. Integrating an LEZ with complementary measures, such as promoting public transport and active mobility while reducing car dependency, will further enhance their impact. Additionally, public educational campaigns can foster acceptance and understanding of LEZs by highlighting their benefits for individuals and communities, thereby ultimately bolstering broader compliance and engagement.

6 Impacts of LEZs on the Local Economy and Urban Development

Low emission zones have been introduced in many cities to address environmental challenges by improving air quality. While their primary goal is ecological, their implications extend far beyond environmental benefits, impacting urban economies, real estate markets, and social dynamics. The analysis presented in this section of the chapter explores the broader economic impacts of LEZs, focussing on the associated costs, effects on property values, and integration with urban logistics.

6.1 Economic Costs and Health Benefits: A Delicate Balance

The implementation of LEZs often comes with significant economic costs, which can overshadow the anticipated benefits. Börjesson et al. (2021) conducted an in-depth analysis of Stockholm's LEZ, which targeted diesel and petrol vehicles that did not comply with the specified Euro standards. Using microeconomic models, they estimated the social costs associated with LEZs and vehicle replacement. The findings revealed substantial economic losses—between EUR 621 and 849 million—attributable to reduced vehicle utility and declining values in the used-car market. In addition, the projected health benefits from improved air quality were limited, as the excluded vehicles contributed only marginally to NO_2 levels. Moreover, since Stockholm already met the EU air quality standards, it was suggested that the natural fleet replacement would have achieved similar results without imposing such high costs on residents. This raises an important policy question: Are the economic burdens of stringent LEZs justifiable in cities already meeting air quality thresholds?

6.2 Impact on Real Estate Markets: Winners and Losers

LEZ policies have also had notable repercussions on property markets, with location and accessibility affecting property values. In Berlin, Aydin and Kürschner Rauck (2023) examined changes in property prices near train stations and motorways following the introduction and tightening of the LEZ. They observed a 5% price increase for homes located within 5–10 min walking distance from train stations, emphasising the importance of proximity to public transport. Interestingly, properties that are located closer or farther from stations did not experience similar gains, highlighting a nuanced impact on property values.

Similarly, Maharani (2024) analysed the LEZ in Amsterdam, particularly after stricter regulations were implemented in 2020. Properties within the LEZ were shown to have experienced a statistically significant increase in value, driven by improved air quality and enhanced living conditions. However, a limitation of the study was that the experimental group had already been covered by LEZ policies before 2018, potentially complicating interpretation. Despite this, the findings collectively suggest that LEZs can boost property values in well-connected urban areas while potentially disadvantaging less accessible locations.

6.3 Urban Logistics and Sustainable Development

LEZs also reshape urban logistics, presenting challenges and opportunities for integrating sustainability into urban planning. Kin et al. (2024) explored contrasting approaches in Paris and Rotterdam. Paris emphasises the integration of logistics within dense urban environments through "logistics hotels" that combine commercial, residential, and logistical functions. This strategy aligns with the city's ambitious climate goals, including a complete ban on diesel vehicles by 2024. Rotterdam, on the other hand, focuses on efficiency by establishing large logistics hubs on the outskirts and creating zero-emission zones. These divergent strategies reflect the distinct geographic and political conditions of each city. While Paris integrates logistics facilities into the urban fabric, Rotterdam relies on peripheral hubs to streamline operations. Both models highlight the adaptability required in balancing urban development with logistical needs in the context of LEZs.

6.4 Area-Specific Conclusions and Recommendations

The introduction of LEZs in urban areas significantly impacts both the local economy and urban development. LEZ policies aim to reduce vehicular emissions, improving air quality and public health, while simultaneously shaping the urban economic landscape. These changes, however, come with complex implications that require a balanced approach. Firstly, LEZ initiatives tend to enhance the value of properties located within or near zones with improved access to public transport and with cleaner environments. By incentivising public transport use and discouraging car dependency, such policies increase the perceived desirability of urban areas with well-connected public transport. This dynamics fosters a shift in real estate trends, with properties near transit hubs or within LEZs gaining in value. Moreover, the role of LEZs in promoting sustainable urban living and encouraging investment in eco-friendly developments becomes more apparent.

However, these economic benefits of improved air quality and appreciation in property value often come at a cost to certain societal groups. Restrictions on vehicle use impose financial burdens, particularly on lower-income residents and small businesses who may struggle with the costs of upgrading to compliant vehicles or adapting to reduced access by car. These measures could inadvertently widen socio-economic disparities, necessitating targeted support mechanisms to ensure equitable transitions. The logistical and infrastructural transformations prompted by LEZs also significantly impact urban planning. Cities like Paris and Rotterdam illustrate diverse strategies in adapting logistics systems to align with LEZ policies. The focus on integrating small-scale logistics into dense urban areas applied in Paris contrasts with the emphasis on centralised hubs and efficiency employed in Rotterdam. These approaches demonstrate how LEZ policies can drive innovation in urban logistics, potentially reducing congestion and emissions.

To maximise the positive effects of LEZs, urban policymakers must adopt a comprehensive, context-sensitive approach. This involves balancing environmental objectives with economic realities, fostering social equity, and ensuring that accompanying measures, such as improved public transport and logistical efficiency, are effectively implemented. In doing so, LEZs can become a pivotal element of sustainable urban development, enhancing not only environmental outcomes but also the economic and social fabric of cities.

7 Broader Impacts of LEZs on Local Communities

The acceptance of low emission zones is a critical factor in determining their impacts on local communities. LEZs, designed to reduce urban air pollution by restricting high-emission vehicles, bring significant changes to daily life, affecting mobility patterns, social inclusion, environmental justice, and public health. The extent to which these zones are embraced by residents depends on multiple factors, including the perceived fairness of the policies, access to alternative modes of transport, and the socioeconomic context of the affected communities. Understanding the determinants of acceptance and the way LEZs shape local dynamics is essential for designing policies that balance environmental goals with community well-being.

7.1 *Environmental and Social Justice in LEZ Policies*

The environmental implications and social justice of LEZ policies are crucial for assessing their broader societal impacts. Charleux (2014) observed that while LEZ policies improved air quality in Grenoble, they simultaneously restricted mobility for lower-income groups with inadequate transport alternatives. These findings underscore the need for a holistic approach to mobility justice that integrates environmental, social, and economic considerations. Similarly, Mullen and Marsden (2016) criticized the promotion of low-emission technologies (e.g., electric vehicles) on exacerbating inequalities. Wealthier individuals benefit more from these policies due to their financial ability to access new technologies, leaving economically disadvantaged groups marginalised.

Verbeek and Hincks (2022) found that the relationship between income and exposure to pollution varies significantly across cities. In Brussels, lower-income households are more exposed to air pollution, while in London wealthier households face higher exposure levels. Furthermore, public transport quality in Brussels is higher in lower-income areas, whereas it is lower in similar areas in London. These dynamics suggests that LEZ boundaries in Brussels may need to be expanded, while the extension of the ULEZ in London disproportionately affects disadvantaged communities with limited public transport access.

7.2 *Impact of LEZs on Mobility and Social Inclusion*

LEZ policies significantly affect individual mobility and can contribute to social exclusion, particularly among those dependent on vehicles which fail to meet the emission standards. In Paris, Poulhès and Proulhac (2021) revealed that while LEZ policies improved air quality, the benefits were disproportionately felt by wealthier residents. Conversely, younger and older residents experienced fewer advantages, which highlights disparities in the distribution of health and mobility benefits. In Germany, Sarmiento et al. (2022) found that the initial introduction of LEZs led to reduced life satisfaction, particularly among diesel vehicle owners, although the health benefits became apparent over time. De Vrij and Vanoutrive (2022) demonstrated that the introduction of an LEZ in Antwerp reduced mobility for lower-income individuals who could not afford compliant vehicles. Furthermore, alternative transport options were often insufficient or inaccessible, amplifying the challenges for vulnerable populations. These studies collectively emphasise the relevance of designing LEZ policies that address the mobility needs of all social groups to prevent exclusion and inequality.

7.3 Determinants of Public Acceptance of LEZs

Public acceptance is a pivotal factor in the successful implementation of LEZs. Sfendonis et al. (2017) examined LEZ acceptance in Thessaloniki (Greece). Although over half of the respondents supported the policy, opposition was found to be higher among residents farther from the city centre. Rizki et al. (2022) found that trust in the government and the perceived fairness of the LEZ restrictions significantly shaped public attitudes in Jakarta (Indonesia). In Cáceres (Spain), Jiménez-Espada et al. (2023) highlighted the importance of enhanced public transport infrastructure and financial support for purchasing clean vehicles as critical factors influencing acceptance. Furthermore, Oltra et al. (2021) demonstrated that the acceptance in Barcelona (Spain) was driven by the perceived health and quality-of-life benefits, alongside the legitimacy of the implementation process. These findings align with those from Paris, where effective communication and transparent policymaking were shown to increase public trust and support for LEZ policies.

Psychological factors, such as moral responsibility and ecological identity, were identified as the key predictors of LEZ acceptability in Bath (England) by Player et al. (2023). Sociodemographic factors, including income and age, played a secondary role, though proximity to the LEZ and family structure (e.g., having children) also influenced acceptance. Overall, many of the reviewed studies suggest that addressing public concerns through information campaigns, participatory decision-making, and financial incentives can improve acceptance and policy effectiveness.

7.4 Area-Specific Conclusions and Recommendations

The implementation of low emission zones requires the balancing of a complex interplay of environmental benefits, social equity challenges, and policy acceptance issues (Fig. 8). While LEZs effectively reduce air pollution and improve public health, their impact on local communities is multifaceted and often disproportionate. A key takeaway is the necessity to balance environmental objectives with social equity to ensure that the benefits of LEZs reach all segments of society. Social acceptance of LEZs hinges on several factors, including perceived fairness, public trust, and the availability of viable alternatives. Acceptance tends to be higher when residents believe that the policy is implemented democratically and delivers tangible improvements to the quality of life, such as better air quality and health. However, the process often highlights pre-existing inequalities, with vulnerable groups—such as low-income households and those dependent on older vehicles—bearing the brunt

Fig. 8 SWOT analysis for the functioning of an LEZ

of mobility restrictions. These groups may face increased social exclusion due to limited access to compliant vehicles and insufficient public transport options.

Recommendations for improving the acceptance and equity of LEZs include conducting extensive community engagement to incorporate diverse perspectives, particularly those of vulnerable groups. Public trust can be fostered through transparent decision-making and the demonstration of the long-term benefits. Investments in sustainable mobility infrastructure, such as enhanced public transport, bicycle lanes, and pedestrian-friendly routes, will provide practical alternatives to private vehicle use. Financial mechanisms, such as subsidies for low-emission vehicles, public transport vouchers, and incentives for car-sharing schemes, can alleviate the economic burden on lower-income residents. Moreover, targeted information campaigns that emphasise the health and environmental benefits of LEZs can increase a public buy-in.

8 Overall Conclusions and Recommendations

This study provided a detailed review of low emission zones (LEZs) implemented in many European cities, their advantages, and implementation challenges. A concise summary of some of the major findings is presented in Table 2. LEZs should be regarded as a component of broader urban policies, and their effectiveness can be significantly enhanced through integration with complementary measures. A key strategy is combining LEZs with congestion charges, as evidenced by the success of Area C in Milan, which has effectively reduced both traffic and emissions. Furthermore, increasing the availability and accessibility of public transport (for instance, by subsidising or providing free fares for LEZ residents) can encourage shifts from private vehicles to mass transit and/or active transport modes. Successful examples can be found in Madrid and Barcelona. Equally important is investment in sustainable mobility infrastructure, including cycle lanes, pedestrian-friendly spaces, and extensive networks of electric vehicle charging stations. LEZ policies should also address emissions from other sources, including residential heating in winter. In cities like Warsaw and Madrid, emissions from heating systems often surpass those from road transport. Introducing emission standards for residential and commercial buildings, combined with financial incentives for upgrading to cleaner heating technologies, could mitigate this seasonal gap in LEZ effectiveness.

Promoting social equity is a critical aspect of implementing LEZs. To prevent the exacerbation of socio-economic disparities, support mechanisms for low-income households and small businesses, such as subsidies or low-interest loans for vehicle upgrades, are essential. The modernisation of the taxi fleet in Lisbon through the national-level collaboration offers a valuable model in this regard. Additionally, engaging local communities, particularly vulnerable groups, in the design of LEZ policies ensures that their mobility needs are met and public trust fostered. The establishment of buffer zones with less stringent requirements and phased implementation timelines can further facilitate compliance transitions. Raising public awareness and

Table 2 Summary of research findings within the selected publications on the impact of low emission zones on air quality, public health, and urban economy. *Source* https://urbanaccessregulations.eu (accessed: 20/09/2024)

Country	City	Access times	Time of implementation	Pollution-related effects	Health-related effects	Socio-economic effects	Authors
Germany	Berlin	24/7	2008	Small but significant reductions in NO_2 and PM_{10} in Berlin	Further efforts to reduce pollutants, particularly those affecting early childhood development, are needed to achieve meaningful health improvements	The findings indicate that implementing an LEZ reduces residents' life satisfaction by approximately 3% compared to the average satisfaction levels prior to the zone's introduction	(Gehrsitz, 2017; Morfeld et al., 2014; Sarmiento et al., 2022)
Italy	Milan	Weekdays 7.30 a.m–7.30 p.m	2008	LEZs reduced PM_{10}, NO_x, and CO_2, but Euro 6 vehicle exceptions limited effectiveness. The Ecopass Program reduced PM_{10} emissions by 15%, and its successor, Area C, achieved an additional reduction of 18% in 2012 compared to 2008	Studies have found that it helps reduce the incidences of heart and lung diseases, benefitting the overall population	The congestion charge for accessing Area C can financially burden low-income individuals with older, more polluting vehicles. Although exemptions and discounts exist, these groups may still struggle to meet the regulations without additional support	(Ku et al., 2020; Marchetti & Antonelli, 2024)

(continued)

Table 2 (continued)

Country	City	Access times	Time of implementation	Pollution-related effects	Health-related effects	Socio-economic effects	Authors
France	Paris	Weekdays 8 a.m–8 p.m	2015	Reduction in NO_2 levels (from 55 $\mu g/m^3$ to 42 $\mu g/m^3$), decreased vehicle emissions in central districts	340 fewer deaths/year, 170 fewer cases of low birth weight, 130 fewer new cases of coronary heart disease, 2,930 fewer asthma cases. Greater benefits observed with larger LEZ coverage	It may deepen social inequalities among low socioeconomic groups	(Bernard et al., 2020; Host et al., 2020)
Sweden	Stockholm	24/7	1996	PM emissions from heavy-duty vehicles across Stockholm dropped by 40% within four years of implementing the LEZ	The health benefits are relatively small	The LEZ for the city centre incurs social costs approximately ten times greater than its health benefits	(Börjesson et al., 2021; Wang et al., 2017)

ensuring compliance are also vital to the success of LEZs. Educational campaigns that highlight the health and environmental benefits of LEZs, coupled with transparent communication of outcomes, can significantly enhance public acceptance. For example, clear and accessible information about the positive impacts of the LEZs in Paris contributed to widespread public support. Regularly publishing data on air quality improvements and health statistics can further strengthen public confidence in these policies.

In the long term, LEZ policies must be dynamic and adaptable. Regular updates to emission standards, informed by technological advancements, are crucial. Transitioning from LEZs to zero emission zones (ZEZs) represents a significant step toward achieving sustained environmental and health benefits. Expanding the monitoring networks, as demonstrated in Rotterdam, captures localised impacts and allows policy measures to be refined. Integrating LEZ strategies with broader urban sustainability initiatives, e.g., climate action plans and the adoption of renewable energy, ensures their alignment with long-term urban development goals.

Future research should focus on addressing knowledge gaps and supporting the evolution of LEZ policies. Longitudinal studies which investigate the long-term health impacts of LEZs, including their impact on life expectancy and the prevalence of chronic diseases would be particularly useful. Additionally, the behavioural and economic effects of LEZs, including changes in mobility patterns and economic activities in zones impacted by traffic displacement, require further exploration. Seasonal variations in the effectiveness of LEZs, particularly during winter months, and comparative studies of LEZ outcomes in diverse geographic and socio-economic contexts, including cities in developing countries, will provide valuable insights and inform scalable best practices. Ultimately, advancing LEZ policies into more comprehensive and equitable instruments will maximise their potential. Integrating LEZs with supportive measures, addressing social equity concerns, and dynamically adapting to the changing conditions will be crucial for fostering cleaner, healthier, and more inclusive urban environments worldwide.

Acknowledgements This study was financed by the National Science Centre in Poland [2023/49/B/HS4/01406, 2024/53/B/HS4/00389].

References

Anderson, M. L. (2020). As the wind blows: The effects of long-term exposure to air pollution on mortality. *Journal of the European Economic Association, 18*, 1886–1927.

Andriolli, A. C., & Silva, L. T. (2024). Are low emission zones truly embraced by the public? *Environments, 11*, 106.

Anenberg, S., Miller, J., Henze, D., Minjares, R. (2019). A global snapshot of the air pollution-related health impacts of transportation sector emissions in 2010 and 2015. Available https://theicct.org/publication/a-global-snapshot-of-the-air-pollution-related-health-impacts-of-transportation-sector-emissions-in-2010-and-2015/

Attia, M. (2020). The effect of the low-emission zones policy and the rotterdam traffic plan on the usage of alternative modes of transport. Available https://thesis.eur.nl/pub/56572

Aydin, E., & Kürschner Rauck, K. (2023). Low-emission zones, modes of transport and house prices: Evidence from Berlin's commuter belt. *Transportation (Amst), 50*, 1847–1895.

Azdad, Z., Stoll, B., Müller, J. (2022). Clean cities the development trends of low-and zero-emission zones in Europe ZONE. *Transport Environment Growth*

Balaguer, J., Pernías, J. C., & Ripollés, J. (2023). Is vehicle scrapping affected by low-emission zones? The case of Madrid. *Transp Res Part A Policy Pract, 172*, Article 103668.

Bergeling, E. (2024). Social aspects of low emission zones: Stockholm case study. https://ieep.eu/wp-content/uploads/2024/06/Social-aspects-of-low-emission-zones-Stockholm-case-study.pdf

Bernard, Y., Miller, J., Wappelhorst, S., Braun, C. (2020). Impacts of the Paris low-emission zone and implications for other cities. https://theicct.org/publication/impacts-of-the-paris-low-emission-zone-and-implications-for-other-cities/

Blandin, L., Bouscasse, H., Mathy, S. (2024). Assessing the ex-ante impacts of a low-emission zone on transport poverty and vulnerability with the VulMob indicator—application to the low-emission zone in Grenoble, France. *SSRN Electronic Journal*

Boogaard, H., Janssen, N. A. H., Fischer, P. H., Kos, G. P. A., Weijers, E. P., Cassee, F. R., van der Zee, S. C., de Hartog, J. J., Meliefste, K., Wang, M., Brunekreef, B., & Hoek, G. (2012). Impact of low emission zones and local traffic policies on ambient air pollution concentrations. *Science of the Total Environment, 435–436*, 132–140.

Boriboonsomsin, K., Tanvir, S., & Barth, M. (2024). Modeling the impacts of low emission zone on route diversion and emissions of heavy-duty trucks: A southern California case study. *Transportation Research Record, 2678*, 794–805.

Börjesson, M., Bastian, A., & Eliasson, J. (2021). The economics of low emission zones. *Transp Res Part A Policy Pract, 153*, 99–114.

Burnett, R., Chen, H., Szyszkowicz, M., Fann, N., Hubbell, B., Pope, C. A., Apte, J. S., Brauer, M., Cohen, A., Weichenthal, S., Coggins, J., Di, Q., Brunekreef, B., Frostad, J., Lim, S. S., Kan, H., Walker, K. D., Thurston, G. D., Hayes, R. B., … Spadaro, J. V. (2018). Global estimates of mortality associated with long-term exposure to outdoor fine particulate matter. *Proceedings of the National Academy of Sciences, 115*, 9592–9597.

Cárdenas-Montes, M. (2020). Evaluation of the impact of low-emission zone: Madrid central as a case study. Available https://arxiv.org/abs/2012.13782

Cesaroni, G., Boogaard, H., Jonkers, S., Porta, D., Badaloni, C., Cattani, G., Forastiere, F., & Hoek, G. (2012). Health benefits of traffic-related air pollution reduction in different socioeconomic groups: The effect of low-emission zoning in Rome. *Occupational and Environmental Medicine, 69*, 133–139.

Chang, T. Y., Graff Zivin, J., Gross, T., & Neidell, M. (2019). The effect of pollution on worker productivity: evidence from call center workers in China. *American Economic Journal: Applied Economics, 11*, 151–172.

Charleux, L. (2014). Contingencies of environmental justice: The case of individual mobility and Grenoble's low-emission zone. *Urban Geography, 35*, 197–218.

Chen, S., Guo, C., & Huang, X. (2018). Air pollution, student health, and school absences: evidence from China. *J Environ Econ Manage, 92*, 465–497.

Currie, J., Hanushek, E. A., Kahn, E. M., Neidell, M., & Rivkin, S. G. (2009). Does pollution increase school absences? *Review of Economics and Statistics, 91*, 682–694.

Cyrys, J., Wichmann, H.-E., Rückerl, R., & Peters, A. (2018). Umweltzonen in Deutschland. *Bundesgesundheitsblatt, Gesundheitsforschung, Gesundheitsschutz, 61*, 645–655.

Dablanc, L., & Montenon, A. (2015). Impacts of environmental access restrictions on freight delivery activities. *Transportation Research Record: Journal of the Transportation Research Board, 2478*, 12–18.

De Vrij, E., & Vanoutrive, T. (2022). 'No-one visits me anymore': Low emission zones and social exclusion via sustainable transport policy. *Journal of Environmental Policy & Planning, 24*, 640–652.

Dias, D., Tchepel, O., & Antunes, A. P. (2016). Integrated modelling approach for the evaluation of low emission zones. *Journal of Environmental Management, 177*, 253–263.

Ding, H., Sze, N. N., Guo, Y., & Lu, Y. (2023). Effect of the ultra-low emission zone on the usage of public bike sharing in London. *Transportation Letters, 15*, 698–706.

Ezeah, C., Finney, K., & Nnajide, C. (2015). A critical review of the effectiveness of low emission zones (LEZ) as a strategy for the management of air quality in major European cities. *Journal of Multidisciplinary Engineering Science and Technology (JMEST), 2*, 1860–1868.

Ferreira, F., Gomes, P., Carvalho, A. C., Tente, H., Monjardino, J., Brás, H., & Pereira, P. (2012). Evaluation of the implementation of a low emission zone in Lisbon. *J Environ Prot (Irvine, Calif), 03*, 1188–1205.

Flanagan, E., Malmqvist, E., Gustafsson, S., & Oudin, A. (2022). Estimated public health benefits of a low-emission zone in Malmö. *Sweden Environ Res, 214*, Article 114124.

Font, A., Guiseppin, L., Blangiardo, M., Ghersi, V., & Fuller, G. W. (2019). A tale of two cities: Is air pollution improving in Paris and London? *Environmental Pollution, 249*, 1–12.

Gehrsitz, M. (2017). The effect of low emission zones on air pollution and infant health. *J Environ Econ Manage, 83*, 121–144.

Griffiths, C., Mudway, I., Wood, H., Marlin, N., Dundas, I., Walton, R., Grigg, J., Sheikh, A., Lee, T., Kelly, F. (2016). P180 impact of the London low emission zone on children's respiratory health: A sequential yearly cross sectional study 2008–2014. Thorax 71, A182.1-A182

Gu, J., Deffner, V., Küchenhoff, H., Pickford, R., Breitner, S., Schneider, A., Kowalski, M., Peters, A., Lutz, M., Kerschbaumer, A., Slama, R., Morelli, X., Wichmann, H.-E., & Cyrys, J. (2022). Low emission zones reduced PM10 but not NO2 concentrations in Berlin and Munich. *Germany J Environ Manage, 302*, Article 114048.

Hajmohammadi, H., & Heydecker, B. (2022). Evaluation of air quality effects of the London ultra-low emission zone by state-space modelling. *Atmospheric Pollution Research, 13*, Article 101514.

He, J., Liu, H., & Salvo, A. (2019). Severe air pollution and labor productivity: Evidence from industrial towns in china. *American Economic Journal: Applied Economics, 11*, 173–201.

Holman, C., Harrison, R., & Querol, X. (2015). Review of the efficacy of low emission zones to improve urban air quality in European cities. *Atmospheric Environment, 111*, 161–169.

Holnicki, P., Kałuszko, A., & Nahorski, Z. (2023). A projection of environmental impact of a low emission zone planned in Warsaw. *Poland. Sustainability, 15*, 16260.

Host, S., Honoré, C., Joly, F., Saunal, A., Le Tertre, A., & Medina, S. (2020). Implementation of various hypothetical low emission zone scenarios in Greater Paris: Assessment of fine-scale reduction in exposure and expected health benefits. *Environmental Research, 185*, Article 109405.

Jiménez-Espada, M., García, F. M. M., & González-Escobar, R. (2023). Citizen perception and ex ante acceptance of a low-emission zone implementation in a medium-sized Spanish City. *Buildings, 13*, 249.

Kang, C., Ota, M., & Ushijima, K. (2024). Benefits of diesel emission regulations: Evidence from the world's largest low emission zone. *J Environ Econ Manage, 125*, Article 102944.

Khomenko, S., Pisoni, E., Thunis, P., Bessagnet, B., Cirach, M., Iungman, T., Barboza, E. P., Khreis, H., Mueller, N., Tonne, C., de Hoogh, K., Hoek, G., Chowdhury, S., Lelieveld, J., & Nieuwenhuijsen, M. (2023). Spatial and sector-specific contributions of emissions to ambient air pollution and mortality in European cities: A health impact assessment. *Lancet Public Health, 8*, e546–e558.

Kin, B., Buldeo Rai, H., Dablanc, L., & Quak, H. (2024). Integrating logistics into urban planning: Best practices from Paris and Rotterdam. *European Planning Studies, 32*, 24–44.

Ku, D., Bencekri, M., Kim, J., Lee, S., & Lee, S. (2020). Review of European low emission zone policy. *Chemical Engineering Transactions, 78*, 241–246.

Lee, K., Bernard, Y., Dallmann, T., Braun, C., Miller, J. (2021). Impacts of a low-emission zone in Sofia. Available https://www.academia.edu/download/92573880/Impacts-of-LEZ-Sofia-TRUE-Report-EN-v4-dec21.pdf

Liu, Y., Yang, S., Li, C., Fang, Q., Han, C., Cheng, C., Cao, Y., Cui, F., & Li, X. (2024). Changes in years of life lost and life expectancy by reducing air pollution. *Air Quality, Atmosphere and Health, 17*, 89–102.

Maharani, F. (2024). Amsterdam's low emission zone (LEZ): A did analysis of their impact on residential property values. Available https://edepot.wur.nl/653342

Marchetti, E., Antonelli, C. (2024). Social aspects of low emission zones: Milan case study. Available https://ieep.eu/wp-content/uploads/2024/06/Social-aspects-of-low-emission-zones-Milan-case-study-IEEP-2024.pdf

Margaryan, S. (2021). Low emission zones and population health. *Journal of Health Economics, 76*, Article 102402.

Mohai, P., Kweon, B.-S., Lee, S., & Ard, K. (2011). Air pollution around schools is linked to poorer student health and academic performance. *Health Affairs, 30*, 852–862.

Moral-Carcedo, J. (2024). Dissuasive effect of low emission zones on traffic: The case of Madrid Central. *Transportation (Amst), 51*, 25–49.

Moreno, E., Schwarz, L., Host, S., Chanel, O., & Benmarhnia, T. (2022). The environmental justice implications of the Paris low emission zone: A health and economic impact assessment. *Air Quality, Atmosphere and Health, 15*, 2171–2184.

Morfeld, P., Groneberg, D. A., & Spallek, M. F. (2014). Effectiveness of low emission zones: large scale analysis of changes in environmental NO2, NO and NOx concentrations in 17 German cities. *PLoS ONE, 9*, Article e102999.

Morillas, C., Alvarez, S., Pires, J. C. M., Garcia, A. J., & Martinez, S. (2024). Impact of the implementation of Madrid's low emission zone on NO2 concentration using Sentinel-5P/TROPOMI data. *Atmospheric Environment, 320*, Article 120326.

Morton, C., Mattioli, G., Anable, J. (2018). A framework for assessing spatial vulnerability to the introduction of low emission zones: a case study of Edinburgh, Scotland. Available https://eprints.whiterose.ac.uk/134945/

Mullen, C., & Marsden, G. (2016). Mobility justice in low carbon energy transitions. *Energy Research & Social Science, 18*, 109–117.

Oltra, C., Sala, R., López-Asensio, S., Germán, S., & Boso, À. (2021). Individual-level determinants of the public acceptance of policy measures to improve urban air quality: The case of the Barcelona low emission zone. *Sustainability, 13*, 1168.

Pasquier, A., André, M. (2016). Decomposition of low emission zone strategies into mechanisms and methodology for assessing their impacts on air pollution. Available https://hal.science/hal-01450251/

Pestel, N., Wozny, F. (2019). Low emission zones for better health: Evidence from German hospitals. Available https://papers.ssrn.com/sol3/papers.cfm?abstract_id=3445811

Peters, J. F., Burguillo, M., & Arranz, J. M. (2021). Low emission zones: Effects on alternative-fuel vehicle uptake and fleet CO_2 emissions. *Transp Res D Transp Environ, 95*, Article 102882.

Player, L., Prosser, A. M. B., Thorman, D., Tirion, A. S. C., Whitmarsh, L., Kurz, T., & Shah, P. (2023). Quantifying the importance of socio-demographic, travel-related, and psychological predictors of public acceptability of low emission zones. *Journal of Environmental Psychology, 88*, Article 101974.

Poulhès, A., & Proulhac, L. (2021). The Paris Region low emission zone, a benefit shared with residents outside the zone. *Transp Res D Transp Environ, 98*, Article 102977.

Quak, H.J. (Hans), de Koster, M. (René) B.M. (2009). Delivering goods in urban areas: how to deal with urban policy restrictions and the environment. *Transportation Science* 43, 211–227

Rizki, M., Irawan, M. Z., Dirgahayani, P., Belgiawan, P. F., & Wihanesta, R. (2022). Low Emission zone (LEZ) expansion in Jakarta: acceptability and restriction preference. *Sustainability, 14*, 12334.

Saldarriaga, L., Patiño, G. (2024). Analysis and characterization of vehicular traffic in a low emission zone. Available https://doi.org/10.3233/ATDE240016

Sarmiento, L., Wägner, N., Zaklan, A. (2022). Standard-Nutzungsbedingungen 12, 99–106

Seter, H., Arnesen, P., & Moscoso, C. (2023). How do drivers' attitudes to low emission zones change after experiencing it? A pilot study in Norway. *Transp Res Interdiscip Perspect, 22*, Article 100934.

Settey, T., Gnap, J., & Beňová, D. (2019). Examining the impact of the deployment of low emission zones in Europe on the technological readiness of road freight transport. *Transportation Research Procedia, 40*, 481–488.

Sfendonis, N., Basbas, S., Mintsis, G., Taxiltaris, C., & Politis, I. (2017). Investigation of the user's acceptance concerning a low emission zone in the center of Thessaloniki, Greece. *Transportation Research Procedia, 24*, 280–287.

da Silva, F. N., Custódio, R. A. L., & Martins, H. (2014). Low emission zone: Lisbon's experience. *Journal of Traffic and Logistics Engineering, 2*, 133–139.

Simeonova, E., Currie, J., Nilsson, P., & Walker, R. (2021). Congestion pricing, air pollution, and children's health. *Journal of Human Resources, 56*, 971–996.

Tartakovsky, D., Kordova Biezuner, L., Berlin, E., Broday, D.M. (2020). Air quality impacts of the low emission zone policy in Haifa. Atmos Environ 232, 117472

Tassinari, F. (2024). Low emission zones and traffic congestion: Evidence from Madrid Central. *Transp Res Part A Policy Pract, 185*, Article 104099.

Treurniet, A. (2018). The effect of the low emission zone in rotterdam on air quality. Available https://thesis.eur.nl/pub/44207/Treurniet-A-384451.pdf

Verbeek, T., & Hincks, S. (2022). The 'just' management of urban air pollution? A geospatial analysis of low emission zones in Brussels and London. *Applied Geography, 140*, Article 102642.

Wang, Y., Song, S., Qiu, S., Lu, L., Ma, Y., Li, X., Hu, Y. (2017). Study on international practices for low emission zone and congestion charging executive summary. Available https://wrimexico.org/sites/default/files/Study_on_International_Practices_for_Low_Emission_Zone_and_Congestion_Charging.pdf

WHO [WWW Document]. (2022). https://iris.who.int/bitstream/handle/10665/368432/9789240047693-eng.pdf?sequence=1. URL https://iris.who.int/bitstream/handle/10665/368432/9789240047693-eng.pdf?sequence=1

Wolff, H. (2014). Keep your clunker in the suburb: low-emission zones and adoption of green vehicles. *The Economic Journal, 124*, F481–F512.

Wood, H. E., Marlin, N., Mudway, I. S., Bremner, S. A., Cross, L., Dundas, I., Grieve, A., Grigg, J., Jamaludin, J. B., Kelly, F. J., Lee, T., Sheikh, A., Walton, R., & Griffiths, C. J. (2015). Effects of Air Pollution and the introduction of the London low emission zone on the prevalence of respiratory and allergic symptoms in schoolchildren in East London: A sequential cross-sectional study. *PLoS ONE, 10*, Article e0109121.

Zhai, M., & Wolff, H. (2021). Air pollution and urban road transport: Evidence from the world's largest low-emission zone in London. *Environmental Economics and Policy Studies, 23*, 721–748.

Zhang, Y., Andre, M., Liu, Y., Wu, L., Jing, B., & Mao, H. (2018). Evaluation of low emission zone policy on vehicle emission reduction in Beijing, China. *IOP Conf Ser Earth Environ Sci, 121*, Article 052070.

Websites

https://www.lez-france.fr/
https://urbanaccessregulations.eu/

Marta Borowska-Stefańska is an Associate Professor and Deputy Director of the Institute of the Built Environment and Spatial Policy at the University of Lodz. Her research interests encompass topics related to flood hazard areas, flood risk management, evacuation, spatial development, daily mobility, transport geography, and GIS. Her academic journey includes a bachelor's degree in Geography (2008) and Spatial Management (2009), a master's degree in Spatial Management (2010), and a Ph.D. in Geographical Sciences (2014). She has also completed postgraduate studies in National Security (2023). Marta Borowska-Stefańska has authored and co-authored over 100 publications, including three academic monographs. She has participated in numerous research projects, serving as the principal investigator for grants funded by the National Science Centre (NCN) and the National Agency for Academic Exchange (NAWA). She actively collaborates with an international network of researchers from countries such as the USA, Germany, Austria, Slovakia, Romania, and Sweden. Her commitment to advancing scientific dialogue is evident in her active participation in over 30 international conferences and her role as a reviewer for prominent journals, including Transportation Research Part D, Safety Science, and Environmental Hazards: Human and Policy Dimensions. Since 2010, she has been teaching at the Faculty of Geographical Sciences of the University of Lodz. Her lectures primarily focus on topics such as Spatial Information Systems and Transport Geography. She has supervised 18 bachelor's theses, 27 engineering projects, and 14 master's theses.

Szymon Wiśniewski is an Associate Professor at the University of Łódź, serving in the Institute of Built Environmental and Spatial Policy within the Faculty of Geographical Sciences. He holds the position of Vice Dean for Development and Promotion of the Discipline of Socio-Economic Geography and Spatial Management. Additionally, he directs the Research Centre for European Spatial Policy and Local Development. His research focuses on transport infrastructure planning and development, transport geography, urban logistics, transport accessibility, spatial mobility, intelligent transport systems, and traffic engineering. His studies have addressed various issues, including the role of low-emission transport zones in shaping urban residents' mobility behavior, the sensitivity of transport systems to urban floods, and changes in travel behavior in Łódź following the implementation of Sunday trade restrictions. As Vice Dean, he oversees the research activities of academic staff, ensuring alignment with institutional goals and disciplinary standards. He monitors research progress, assesses publication and project activities, and documents the societal and economic impact of scientific endeavors. He also supervises the preparation of reports demonstrating the societal relevance of research, assisting researchers in compiling the necessary evidence. Szymon Wiśniewski has authored and co-authored over 130 publications, including four academic monographs. He has participated in numerous research projects, serving as the principal investigator in grants funded by the National Science Centre (NCN) and the National Agency for Academic Exchange (NAWA). His academic journey includes a bachelor's degree in Spatial Management (2009) and Logistics (2011), a master's degree in Spatial Economy (2011), a Ph.D. in Geographical Sciences (2014), and a habilitation in Social Sciences (2022). He has also completed postgraduate studies in Traffic Engineering and Transport Planning (2019) and National Security (2023). As an educator, Szymon Wiśniewski delivers courses aligned with his research interests, equipping students with knowledge to address contemporary challenges in transport planning and spatial development.

Magdalena Magiera holds a master's degree in space planning and organization from the University of Lodz. She has a bachelor's degree in spatial economy from the University of Lodz. Her research interest include flood vulnerability assessment, which she investigates in her doctoral dissertation. She completed internship named "Students' Power" which has allowed her to broaden her knowledge in urban planning and regeneration of urban areas in European cities, working alongside one of her professors. During her studies, she focused on sustainable spatial economy, environmentally friendly solutions in modern cities, and development of housing estates. Ms. Magiera was a member of students' scientific association SKN Civitas.

Strategic Transport Management as a Tool for Climate Change Mitigation

Miroslava Mikusova

Abstract The transportation sector significantly contributes to climate change through greenhouse gas (GHG) emissions, yet it often lacks comprehensive strategies for effective mitigation. Current approaches tend to focus narrowly on improving vehicle efficiency and reducing traffic volumes, which oversimplifies the complex challenges faced by this sector. This chapter argues that a more integrated and systematic proposal for mitigation measures is essential, particularly within the framework of environmental impact assessments. The proposed approach is global, referencing researches and key studies from all around the word. The central point is that the development of a clear typology of mitigation measures—encompassing preventive, corrective, and compensatory strategies—is crucial for addressing the climate impacts of transportation projects. By prioritizing prevention and incorporating actionable strategies at the project level, this study seeks to fill a significant gap in existing methodologies. Through a thorough analysis of current practices, literature review, and insights from international best practices, the chapter proposes a strategic management framework aimed at enhancing the environmental performance of transport systems. It emphasizes the need for local and regional authorities to take ownership of implementation processes, ensuring that interventions are contextually relevant and effective. Ultimately, this research aims to contribute to the sustainable development of transport systems that are resilient to climate change and aligned with broader environmental goals.

Keywords Climate change · Mitigation · Transport · Projects · Environmental impact assessment

---

M. Mikusova (✉)
Department of Road and Urban Transport, Faculty of Operation and Economics of Transport and Communications, University of Zilina, Zilina, Slovakia
e-mail: miroslava.mikusova@uniza.sk

M. A. Dulebenets et al. (eds.), *Transport and Climate Change*,
https://doi.org/10.1007/978-3-032-00563-2_12

1 Introduction

The transportation sector's contribution to climate change is predominantly linked to the emission of greenhouse gases (GHGs) from traffic (Pal et al., 2023). Mitigation efforts are typically focused on reducing these emissions through improved vehicle efficiency or reduced traffic volumes (SLOCAT, 2022). While this approach is valid, it represents an oversimplification that has hindered broader, more integrated strategies. For instance, in the context of Environmental Impact Assessments (EIAs), such narrow perspectives often result in the omission of meaningful mitigation measures when these extend beyond the traditional scope of EIA applications (Gharehbaghi et al., 2020; Soria-Lara et al., 2020). A significant research gap exists regarding the systematic integration of specific, actionable mitigation strategies within the EIA framework for transport projects. While studies highlight the need for broader considerations, there is currently no formalized or comprehensive approach to incorporating these strategies into EIA processes (Mayembe et al., 2023).

Hands and Hudson (2016) presented a new criteria-based review package to evaluate climate change adaptation and mitigation in environmental statements. After the application to a sample of 10 transport developments in England, their study revealed that many projects lacked detailed climate change adaptation and mitigation measures, with the evidence of commitment to these measures being poor or non-existent. While air quality and noise pollution were typically monitored, broader climate change mitigation measures were not adequately addressed. Similarly, the report by OECD (2024) underscores the absence of formalized guidelines and the need for a more holistic approach to incorporating climate change mitigation into EIAs. The case study by Ambade and Ambade (2023) discusses the lack of comprehensive mitigation strategies in the EIA process for metro rail construction projects.

Additionally, the Gautrain Rapid Rail Link case study (Gautrain Management Agency, 2023) highlights the difficulties in implementing comprehensive mitigation strategies within the EIA framework, particularly in the context of newly emerging EIA legislation. Furthermore, a report by the International Energy Agency (IEA, 2019) underscores the need for comprehensive strategies that go beyond current EIA practices, especially in harder-to-abate modes like heavy-duty vehicles, maritime, and aviation. Current practices often lack detailed climate change mitigation and adaptation measures. The study by Safitri (2023) identifies a research gap in integrating macro-ergonomics approaches to improve overall system efficiency and safety in public transportation, further emphasizing the need for comprehensive mitigation strategies within EIAs. Moreover, the Gaps Analysis Report in the Smart Mobility Field (Shiftan et al., 2021) identifies fundamental challenges and research gaps in smart mobility, including the need for better integration of climate change mitigation strategies within transport planning and EIA processes. Addressing this issue requires the development of a structured methodology to ensure that transport project assessments effectively contribute to climate change mitigation.

This chapter aims to present a systematic proposal for mitigation measures to address the climate change impacts of transportation projects, addressing the identified gap where such measures are currently absent in an integrated form. The first section delineates the typology of mitigation measures, categorizing them into preventive, corrective, and compensatory strategies, aligned with the mitigation hierarchy. After that, the chapter reviews existing methodologies and identifies the best practices through case studies, demonstrating the potential for effective integration of these measures in project planning and execution. The subsequent sections explore the challenges and opportunities associated with applying this framework in different transport contexts, with a particular focus on how local and regional authorities can implement these strategies effectively. The last section concludes with a discussion on the state of the art, assessing mitigation options and their potential for application, and the reasons that have led to their limited consideration in the EIA to date. After presenting the possibilities of climate change mitigation in transport infrastructure, the main challenges for its incorporation into both the EIA and existing infrastructure and the financing possibilities are analyzed. This research aims to provide actionable recommendations for stakeholders in the transportation sector to enhance environmental performance and resilience against climate change.

2 General Considerations

There are differences in the interpretation of the term mitigation between European countries, where mitigation covers prevention and correction measures, and the United States, where this term also includes compensation, a dominant view in the field of climate change (Soria-Lara et al., 2016). When analyzing the impacts on biodiversity and the applicable mitigation measures, it is common to refer to the mitigation hierarchy, a concept that is rarely applied to climate change. This hierarchy establishes an order of priority when applying mitigation measures, first to avoid impacts, then to minimize them, and finally to compensate for the unavoidable ones. This hierarchy is presented on Fig. 1.

Mitigation hierarchy gives rise to three types of mitigation measures to be applied, following descending hierarchy, which are:

- Preventive mitigation (or preventive or protective measures) that is applied with the aim of avoiding impacts before they occur. In case of climate change, the main aim is to avoid or reduce GHG emissions or to avoid the destruction of carbon sinks. Due to its typology, it is the most desirable type of mitigation, since it avoids the generation of impacts, instead of acting to correct the damage caused.
- Corrective mitigation (or corrective measures) that is applied with the objective to correct (in the ideal situation) or more realistically to alleviate, attenuate, or minimize impacts that will certainly and inevitably occur. In case of climate change, this type of mitigation is very limited, since once the GHGs have been

Fig. 1 Mitigation hierarchy

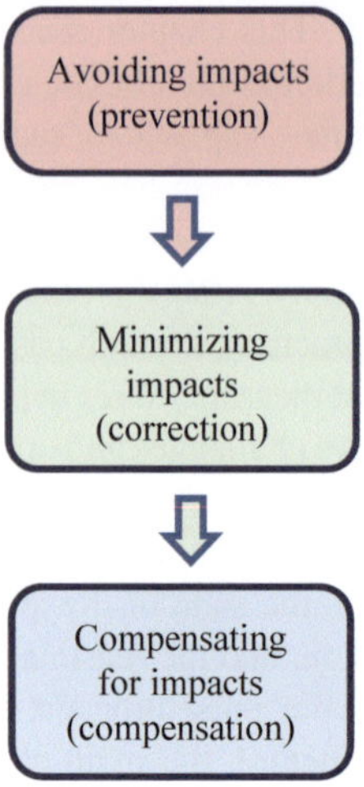

emitted, no correction is possible. Its application focuses on the recovery of live sinks affected by the construction of transport infrastructures.

- Compensatory mitigation (or compensatory measures), which aims to create new values or adopt positive environmental measures for a resource that compensate or balance an inevitable loss of that resource. In case of climate change, GHG emissions can be offset by actions in other areas aimed at reducing their concentration in the atmosphere or preventing their increase, and the irremediable loss of sinks can be offset by the creation of new ones.

3 Preventive Mitigation

Prevention to avoid GHG emissions is the preferred line of action against climate change, and the greatest efforts should be focused on it. The main measures proposed by the Intergovernmental Panel on Climate Change (IPCC) for transport are behavioral changes, such as avoiding unnecessary trips or driving more efficiently, shifting to more efficient modes of transport (from private vehicles to public transport, walking, or cycling), improving the performance of planes, trains, ships, and vehicles and using low-carbon fuels. Preventive mitigation includes measures intended to prevent the occurrence of environmental impacts, which in case of climate change are the emission of GHGs or the destruction of sinks. Its main characteristic is that the associated measures are anticipatory in nature and normally applied before the impact occurs unlike responsive measures (e.g., corrective or compensatory mitigation measures, which are applied when the impact already exists) (Fawzy et al., 2020). The preventive mitigation measures can be categorized according to their relationship to the environmental assessment into one of the following groups:

- Preventive mitigation outside of the Environmental Assessment (EA). Measures that go beyond the EA and therefore cannot be incorporated specifically into it, although they can be cited or recommended. Examples include: the energy efficiency of vehicles, the types of fuels used, the implementation of recharging points for specific fuels (biofuels, electricity, or hydrogen), increased vehicle occupancy (although this could be encouraged by designing lanes for high-occupancy vehicles), road management measures, the application of fees and taxes to penalize certain modes of transport or fuels, or the application of subsidies with the opposite aim.
- Preventive mitigation specific to the Strategic Environmental Assessment (SEA). This includes measures specific to strategic planning, such as urban planning, sustainable mobility, modal connections, low climate impact transport modes and public transport, or design and extension of environmentally friendly transport networks.
- Preventive mitigation specific to the Environmental Impact Assessment (EIA). This includes measures specific to the project design and the EIA.

Focusing on the EIA, preventive mitigation measures can be applied in three different phases of infrastructure development: design, construction, and operation. Most preventive measures should be adopted in the design, such as the selection of locations, geometry of the routes, typologies, materials, or avoided deforestation. In the construction phase, the machinery and work methods used, as well as the organization of work and transport routes, are particularly influential. There is also a group of preventive measures related to the operation of the infrastructure, which often go beyond the scope of the EIA, as they can be decided after the evaluation, but which can sometimes be incorporated into the decision-making process, having a certain significance.

Within the context of climate change mitigation various studies underscore the importance of the Polluter Pays Principle (PPP) in addressing social exclusion and spatial justice. Khan (2015) highlights PPP as a crucial economic, ethical, and legal instrument that promotes global responsibility for climate change adaptation and mitigation. By ensuring that polluters bear the costs of their actions, PPP can generate reliable funding for climate initiatives and address the global common issue. This principle supports the equitable treatment of disadvantaged communities and helps address inequalities exacerbated by environmental degradation. Bugge (2009) discusses the dilemmas of justice in national and international contexts related to PPP. The study highlights that PPP is not only an effective approach for internalizing environmental costs but also a legal principle for allocating the costs of pollution prevention and liability for environmental damage. This dual role of PPP is crucial for promoting environmental justice and ensuring that polluters are held accountable for their actions. The International Transport Forum (ITF, 2024) emphasizes the need for effective data reporting and monitoring frameworks to understand the environmental impacts of new mobility services. By implementing these frameworks, local authorities can better manage the benefits and negative impacts of new mobility, ensuring that the costs of pollution are borne by those responsible.

Recent studies also highlight several key strategies for mitigating climate change through effective urban planning and transportation policies. Coordination and integration of spatial and transport planning are essential for reducing CO_2 emissions and promoting sustainable urban development. By aligning transportation systems with land use planning, cities can encourage the use of public transit, walking, and cycling, which are lower-emission modes of transportation compared to private vehicles (U.S. Department of Transportation, 2025). Another significant strategy is Transit Oriented Development (TOD). It consists in creating compact and walkable communities that are centered around high-quality public transportation. TOD reduces the need for private vehicle use, thereby it is lowering emissions and promoting sustainable living. A report by the Institute for Transportation and Development Policy (2017) emphasizes that TOD can significantly reduce emissions by creating walkable communities with easy access to public transit. This model integrates housing, retail, and commercial spaces within close proximity to transit hubs, encouraging the use of public transportation over private vehicles.

The research of Center for Climate-Smart Transportation (2025) highlights the impact of mixed-use developments on vehicle miles travelled (VMT) and greenhouse gas emissions. Their research indicates that TOD can effectively reduce VMT, leading to lower emissions and improved air quality. This is particularly evident in projects that promote active transportation, such as walking and cycling, alongside public transit. The Resilient Connecticut (2024) project underscores the importance of TOD in enhancing climate resilience. By steering the development away from environmentally sensitive areas and reducing runoff, TOD can mitigate the impacts of climate change while promoting sustainable growth. This approach also supports the creation of affordable housing, which is crucial for equitable urban development. The Pace Environmental Law Review (2023) discusses the benefits of TOD in improving mobility and access to essential services. By providing affordable and accessible transit options, TOD can reduce overall commuting expenses and decrease the carbon footprint of urban areas. The study also highlights the need for equitable TOD to ensure that low- and moderate-income communities benefit from these developments.

3.1 Preventive Mitigation in Design of Transport Projects

Preventive mitigation in the design of transport projects focuses on implementing measures that prevent environmental impacts, particularly the emission of greenhouse gases (GHGs), before they occur. This section explores various aspects of preventive mitigation, emphasizing the importance of anticipatory actions over reactive measures. The key topics include the influence of route geometry on GHG emissions, the selection of construction typologies and materials, and the strategic location of transport infrastructure. The discussion also covers optimizing transport effectiveness, avoiding deforestation, and designing reserved platforms for public transport. By prioritizing shorter routes, efficient construction methods, and environmentally friendly materials, transport projects can significantly reduce their carbon footprint and contribute to sustainable development.

3.1.1 Aspects Associated with the Geometry of Layouts

The design of routes will have repercussions on GHG emissions during operation. Although many environmental and social factors influence the situation, with equal impacts on other aspects, priority should be given to shorter routes. However, this depends on a large number of factors such as relief, physical or geotechnical limitations, environmental values of the territory or the population to be served. If the layout conditions are similar for the options, the longer they are, the greater investment is required, so it is not likely that longer options will be adopted unjustifiably. If the conditions are not equal, a longer option may have economic advantages, for example, if it requires fewer structures or tunnels.

In case of roads, the lower the gradient of the routes, the lower the fuel consumption of the vehicles that circulate on them. But under other equal conditions, the better the route, the higher the average speed, which also influences fuel consumption. However, many other factors influence this pattern, such as the relief and environmental values of the territory. Feijoo and Moreira (2020) analyzed the emissions generated on a stretch of motorway with different geometric solutions, referring to the gradient, the section of the road, and the design speed, detecting significant differences in CO_2 emissions. Variations in gradient can bring significant increases in CO_2 emissions, since heavy vehicles and buses depend largely on the inclination of the gradient, not only on the speed. In case of railways, the layout also determines the speed and, therefore, the travel times. This is an essential factor for their competitiveness, especially in commuter railways, where there is competition between modes of transport for minutes or tens of minutes of savings in travel time.

3.1.2 Selection of Typologies and Construction Solutions

Different construction typologies for the same project can have important repercussions in terms of GHG emissions in construction. In decision-making, the GHG emissions of the alternatives should be considered among the aspects that influence decision-making. Some examples include the following:

- **Tunnel typology**. Underground sections can be constructed by means of excavation in a mine or by means of cut-and-cover tunnels. The decision is usually conditioned by geotechnical, environmental, urban or economic aspects, but the GHG emissions of each construction alternative should also be incorporated into the decision-making process.
- **Types of viaducts**. The type of viaduct, both its design and its form of construction, is usually conditioned by hydrological, environmental, functional, and economic

aspects. Depending on the type and quantity of materials used and the form of construction, the greenhouse gas emissions will vary.

- **Types of walls and defenses**. The construction of retaining walls is primarily affected by geotechnical, functional, safety, and economic aspects. However, depending on the wall type (in situ concrete, prefabricated, gabions, breakwaters, blocks, green walls, etc.), GHG emissions can vary considerably.
- **Types of sections**. A section of infrastructure, whether it runs on an embankment or a cut, can be designed with slopes or with walls to reduce the occupation or volume of excavated or supplied earth. Both options have a direct influence on the associated emissions.
- **Types of interchanges**. The design of interchanges is determined primarily by their functionality, land availability, and costs. However, construction solutions, with one or several levels and longer or shorter structures, have repercussions on GHG emissions. In general, the greater the complexity of the interchange and the structures planned, the higher the emissions.

3.1.3 Material Selection in the Design Phase

The decision to use one or another material can be made in the design phase or during the construction itself, depending on the specific material in question. Mitigation measures can be adopted in the design of transport infrastructures, taking into account the GHG emissions of construction materials, so that, with equal operability or functionality, the materials with the lowest impact are preferred. Some examples of decisions on construction materials in the design phase that may have an impact on GHG emissions are related to type of road surface. The choice of the type of road surface (conventional aggregate, microaggregate, drainage, concrete, etc.) has a clear impact on GHG emissions as well as on adaptation to climate change, as it is more or less resistant to climatic influences.

Numerous works analyze the life cycle of pavements (Giani et al., 2015; Keijzer et al., 2015; Mahdi et al., 2023; Plati & Tsakoumaki, 2023; Zheng et al., 2019). For example, Keijzer et al. (2015) propose replacing brick pavements, which consume much more energy than other types, with concrete blocks for infrastructure projects in the Netherlands. For bituminous pavements, the use of low-energy asphalt and the application of rejuvenators have the greatest potential to improve the sustainability of pavement materials (Keijzer et al., 2015). The use of end-of-life tires (ELT) is also very interesting. It is possible to use pavements with lower friction in order to reduce fuel consumption (Maga et al., 2023). Selecting construction materials with lower greenhouse gas emissions during the design phase, such as low-energy asphalt or concrete blocks instead of brick pavements, can significantly reduce the environmental impact of transport infrastructure.

3.1.4 Location Selection

Alternative locations for railway stations and the type of transport used to access them influence GHG emissions and the effectiveness of the transport mode, its social acceptance, and use. Adequate consideration of these aspects in the design can contribute to mitigation of GHG emissions during operation. On roads, this influence is much less significant, as it does not depend on specific transport connection points such as railway stations. However, the location of links on dual carriageways and motorways, for example, can lengthen or shorten journeys to access these roads from the region crossed.

3.1.5 Optimizing Transport Effectiveness

In case of rail transport, the greater the number of passengers, the greater the reduction in the use of other modes of transport with higher emissions, which leads to a reduction in overall emissions. The design of the routes, especially their length, geometry, and connected points, determine the speed of circulation and travel times, and the population reached, thus determining the number of passengers and the effectiveness of transport. Generally speaking, the shorter a railway route is and the fewer stops it has, the faster the transport will be, and the more competitive it will be with respect to other modes of transport. However, both factors, length and stops, can reduce the number of points served, and thus the potential user population. Conversely, long routes with many stops may potentially attract a larger population, but at the cost of slow transport that is not competitive in terms of time with other modes. A clear example of this is the commuter rail network, where the essential design factor is to provide service to certain essential nodes, which completely condition the layout.

3.1.6 Avoiding Deforestation

The construction of transport infrastructures involves the temporary or permanent occupation of land, except for bare soil, infrastructure or urbanized areas, the destruction of the existing vegetation cover. The more carbon sinks an alternative has, the less its contribution to climate change. Minimizing the impact on vegetation is an essential preventive measure that must be adopted from the early stages of infrastructure design. From the perspective of climate change, it is desirable to affect plant communities with a high carbon stock and sequestration capacity as little as possible. However, decisions should not be based solely on this criterion, and biodiversity should be considered, since there is no direct relationship between stock/sequestration and the ecological value of vegetation. Numerous authors have highlighted the risk to biodiversity due to maximizing the production of carbon credits (Andersen & Jang, 2021; Andres et al., 2023; Kedward et al., 2023; Prodanovic et al., 2024; Ward et al., 2015), an aspect that should be carefully taken into account.

Consequently, efforts must be made to minimize the impact on vegetation, as indicated by good practice in the EIA and in design with environmental criteria, but with a vision where the criteria of biodiversity and naturalness prevail over mere carbon storage. In order to achieve an adequate design that minimizes the impact on vegetation, it is necessary to have detailed vegetation mapping from the start of planning, prepared in the stages prior to the EIA, so that the initial planning of corridors and subsequent planning of specific alternatives is done taking into account the possible impacts. Currently, informative studies for major roads and railway lines follow a phased planning, where in the first stage the capacity of the territory is assessed from different points of view, looking for the most suitable corridors. This phase prior to the EIA is the basis for the preparation of the initial project documentation, which is used to carry out the scoping phase and start the regulated EIA procedure. In smaller projects, this prior phase is often not taken into account, so the EIA acts in a more reactive manner, assessing the impacts that will occur in the environment without having the anticipatory capacity as would be desirable, in order to avoid the impacts before they occur.

3.1.7 Induced Mobility and Accessibility

On highways, it is logical to look for alternatives that attract the most traffic, since that is their objective, but this attraction should translate into a reduction in traffic on other routes; that is, the new highways, especially if they have better performance, should attract traffic from other lower-quality roads. The differences in traffic attraction are closely related to the layout, the distance to urban centers, or the existence of links and may differ significantly between alternatives. The choice of road type will also determine the type of traffic and travel. Conventional roads provide full accessibility by having direct connections, but they are slower and more dangerous. Dual carriageways also provide good accessibility, with a large number of junctions. Motorways have fewer junctions and are focused on attracting long-distance traffic, not local traffic.

3.1.8 Design of Reserved Platforms

A highly desirable measure on access roads to large cities is the construction of reserved platforms for public transport and sometimes for high-occupancy vehicles (HOV). These restricted-use platforms have less traffic than the rest of the road and therefore less congestion during rush hours, which makes traffic flow more fluid, travel times shorter and public transport or high-occupancy vehicles more efficient. The decision to build reserved platforms on a roadway segment is often strategic, prior to the EIA. They are usually designed on existing roads and developed through specific projects, rather than alternatives in the design of a road.

However, in the detailed planning process and in the EIA, different operational alternatives can be considered, which influence its design, especially in terms of the type of authorized users (e.g., platforms reserved solely for public transport). This is the case for bus or bus-taxi lanes in cities as well as trams in some places. However, this is also applicable to roads. Reserved lanes can be reversible, with a variable direction of traffic depending on the time of day or fixed with a single direction of traffic. The decision to build one type or another depends on the time and direction of the congestion. On roads with congestion marked in time and in the direction of traffic, usually entering the city in the early morning and leaving in the afternoon, it is possible to build reversible lanes. However, when there is no marked time, and the congestion is repeated throughout the day and in both directions, the reversible lanes are not operational, requiring two reserved lanes, one in each direction.

Another decision that may generate different operational alternatives is the type of traffic accepted on the reserved platform, whether it will be a bus lane or a bus-HOV lane, and in the latter case when a roadway is considered to be an HOV (what should be the vehicle occupancy). This decision depends on the traffic captured in each option, and the capacity of the reserved platform (one or more lanes). Regarding the HOV concept, there is some disagreement as to whether vehicles with more than two or more than three passengers should be considered as high-occupancy vehicles. If the traffic captured by vehicles with three or more passengers is very low, it can be decided to consider vehicles with two or more passengers as HOV, but this decision may vary depending on the specific case. Decision-making of this type is determined by its functionality and the traffic captured. It would be desirable to also incorporate consideration of the contribution to climate change in each of the operational alternatives, according to the type of platform and traffic captured, since they are usually practically the same in terms of the layout.

3.2 Preventive Mitigation in Construction of Transport Projects

Preventive mitigation in construction is most effective when the key decisions are made during the design stage of a transport project. This includes choices about the project's layout, construction methods, types of structures, and materials. These early decisions have a significant impact on the project's overall environmental footprint and offer the best opportunity to reduce greenhouse gas emissions.

One of the most important aspects of prevention during the construction phase of a transport project is the selection of materials. While the essential characteristics of the materials are typically outlined in the project's technical specifications, there is often some flexibility for contractors to make specific choices. Within these established guidelines, contractors can opt for more sustainable options, such as recycled materials. Using recycled materials is a particularly promising strategy for reducing the environmental impact of construction. Whether this possibility is explicitly defined in the project or left to the discretion of the contractor, it represents a practical and impactful way to incorporate preventive mitigation into the construction process. The results of Hoy et al. (2024) show that the use of recycled materials instead of primary resources greatly reduces the environmental impact of road material manufacturing. There are various recycled materials currently being used for roadway construction, such as end-of-life tires (ELT), recycled aggregates from construction and demolition waste (CDW), roadbed waste, biomass ash or slag from thermal power plants. Possibly, the recycled material that has aroused the most interest is used tires, as their disposal is a major environmental problem. From a climate change perspective, the application of Non-Forest Use (NFU) means a saving in emissions compared to the production of new materials and avoids the generation of GHG that could be associated with the disposal of these tires.

Another important aspect of preventive measures during the construction phase is the use of low-impact materials. There is increasing knowledge about the life cycle of materials, which allows for a more environmentally friendly selection. Papasavva et al. (2022) provides a detailed life cycle assessment (LCA) of various paint processes, including those used for road markings, highlighting their environmental impacts. Burghardt and Pashkevich (2020) discuss the financial and environmental impacts of different road marking materials, including acrylic paints, over their life cycle. Xu et al. (2021) analyze the life cycle of acrylic paints for road markings. According to these authors it is a good environmental practice to check the carbon footprint of the different products when deciding which ones to use. This aspect should be integrated into the Environmental Management Systems (EMS) that practically all construction companies use, but its voluntary nature and the fact that significant economic differences between suppliers or types of materials may be at stake make the effectiveness of this option very moderate. Consequently, it is possible to promote the use of materials with a lower carbon footprint by including it as a valuable criterion in tenders.

Regarding the origin of materials used in the construction phase of any transport project, the contractor usually has a wide freedom to choose suppliers and places of origin. Except for some materials such as soil, the transport of which is considered within the projects as an essential unit of work, most materials are paid for on site. That means that the price of the material includes transport from the place of production and distribution to the construction place. Obviously, the most important criterion when choosing suppliers is the cost of the materials used, rather than the place of origin. However, the transport of materials can cause their carbon footprint to differ greatly, depending on whether the source is nearby or not. In this sense, the considerations made in the previous section are applicable, regarding the convenience of considering the carbon footprint, in this case associated with transport, its integration into the EMS, and its assessment in tenders.

3.3 Preventive Mitigation During the Operation Phase of Transport Projects

Preventive mitigation during the operation phase of transport projects focuses on managing traffic to create conditions that result in lower GHG emissions. Although opportunities for mitigation in this phase are more limited compared to the design stage, they are still significant. The greatest potential lies in improving the energy efficiency of vehicles and promoting the use of clean fuels, which can substantially reduce emissions during the life of a transport system. Traffic regulation and control measures are the two key strategies for lowering emissions, particularly through speed management. A study by Singh et al. (2015) examined three approaches to reducing GHG emissions in Norway's transport sector: increasing the use of electric vehicles to 39%, reducing car trips or kilometres travelled by 25%, or combining these approaches with 15% electric vehicle adoption and a 19% reduction in mobility. All three strategies achieved a 25% reduction in emissions, meeting Norway's 2020 climate targets. Promoting electric vehicles requires national-level strategic planning and investment, while reducing mobility depends on broader measures such as shifting transport modes, redesigning urban areas, and redistributing work locations. When designing transport systems, priority should be given to options that minimize emissions and reduce overall travel distances.

Traffic regulation and control measures also play a role in emission reductions, particularly through speed management. The research of Perez-Prada et al. (2017) confirmed that decreasing the speed limit is associated with substantial environmental benefits and insignificant impacts on traffic performance and shows great potential to alleviate atmospheric pollution in highly-populated urban locations situated next to busy highway segments. Fondzenyuy et al. (2024) performed a systematic review investigating the impacts of speed limit changes on emissions, highlighting the potential environmental benefits of speed management policies. This study investigated how different speed management strategies in urban areas can influence CO_2 emissions, emphasizing the benefits of smoother traffic flow.

Numerous studies have shown a clear link between speed reductions and lower GHG emissions (Albuquerque et al., 2020; Bigazzi & Rouleau, 2017; EEA, 2021; Folgerø et al., 2020; Fondzenyuy et al., 2024; Gao et al., 2024; Metcalfe, 2023). Comprehensive data for various vehicle categories and speed conditions is provided in the Handbook Emission Factors for Road Transport (HBEFA) (Table 1). For example, reductions in road speed by 10 to 20 km/h can decrease CO_2 emissions by 10% to 23%, depending on the specific conditions and types of vehicles involved. These findings highlight the importance of implementing policies that regulate traffic speeds to improve both energy efficiency and environmental outcomes in transport systems.

At low speeds, vehicles generally produce higher CO_2 emissions due to stop-and-go traffic and inefficient fuel use, especially for passenger cars, buses, and heavy-duty trucks operating in urban areas. Emissions tend to decrease at moderate speeds (50-90 km/h) for most vehicle categories, as engines operate more efficiently, resulting

Table 1 Relationship between speed and CO_2 emission factor for different types of vehicles (INFRAS, 2023)

Vehicle category	Speed (km/h)	CO_2 emissions (g/km)
Passenger cars (petrol)	30	180–200
	50–90	140–160
	120+	180–220
Passenger cars (diesel)	30	150–170
	50–90	120–140
	120+	160–200
Electric vehicles	All speeds	0
Buses	30	900–1100
	50–70	800–900
	90+	1000–1200
Light commercial vehicles	30	180–220
	50-80	160–180
	100+	200–240
Heavy-duty trucks	30	800–1000
	60–80	500–700
	90+	1000–1200
Motorcycles	30	100–140
	50–70	80–100
	90+	120–150
Trains (electric)	All speeds	0
Trains (diesel)	30–60	80–100
	100–160	100–120

in lower fuel consumption. However, at higher speeds, such as on highways, aerodynamic drag increases fuel use, causing emissions to rise again, particularly for passenger cars, motorcycles, and heavy vehicles. Electric vehicles (EVs) stand out for producing zero direct emissions, though their overall environmental impact depends on the electricity source used for charging. Trains also demonstrate variability: electric trains produce no emissions, while diesel trains contribute CO_2, especially at higher speeds. In general, heavy-duty vehicles and buses emit significantly more CO_2 compared to lighter vehicles due to their larger size and weight, which require greater energy to operate efficiently.

Another advantage associated with reducing road speeds is reducing the need for high-powered vehicles. Despite the increasing energy efficiency of engines and vehicles, EU-15 data (EPA, 2022) indicate a progressive increase in the power, cylinder capacity, and weight of cars since the 1990s, which implies higher fuel consumption for their operation. The use of lighter and lower-powered vehicles allows lower emission ratios to be achieved. As a consequence, reducing the speed of traffic on roads would lead to a reduction in CO_2 emissions, being a clear preventive mitigation measure. Various countries, such the Netherlands, France, Sweden, Finland, Germany, Denmark, the United Kingdom, Australia, and the United States, have adopted speed control measures (EPA, 2023). In Rotterdam, for example, reducing the speed on motorways to 80 km/h allowed for better use of the road, and despite a 3% increase in traffic, congestion was reduced by 2 km and its duration by 30 minutes. Different EU countries have decided to reduce the speed of trucks, and it is considered advisable to apply speed limits to vans as well.

A more drastic group of measures than speed reduction is the introduction of road traffic restrictions, which are designed to forcibly reduce traffic. These measures are often associated with acute episodes of air pollution, but they similarly reduce GHG emissions. Li and Jones (2015) analyze different scenarios of restrictions on private vehicle traffic in Beijing, a city with severe pollution problems, and their influence on CO_2 emissions. Another possible measure to reduce traffic is the introduction of tolls. This does not fall under the previous group of traffic restrictions, since in reality the passage of vehicles is not restricted, but rather a toll is levied, which may have a different underlying approach. Tolls are most commonly associated with administrative concessions for the construction and operation of roads, especially motorways. The toll, paid by the drivers who use the road, covers the costs of construction, maintenance, and operation, and the profit of the concessionaire company, all for a set period of time.

Recent studies highlight the effectiveness of tools, taxes, and restrictions in managing urban traffic and reducing emissions. Lehe (2019) and Eliasson (2014) demonstrate that congestion pricing in cities like Stockholm and London significantly reduces traffic volume and emissions. Gibson and Carnovale (2015) show that road pricing in Milan improves air quality by altering driver behavior. De Palma et al. (2018) discuss the economic benefits of urban road pricing, emphasizing its role in mitigating congestion and emissions.

All these findings underscore the potential of fiscal and regulatory measures to enhance urban sustainability. An interesting example of toll management is the Lyndon B. Johnson Expressway (LBJ) located north of Dallas (Texas). On this highway, managed toll lanes are established in which the toll rate is adjusted in real time, constantly changing based on the average speed or the number of vehicles that want to use these lanes. In this way, as traffic increases, the price goes up, which discourages the use, helping to reduce traffic on the express lanes, and thus maintaining the fluidity of traffic. Drivers can choose to use the toll lanes by paying or drive on the general lanes of the highway. This model is focused on guaranteeing the fluidity of traffic, at least on the toll lanes, and not on environmental aspects, but the idea of penalizing the moments of greatest traffic intensity is really interesting and would be applicable to metropolitan traffic management models at peak times.

The use of tolls has a significant national, social and political influence. For example, France's high-capacity road network is dominated by toll motorways. In Germany, there are no tolls. In Switzerland and Slovenia, a fixed weekly, monthly, or annual fee is paid when entering the country. Politically, there are arguments because tolls are considered discriminatory, favoring the wealthiest classes, or even being a double taxation. However, many of these criticisms are associated with administrative concessions (toll motorways). Beyond these discussions, the establishment of tolls that act as a Pigouvian tax to curb the emission of pollutants seems reasonable and also an application of the "polluter pays" principle, which is the basis of the EU environmental regulations.

3.4 Behavioral Preventive Mitigation

There is a type of preventive mitigation that is based on modifying the behavior of a group, or of society in general, to achieve a lower contribution to climate change. Energy-saving campaigns aimed at achieving greater awareness and reducing consumption have been common for years. Focusing on transport infrastructure, the main behavioral measures that promote mitigation are the use of public transport instead of private vehicles and efficient driving.

3.4.1 Use of Public Transport

Promoting public transport goes beyond planning, depending on political will. The expansion of public transport networks, the cost to users, or the disincentives to use private vehicles are political measures conditioned by social acceptance. To define the possibilities for action, it is recommendable to start from the differences between mobility by private vehicles and public transport (Fig. 2). Starting from the different stages that make up a journey and adding the overall cost involved, measures can be established to encourage the use of public transport and discourage the use of private vehicles (Table 2). The promotion of public transport will be more successful with more measures taken in both directions. While encouraging public transport has broad social acceptance, discouraging private vehicles is rejected by the majority, and these trends will depend on the environmental awareness and education of society.

The comfort of public transport has improved a lot, except during rush hours in urban areas for short journeys. Its disadvantage is a longer journey time if many stops are to be made, connections are required, or traffic jams occur. For private vehicles, the disadvantages are traffic jams, parking and higher operational costs. The expansion of networks, an increase in frequencies and interchanges, with acceptable costs for the user, should be encouraged, whereas higher parking costs and even urban tolls make the private vehicle an unattractive alternative.

3.4.2 Efficient Driving

Another aspect of preventive behavioral mitigation is efficient driving. There is a consensus on the need to adopt efficient driving to reduce fuel consumption and emissions, and there are several guides dedicated to this (Porter et al., 2013). Among

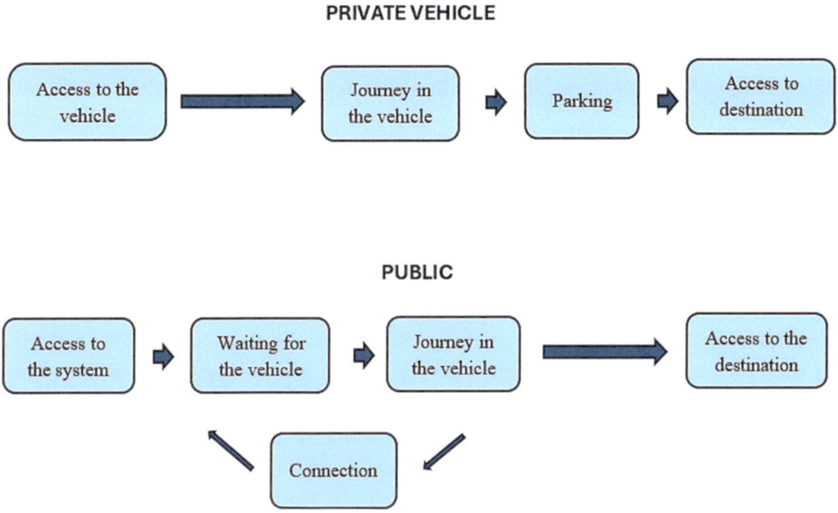

Fig. 2 Differences in mobility by private vehicles and public transport

Table 2 Measures to encourage the use of public transport and discourage the use of private vehicles

Stage	Public transport incentive	Discouragement of private vehicles
Access to the vehicle	Extension of networks, proximity to stops	–
Waiting	Higher frequencies, punctuality	–
Replacement	Speed, reserved lanes, comfort	Speed limit, traffic jams
Connection	Modal interchanges	–
Parking	–	Paid parking, high rates
Access to destination	Extension of networks, proximity of stops	Tolls for access to cities
Cost	Lower costs than private vehicles	Costs higher than public transport

the most common measures for efficient driving are gradual acceleration, maintaining a stable speed, anticipating traffic, avoiding high speeds, or decelerating instead of braking. Other measures include keeping the windows closed or properly regulating the air conditioning. Proper driving training can significantly reduce fuel consumption and CO_2 emissions.

Several studies highlight the significant impact of efficient driving on reducing CO_2 emissions. For example, Barla et al. (2017) provided a comprehensive review showing that eco-driving practices can substantially lower fuel consumption and CO_2 emissions. Wang and Boggio-Marzet (2018) presented field study data confirming the positive impact of eco-driving on reducing vehicle emissions. Coloma et al. (2020) explored the role of eco-driving in urban settings, highlighting its potential to significantly reduce traffic emissions. Chawla et al. (2021) demonstrated that driving speed and acceleration patterns are crucial factors that are influencing urban air quality and GHG emissions. Shahariar et al. (2022) evaluated various driving styles, concluding that aggressive driving leads to higher GHG emissions compared to smooth driving.

Studies show that immediately after training, private drivers can achieve a 25% reduction in fuel use, with a sustained 10% reduction in the long term. For employed drivers, these reductions are 25% initially and 6.5% over time (Department for Transport, 2016). Despite these benefits, public awareness of efficient driving remains low. In the United Kingdom, the report by Department for Transport (2016) noted that efficient driving is not widely emphasized in driver training or licensing exams. Additionally, there is little interest from private drivers and companies in paying for such courses, as the benefits are not seen as directly profitable. Beyond training, technological improvements of vehicles can also help reduce emissions by providing drivers with real-time information on fuel consumption. These features are becoming more common and can be effective in promoting efficient driving, as long as they do not increase the cost of the vehicle significantly. Both training and technology, if implemented more widely, have the potential to make driving more sustainable and environmentally friendly.

4 Corrective Measures

Corrective measures are intended to correct, alleviate, attenuate, or minimize impacts that will occur in a certain and inevitable way. The possibilities of corrective mitigation of the contribution to climate change are very limited. It is possible to achieve a minimization of emissions by applying the preventive measures indicated in the previous section, but if GHGs are emitted, the only alternative is the compensation. There are only limited cases when it is possible to speak of corrective mitigation: the restoration of the vegetation cover affected by the works, since when the implantation of vegetation is done in another place, as a replacement for the destroyed vegetation, it would be a compensatory measure.

During the works, vegetation is destroyed, which leads to a loss of the accumulated carbon stock and its sequestration capacity. Vegetation is destroyed as a result of clearing, but the affected land may or may not be permanently occupied, and therefore may or may not be suitable for re-vegetation. Sometimes, the works affect land that is necessary during construction, but not permanently, such as auxiliary installation areas, storage areas, areas under viaducts or temporary construction roads. On other occasions, the work occupying the land permanently may generate new surfaces that are capable of supporting vegetation, such as cut and embankment slopes, false tunnel covers, or connecting islands. There are also areas that are permanently occupied and altered, but they are capable of being re-vegetated, such as borrow pits and landfills. For each of the affected areas listed in Table 3, the possibility of recovering the stock and sequestration by planting new vegetation should be assessed. This restoration of the vegetation cover allows for a partial correction of the impact generated by the

Table 3 Re-vegetation potential for different zone types during transport projects

Zone	Occupation	Re-vegetation potential	Observations
Platform	Strict platform	Permanent	None
Slopes	Excavation slope	Permanent	Low
Embankment slope	Permanent	Medium	Terrain with slope but less compact
Tunnel entrances	Permanent	Very low	Very difficult terrain, often gunited
False tunnel cover	Permanent/Temporary	Medium/High	Generally can be returned to the previous use
Areas under viaducts	Permanent/Temporary	Medium/High	Limited water and light under low viaducts
Stations and parking lots	Permanent	None	Permanently occupied
Electrical substations	Permanent	None	Permanently occupied
Service areas	Permanent	None	Permanently occupied
Margins (public domain)	Temporary	Low/High	Highly variable terrain, from flat to steep
Auxiliary facilities zones	Temporary	Medium/High	Generally flat areas, recoverable
Earth stockpiles	Temporary	Medium/High	Generally moderate slopes, recoverable
Borrow areas	Permanent	Very low	Often with difficult excavation fronts
Landfills	Permanent	Low	Easier to restore than borrow areas

works (total correction is impossible due to the existence of areas with permanent occupation) and for the reestablishment of a vegetation cover that partially mitigates the loss of stock and sequestration.

The recovery of vegetation cover through sowing and planting is a slow process. It starts with the implantation of seeds or young plants, which will develop and will also suffer losses. Therefore, its evaluation requires a study over a period of time, which must coincide with that established for the evaluation of the loss of sequestration, and the planning period should correspond with the one used for the infrastructure design. During re-vegetation, the carbon stock of the planted vegetation is practically negligible, as it usually consists of seeds and young plants. Therefore, the assessments of sequestration and carbon stock are the same. The CO_2 sequestration produced in the reference period (which is derived from the annual growth of the vegetation) can be assessed, or the stock of planted vegetation at the end of the reference period can be assessed, understanding that the said stock is associated with an equivalent sequestration of CO_2.

5 Compensatory Mitigation

There are numerous definitions of environmental compensation in technical, academic, and legal fields. Cowell (2000) defines environmental compensation as the adoption of positive environmental measures to correct, balance, or remedy the loss of environmental resources. In the field of EIA, it is a concept established since the first EIA Directive in 1985. Compensatory measures can be divided into following categories, not exclusive of each other (Cole et al., 2021; Cuperus et al., 1996, 2001; Penca, 2024):

- Compensatory measures in situ (on-site) where the impacts are generated.
- Compensatory measures ex situ (off-site) outside the affected area.
- In-kind compensation, related to the affected resource, compensating with a value equal to or similar to that affected.
- Out-of-kind compensation, without a direct relationship to the affected resource.
- Monetary compensation referring to direct monetary payments, where it is difficult to establish the relationship between affected resources and measures adopted.

There is a broad consensus that compensation measures should be direct, in situ, and in kind. This translates into specific actions to restore, improve, or create habitats in the affected area or its surroundings. Two key aspects to understand compensation are additionality and permanence. Additionality refers to the compensatory measures that must imply a change in the current situation in terms of environmental quality or risk reduction. Permanence refers to the fact that if the impact is permanent, compensatory measures must also be permanent. Compensation through temporary mechanisms does not guarantee permanence, unless they include specific guarantees.

There is an urgent need to intensely reduce GHG levels in the atmosphere if we want to achieve the objective of not exceeding 2 ºC of temperature increase. Considering current trends, it does not seem that this objective will be met, unless major additional efforts are made at the global level. The main efforts should focus on preventive mitigation, drastically reducing GHG emissions from all countries, and especially the most developed ones. Reducing the concentration of GHG in the atmosphere through their sequestration and storage can also play an important role. Little interest is being paid to offsetting emissions generated by transport infrastructure, probably due to their magnitude and liability issues, especially in case of roads. Applying the mitigation hierarchy, once all possible prevention measures have been applied, the unavoidable residual impacts, in this case GHG emissions, should be offset. Compensatory mitigation associated with climate change can be undertaken

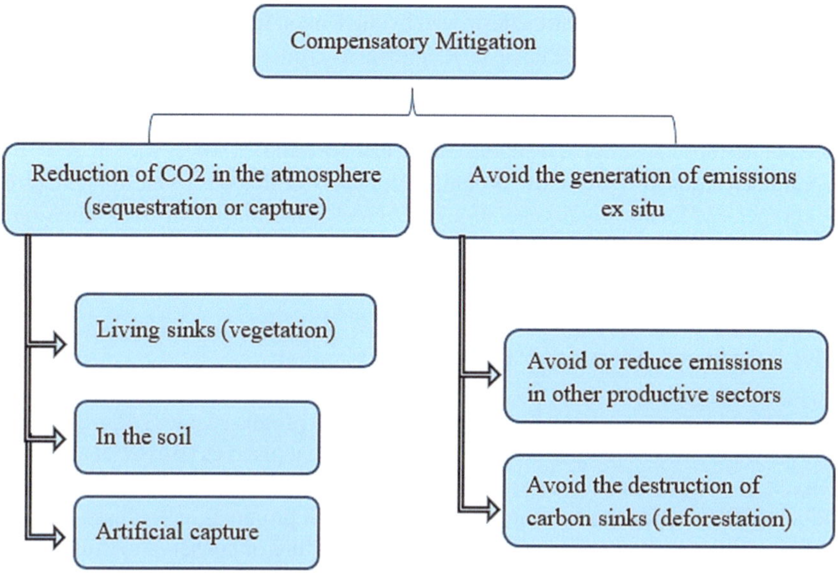

Fig. 3 Types of compensatory mitigation against climate change

from two major perspectives, sequestration of emissions or avoiding creation of new emissions (Fig. 3).

5.1 Carbon Sequestration

5.1.1 Carbon Sequestration Using Living Sinks

Carbon sequestration using living sinks is based on promoting the development of vegetation, which absorbs atmospheric CO_2 through photosynthesis, transforming it into plant matter. The use of sinks is a complementary tool to the emission reduction policies that states can use (Keenan & Williams, 2018). During the construction of transport infrastructure, the destruction of carbon sinks is frequent, which implies the loss of the carbon stock they have stored, as well as the decrease of their sequestration capacity. In some areas, it is possible to restore or recover the affected vegetation, which is considered a corrective measure. Vegetation can also be used as a compensatory measure, to create new sinks, or to enhance vegetation in existing sinks by increasing density or varying its composition. Unlike corrective mitigation, offsetting does not restore damaged vegetation, but rather implants new vegetation in a different location to compensate for the vegetation destroyed, or create a new carbon sink where none existed, thereby offsetting emissions from transport infrastructure. Two types of compensation for re-vegetation can be proposed:

- Compensation for vegetation destroyed by the works. In this case, a double objective is sought: to avoid the net loss of biodiversity *(no-net-loss)* due to the destruction of vegetation, especially if they are habitats of interest, and to compensate for the loss of carbon sinks. Priority will be given to recovering plant communities equal to those destroyed, using an ecological criterion.
- Compensation of emissions through new sinks. In this case, the aim is not to compensate for destroyed vegetation but to create sinks that offset GHG emissions. In this sense, the greater the sequestration capacity of the sink, the more effective this compensation will be.

5.1.2 Carbon Sequestration Through Soil Fixation

Agricultural land use, especially tillage, can lead to a progressive loss of soil organic carbon (SOC), reaching up to two-thirds of the original reserves (Bossio et al., 2020; Feng et al., 2024; Galluzzi et al., 2024; Jiang et al., 2022; Lal, 2004). In contrast, reducing tillage and maintaining crop residues in soil promotes an increase in SOC. These techniques are known as conservation agriculture or conservation tillage and mainly include:

- No-tillage (NT), complete elimination of tillage on land subjected to conventional tillage (CT, conventional tillage).
- Reduced tillage (RT), elimination of some of the tillage passes applied to CT crop fields, but not all as in NT.
- Maintaining crop residues on the ground, without collecting them, such as cereal straw.

Early studies on conservation tillage provided encouraging results on the potential of this technique for carbon sequestration (Powlson et al., 2011), mainly because they focused on the topsoil. The study of Ariluoma et al. (2021) found that integration of trees and biochar into urban residential yards in Helsinki could significantly enhance carbon sequestration and storage, with biochar contributing 65% and tree biomass 35% of the total potential, amounting to 520 kg CO_2 per resident over 50 years. NT results in an increase in SOC in the upper soil layer and a vertical gradient, but below the tillage depth the concentration of SOC is higher with CT. This is because CT buries the plant remains on the soil, all the residue or the part that remains after removing them, which favors their degradation, and a higher concentration of SOC at depth. Consequently, NT produces a more natural vertical gradation of SOC, with maxima at the surface and a reduction with depth, while CT produces a more homogeneous concentration throughout the tillage depth.

5.2 Carbon Capture and Storage

Since the end of the 20th century, artificial methods and technologies have been developed that allow the capture of atmospheric CO_2 and its fixation in some material or its storage, generally in deep geological layers. To avoid a temperature increase of more than 2 °C compared to pre-industrial levels, it is necessary to maintain an atmospheric concentration close to 450 ppm CO_2-eq in 2100, which implies scenarios with GHG emissions in 2050 that will be 40–70% lower than in 2010 and emission levels close to zero Gt CO_2 eq in 2100, associated with rapid improvements in energy efficiency, increased supply of renewable and nuclear alternatives, and CO_2 capture and storage (CCS) (Edenhofer et al., 2014). The IPCC has published a report on CCS, concluding that it has considerable potential and that the cost of climate change mitigation could be reduced (IPCC, 2005). Dooley (2013) notes that the demand for geological storage of CO_2 is quite small compared to the capacity of the earth and is not an impediment. Captured CO_2 can be managed in several ways:

- **Fixation in materials** (for example during construction). There are experiences with materials and cements capable of fixing CO_2, but with limited potential, not being a widespread alternative for reducing CO_2 concentration.
- **Transformation into other products**. There are studies on the transformation of CO_2 into synthesis gas, a viable but not very energy efficient process, as it is generating more carbon than it eliminates. Perhaps, the use of solar energy in the process could open up new expectations.
- **Deep geological storage**. This technique is not widely used due to its low efficiency in the mitigation hierarchy, as CO_2 emissions are not reduced but are "hidden". It also has potential seismic risks.

Recent studies on CCS highlight significant advancements and challenges in the field. Rubin et al. (2015) analyze the economic and performance aspects of CCS technologies in the power sector, identifying cost and efficiency as critical factors. Mac Dowell et al. (2017) explore the potential of CO_2 capture and utilization technologies to mitigate climate change, stressing the importance of integrating these technologies into broader climate strategies. Bui et al. (2018) provide a comprehensive review of the current status, challenges, and future directions of CCS technologies, advocating for continued research and policy support to overcome existing barriers. The future importance of CCS in mitigating climate change will depend on the financial incentives provided for its implementation and whether the risks posed by storage can be adequately managed. Osman et al. (2021) evaluate major CCS technologies, including pre-combustion, post-combustion, and oxyfuel combustion, and discuss recent advances in carbon storage and utilization. Wang et al. (2022) focus on a detailed life cycle assessment of combustion-based electricity generation technologies that are directly integrated with the CCS concept.

6 Mechanisms for the Application of Environmental Compensation of Transport Projects

There are various mechanisms for applying environmental compensation measures. In general, diverse forms of compensation implementation can be established which are discussed in Table 4. More details regarding different environmental compensation mechanisms are presented in the following sections.

6.1 Direct Compensation by the Developer

The first solution to carry out environmental compensation, whether on biodiversity or for contribution to climate change, is its execution directly by the developer or person responsible for the transport project. The developer may execute the compensation directly, or through a company contracted for the specific execution of some actions, but this is not considered execution through third parties as it is a mere

Table 4 Methods of implementing environmental compensation

<table>
<tr><td rowspan="5">Non-financial compensation</td><td colspan="3">Carried out by the promoter. The promoter is legally responsible for the compensation, assuming it directly or through contracts with third parties under their responsibility.</td></tr>
<tr><td rowspan="4">Delegated to third parties. The promoter entrusts compensation to a third party.</td><td rowspan="4">Transfer of legal responsibility. The promoter transfers the legal responsibility to execute the compensation to a third party through an agreement and payment.</td><td>Regulated carbon markets. The promoter purchases carbon credits in regulated markets. Applicable to GHG emission compensation.</td></tr>
<tr><td>Mitigation banks. The promoter buys mitigation credits from a bank that assumes responsibility for executing compensation.</td></tr>
<tr><td>Agreements with authorized third parties. The promoter entrusts compensation to a third party authorized to assume legal responsibility for the compensation.</td></tr>
<tr><td>Agreements with environmental authorities. The promoter funds compensation carried out by environmental authorities.</td></tr>
</table>

(continued)

Table 4 (continued)

		Without transfer of legal responsibility. The promoter entrusts the execution of measures to NGOs or foundations, generally without transferring legal responsibility.	**Agreements with land trustees or NGOs.** Measures are entrusted to these entities, but the promoter maintains responsibility.
			Voluntary carbon markets. The promoter buys carbon credits on voluntary markets. Applicable to GHG emissions.
Financial compensation	**Impossible direct financial compensation.** Financial compensation is used as a last resort. It is feasible in many countries but requires a legal mechanism.		
	Legally established financial compensation. The promoter pays amounts into a compensation fund managed by public authorities to restore and improve the environment.		
	Direct or financial compensation at the promoter's discretion. The promoter pays amounts into a compensation fund managed by public authorities to restore and improve the environment.		
	Financial compensation in addition to direct compensation. Additional payment is required to cover uncertainties in achieving ecological quality objectives sought through compensation.		

contract. One of the main limitations that it presents is the difficulty in acquiring the necessary land due to the restrictive interpretation that many public administrations make of the powers of compulsory acquisition. Once the measures have been executed, such as the planting of sinks, it is undesirable for the developer to deal with its conservation and management, as this is often not among its objectives. Therefore, it is preferable to opt for transfer to a third party, with adequate financial support and transfer of legal responsibility. These options, generally applied in the international sphere, are transferred to a public environmental administration or to non-governmental organizations for conservation purposes.

6.2 Market-Based Instruments (MBI)

Since the late 1980s, environmental agencies, governments and organizations have adopted market-based mechanisms, or *Market Based Instruments* (MBI), in environmental management, which include, among others, carbon markets, habitat banks, biodiversity compensation, or Payment for Environmental Services (PES). These instruments are claimed by various authors to be flexible and cost-effective mechanisms capable of mobilizing resources from the private sector (Gómez-Baggethun & Muradian, 2015; Pirard, 2012). The MBI mechanisms have their detractors and supporters. A number of researchers believe that the environment is being commoditized, that the anticipated ecological restoration is not supported by evidence, and that the promise of compensation can increase the permissiveness of damages mainly to biodiversity, but also less efforts to mitigate GHG emissions (Maron et al., 2012; McAfee, 2012; Robertson, 2004; Walker et al., 2015).

Other authors highlight the limitations of compensations on biodiversity, especially associated with the effectiveness of ecological restoration. Öckinger et al. (2021) emphasize the challenges in achieving no net loss of biodiversity due to knowledge gaps and the need for systematic, long-term monitoring. Yirdaw et al. (2023) discuss the difficulties in implementing biodiversity offsets, particularly in measuring ecological equivalence and the extended time required for effective restoration. The study by Salès et al. (2023) compares various offsetting schemes, pointing out limitations related to ecological equivalence, spatial and temporal considerations, and uncertainties in achieving biodiversity gains. These findings underscore the complexities and challenges in using ecological compensation as a reliable tool for biodiversity conservation.

Vaissière and Levrel (2015) analyze mitigation banks in Florida concluding that they are a hybrid form halfway between the market and the hierarchy because the system is based on rules imposed by public agents. Boisvert (2015) indicates that the discourses on mitigation banks are based on the fact that they are not a market, but rather a hierarchy. Pros and cons of conservation banks, caricatured as market tools or licenses to kill endangered species and destroy their habitats, tend to obscure their true characteristics. Criticism is particularly virulent towards conservation or habitat banks. They are less so compared to carbon markets, although the "delocalization" between emissions and compensation is undoubtedly a cause for controversy. Finally, PES, despite also being an MBI, are much more widely accepted.

6.2.1 Carbon Markets

Carbon markets are compensation mechanisms applied to the contribution to climate change, in which a buyer who wishes to offset a certain amount of GHG emissions buys carbon credits equivalent to their emissions from a seller, and for which there are guarantees of performance and additionality. There are two types of carbon markets, the regulated market, derived from the Kyoto Protocol, and voluntary markets.

The Kyoto Protocol provides additional means to achieve the targets set through market-based mechanisms. One such mechanism is emissions trading, or carbon market, which the EU has created in the EU Emissions Trading Scheme, the world's first carbon market. Targets agreed by countries are expressed in allowable emissions levels or assigned amounts, divided into assigned amount units. Emissions trading allows countries with surplus emissions units to sell these excess units to countries that exceed their targets. Other flexibility mechanisms under the Kyoto Protocol include the Clean Development Mechanism (CDM), the Joint Implementation (JI) mechanism, and the Adaptation Fund. The regulated market and flexibility mechanisms apply to industrial sectors but not to diffused sectors such as transport.

The voluntary market is also based on the supply and demand of offset units or carbon credits, but it is not regulated by the Kyoto Protocol or any other standard. The buyer voluntarily comes to offset his GHG emissions, and the seller is any entity that offers carbon credits. Since these markets do not have a legal basis, an essential aspect in voluntary carbon markets is their credibility, for which various norms and standards have been developed, seeking to provide legitimacy to the clearing units and avoid double accounting (when the same credit is sold twice). Much of the supply of voluntary carbon markets is based on avoided deforestation projects in developing countries, generally through REDD+ mechanisms, a mechanism developed by the Parties to the United Nations Framework Convention on Climate Change, intended to incentivize developing countries to decrease emissions of carbon caused by forest degradation and deforestation. However, as noted, there is no additionality in this offset mechanism in developed nations, where there are strong regulatory mechanisms for forest protection. In these nations, additionality is associated with obvious carbon sequestration.

6.2.2 Payments for Environmental Services

A market-based instrument that can add value to both biodiversity and carbon sequestration, thereby facilitating the development of environmental offsets, with Payments for Environmental Services (PES), aims to achieve sustainable development via significant recognition of ecosystem values. Consideration of carbon sequestration is an approach of enhancing the value of ecosystems, so there is an obvious connection between PES and carbon markets. The link between PES, including carbon markets, and poverty decrease has also been widely studied.

Jack et al. (2008) review the design of PES schemes, underscoring the importance of well-designed incentive mechanisms for achieving both environmental and socio-economic outcomes. Pagiola et al. (2008) find that poor households can participate in and benefit from PES schemes, as demonstrated by the Silvopastoral Project in Nicaragua. Clements et al. (2010) compare PES programs in Cambodia, showing their effectiveness in biodiversity conservation and poverty reduction despite weak institutional capacities in this country. Muradian et al. (2010) provide a conceptual framework for understanding the socio-economic impacts of PES, highlighting the importance of reconciling theory and practice. Tacconi et al. (2013) emphasize the positive impacts of PES on livelihoods, particularly in the context of REDD+ initiatives. Wunder et al. (2020) review the effectiveness of PES schemes, noting their potential to deliver environmental benefits while also improving livelihoods.

6.2.3 Conservation Banks

Conservation banks were created in the United States in connection with the Clean Water Act of 1972 to compensate for damage to wetlands. They are focused on environmental compensation for damage to biodiversity, species, and habitats. Conservation, mitigation, or habitat banks are permanently protected lands where restoration and improvement actions are carried out and are managed with conservationist objectives. Their particularity is that these restoration and improvement actions are translated into compensation titles or credits, which can be sold to developers who need to compensate for the impacts on habitats or species derived from their projects, functioning as a banking mechanism. A key aspect for a successful integration of conservation banks and carbon markets is to allow double accounting of credits, for biodiversity and for carbon, which would provide additional value to the bank, reducing the risk of drifting into pure carbon offsets. In the forestry sector, there are management measures that can favor an increase in carbon absorption and stock, with added ecological advantages.

7 Challenges in Incorporating Climate Change Mitigation into Transport Projects

7.1 Climate Change Mitigation in Projects Subjected to EIA

Consideration of climate change in EIAs is quite limited, especially in the area of transport infrastructure, where the mitigation is practically non-existent. Therefore, climate change mitigation in the EIA, both preventive and compensatory, can be considered practically non-existent, and its effective incorporation in the future will be a great challenge. An adequate incorporation of mitigation and compensation of the contribution to climate change is not only desirable to reduce the concentration of GHG in the atmosphere but also should have a significant influence on decision-making. Mitigation costs may be so significant that it is advisable to seek alternatives with lower emissions, within a multi-criteria analysis process, since there may be other reasons that justify these solutions.

As the first step, all preventive mitigation possibilities should be analyzed in the design phase of the transport project, and the best possible solutions should be adopted that are technically and environmentally reasonable. Corrective mitigation of affected vegetation and compensation for lost sinks should also be considered. The second phase should consider offsetting GHG emissions associated with construction, so that the project has a zero-carbon balance. The third, more ambitious phase should be mitigation during operation, both with preventive measures to reduce emissions and through offsetting emissions, which requires a higher level of planning and political and administrative involvement for its financing and implementation.

7.2 *Climate Change Mitigation in Already Operating Projects*

A huge challenge in relation to climate change is the mitigation of the contribution to climate change in transport infrastructures that have already been built, and therefore where no environmental assessment mechanism is applicable. The above presented considerations regarding the consideration of climate change in the EIA are not applicable to these existing networks, as they are already in service, but they are the bulk of the transport sector's contribution to climate change. If action is not taken to mitigate their contribution to climate change, it will be very difficult to achieve the reduction targets agreed upon the Kyoto Protocol. Three possible lines of action include the following: (i) preventive mitigation in operation by reducing and regulating traffic; (ii) preventive mitigation in operation through changes in lighting and the source of energy along with optimization of maintenance frequency and efficiency; (iii) compensatory mitigation through CO_2 sequestration in sinks or avoiding *ex situ* emissions.

Among these measures, there are two actions that are particularly important due to their potential to combat global warming: reducing the use of private vehicles as a means of transport, discouraging their utilization and achieving a modal shift to rail, and compensatory mitigation through carbon sequestration. A tool to achieve the GHG emission reduction targets can also be an increasing use of electric vehicles (Mikusova et al., 2018). This modal shift can be intense in urban agglomerations, where the highest traffic intensities and also many congested areas are concentrated, so that effective public transport is an essential measure in terms of emissions and attractiveness for the user in terms of transport times and costs. It is therefore necessary to continue investing in public transport, both in the construction of new transport possibilities and in the adaptation of infrastructure for preferential use, as is the case of reserved platforms or bus-HOV lanes. This investment, however, must be supported by clear data on passenger attraction and benefits in terms of the overall balance of emissions in order to avoid costly underused infrastructure.

Regarding the second measure, it is necessary to assume that it is possible to offset GHG emissions, at least in part. Offsetting is not a theoretical concept, nor a utopia. There is a real possibility of creating new sinks, both through new vegetation and by promoting fixation in the soil. The application of this offset would have enormous ecological and landscape advantages. Whatever the measures adopted, the EIA cannot be considered the end of the road, a milestone that, once passed, makes consideration of the environmental problems of transport infrastructure unnecessary.

8 Mitigation Responsibilities and Financing

To establish the possible financing of mitigation measures, especially compensatory mitigation, the "polluter pays" principle should be applied. In this case, the person responsible for covering the costs of mitigation and compensation is the person who generates or induces the generation of GHG emissions. In the construction phase of transport infrastructure, the developer is responsible for GHG emissions and for implementing and financing the necessary mitigation and compensation measures. The construction project should include a detailed calculation of the carbon footprint, both for the construction work and for the destruction of sinks. It should include the compensation in the project itself, as an inherent part of it, and therefore be financed in a similar way to the rest of the infrastructure. In the operational phase, the road manager is responsible for emissions associated with lighting, cleaning or maintenance, but the main contribution comes from users. The bulk of GHG emissions come from road traffic, made up of many users, whose contribution varies according to the use of transport infrastructure that they make.

9 Discussion and Conclusions

Determining the contribution of a transport project to climate change is an essential step that must be incorporated into decision-making to influence the choice of the least unfavorable alternatives. However, this is not enough. Appropriate mitigation measures should be adopted to reduce the impact of each project on climate change. The main mitigation measures proposed for transport are behavioral changes (e.g., avoiding unnecessary journeys or driving more efficiently), shifting to more efficient modes of transport (especially rail), improving the energy efficiency of trains and vehicles, using low-carbon fuels or those with lower GHG emissions, establishment of fuel taxes, inclusion of aspects of sustainable mobility in urban planning, increase of vehicle occupancy, or application of environmental criteria and fees in urban parking management. Most of these measures exceed the scope of the environmental assessment, as they are decisions of a political nature. In some cases, they could be incorporated into the SEA of infrastructure plans, such as the promotion of railways, but this also has an undoubted political burden in decision-making.

There is a notable lack of possible mitigation measures applicable at the project level, and therefore susceptible to being incorporated into the EIA of transport infrastructure. However, mitigation applied to specific projects is possible and necessary. There are numerous preventive mitigation measures that should be incorporated into the design of infrastructure components to reduce GHG emissions, both in the construction and operation phases. The concept of preventive mitigation, based on avoiding impacts before they occur, should lead to preventive designs, rather than reactive mitigation proposals. The same principle should be applied to the protection of other values, such as biodiversity, where the best prevention is an adequate selection of routes that are minimizing the environmental damage. A preventive design allows for the proposal of alternatives with a lower contribution to climate change.

One of the great options of mitigation in the EIA of transport projects is the possibility of considering compensation for their impact on climate change, an aspect that is currently often ignored. Following the mitigation hierarchy, firstly the impacts must be avoided, and if they are unavoidable, they must be corrected or minimized and, finally, if this is not possible, they must be compensated. Compensation for the contribution of transport infrastructure to climate change has two facets: compensation for destroyed sinks, which is easily assumed, and compensation for GHG emissions, which is a much more complex and ambitious objective. Neither of these aspects is currently considered in the EIA of transport infrastructure projects, and compensation for sinks has only been proposed very occasionally in some other type of projects, without specifying it in detail.

Both types of compensation, sinks and GHG emissions, are technically feasible at the global level simply through "land use, land-use change and forestry" (LULUCF) activities, associated with land use and forestry, and can be undertaken without causing social impacts. On the contrary, in addition to the climatic advantages, they can help improve the landscape, biodiversity, and rural development, mobilizing investments in the countryside. The great challenge is financing. Compensation requires significant investments, and that in turn requires political decisions. Among the sources of financing, it seems logical that at least part of the funds should come from fuel taxes. This is an application of the "polluter pays" principle, which is basic to the EU regulations, so that the user of the vehicle that consumes fuel and emits GHG is responsible for its compensation. However, given the great social and environmental benefits of this compensation, it also seems reasonable that there should be more sources of financing and direct investments made by public administrations.

Incorporating compensation for the contribution to climate change in the EIA will be complex until the challenge of financing is resolved. However, this should be implemented progressively. The first step is to compensate for the destroyed sinks, something perfectly acceptable within the costs of the project. In the second phase, the emission of GHG in the works should be compensated, both for the transport infrastructure and for any other activity, so that their construction is carbon neutral. The final challenge would be compensation in operation, especially derived from road traffic, which will require more ambitious political decisions. Climate change compensation in transport infrastructure opens a hopeful horizon for more sustainable transport, environmental improvement and more equitable rural development. It can be an essential tool to meet the GHG emissions targets.

From 2030, the LULUCF sector will be included in the national emissions accounting. It is currently included in the inventories, but not in the global calculations. The integration of climate change considerations into the EIA for transport projects remains a significant challenge. This chapter highlighted the necessity of systematically addressing the contribution of transport infrastructure to climate change throughout all project phases, from construction to operation and maintenance. In this context, the critical direction for future research is the development and refinement of methodologies to calculate the carbon footprint of transport projects. The first step in this process is to establish a reliable measure of GHG associated with both the construction and operational phases of a project. While there are numerous practical examples of carbon footprint calculations applied to specific projects, these methodologies are not typically utilized within the EIA framework. One reason for this disconnect is that detailed project knowledge is often unavailable at the time of assessment.

To better incorporate carbon calculations into EIAs, it is essential to adapt existing methodologies to align with the level of detail typically found in EIAs. Construction of transport infrastructure inherently contributes to climate change due to GHG generated from machinery use and material consumption. Although the available detail at the EIA stage may be limited, fundamental data, such as the length and cross-section of the route, estimates of earthworks, drainage, and structural passages, are generally accessible. By applying standard emission factors to these principal components, it should be possible to produce sufficiently accurate estimates of GHG emissions during the construction project phase. While more comprehensive carbon footprint models could yield better precision, this is often impractical due to the lack of detailed information. Moreover, the destruction of carbon sinks during the construction phase needs to be acknowledged. Clearing vegetation results in not only the immediate release of stored carbon but also the subsequent loss of ongoing sequestration capacity. EIAs should quantify these effects on vegetation as part of the assessment process. By applying carbon stock and sequestration data to identified vegetation types, it becomes feasible to estimate a more accurate project's contribution to climate change.

During the operational phase, the contribution of transport projects to climate change varies significantly by transport mode. For road networks, emissions stem primarily from vehicular traffic, which is ultimately the responsibility of road users. GHG emissions are calculable based on anticipated traffic on new routes and residual traffic on pre-existing roads using CO_2 emission factors that evolve over time. Comparison of these findings with a "do-nothing" alternative is vital for determining net emissions. In public transport scenarios, while the transporter typically holds responsibility for emissions, these can also affect users indirectly. Supposing that public transport options maintain a viable rate of use, they usually present lower GHG emissions per user compared to road transport. Investigating user attraction to more polluting transport options, such as aviation or private vehicles, is another aspect that warrants attention, along with the potential GHG reduction benefits associated with modal shifts. Despite the existence of tools and methodologies capable of evaluating the climate change contributions from transport projects, their actual incorporation into EIAs remains minimal. It is important that these tools are implemented effectively, allowing the integration of the climate impact of transport projects as a criterion in alternative selection processes within EIA. For meaningful progress, climate change considerations must be systematically addressed within EIAs, not merely contingent upon significance thresholds that can lead to inadequate assessments being dismissed or overlooked.

In summary, addressing these concerns presents not only challenges but also substantial opportunities for future research to enhance the integration of climate change considerations into transport project planning and assessment. Establishing a framework that enshrines this type of evaluation within the EIA process is crucial for making informed decisions that favour sustainable development and climate resilience in transport systems.

Glossary of Terms and Acronyms

CDM	Clean Development Mechanism	A mechanism defined in the Kyoto Protocol allowing industrialized countries to invest in greenhouse gas emission reductions in developing countries as a way to meet their own emission reduction targets.
CCS	Carbon Capture and Storage	A technology that captures carbon dioxide emissions at their source and stores it underground or uses it in some way to prevent its release into the atmosphere.
CDV	Circular Development Value	A concept related to creating economic value through sustainable practices that support circular economy principles.
CO_2	Carbon Dioxide	Greenhouse gas that contributes to global warming.
CT	Carbon Tax	A tax levied on the carbon content in fuels, with the objective of decreasing greenhouse gas emissions.
EIA	Environmental Impact Assessment	A process used to evaluate the potential environmental effects of a proposed project before it is allowed to proceed.
ELT	End-of-Life Tires	Used tires that can be repurposed or recycled, reflecting waste management and environmental concerns.
EMS	Environmental Management Systems	A part of an organizational approach to address governmental regulations and apply relevant environmental policies.
GHG	Greenhouse Gas	Gases that trap heat in the atmosphere, contributing to global warming, including methane (CH_4), carbon dioxide (CO_2), and nitrous oxide (N_2O).
Gt CO_2	Gigatonnes of Carbon Dioxide	A metric used to quantify large amounts of carbon dioxide emissions (1 Gt CO_2 = 1 billion tonnes of CO_2).
HOV	High-Occupancy Vehicles	Vehicles that carry multiple passengers, typically two or more, and are often given access to special lanes on highways to reduce traffic congestion and promote carpooling.
IPCC	Intergovernmental Panel on Climate Change	A scientific body under the United Nations that assesses the climate change risk due to by human activities.
JI	Joint Implementation	A mechanism under the Kyoto Protocol where countries with emission reduction commitments can invest in emission-reduction projects in other developed countries.

LCA	Life Cycle Assessment	An analysis used to evaluate the environmental effects caused by all the stages of a product's life cycle.
LULUCF	Land Use, Land-Use Change and Forestry	A greenhouse gas inventory sector that oversees emissions and reduction of greenhouse gases caused by direct human-induced land-use, land-use change, and forestry actions.
MBI	Market-Based Instrument	Economic tool used to encourage environmentally friendly practices, such as carbon trading or pollution taxes.
NFU	Non-Forest Use	Areas that do not comprise forest land but may influence environmental and land-use policies.
NGO	Non-Governmental Organization	A non-profit group that operates independently of any government, often focused on social or environmental issues.
NT	Nature-Based Target	Goal related to the use of natural processes and ecosystems to achieve conservation and sustainability objectives.
PES	Payment for Ecosystem Services	A financial mechanism to incentivize the conservation and sustainable use of ecosystem services.
PPP	Polluter Pays Principle	A fundamental concept in environmental economics which stipulates that those who produce pollution should bear the costs associated with managing it.
REDDT	Reducing Emissions from Deforestation and Forest Degradation	A concept aimed at encouraging conservation and sustainable management of forests to reduce emissions.
RT	Renewable Technology	Technologies that utilize renewable resources for energy generation, including solar, wind, and hydro technologies.
SOC	Soil Organic Carbon	The carbon stored in soil organic matter, important for soil health and carbon sequestration.
TOD	Transit Oriented Development	An urban planning strategy that focuses on creating compact, walkable neighbourhoods located in the vicinity of high-quality public transport networks.

References

Albuquerque, F. D. B., Maraqa, M. A., Chowdhury, R., Mauga, T., & Alzard, M. (2020). Greenhouse gas emissions associated with road transport projects: Current status, benchmarking, and assessment tools. *Transportation Research Procedia, 48*, 2018–2030. https://doi.org/10.1016/j.trpro.2020.08.261

Ambade, B. R., & Ambade, C. B. (2023). Environmental impact assessment: A Case study of construction of metro rail Nagpur, Maharashtra (India). *SSRG International Journal of Recent Engineering Science, 10*(3), 39–43.

Andersen, D., & Jang, Y. (2021). Biodiversity and transportation infrastructure in the republic of Korea: A review on impacts and mitigation in developing the country. *Diversity, 13*(11), 519. https://doi.org/10.3390/d13110519

Andres, S. E., Standish, R. J., Lieurance, P. E., Mills, C. H., Harper, R. J., Butler, D. W., Adams, V. M., Lehmann, C., Tetu, S. G., Cuneo, P., Offord, C. A., & Gallagher, R. V. (2023). Defining biodiverse reforestation: Why it matters for climate change mitigation and biodiversity. *Plants People Planet, 5*, 27–38. https://doi.org/10.1002/ppp3.10329

Ariluoma, M., Ottelin, J., Hautamäki, R., Tuhkanen, E.-M., & Mänttäri, M. (2021). Carbon sequestration and storage potential of urban green in residential yards: A case study from Helsinki. *Urban Forestry and Urban Greening, 57*, Article 126939. https://doi.org/10.1016/j.ufug.2020.126939

Barla, P., Gilbert-Gonthier, M., Lopez Castro, M. A., & Miranda-Moreno, L. (2017). Eco-driving training and fuel consumption: Impact, heterogeneity and sustainability. *Energy Economics, 62*, 187–194. https://doi.org/10.1016/j.eneco.2016.12.018

Bigazzi, A. Y., & Rouleau, M. (2017). Can traffic management strategies improve urban air quality? A review of the evidence. *Journal of Transport and Health, 7*, 111–124. https://doi.org/10.1016/j.jth.2017.08.001

Boisvert, V. (2015). Conservation banking mechanisms and the economization of nature: An institutional analysis. *Ecosystem Services, 15*, 134–142. https://doi.org/10.1016/j.ecoser.2015.02.004

Bossio, D. A., Cook-Patton, S. C., Ellis, P. W., Fargione, J., Sanderman, J., Smith, P., Wood, S., Zomer, R. J., Von Unger, M., Emmer, I. M., & Griscom, B. W. (2020). The role of soil carbon in natural climate solutions. *Nature Sustainability, 3*(5), 391–398. https://doi.org/10.1038/s41893-020-0491-z

Bugge, H. C. (2009). The polluter pays principle: Dilemmas of justice in national and international contexts. In: J. Ebbesson & P. Okowa (Eds.), *Environmental Law and Justice in Context*. Cambridge: Cambridge

Bui, M., Adjiman, C. S., Bardow, A., Anthony, E. J., Boston, A., Brown, S., Fennell, P. S., Fuss, S., Galindo, A., Hackett, L. A., & Hallett, J. P. (2018). Carbon capture and storage (CCS): The way forward. *Energy and Environmental Science, 11*(5), 1062–1176. https://doi.org/10.1039/C7EE02342A

Burghardt, T. E., & Pashkevich, A. (2020). Materials selection for structured horizontal road markings: Financial and environmental case studies. *European Transport Research Review, 12*, 11. https://doi.org/10.1186/s12544-020-0397-x

Center for Climate-Smart Transportation. (2025). *Research on mixed-use development and VMT*. Retreived February 6, 2025 from https://publichealth.jhu.edu/center-for-climate-smart-transportation/research

Chawla, A., Khare, M., & Khan, S. (2021). Evaluating the effect of speed variation on vehicular emission using an integrated modelling approach. In: D.K. Ashish, J. De Brito & S.K. Sharma, (Eds.), *3rd International Conference on Innovative Technologies for Clean and Sustainable Development* (pp. 299–315). Cham: Springer International Publishing. https://doi.org/10.1007/978-3-030-51485-3_20

Clements, T., John, A., Nielsen, K., An, D., Tan, S., & Milner-Gulland, E. J. (2010). Payments for biodiversity conservation in the context of weak institutions: Comparison of three programs from Cambodia. *Ecological Economics, 69*(6), 1283–1291. https://doi.org/10.1016/j.ecolecon.2009.11.010

Cole, S., Moksnes, P.-O., Söderqvist, T., Wikström, S. A., Sundblad, G., Hasselström, L., Bergström, U., Kraufvelin, P., & Bergström, L. (2021). Environmental compensation for biodiversity and ecosystem services: A flexible framework that addresses human wellbeing. *Ecosystem Services, 50*, Article 101319. https://doi.org/10.1016/j.ecoser.2021.101319

Coloma, J. F., Garcia, M., Boggio-Marzet, A., & Monzón, A. (2020). Developing eco-driving strategies considering city characteristics. *Journal of Advanced Transportation, 2020*, 1–13. https://doi.org/10.1155/2020/2083074

Cowell, R. (2000). Environmental compensation and the mediation of environmental change: Making capital out of Cardiff Bay. *Journal of Environmental Planning and Management, 43*(5), 689–710. https://doi.org/10.1080/713676580

Cuperus, R., Canters, K. J., & Piepers, A. A. G. (1996). Ecological compensation of the impacts of a road: Preliminary method for the A50 road link (Eindhoven-Oss, The Netherlands). *Ecological Engineering, 7*(4), 327–349. https://doi.org/10.1016/S0925-8574(96)00024-9

Cuperus, R., Bakermans, M. M. G. J., Haes, H. A. U. D., & Canters, K. J. (2001). Ecological compensation in dutch highway planning. *Environmental Management, 27*, 75–89. https://doi.org/10.1007/s002670010135

De Palma, A., Lindsey, R., & Monchambert, G. (2017). The economics of crowding in rail transit. *Journal of Urban Economics, 101*, 106–122. https://doi.org/10.1016/j.jue.2017.06.003

Department for Transport. (2016). Efficient driving: A rapid evidence assessment for the department for transport

Dooley, J. J. (2013). Estimating the supply and demand for deep geologic CO_2 storage capacity over the course of the 21st century: A meta-analysis of the literature. *Energy Procedia, 37*, 5141–5150. https://doi.org/10.1016/j.egypro.2013.06.429

Edenhofer, O., et al. (2014). Climate change 2014: Mitigation of climate change. Contribution of working group III to the fifth assessment report of the intergovernmental panel on climate change

EEA. (2021). Transport and environment report 2021. *Decarbonising Road Transport—The Role of Vehicles, Fuels and Transport Demand*. European Environment Agency

Eliasson, J. (2014). The Stockholm congestion charges: an overview. Working papers in Transport Economics. Centre for Transport Studies Stockholm (KTH and VTI)

EPA. (2022). *The 2022 EPA Automotive Trends Report: Greenhouse Gas Emissions, Fuel Economy, and Technology Since 1975*. U.S. Environmental Protection Agency.

EPA. (2023). *Highlights of the automotive trends report*. Retreived January 21, 2025 from https://www.epa.gov/automotive-trends

Fawzy, S., Osman, A. I., Doran, J., & Rooney, D. W. (2020). Strategies for mitigation of climate change: a review. *Environmental Chemistry Letters, 18*, 2069–2094. https://doi.org/10.1007/s10311-020-01059-w

Feijoo, G., & Moreira, M. T. (2020). Análisis de Ciclo de Vida y Huella de Carbono: Casos Prácticos. https://doi.org/10.13140/RG.2.2.11030.50240/1

Feng, X., Dai, G., Liu, T., Jia, J., Zhu, E., Liu, C., Zhao, Y., Wang, Y., Kang, E., Xiao, J., & Li, W. (2024). Understanding the mechanisms and potential pathways of soil carbon sequestration from the biogeochemistry perspective. *Science China Earth Sciences, 67*, 3386–3396. https://doi.org/10.1007/s11430-024-1359-9

Folgerø, I. K., Harding, T., & Westby, B. S. (2020). Going fast or going green? Evidence from environmental speed limits in Norway. *Transportation Research Part D: Transport and Environment, 82*, 102261. https://doi.org/10.1016/j.trd.2020.102261

Fondzenyuy, S. K., Turner, B. M., Burlacu, A. F., Jurewicz, C., Usami, D. S., Feudjio, S. L. T., & Persia, L. (2024). The impact of speed limit change on emissions: A systematic review of literature. *Sustainability, 16*, 7712. https://doi.org/10.3390/su16177712

Galluzzi, G., Plaza, C., Priori, S., Giannetta, B., & Zaccone, C. (2024). Soil organic matter dynamics and stability: Climate versus time. *Science of the Total Environment, 929*, 172441. https://doi.org/10.1016/j.scitotenv.2024.172441

Gao, C., Xu, J., Jia, M., & Sun, Z. (2024). Correlation between carbon emissions, fuel consumption of vehicles and speed limit on expressway. *Journal of Traffic and Transportation Engineering (English Edition), 11*, 631–642. https://doi.org/10.1016/j.jtte.2023.02.007

Gautrain Management Agency. (2023). *Gautrain rapid rail link project*. Retreived February 2, 2025, from https://managingppp.gihub.org/case-studies/gautrain-rapid-rail-link/

Gharehbaghi, K., Hosseinian-Far, A., & Hilletofth, P. (2020). The predicaments of environmental impact assessment (EIA) for transport infrastructure: An examination of policy stagnation and progress. *Transforming Government: People, Process and Policy.* https://doi.org/10.1108/TG-01-2020-0010

Giani, M. I., Dotelli, G., Brandini, N., & Zampori, L. (2015). Comparative life cycle assessment of asphalt pavements using reclaimed asphalt, warm mix technology and cold in-place recycling. *Resources, Conservation and Recycling, 104*, 224–238. https://doi.org/10.1016/j.resconrec.2015.08.006

Gibson, M., & Carnovale, M. (2015). The effects of road pricing on driver behavior and air pollution. *Journal of Urban Economics, 89*, 62–73. https://doi.org/10.1016/j.jue.2015.06.005

Gómez-Baggethun, E., & Muradian, R. (2015). In markets we trust? Setting the boundaries of Market-Based Instruments in ecosystem services governance. *Ecological Economics, 117*, 217–224. https://doi.org/10.1016/j.ecolecon.2015.03.016.

Hands, S., & Hudson, M. D. (2016). Incorporating climate change mitigation and adaptation into environmental impact assessment: A review of current practice within transport projects in England. *Impact Assessment and Project Appraisal, 34*(4), 330–345. https://doi.org/10.1080/14615517.2016.1228340

Hoy, M., Horpibulsuk, S., Chinkulkijniwat, A., Suddeepong, A., Buritatum, A., Yaowarat, T., Choenklang, P., Udomchai, A., & Kantatham, K. (2024). Innovations in recycled construction materials: Paving the way towards sustainable road infrastructure. *Frontiers in Built Environment, 10*, 1449970. https://doi.org/10.3389/fbuil.2024.1449970

IEA. (2019). *Innovation gaps.* IEA, Paris. Retrieved February 4, 2025, from https://www.iea.org/reports/innovation-gaps

INFRAS. (2023). *Handbook emission factors for road transport (HBEFA).* Retrieved January 21, 2025, from https://www.hbefa.net/

Institute for Transportation and Development Policy. (2017). *TOD Standard*. Retrieved January 20, 2025, from https://itdp.org/2017/06/23/tod-standard/

International Transport Forum (ITF). (2024). *Measuring new mobility: Case studies and best practices*

IPCC. (2005). La captación y el almacenamiento de dióxido de carbono: Resumen para responsables de políticas y resumen técnico. Retrieved from https://archive.ipcc.ch/pdf/special-reports/srccs/srccs_spm_ts_sp.pdf

Jack, B. K., Kousky, C., & Sims, K. R. E. (2008). Designing payments for ecosystem services: Lessons from previous experience with incentive-based mechanisms. *Proceedings of the National Academy of Sciences of the United States of America, 105*, 9465–9470. https://doi.org/10.1073/pnas.0705503104

Jiang, P., Xiao, L. Q., Wan, X., Yu, T., Liu, Y. F., & Liu, M. X. (2022). Research progress on microbial carbon sequestration in soil: A review. *Eurasian Soil Science, 55*, 1395–1404. https://doi.org/10.1134/S1064229322100064

Kedward, K., Ryan-Collins, J., & Chenet, H. (2023). Biodiversity loss and climate change interactions: Financial stability implications for central banks and financial supervisors. *Climate Policy, 23*, 763–781. https://doi.org/10.1080/14693062.2022.2107475

Keenan, T. F., & Williams, C. A. (2018). The terrestrial carbon sink. *Annual Review of Environment and Resources, 43*, 219–243. https://doi.org/10.1146/annurev-environ-102017-030204

Keijzer, E. E., Leegwater, G. A., De Vos-Effting, S. E., & De Wit, M. S. (2015). Carbon footprint comparison of innovative techniques in the construction and maintenance of road infrastructure in The Netherlands. *Environmental Science and Policy, 54*, 218–225. https://doi.org/10.1016/j.envsci.2015.06.010

Khan, M. R. (2015). Polluter-pays-principle: The cardinal instrument for addressing climate change. *Laws, 4*(3), 638–653. https://doi.org/10.3390/laws4030638

Lal, R. (2004). Soil carbon sequestration to mitigate climate change. *Geoderma, 123*, 1–22. https://doi.org/10.1016/j.geoderma.2004.01.032

Lehe, L. (2019). Downtown congestion pricing in practice. *Transportation Research Part c: Emerging Technologies, 100*, 200–223. https://doi.org/10.1016/j.trc.2019.01.020

Li, P., & Jones, S. (2015). Vehicle restrictions and CO_2 emissions in Beijing—A simple projection using available data. *Transportation Research Part D: Transport and Environment, 41*, 467–476. https://doi.org/10.1016/j.trd.2015.09.020

Mac Dowell, N., Fennell, P. S., Shah, N., & Maitland, G. C. (2017). The role of CO_2 capture and utilization in mitigating climate change. *Nature Climate Change, 7*, 243–249. https://doi.org/10.1038/nclimate3231

Maga, D., Aryan, V., & Blömer, J. (2023). A comparative life cycle assessment of tyre recycling using pyrolysis compared to conventional end-of-life pathways. *Resources, Conservation and Recycling, 199*, Article 107255. https://doi.org/10.1016/j.resconrec.2023.107255

Mahdi, S. N., Imjai, T., Wattanapanich, C., Garcia, R., Kaur, H., & Musarat, M. A. (2023). Life cycle cost analysis of flexible pavements reinforced with geo-synthetics: A case study of new construction or repair overlays in Thailand's roads. *Engineering Science.* https://doi.org/10.30919/es1071

Maron, M., Hobbs, R. J., Moilanen, A., Matthews, J. W., Christie, K., Gardner, T. A., Keith, D. A., Lindenmayer, D. B., & McAlpine, C. A. (2012). Faustian bargains? Restoration realities in the context of biodiversity offset policies. *Biological Conservation, 155*, 141–148. https://doi.org/10.1016/j.biocon.2012.06.003

Mayembe, R., Simpson, N. P., Rumble, O., & Norton, M. (2023). Integrating climate change in environmental impact assessment: A review of requirements across 19 EIA regimes. *Science of The Total Environment, 869*,161850. https://doi.org/10.1016/j.scitotenv.2023.161850.

McAfee, K. (2012). The contradictory logic of global ecosystem services markets. *Development and Change, 43*, 105–131. https://doi.org/10.1111/j.1467-7660.2011.01745.x

Metcalfe, J. (2023). Effect of speed on emissions and air quality. Waka Kotahi NZ Transport Agency and Auckland Transport

Mikusova, M., Zukowska, J., & Torok, A. (2018). Community road safety strategies in the context of sustainable mobility. In: J. Mikulski (Ed.) *Management Perspective for Transport Telematics* (pp.115–128). Cham: Springer International Publishing. https://doi.org/10.1007/978-3-319-97955-7_8.

Muradian, R., Corbera, E., Pascual, U., Kosoy, N., & May, P. H. (2010). Reconciling theory and practice: An alternative conceptual framework for understanding payments for environmental services. *Ecological Economics, 69*, 1202–1208. https://doi.org/10.1016/j.ecolecon.2009.11.006

Öckinger, E., et al. (2021). When can ecological compensation contribute to sustaining biological diversity and ecosystem services? https://www.naturvardsverket.se/en/publications/6900/978-91-620-6996-4/

OECD. (2024). *Measuring new mobility: Case studies and best practices.* Retrieved January 21, 2025, from https://www.itf-oecd.org/measuring-new-mobility-case-studies

Osman, A. I., Hefny, M., Abdel Maksoud, M. I. A., Elgarahy, A. M., & Rooney, D. W. (2021). Recent advances in carbon capture storage and utilisation technologies: A review. *Environmental Chemistry Letters, 19*, 797–849. https://doi.org/10.1007/s10311-020-01133-3

Pagiola, S., Rios, A. R., & Arcenas, A. (2008). Can the poor participate in payments for environmental services? Lessons from the Silvopastoral Project in Nicaragua. *Environmental Development Economics, 13*, 299–325. https://doi.org/10.1017/S1355770X08004270

Pal, P., Gopal, P. R. C., & Ramkumar, M. (2023). Impact of transportation on climate change: An ecological modernization theoretical perspective. *Transport Policy, 130*, 167–183. https://doi.org/10.1016/j.tranpol.2022.11.008

Papasavva, S., Kia, S., Claya, J., & Gunther, R. (2002). Life cycle environmental assessment of paint processes. *Journal of Coatings Technology, 74*, 65–76. https://doi.org/10.1007/BF02720151

Partnership on Sustainable, Low Carbon Transport (SLOCAT). (2022). Transport and Climate Change Global Status Report—2nd edn. https://unfccc.int/documents/461584

Penca, J. (2024). Biodiversity offset mechanisms and compensation for loss from exceptional to popular: Rediscovering environmental law. *Journal of International Wildlife Law and Policy*, pp. 1–24. https://doi.org/10.1080/13880292.2024.2375862

Perez-Prada, F., & Monzon, A. (2017). Ex-post environmental and traffic assessment of a speed reduction strategy in Madrid's inner ring-road. *Journal of Transport Geography, 58*, 256–268. https://doi.org/10.1016/j.jtrangeo.2016.12.018

Pirard, R. (2012). Market-based instruments for biodiversity and ecosystem services: A lexicon. *Environmental Science and Policy, 19–20*, 59–68. https://doi.org/10.1016/j.envsci.2012.02.001

Plati, C., & Tsakoumaki, M. (2023). Life cycle assessment (LCA) of alternative pavement rehabilitation solutions: A case study. *Sustainability, 15*, 2129. https://doi.org/10.3390/su1503 2129

Porter, C., Wang, M., Weber, S. & Daigneault, A. (2013). Effects of travel reduction and efficient driving on transportation: Energy use and greenhouse gas emissions. *Transportation Energy Futures Series (DOE/GO-102013-3704)*

Powlson, D. S., Whitmore, A. P., & Goulding, K. W. T. (2011). Soil carbon sequestration to mitigate climate change: A critical re-examination to identify the true and the false. *European Journal of Soil Science, 62*, 42–55. https://doi.org/10.1111/j.1365-2389.2010.01342.x

Prodanovic, V., Bach, P. M., & Stojkovic, M. (2024). Urban nature-based solutions planning for biodiversity outcomes: Human, ecological, and artificial intelligence perspectives. *Urban Ecosystems, 27*, 1795–1806. https://doi.org/10.1007/s11252-024-01558-6

Resilient Connecticut. (2024). *Transit oriented development for a more resilient CT final report*. Retrieved February 6, 2025, from https://resilientconnecticut.uconn.edu/resilient_ct_res/transit-oriented-development-for-a-more-resilient-ct-final-report/

Pace Environmental Law Review. (2023). *Climate resilient development at the local level: benefits of transit-oriented development*. Retrieved February 6, 2025, from https://pelr.blogs.pace.edu/2023/05/07/climate-resilient-development-at-the-local-level-benefits-of-transit-oriented-dev elopment/

Robertson, M. M. (2004). The neoliberalization of ecosystem services: Wetland mitigation banking and problems in environmental governance. *Geoforum, 35*, 361–373. https://doi.org/10.1016/j.geoforum.2003.06.002

Rubin, E. S., Davison, J. E., & Herzog, H. J. (2015). The cost of CO_2 capture and storage. *International Journal of Greenhouse Gas Control, 40*, 378–400. https://doi.org/10.1016/j.ijggc.2015.05.018

Safitri, D. M. (2023). Finding a research gap on service quality and safety improvement in public transportation. In *The 2nd National Conference on Mathematics Education (NACOME) 2021: Mathematical Proof as a Tool for Learning Mathematics*

Salès, K., Marty, P., & Frascaria-Lacoste, N. (2023). Tackling limitations in biodiversity offsetting? A comparison of the Peruvian and French approaches. *Regional Environmental Change, 23*, 145. https://doi.org/10.1007/s10113-023-02143-x

Shahariar, G. M. H., Bodisco, T. A., Zare, A., Sajjad, M., Jahirul, M. I., Van Chu, T., Bartlett, H., Ristovski, Z., & Brown, R. J. (2022). Impact of driving style and traffic condition on emissions and fuel consumption during real-world transient operation. *Fuel, 319*, Article 123874. https://doi.org/10.1016/j.fuel.2022.123874

Shiftan, Y., Israel, F., & Reuveni, E. (2021). *Gaps analysis report in the smart mobility field*. Retreived February 2, 2025, from https://istrc.net.technion.ac.il/files/2021/11/Gaps-Analysis-Report-in-the-Smart-Mobility-Field.pdf

Singh, B., Ellingsen, L. A. W., & Strømman, A. H. (2015). Pathways for GHG emission reduction in Norwegian road transport sector: Perspective on consumption of passenger car transport and electricity mix. *Transportation Research Part d: Transport and Environment, 41*, 160–164. https://doi.org/10.1016/j.trd.2015.09.028

Soria-Lara, J. A., Bertolini, L., & Te Brömmelstroet, M. (2016). Towards a more effective EIA in transport planning: A literature review to derive interventions and mechanisms to improve knowledge integration. *Journal of Environmental Planning and Management, 59*(8), 1537–1559. https://doi.org/10.1080/09640568.2016.1180282

Soria-Lara, J. A., Arce-Ruiz, R. M., Arranz-López, A., & Ariza-Álvarez, A. (2020). Environmental Impact Assessment for transport projects: A review of technical and process-related issues. In: N. Mouter (Ed.), *Advances in Transport Policy and Planning*, (vol 6, pp. 255–285). Academic Press. https://doi.org/10.1016/bs.atpp.2020.07.002

Tacconi, L., Mahanty, S., & Suich, H. (2013). The livelihood impacts of payments for environmental services and implications for REDD+. *Society and Natural Resources, 26*, 733–744. https://doi.org/10.1080/08941920.2012.724151

U.S. Department of Transportation. (2025). *Coordinated transportation planning—climate strategies that work*. Retrieved January 20, 2025, from https://www.transportation.gov/sites/dot.gov/files/2025-01/Coordinated%20Transportation%20Planning%20-%20Climate%20Strategies%20that%20Work.pdf

Vaissière, A.-C., & Levrel, H. (2015). Biodiversity offset markets: What are they really? An empirical approach to wetland mitigation banking. *Ecological Economics, 110*, 81–88. https://doi.org/10.1016/j.ecolecon.2015.01.002

Walker, M., Gibbard, P., & Lowe, J. (2015). Comment on "When did the Anthropocene begin? A mid-twentieth century boundary is stratigraphically optimal. Jan Zalasiewicz et al. (2015). *Quaternary International, 383*, 196–203. https://doi.org/10.1016/j.quaint.2015.04.007

Wang, Y., & Boggio-Marzet, A. (2018). Evaluation of eco-driving training for fuel efficiency and emissions reduction according to road type. *Sustainability, 10*, 3891. https://doi.org/10.3390/su10113891

Wang, Y., Pan, Z., Zhang, W., Borhani, T. N., Li, R., & Zhang, Z. (2022). Life cycle assessment of combustion-based electricity generation technologies integrated with carbon capture and storage: A review. *Environmental Research, 207*, Article 112219. https://doi.org/10.1016/j.envres.2021.112219

Ward, A. I., Dendy, J., & Cowan, D. P. (2015). Mitigating impacts of roads on wildlife: An agenda for the conservation of priority European protected species in Great Britain. *European Journal of Wildlife Research, 61*, 199–211. https://doi.org/10.1007/s10344-015-0901-0

Wunder, S., Börner, J., Ezzine-de-Blas, D., Feder, S., & Pagiola, S. (2020). Payments for environmental services: Past performance and pending potentials. *Annual Review of Resource Economics, 12*, 209–234. https://doi.org/10.1146/annurev-resource-100518-094206

Xu, L., Chen, Z., Li, X., & Xiao, F. (2021). Performance, environmental impact and cost analysis of marking materials in pavement engineering: The state of the art. *Journal of Cleaner Production, 294*, Article 126302. https://doi.org/10.1016/j.jclepro.2021.126302

Yirdaw, E., Kanninen, M., & Monge, A. (2023). Synergies and trade-offs between biodiversity and carbon in ecological compensation. *Sustainability, 15*, 11930. https://doi.org/10.3390/su151511930

Zheng, X., Easa, S. M., Yang, Z., Ji, T., & Jiang, Z. (2019). Life-cycle sustainability assessment of pavement maintenance alternatives: Methodology and case study. *Journal of Cleaner Production, 213*, 659–672. https://doi.org/10.1016/j.jclepro.2018.12.227

Miroslava Mikusova is a transport engineer, lecturer, and researcher in the Department of Road and Urban Transport at the University of Zilina, Slovakia. She specializes in mobility management, road safety management, intelligent transport systems (ITS), and sustainable urban mobility. Her research focuses on reducing the environmental impact of transportation through innovative mobility solutions, artificial intelligence, and smart city applications. She has extensive international experience, having collaborated with leading transport research institutions across Europe and Latin America. She has also contributed to on-site projects in Chile, Colombia, Brazil, and Cuba. As an ERA Chair researcher for Intelligent Transport Systems, she played a key role in advancing research excellence in ITS at the University of Žilina. She has led 13 international research projects on ITS, road safety, sustainable transport, and smart cities. As a member of ECTRI's Transport Safety Working Group, she actively contributes to discussions on transport and climate change. She has served as an editor of three scientific books and as a guest editor for the special issue "Towards Safe Horizons: Redefining Mobility in Future Transport" and the topical collection "Methods and Measures to Improve Road Safety and Travel Efficiency in Sustainable Urban Transport Management" in the journal Sustainability (MDPI). Dr. Mikusova is also an active scientific reviewer, having completed 109 verified peer reviews for 12 international scientific journals. Additionally, she is a committee member for over 12 international scientific conferences. Beyond her academic roles, she is a strong advocate for inclusive road infrastructure, particularly for people with disabilities. Her research and expertise provide valuable insights into innovative transport solutions aimed at mitigating climate change while enhancing safety and efficiency in urban mobility systems.

Sectorial Climate Policy Targets in Germany: Recent Developments and Opportunities

Christian Diller

Abstract In April 2024, Germany enacted a new law that deviated from its previous policy of formulating sectorial climate targets for individual policy areas. The primary rationale for this decision was the failure to achieve the targets set for the transport sector for an extended period of time. Achieving quite ambitious targets for the transport sector would have resulted in immediate measures, such as driving bans. This chapter examines the question of sectorial differentiation of policy objectives using three representative examples: (1) the targets for designating areas for wind energy; (2) the quantitative target of new housing development; and (3) the 30-hectare land-saving target. It is evident that there are differences between the policy fields in terms of the disaggregation of targets, which also depends on the political context. The question of whether disaggregating targets is an effective approach to achieve them better in a scientific view cannot be answered with certainty. At the end, this can be viewed as a question that is associated with a political risk.

Keywords Transport policy · Sectorial and spatial disaggregated targets · Climate policy · Policy change · Germany

1 Introduction

1.1 Climate Protection Act Considerations in Germany

Transport policy is an important part of climate policy. Science-based targets are considered an important climate policy goal for countries and companies in order to contribute to reducing global warming to 1.5 degrees (SBT, 2024). The German government has also formulated such quantitative targets, with the initial Climate Protection Act being passed in 2019. The legislation was reformed in 2021 under the

C. Diller (✉)
Urban Geography and Spatial Planning, Institute for Geography, Justus-Liebig-University Gießen, Gießen, Germany
e-mail: christian.diller@geogr.uni-giessen.de

M. A. Dulebenets et al. (eds.), *Transport and Climate Change*,
https://doi.org/10.1007/978-3-032-00563-2_13

new federal government, which consisted of the Sozialdemokratische Partei Deutschland (SPD, the social democrats), the Green Party, and the Freie Demokratische Partei Deutschland (FDP, the liberals).

The law commits Germany to reduce its greenhouse gas emissions by at least 65 per cent by 2030 in comparison to 1990 levels, with the ultimate objective of achieving climate neutrality by 2045. In comparison to the targets set by previous federal administrations, this target is arguably more ambitious. The Climate Act 2021 was primarily intended to reaffirm the goal of climate neutrality by formulating not only an overall target, but also sectorial targets for individual areas: energy, industry, buildings, transport, agriculture, waste and land use. Various federal ministries were involved, including the Ministry of Economic Affairs, the Ministry of Transport, the Ministry of Construction, and the Ministry of Agriculture. Table 1 shows the annual CO_2 emission targets formulated in the Climate Protection Act 2021 up to 2030. The sectors are classified in accordance with the Common Reporting Tables (CRT) set forth in the European Climate Reporting Regulation or a subsequent regulation enacted on the basis of Article 26(7) of the European Governance Regulation. For the transport sector, a CO_2 reduction target was set to decrease the annual emission volume in million tonnes of CO_2 equivalent from 150 million tonnes in 2020 to 85 million tonnes in 2030. A quite ambitious CO_2 reduction target was established for the energy industry as well.

The regulation also requires a strict controlling. In case of two consecutive years of evident failure by the German government to achieve its climate target for 2030, a corrective action must be taken. If individual sectors, such as transport or buildings, fail to meet the statutory targets for CO_2 emissions, the responsible ministries are obliged to present immediate action programmes in the following year. The most recent assessment of the German government's climate targets indicates a positive overall outcome. In 2023, greenhouse gas emissions decreased by over 10% compared to the previous year, placing the climate targets for 2030 within a reach. In the energy and industrial sectors, a greater reduction in greenhouse gas (GHG)

Table 1 Annual sectorial CO_2 emission targets for Germany until 2030 according to the climate protection act 2021. *Source* BKSG (2021)

Annual emission volume in million tonnes of CO_2 equivalent	2020	2021	2022	2023	2024	2025	2026	2027	2028	2029	2030
Energy industry	280		257						108		108
Industry	186	182	177	172	165	157	149	140	132	125	118
Buildings	118	113	108	102	97	92	87	82	77	72	67
Transport	150	145	139	134	128	123	117	112	105	96	85
Agriculture	70	68	67	66	65	63	62	61	59	57	56
Waste management and other	9	9	8	8	7	7	6	6	5	5	4

emissions was achieved than required by the climate targets (Die Bundesregierung, 2024a). However, it became evident that the sectorial targets would not be met in all sectors, with the transport sector exhibiting particularly evident deficiencies in implementation. Consequently, efforts commenced on a revised Climate Protection Act halfway through the legislative period, with the finalised version being passed in June 2023. Since then, the legislation has been subject of debates within the Bundestag. Despite the government's initial disagreement, a law was presented that no longer monitors compliance with climate targets retrospectively by sector. Instead, it focuses on the future, with a view to achieving the targets over several years and, most importantly, across sectors. This approach means that only the overall value needs to be achieved. In the case of a sector failing to meet the prescribed climate targets, its shortfall can be offset by the other sectors (Turczynski, 2024).

The primary rationale of underlying this newly proposed regulatory framework is the fact that, in 2023, the transportation sector—for which the FDP assumed responsibility in 2021 (previously from the Christlich Soziale Union, CSU since 2010)—surpassed the permissible annual emissions levels for the third consecutive year. In lieu of the permissible 134 million tonnes of CO_2 equivalents, 146 million tonnes of greenhouse gases were emitted (Expertenrat Klima, 2024). In consequence, an emergency plan should have been implemented in accordance with the prevailing legal framework. This might have entailed nationwide driving bans, a measure that was last implemented in Germany in 1974 in the context of the oil crisis. Consequently, Transport Minister Volker Wissing (FDP) would have been required to present an emergency plan.

However, the new legislation permits the overruns in transport to be compensated for by other sectors. The law passed in April 2024 has received criticism from climate experts, who have expressed concerns that without the establishment of sector-specific targets, the country could face the discrepancy between the targets and rising actual emissions of the transport sector (Die Bundesregierung, 2024a). Environmental organisations Greenpeace and Germanwatch, in collaboration with over 54,000 co-plaintiffs, have recently initiated a constitutional complaint against the Federal Government's inadequate climate protection policy. Additionally, they have expressed the necessity for the implementation of concrete measures aimed at reducing CO_2 emissions in the transport sector (ZEIT, 2024). The question thus arises as to why the sectorial differentiation was abandoned due to the transport sector. This necessitates a more detailed examination of the German transport policy.

1.2 Lack of Innovative German Transport Policy

The German transport policy in recent decades can be regarded as less innovative in light of the formulated climate protection targets. It is still strongly focused on motorized private transport and, specifically, on the conventional CO_2-emitting combustion engines. The motorization rate is one of the indicators for which Germany holds a top position in a global comparison. In 2024, it ranked eighth in the European Union

(EU) with 583 cars per 1,000 inhabitants (DESTATIS, 2024). The density is continuously increasing. Almost the same finding shows the statistic for the modal split. In 2021, over 80% of the passenger kilometers associated with the transport sector were completed by car in Germany. Among 30 European states in the statistics, that is position No. 6 and one of the highest levels (EUROSTAT, 2023). Although having position No. 6 and above the European average can be partially attributed to railway passenger kilometers (EUROSTAT, 2023), Germany with its "Autobahn" can be titled as a car-oriented country. One reason for this focus is undoubtedly the significant role of the automotive industry. For an extended period, Volkswagen was regarded as the largest automotive group in the world, while Daimler-Benz and BMW are among the leading global players in the luxury sedan segment. The influence of automobile lobbyists on land and state politics has been accordingly considerable.

The preference for the automobile over passenger rail transport was particularly evident in the respective federal transportation plans, in which the prioritization of investments became apparent. For decades, investments in road infrastructure were prioritized over railway investments. The current Federal Transport Infrastructure Plan 2030, adopted in 2016, allocates 49.3% of the total volume of €269.6 billion for the maintenance and expansion of transport networks to road transport, 41.6% to rail, and 9.1% to waterways (BMVI, 2016). Despite the considerable increase in funding compared to previous plans and the growing recognition of the importance of maintaining and expanding rail infrastructure, it is evident that infrastructure for motorized private transport remains a priority compared to rail transport.

Considering the aim of decarbonizing the transport system, most of the parties can be named as indifferent and conservative. The parties Christlich-Demokratische Union (CDU), Christlich-Soziale Union (CSU), and Sozialdemokratische Partei Deutschland (SPD) have a lot of respect for the car industry in the states of Lower Saxony and Saxonia (Volkswagen), Bavaria (BMW, Audi), and Northrine-Westfalia (Opel, Ford). The FDP in particular has traditionally emphasized the individual right to drive according to the slogan "free travel for free citizens" (ZIN, 2024), which already emerged during the 1974 debate. The emerging right-wing party Alternative für Deutschland (AFD) even denies the man-made component in climate change and doesn't see any requirement of a modification in transport policy. Only the Green Party (Die Grünen) continuously follows a consequent traffic policy towards CO_2 reduction.

The discussion about transport policy measures in Germany at the federal level has been very much focused on symbolic issues in recent decades. The introduction of tolls for foreign trucks on German freeways was the central political project of CSU's Minister Scheuer, who served from 2018 to 2021, which ultimately ended in a political fiasco. To date, Germany is one of few countries in the world without such a limit. Despite the existence of speed restrictions on a significant portion of the German freeway network, which spans approximately 13,200 km (Statista, 2024a), a uniform speed limit has yet to be established. In particular, the Free Democratic Party (FDP) and the Christian Social Union (CSU) of the Christian Democratic Union (CDU) have repeatedly articulated opposing positions to the demand for a speed limit on freeways. The debate has been characterized by a strong ideological dimension. The focus on debates with a rather symbolic effect was accompanied by an insignificant impetus in favor of an innovative CO_2-reducing transport policy.

Far more dramatic, however, years of underinvestment in maintaining the network infrastructure were observed, both on roads and on railways. This was evidenced by the occurrence of several notable bridge collapses, including the recent incident in Dresden (ZDF, 2024a), as well as the frequent closure of bridges and other transportation infrastructure. The results of this policy are even more dramatic for the German rail network. It would not be an exaggeration to describe this network as being in a state of severe disrepair. The resulting increase in number of delays has been well-documented, with particularly problematic instances on long-distance routes (Statista, 2024b).

The new federal government that took office in 2021 did formulate more ambitious goals for a climate-friendly transport transition in the coalition agreement than the preceding governments. However, the Green Party, which formulated these goals most consistently, neglected to secure the Ministry of Transport. Consequently, the responsibility for achieving this goal falls upon the shoulders of the Free Democratic Party (Freie Demokratische Partei Deutschlands, FDP), a party that is traditionally regarded as an advocate of motorized private transport. The most significant governmental objective at present is the consolidation of public finances, which contradicts the goal of maintaining and expanding infrastructure. In recent years, more investment has been made in maintaining and expanding the rail network and, in response to the drastic decline in rail user numbers in times of coronavirus, a nationwide rail ticket was introduced, which also led to a short-term increase in the number of users. However, the fundamental refurbishment of the rail network still appears to be a challenge that has only just begun to be tackled.

It is becoming increasingly evident within the automotive industry, which is of central importance to Germany, that both German companies and politicians have failed to keep pace with developments, particularly in the low-cost electric car segment. This has resulted in a corresponding increase in market share for Chinese products. The policy for subsidies for the purchase of electric cars was inconsistent and the expansion of the charging infrastructure also stagnated. This provides an explanation as to why the proportion of electric cars in the new registrations in Germany in 2024 experienced a significant decline in comparison to the previous years, reaching a mere 13% in August (ADAC, 2024). Thus, Germany may lose its good position in the share of electric cars, which was No. 6 in Europe in 2022 (EUROSTAT, 2023). Consequently, the German government has convened a crisis summit with the German automotive industry with the objective of enhancing innovation and sales in the electric car sector (NTV, 2024).

2 Quantitative Targets: Examples from Other Spatially Relevant Policy Fields in Germany

Against the backdrop of the abandonment of sectorial climate protection targets in the new Climate Protection Act, the question of sectorial disaggregation of quantified targets is examined using the examples of other policy fields, supplemented by the interesting question of spatial disaggregation from a geographical perspective. Important questions are addressed through the examination of the following case studies: (1) the designation of areas for wind energy; (2) the quantitative target for the development of new housing; and (3) the 30-hectare land-saving target. To what extent is a sectorial and spatial disaggregation of the targets appropriate, and has it been in fact undertaken? How can the success of such targets be evaluated? And what would happen if these targets were not achieved?

2.1 *Areas for Wind Energy*

In the field of preventive climate protection, the central task for spatial planning is the expansion of renewable wind energy. The objective is to "significantly accelerate the expansion of renewable energies and remove all obstacles in the process" (Koalitionsvertrag, 2021). It is intended that 80% of the gross electricity demand of 680–750 TWh, which is to be met in 2030, shall be provided for from renewable energy sources. Consequently, grid expansion is to be accelerated. The capacities for offshore wind energy shall be increased to at least 30 GW in 2030, 40 GW in 2035, and 70 GW in 2045 (Koalitionsvertrag, 2021). The Erneuerbare Energien-Gesetz (EEG) and the Offshore Wind Energie-Gesetz have set forth fixed expansion targets for the coming years. In 2030, the output of wind power plants is to more than double to 145 gigawatts, and wind power is to reach 230 gigawatts by 2045 (Sander, 2024).

The proportion of land dedicated to wind energy constitutes a sectorial target for a renewable energy provider, derived from the overarching goal of increasing the proportion of renewable energy. In accordance with the recently introduced regulations included in the Baugesetzbuch (Building Code), a minimum of two per cent of federal state land is to be designated for onshore wind energy. In order to achieve this, concerted actions by the land and state governments are planned in the near future. Land owned by the federal government in the external area is prioritized for climate protection as potential sites for wind power and photovoltaic (PVC) systems. This sectorial target for wind energy plants has also been spatially disaggregated in the Act on the Determination of Land Requirements for Onshore Wind Energy Plants (Wind Energy Land Requirements Act—WindBG) of 2022 (see Table 2). A binding target was established for each federal state, specifying the percentage of state territory designated for wind energy by 2027 and 2032. The range of percentages established for each state varied considerably. In Berlin, for instance, the target was 0.5%, while

Table 2 Area contribution values of the federal states for wind energy plants. *Source* BfJ (2024)

Federal state	Area contribution value to be achieved by 31 Dec. 2027 (percentage of federal state area)	Area contribution value to be achieved by 31 Dec. 2032 (percentage of federal state area)
Baden-Württemberg	1.1	1.8
Bavaria	1.1	1.8
Berlin	0.25	0.5
Brandenburg	1.8	2.2
Bremen	0.25	0.5
Hamburg	0.25	0.5
Hesse	1.8	2.2
Mecklenburg-Western-Pomerania	1.4	2.1
Lower Saxony	1.7	2.2
North Rhine-Westphalia	1.1	1.8
Rhineland-Palatinate	1.4	2.2
Saarland	1.1	1.8
Saxony	1.3	2.0
Saxony-Anhalt	1.8	2.2
Schleswig-Holstein	1.3	2.0
Thuringia	1.8	2.2

in the largely rural states of Hesse, Rhineland-Palatinate, Thuringia, Saxony-Anhalt, and Brandenburg, it was 2.2%.

However, the simultaneous disaggregation of the sectorial data set to the individual energy sources and the disaggregation of the spatial data set to the federal states resulted in conflicts. Some federal states, such as Bavaria, advanced the argument that their potential in wind energy was limited but they had significant potential in other renewable energy sources, such as biomass and photovoltaic systems. In fact, their performance in these areas exceeded the federal average. In regard to renewable energy, however, the State of Bavaria did not consider itself constrained by the targets for wind energy areas. At times, even legal provisions were enacted, such as the wind turbine distance decree, which rendered this target unattainable (BSWBV, 2023).

Nevertheless, the measures implemented to accelerate the expansion in Germany appear to be successful overall (Die Bundesregierung, 2024b). As reported by the Federal Statistical Office, 61.5% of the electricity generated in this country during the initial six-month period of 2024 was derived from wind, solar, hydropower, and biomass sources. A total of 57 per cent of the electricity consumed was derived from renewable energy sources. This represents a nine per cent increase in climate-friendly electricity generation compared to the first half of 2023, representing the highest level of production in the first half of a year to date. Wind energy was the predominant source of electricity generation, accounting for one-third of domestic electricity production in the first half of 2024.

2.2 New Housing

As in many other European countries, the problem of housing shortages has worsened dramatically in Germany in recent years, particularly in the lower price segments, despite intensified construction activity (BfWSB, 2023). The cost of renting and purchasing property has risen significantly, particularly in major cities, such as Hamburg, Berlin, Munich, Frankfurt, and Cologne, but also in smaller urban centers. One of the causes of this increasingly dramatic housing shortage was the drastic reduction in social housing construction in Germany over the years, which resulted in an insufficient number of affordable housing units being created. The issue had consistently been overlooked by preceding administrations, but it was the new federal government, in 2021, that first established quantitative targets for housing construction in Germany. According to the coalition agreement, the goal was to construct a total of 400,000 new homes annually. This would equate to 1.6 million new units over five years. The most notable shift in emphasis compared to the previous government is the markedly greater importance attached to publicly subsidized housing construction. Accordingly, the Coalition Agreement 2021 (p. 88) stipulates the construction of 100,000 new publicly subsidized flats on an annual basis. In this regard, a differentiation is to be made between sectors, although this will not apply to political departments. In theory, this target could be disaggregated to the level of federal states and even municipalities. However, this approach would not adequately address the spatially disparate needs of the markets and was therefore excluded.

In light of the pressing necessity for housing, this objective garnered considerable public interest. As part of the Alliance for More Affordable Housing, initiated by the German government, a comprehensive package of measures was defined that pursued the main objectives: increased investment and deceleration of cost containment in new construction (BfWSB, 2024). To achieve this, various alternatives were implemented to strengthen social housing construction, which include the following: mobilize building land, facilitate inner-city development, and reduce cost-driven construction standards in planning procedures.

Despite the implementation of the aforementioned measures, the preliminary evaluation for 2023 evidently demonstrates that this objective was not achieved during the initial years. In contrast to the 400,000 new homes initially projected, only 295,000 were completed. Therefore, even the preceding figures were not attained, with the number of new flats declining by 0.3% in 2023 in comparison to 2022 (ZDF, 2024b). The reasons for this decline were numerous, with the most significant being the supply difficulties experienced by the construction industry during the pandemic and the subsequent conflict in Ukraine. Since the end of 2022, the real estate market has also slowed down considerably due to interest rate hikes and construction costs. Consequently, it can already be expected that the figures will not be reached in the coming years despite the measures taken to incentivize residential construction. In view of the 2025 federal elections, the federal government is under increasing pressure to justify its actions. Thus far, the political pressure on this issue has been relatively limited, with the delayed effect of measures. However, it will increase. Nevertheless, the current federal government remains committed to the housing construction targets that have been set.

2.3 The 30-Hectare Land-Saving Target

The so-called 30-ha target is one of the oldest quantitative targets of spatially relevant policy in Germany. The 30-hectare target was formulated with a central goal of reducing the consumption of the finite resource of land and soil in the German government's sustainability strategy as early as 2002. It was adopted in the coalition agreement of the new federal government in 2021 and underpinned by further measures (Diller, 2022). Specifically, the goal is to significantly reduce the consumption of new settlement areas in Germany from 30 hectares per day. The most important policy areas for achieving this goal are the urban development policy, spatial development policy, settlement policy, and housing policy. Transport policy is also affected, as a considerable proportion of new land is used for transport. From a technical point of view, this target can theoretically be broken down by sector by formulating specific targets for individual types of land (e.g., residential land, commercial land, transport land, retail land).

However, this sectorial disaggregation has not been implemented politically. Spatial disaggregation would be possible, for example, by using a municipality's share of the total settlement area in Germany as the basis for calculating the permissible share of new areas with the maximum target value of 30 hectares per day, as is practised in cities such as Heidelberg. However, a prescribed disaggregated value would not do justice to the different structures and development dynamics of rural areas on the one hand and urban areas on the other. The state and regional plans therefore refer to the general target of 30 ha. However, this target value is not simply reduced. At best, target values for individual municipalities are formulated in the regional plans, in which an attempt is made to find a compromise between the past settlement development and the land reduction target. There are no precise, scientifically-based values. Differentiated targets are ultimately decided politically between municipalities in regional assemblies. Comparative studies indicate differences between federal states and regions in the stringency of the formulation and implementation of settlement development targets (Pehlke et al., 2021).

A key starting point for saving space is the internal development approach. Already developed areas are used more intensively and revitalised when they fall into disuse. Small-scale, differentiated studies on the realisation of internal development potential are not yet available for Germany as a whole. A study for North Rhine-Westphalia concluded that internal development potential has been realised to a relevant extent over time, but that there are significant differences between larger cities and small municipalities in suburban and rural regions (Eichhorn & Siedentop, 2022). Brownfield development is often a successful instrument in attractive locations, particularly in large cities, and its potential has not yet been fully exploited. Changing the framework conditions could also favour the achievement of the 30-ha target, taking into account the trend towards compact cities, rising land prices, and declining populations, at least in some peripheral regions. In addition, there have been changes in planning laws that at least partially counteract the consumption of greenfield land.

A differentiated assessment must be made of whether this goal has been achieved. Current statistics shows that land consumption in Germany has fallen continuously between 2006 and 2019 (UBA, 2023). It should be noted that Germany was in a continuous economic upswing in the 2010s and that the population in Germany has been growing overall since 2011, mainly as a result of immigration (approx. 80 million in 2011 and 84 million in 2021). In this respect, it can be assumed that the measures to implement the 30-ha target have been at least partially successful. However, the 30-ha target has not yet been achieved within the set frame. In the period 2019–2022, 52 ha of new housing areas were designated per day. In previous years, the target had already been partially achieved, but not to the desired extent. The political response in the 2018 revision of the sustainability strategy was the retention of the target and an extension of the time horizon for its achievement.

3 Conclusion and Prospects

By taking Germany as a case study, the preceding observations firstly demonstrate that quantitative targets do not universally apply across all policy domains. Nevertheless, in certain spatially relevant policy areas, such as climate and energy policies, as well as transport policy at the federal level, there are notable instances of quantitative targets. These have also been formulated partly in a sectorial and spatially disaggregated manner. The following findings can be summarized for these policy areas. To what extent is a sectorial and spatial disaggregation of the targets useful and has this actually been implemented? The example of the area share for wind turbines represents the only target in which both sectorial and spatial disaggregation was undertaken. On the one hand, a target value was set for the total proportion of renewable energy to be provided by wind turbines.

On the other hand, a nationwide target for the proportion of land used for wind turbines was established, although no clear scientific link was made to the capacity target. This target was also disaggregated to the level of individual federal states in the corresponding law, reflecting the outcome of negotiation processes between the federal government and the federal states. The CO_2 reduction target was initially disaggregated by department. However, this sectorialisation was discontinued following the amendment to the law adopted in 2024. Sectorial disaggregation was, at the very least, applied in a basic manner to the housing development target by quantifying the proportion of social housing within the overall target. In the case of the land-saving target, disaggregation by individual land types, which is entirely feasible from a methodological standpoint, was omitted. Spatial disaggregation in the form of statutory requirements was not applied to the targets for housing development and land saving. The decision to break down the federal targets to their own levels lies with the federal states, regions, and local authorities.

As for transport policy, a combination of sectorial disaggregation with spatial disaggregation was never discussed. The former approach of sectorial disaggregation was actually given up by the federal government. The examples of land use policy and renewable energy policy show that sectorial disaggregation could be a useful instrument. But it is not suitable to correct long-term policy paths in short periods. In Germany, the change to a decarbonized traffic system has to be better prepared. If formulated sectorial targets cannot be reached because of structural conditions, they lead to considerable political legitimation deficits. In the field of regenerative energy, the sectorial and spatial disaggregated targets were formulated after over 20 years of experience to accelerate a process which was basically advanced. When it comes to transport policy, this process seems to be merely at the beginning. A good illustrative example of this tendency could be the abrupt abolition of the subsidiaries for electric vehicles by the Federal government. Perhaps, sectorial targets in transport policy as part of decarbonization policy will make a revival in the future on a more solid policy base.

How are these targets being monitored and what will happen if they are not achieved? All of the quantitative targets described here, particularly in transport policy, are subject to review at the federal level as part of the periodic reporting process based on official statistics. Interesting differences between the targets can be seen with regard to the question of what happens if the targets are not met. It is premature to evaluate the quantitative targets for the expansion of wind energy. However, it is probable that the stagnation observed in previous years has been at least partially overcome. In contrast, it is already foreseeable that the housing construction target will not be achieved within the current legislative period. Nevertheless, the target is still being adhered to, with public political discourse emphasizing the numerous and diverse activities aimed at achieving the target. Of a particular note is the emphasis placed on the general conditions that are impeding their fulfillment. A similar approach has been taken with regard to the land-saving target, whereby even if the target for the envisaged period has not been met, the time horizon has been extended.

In light of the aforementioned circumstances, how might the amendment to the legislation regarding the sectorial CO_2 reduction targets be evaluated? It is not a common practice in politics to set quantitative targets. In comparison to qualitative targets, quantitative targets demonstrate a greater degree of commitment, which appears to indicate a stronger political resolution to address the issue stringently. The conciseness of these targets renders them an invaluable instrument in the political discourse. Nevertheless, the formulation of quantitative targets carries a certain degree of political risk. In particular, politicians are exposed to the risk of being judged in the public debate for something that they cannot actually ensure. In the majority of policy areas, the effects of measures do not manifest until a considerable period after they have been formulated. In many cases, this delay extends to several legislative periods, during which the political environment has often undergone significant changes. The starting level of the key performance indicators have to be respected. As an example, the German climate policy seemed to be successful for a long time because the indicator values before 1990 were taken as a base, but the targets were reached not because of the climate policy but due to the breakdown of the Eastern German Industry soon after 1990.

The implementation of legal changes across various levels of government, particularly in a federal state like Germany, where the influence of overarching European Union (EU) requirements must also be considered, is a lengthy process. The difficulty of influencing global context factors, such as energy prices and inflation in the aforementioned policy areas, has a significant impact on housing markets, construction, and energy prices. Consequently, the effectiveness of measures is influenced by these factors. In the context of day-to-day politics, governments are primarily concerned with justifying outcomes for which they have limited responsibility and which they are unable to influence immediately. Furthermore, the German political landscape has become more diverse, and the formation of governments is increasingly volatile, which supports the logic of focusing politics on one legislative period rather than the long term. In this context, the criticism of the amendment to the law to no longer disaggregate climate targets by sector is understandable. It is likely to result in a continuation of a less innovative German transport policy, which will not contribute to CO_2 reduction in the future.

Conversely, the assertion that the overarching equilibrium is of paramount importance is also a valid position. It is understandable that the Federal Government, which has suffered considerable political damage and currently has very low satisfaction ratings in the population (Statista, 2024b), tends to emphasize successful efforts in its public image. Adherence to the sectorial climate targets would have necessitated the implementation of immediate measures within the transport sector, such as driving bans. Such measures would have resulted in a further reduction in the government's acceptance among the population. More significantly, they would have diminished the likelihood of success in the 2025 elections for the FDP, which is responsible for the transport department, and the Greens, who are particularly committed to a CO_2-reducing transport policy. The implementation of quantitative, sectorial and spatially disaggregated policy targets has the potential to serve as an effective instrument for a successful policy, as evidenced by the current approach to wind energy areas. However, the efficacy of this instrument is contingent upon a comprehensive situational context. It is rather difficult to give a definitive answer to the question of whether the disaggregation of targets facilitates the achievement of the desired outcomes.

References

ADAC. (2024). Pkw-Neuzulassungen August 2024: E-Auto-Absatz bricht dramatisch ein. https://www.adac.de/news/neuzulassungen-kba/

BfJ. (2024). Bundesministerium der Justiz: Windenergieflächenbedarfsgesetz vom 20. Juli 2022 (BGBl. I S. 1353), das zuletzt durch Artikel 12 des Gesetzes vom 8. Mai 2024 (BGBl. 2024 I Nr. 151) geändert worden ist. https://www.gesetze-im-internet.de/windbg/BJNR135310022.html

BfWSB. (2023). Bundesministerium für Wohnen, Stadtentwicklung und Bauwesen: Faktenblätter zum deutschen Wohnungsmarkt 2023, Stand: Dezember 2023, Berlin. https://www.bmwsb.bund.de/SharedDocs/topthemen/Webs/BMWSB/DE/fakten-zum-wohnungsmarkt/faktenblaetter.pdf;jsessionid=1FE210115B1C8B0714C2C38A61B0CEBB.live871?__blob=publicationFile&v=2

BfWSB. (2024). Bundesministerium für Wohnen, Stadtentwicklung und Bauwesen: Bündnis bezahlbarer Wohnraum. https://www.bmwsb.bund.de/SharedDocs/topthemen/Webs/BMWSB/DE/buendnis-bezahlbarer-wohnraum/buendnis-Artikel.html

BKSG. (2021). Bundes-Klimaschutzgesetz vom 12. Dezember 2019 (BGBl. I S. 2513), das durch Artikel 1 des Gesetzes vom 15. Juli 2024 (BGBl. 2024 I Nr. 235) geändert worden ist Geändert durch Art. 1 G v. 18.8.2021 I 3905 Änderung durch Art. 1 G v. 15.7.2024 I Nr. 235 textlich nachgewiesen, dokumentarisch noch nicht abschließend bearbeitet.

BMVI. (2016). Bundesministerium für Verkehr und digitale Infrastruktur: Bundesverkehrswegeplan 2030, Berlin, https://bmdv.bund.de/SharedDocs/DE/Publikationen/G/bundesverkehrswegeplan-2030-gesamtplan.pdf?__blob=publicationFile

BSWBV. (2023). Bayerisches Staatsministerium für Wohnen, Bau und Verkehr (Bauleitplanung für Windenergieanlagen, insbes. Repowering-Bebauungsplan. Ein Merkblatt für Städte und Gemeinden, Behörden und Träger öffentlicher Belange, Planerinnen und Planer, Projektträger sowie Bürgerinnen und Bürger. Überarbeitete Auflage Stand: 06.04.2023

DESTATIS. (2024). Pkw-Dichte im Jahr 2022 erneut auf Rekordhoch. https://www.destatis.de/DE/Presse/Pressemitteilungen/2023/09/PD23_N048_46.html

Die Bundesregierung. (2024a). Treibhausgasemissionen sinken deutlich. https://www.bundesregierung.de/breg-de/aktuelles/treibhausgasbilanz-2023-2265440#:~:text=In%20allen%20Sektoren%20werden%20Treibhausgase%20eingespart.%20

Die Bundesregierung. (2024b). Energiewende. So läuft der Ausbau der Erneuerbaren Energien in Deutschland. https://www.bundesnetzagentur.de/SharedDocs/Pressemitteilungen/DE/2024/20240719_EEZubauHJ1.html?nn=659670

Diller, C. (2022). Mehr Fortschritt durch raumbedeutsame Politik? Eine vertiefende Betrachtung ausgewählter Handlungsfelder des Koalitionsvertrags der neuen Bundesregierung, in: Raumplanung 126, 2/2022, S. 66–72

Eichhorn, S., Siedentop, S. (2022). Innenentwicklung in Deutschland - ein Klassifizierungsansatz. In: Flächennutzungsmonitoring XIV: Beiträge zu Flächenmanagement, Daten, Methoden und Analysen (S. 139–148). Berlin: Rhombos-Verlag. https://doi.org/10.26084/14dfns-p015

EUROSTAT. (2023). Key figures on European transport 2023 edition. https://ec.europa.eu/eurostat/documents/15216629/18384997/KS-HE-23-001-EN-N.pdf/65eb90bc-4856-f6a5-b12f-cf87854587f7?version=4.0&t=1707145038133

Expertenrat Klima. (2024). Prüfung Projektionsdaten: Einhaltung des Klimaziels für 2021 bis 2030 nicht bestätigt. https://www.expertenrat-klima.de/

Koalitionsvertrag. (2021). Mehr Fortschritt wagen. Bündnis für Freiheit, Gerechtigkeit und Nachhaltigkeit. Koalitionsvertrag 2021–2025 zwischen der Sozialdemokratischen Partei Deutschlands (SPD), BÜNDNIS 90/DIE GRÜNEN und den Freien Demokraten (FDP), Berlin

NTV. (2024). Belebung der E-Mobilität. Jede Menge Vorschläge für Habecks Autogipfel. https://www.n-tv.de/wirtschaft/Jede-Menge-Vorschlaege-fuer-Habecks-Autogipfel-article25243894.html

Pehlke, D., Eichhorn, S., Diller, C. (2021). Hat die Trägerschaft der Regionalplanung Einfluss auf ihre Planinhalte und die regionale Siedlungsentwicklung? Theoretische Überlegungen und eine

empirische Analyse für Deutschland, in: Raumforschung und Raumordnung|Spatial Research and Planning, 2021, 1–17. rur.oekom.de. https://doi.org/10.14512/rur.74
Sander, L. (2024). Wie läuft der Ausbau erneuerbarer Energien in Deutschland? https://www.ndr.de/nachrichten/ndrdata/Wie-laeuft-der-Ausbau-von-Solar-Windkraft-Batteriespeicher-Erneuerbare-Energien-in-Deutschland,erneuerbare104.html
SBT. (2024). Science Based Targets: Ambitious corporate climate action. https://sciencebasedtargets.org/#:~:text=The%20framework%20enables%20businesses%20in%20the%20buildings%20value,to%20prevent%20the%20worst%20effects%20of%20climate%20change
STATISTA. (2024a). Länge der Autobahnen in Deutschland von 1995 bis 2023. https://de.statista.com/statistik/daten/studie/2972/umfrage/entwicklung-der-gesamtlaenge-des-autobahnnetzes/
STATISTA. (2024b). Anteil pünktlicher Züge der Deutschen Bahn im Personenverkehr (Nah- und Fernverkehr) von Januar 2019 bis August 2024. https://de.statista.com/statistik/daten/studie/202122/umfrage/puenktlichkeit-der-zuege-der-deutschen-bahn/
STATISTA. (2024c). Wie zufrieden sind Sie mit der Arbeit der Bundesregierung? https://de.statista.com/statistik/daten/studie/2953/umfrage/zufriedenheit-mit-der-arbeit-der-bundesregierung/
Turczynski, J. (2024). Weg mit den Sektorzielen - Was bedeutet das neue Klimagesetz? BR-Online. https://www.br.de/nachrichten/deutschland-welt/klimaschutzgesetz-was-bedeutet-der-wegfall-der-sektorziele,UAHdT9w
UBA. (2023). Umweltbundesamt: Flächensparen – Böden und Landschaften erhalten. https://www.umweltbundesamt.de/themen/boden-landwirtschaft/flaechensparen-boeden-landschaften-erhalten#flachenverbrauch-in-deutschland-und-strategien-zum-flachensparen
ZDF. (2024a). Abrissarbeiten in Dresden. Carolabrücke: Weiterer Abschnitt eingestürzt. https://www.zdf.de/nachrichten/panorama/dresden-carolabruecke-einsturz-elbe-100.html
ZDF. (2024b). Bundesregierung verfehlt Wohnungsbauziel deutlich. https://www.zdf.de/nachrichten/wirtschaft/weniger-neue-wohnungen-deutschland-100.html
ZEIT. (2024). ZEIT-Online: Umweltverbände klagen gegen Regierung für besseren Klimaschutz. https://www.zeit.de/politik/deutschland/2024-09/klage-klimaschutz-greenpeace-germanwatch-klimaschutzgesetz
ZIN. (2024). Zentrum für Interdisziplinäre Nachhaltigkeitsforschung: Freie Bürger, Freie Fahrt. https://nach-haltig-gedacht.de/glossary/freie-buerger-freie-fahrt/

Christian Diller has been a faculty member of Urban Geography and Spatial Planning in the Institute for Geography at the University of Gießen in Germany since 2007. His research interests focus on urban geography and spatial planning in all scales with an accent on regional planning. Further research interests are regional development and regional government as well as evaluation research. Until now, he has supervised over 230 bachelor's theses and 115 master's theses. After his diploma in spatial planning at the Technical University of Berlin in 1989, he worked at the private Institut für Stadtforschung in Berlin until 1995. Afterwards, he finished his Ph.D. in spatial planning in 2000 at the Technical University of Berlin. He worked in planning administrations in Hamburg and the Federal State Schleswig-Holstein. Dr. Diller finished his habilitation in 2005 again at the Technical University of Berlin. He is a regular member of the ARL Akademie für Raumentwicklung der Leibnitz Gemeinschaft, the DEGEVAL Deutsche Gesellschaft für Evaluation, and the DVAG Deutscher Verband für Angewandte Geographie.

Harnessing Digital Twins in Transportation: Adaptation and Mitigation Strategies for Sustainability and Resilience

Guilhermina Torrao, Ahmet Onur Agca, and Reem Hadeed

Abstract The transportation sector is essential for the global economic and social progress but also significantly contributes to greenhouse gas emissions and climate change. While the application of digital twin technology has been noted in various sectors, such as aerospace, it has only recently begun to expand into the transportation field. This chapter explores the transformative potential of digital twin technology in developing sustainable and resilient transportation systems. Digital twins are dynamic virtual replicas of physical systems that incorporate real-time data and advanced analytics to enhance transportation networks' performance, adaptability, and efficiency. This study is grounded in a systematic literature review that analyzes the latest research, specifically, peer-reviewed publications from 2018 onwards, focused on the use of digital twin technology in future mobility and optimizing transport network management and supply chains. Real-world applications, such as the Thameslink Railway Program and the Port of Rotterdam, demonstrate how digital twins can enhance energy consumption and streamline logistics operations. The research presented in this chapter contributes to the current understanding of digital twin technology by critically examining its dual benefits and highlighting its role in adaptation and mitigation strategies to improve transportation resilience against climate change and extreme weather conditions. Furthermore, it addresses the implementation challenges and provides recommendations for future research.

Keywords Climate change · Adaptation · Mitigation · Digital twin · Supply chains · Transport systems

G. Torrao (✉) · A. O. Agca · R. Hadeed
Engineering Management, College of Engineering and Physical Sciences, Aston University, Birmingham, UK
e-mail: g.torrao@aston.ac.uk

A. O. Agca
e-mail: a.agca@aston.ac.uk

R. Hadeed
e-mail: r.hadeed@aston.ac.uk

M. A. Dulebenets et al. (eds.), *Transport and Climate Change*,
https://doi.org/10.1007/978-3-032-00563-2_14

1 Introduction

1.1 Transport Pollution and Contribution to Climate Change

Transport is a crucial driver of society and economic growth, and the transport system is critical to worldwide businesses and global supply chains. Transport is fundamental to achieving the United Nations (UN) Millennium Development Goals, in particular to ensure environmental sustainability and respond to the climate change emergency. At the same time, transport generates externalities, such as energy consumption, pollutant emissions, noise, congestion, and accidents. According to UN (2024), the transport sector relies heavily on fossil fuel combustion, making it a leading source of urban and regional air pollution. On a global scale, this sector accounts for almost 60% of oil demand and 30% of energy consumption (UN, 2024). The International Energy Agency (IEA) reports that from 1990 to 2022, transport emissions grew at an annual rate of nearly 2%, outpacing increases in all other end-use sectors, except for industry (IEA, 2024a). On the other hand, the European Commission (EC) states that to achieve Net Zero Emissions (NZE) by 2050, it is crucial to reduce transport emissions worldwide by 90% by that year (EC, 2024).

In 2023, transportation accounted for about 16% of global greenhouse gas (GHG) emissions, making it a major contributor to global warming and climate change (Statista, 2024). This significantly increases the risks of rising temperatures, rising sea levels, melting glaciers, and severe weather occurrences. These conditions lead to more frequent and intense hurricanes, heat waves, and heavy rainfall, all of which can cause substantial disruptions on both regional and global scales. Consequently, they pose heightened risks to human health, agricultural stability, energy supply, supply chains, as well as the operations of businesses and transportation systems. This chapter explores the potential of innovative solutions to facilitate the transition to energy-efficient transport and mobility, as well as intelligent supply chains, through the integration of advanced technologies. In light of the pressing environmental challenges facing the transport sector, including congestion and GHGs emissions, as mentioned previously, we embark on an inspiring exploration of how digital technology can empower adaptation and resilience. By delving into the transformative potential of digital twins, this chapter presents a compelling vision for a robust response to climate change.

A digital twin can be understood as a dynamic virtual representation of a physical entity, such as an object, a system element, an entire system, a network, a process, or even a human or group of individuals alongside its contextual environment (Krmac & Djordjevic, 2024; Xiao et al., 2024). The basic concept of a digital twin is illustrated in Fig. 1. Bhatti et al. (2021) define digital twin technology as a simulation that integrates various physical processes and scales to represent a complex product, reflecting the life of its real counterpart or "corresponding twin" as called in this study.

The digital twin automatically gathers raw data from the real world and processes this information to fulfil its intended purpose: to enhance the performance of or

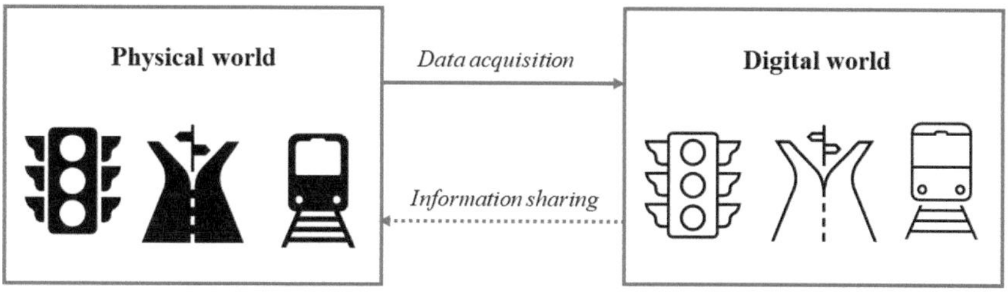

Fig. 1 Conceptual representation of a digital twin

deepen our understanding of the physical entity. It achieves this through online visualization, making predictions or forecasts, detecting anomalies, and assisting in decision-making (Krmac & Djordjevic, 2024). This innovative approach is intricately linked to the Internet of Things (IoT) and current Fourth Industrial Revolution (Industry 4.0), showcasing its exceptional versatility and adaptability. The digital twin functionality significantly enhances IoT's capabilities by enabling two-way communication between the physical object and its virtual representation. While the IoT connects physical devices to cloud services, the digital twin harnesses and interprets real-time data streams from these devices, unlocking new possibilities. Notably, the digital twin technology enables concurrent data analysis derived from IoT sources and the specific real-world system it mirrors. It uses advanced technologies to create real-time representations of complex systems, including the physical and social infrastructures of urban environments (Shulajkovska et al., 2024). The foundational elements contributing to forming a digital twin include sensor data and artificial intelligence.

Recognized as a leading emerging innovation, digital twin technology is expected to transform various industries by providing significant benefits that improve the functionality and reach of IoT, a fundamental component of future Industry 5.0. The application of digital twins has been extensively observed in sectors such as the biomedical and aerospace industries (Krmac & Djordjevic, 2024). Recently, the concept of digital twin technology has expanded into the transportation sector, highlighting its potential to address environmental challenges.

1.2 Aim of This Study

This chapter explores the role of digital twin technology in transportation, specifically focusing on enhancing sustainable mobility and resilience of transportation systems against the impacts of climate change. It guides readers through research studies highlighting the dual benefits of digital twins in incorporating adaptive and mitigative strategies. Real-world case studies are included to show how digital twins are applied in transportation, providing practical examples to help readers understand their potential to create smarter and more sustainable systems. This chapter reviews the latest research on digital twins, specifically their application in optimizing transportation network alternatives and improving intelligent supply chain management solutions. The studies discussed in this chapter result from a systematic literature review of peer-reviewed articles published since 2018, utilizing a targeted search with the following keywords in the Scopus database: "sustainable" AND "digital twin" AND "transport".

2 The Role of Digital Twins for Sustainable and Resilient Transport

2.1 Strategies for Future Mobility and Transportation Network Management

The 2030 Agenda for Sustainable Development connects sustainable transport and urban mobility to several Sustainable Development Goals (SDGs), in particular, SDG 11 titled "Make cities and human settlements inclusive, safe, resilient, and sustainable". This goal emphasizes the need to provide access to accessible, affordable, safe, and sustainable transportation systems by 2030. Additionally, SDG 13 titled "Take urgent action to combat climate change and its impacts" calls for immediate measures to mitigate climate change effects (UN, 2015) Digital twin technology can enhance sustainable transport solutions by integrating advanced technologies and eco-friendly alternatives, evaluating transport planning scenarios that facilitate the transition towards energy-efficient sustainable urban mobility and smart cities, and resilient transport systems, which will be explored further.

2.1.1 Future Mobility Solutions

Advancements in technology and the prioritization of environmental sustainability significantly influence future mobility solutions. These solutions, such as autonomous vehicles, electric vehicles and drones, can address critical urban challenges, including congestion, pollution, and accessibility.

Urban air mobility and drone technology are expected to emerge as sustainable alternatives for urban transportation in the near future. Urban Air Mobility (UAM) refers to the utilization of small electric or hybrid-electric vertical aircraft and other innovative aerial technologies specifically designed for operation in urban environments. Examples of UAM include air taxis and drones intended for short-distance transportation, emergency services, and the expedited delivery of goods, especially in remote or congested urban areas. This innovative mode of transportation is projected to experience significant growth in the forthcoming years, driven by an increasing demand for efficient urban mobility solutions. Integrating digital twin technology offers an innovative and constructive approach to urban air transportation planning. It enables the development of precise digital models that accurately represent different aspects of airspace-related infrastructure. These elements include aircraft routes, existing airports, and the crucial potential of future vertiports, whose significance will grow as UAM expands. Within the UAM landscape, it is helpful to define a vertiport: a designated area engineered explicitly for the landing, takeoff, and servicing of vertical takeoff and landing (VTOL) aircraft. Hamburg, Germany, which features two city airports and a prototype vertiport is showcased as an example of a UAM case study (Bubalo, 2024).

Bubalo (2024) study focused on Hamburg airports and utilized a digital twin model to create a SIMMOD simulation. This model effectively integrated airport information, such as flight routes and schedules, to facilitate demand forecasting for UAM systems. This study demonstrated that the capacity of vertiports is significantly impacted by several key factors. These include the availability of charging stations, UAM vehicles parking, drone operational speeds, turnaround times, autonomy in operations, and the characteristics of service routes. Furthermore, Bubalo (2024) indicated that the main airport's operations predominantly influence the overall capacity of Hamburg's airspace. In this study, Bubalo (2024) presents an inspiring vision for the Airbus production facility in Hamburg-Finkenwerder by suggesting the construction of a parallel taxiway alongside its existing runway to elevate its capacity and efficiency. This infrastructural expansion would alleviate some capacity constraints and support the anticipated growth of UAM.

Across the globe, Electric Vehicles (EVs) have emerged as a sustainable way to achieve transport decarbonization and to help governments meet their strategic goals for reaching net-zero emissions by 2050, which demands that our economy either eliminates GHG emissions or offsets them. In this vision, the UK's "Build Back Greener" strategy and the European Green Deal represent proactive efforts to eliminate net greenhouse gas emissions by the targeted year. These initiatives are vital to combat climate change and environmental degradation. In 2023, EV sales reached 14 million, with the proportion of EVs in total car sales increasing from 4% in 2020 to 18% in 2023 (IEA, 2024b). However, to meet 2050, EV uptake faces challenges that impact their adoption, performance, and integration into existing transport systems.

In the context of the next generation of smart vehicles, such as autonomous vehicles and EVs, digital twin technology applications effectively enhance vehicle health management systems. Digital twin models enable real-time monitoring of EV motors through cloud-based interconnections, providing up-to-date health statistics and in-house prognosis for electric motor operations. The study by Li et al. (2024) proposed a digital twin framework for e-mobility utilizing a self-learning algorithm. This framework empowers EVs to simulate future multistep developments based on historical data. The authors claim that this method can enhance real-time communication by reducing delays and ensuring the accuracy and timeliness of planned driving strategies for autonomous vehicles (Li et al., 2024).

The emergence of Smart Electric Vehicles (SEV) represents a significant advancement in the automotive sector, characterized by enhanced connectivity and automated driving capabilities. The research conducted by Bhatti et al. (2021) indicated that these vehicles have the potential to reduce CO_2 emissions by as much as 43% compared to traditional diesel engines. However, the successful market adoption of SEV necessitates substantial upgrades to existing infrastructure and architecture. In parallel, the growing salles of EV introduces considerable challenges to both power grid stability and the adequacy of charging station infrastructure. A difficulty in representing the SEV lies in the intricate interdependencies among their various subsystems. The application of digital twin technology emerges as a promising solution, facilitating resource-efficient testing and optimization. Bhatti et al. (2021) suggested that SEV should be equipped with extensive sensory feedback mechanisms and the capability to generate real-time data, rendering them particularly suitable for digital replication.

Digital twin systems are anticipated to be increasingly utilized in real-time scenarios, revolutionizing areas such as SEV's health monitoring, SEV's battery management, and intelligent charging systems. This innovative technology enables a comprehensive examination of the lifecycle of SEV within a digital framework, thereby optimizing the supporting infrastructure. Furthermore, integrating charging stations with smart distribution systems can be explored through digital twin applications, ultimately enhancing the efficiency and effectiveness of the overall electric vehicle ecosystem (Bhatti et al., 2021; Verma et al., 2024).

Sustainable transport strategies are essential for shifting travel behavior from private car reliance to greater use of active transport and public transit options, such as buses, trams, and rail systems. In the context of medium to long-distance travel, rail is a particularly sustainable choice for passenger and freight transport due to its lower environmental impact. However, the complexity of railway systems necessitates regular maintenance to ensure the secure and effective operation of train services. Moreover, rail infrastructure is susceptible to disruptions from unforeseen events, which can result in significant adverse outcomes for operators and users. To address these challenges, integrating digital twin technology, along with advancements in artificial intelligence (AI), blockchain technology, and 5G communications, has been identified as a highly promising innovation within the rail sector. This combination is seen as vital for advancing the goals of Rail 4.0, as discussed by Moloney (2020) and Krmac and Djordjevic (2024). Both passenger and freight rail services can leverage digital twin technology for enhanced operational efficiency. Krmac and Djordjevic (2024) suggested that by utilizing data from ticketing systems, traffic flow sensors, and train schedules, operators can predict overcrowding situations, offer real-time updates to passengers about delays or disruptions, and optimize the use of their fleets. This transformative approach can improve user travel experience and is critical in promoting sustainability and resilience within the rail industry.

The expansion of high-speed rail networks, alongside the anticipated full automation of railway operations and upgrades to railway infrastructure, necessitates advancing digital twin technology. This development is critical for monitoring operational integrity, risk management, cost reduction, and enhancing predictive capabilities (Krmac & Djordjevic, 2024). Nevertheless, implementing digital twin models in the railway sector encounters significant challenges. These models are often developed in isolation by various stakeholders, including operators, infrastructure managers, and manufacturers. Such a lack of collaboration hinders the sharing of concepts and models, thereby creating obstacles to standardization and scalability (Krmac & Djordjevic, 2024). This highlights the need for a cohesive strategy to foster collaboration among all parties involved, ensuring that advancements in digital twin technology can be effectively integrated and utilized across the entire railway ecosystem.

2.1.2 Smart Infrastructure and Route Optimization

Road transport, particularly for passenger movement, experiences higher demand compared to rail or air transport. As a result, road infrastructure must support the transition to sustainable transport by reducing emissions, enhancing accessibility, and optimizing resource resilience against unexpected events and extreme weather conditions. Smart infrastructure requires dynamic traffic management to optimize traffic signals, reduce congestion, and efficiently manage emergency scenarios. Adaptive traffic signals adjust their timing in real-time based on current traffic conditions. This advanced traffic management technology is crucial for sustainable, intelligent transport management through several means: it enhances safety by preventing accidents, improves efficiency by minimizing congestion and enhancing traffic flow, and promotes environmental sustainability by reducing idling time and unnecessary acceleration or deceleration. Additionally, it prioritizes non-motorized transport options, such as bicycles and e-scooters, as alternatives to private car use. However, such systems face challenges due to stochastic events and irregular and unpredictable variations in traffic patterns caused by various random factors.

Digital twin technology represents an inspiring advancement to successfully navigate the challenges of fluctuating traffic demands in urban transportation. For instance, Zhu et al. (2024) introduced an innovative digital twin for an adaptive traffic signal control system. It was specifically designed to effectively manage signal control challenges at intersections with varying traffic demands. This system aimed to reduce delays at individual intersections under various synchronization conditions. The authors estimated that the digital twin could reduce delays by up to 54% compared to traditional methods. Applying digital twin technology in the context of time-varying traffic demand offers significant advantages for enhancing traffic signal control. This technology can reduce the number of stops and starts, minimize idling time, and limit unnecessary acceleration or deceleration. As a result, these improvements can lead to energy savings and a reduction in emissions, contributing to the development of more sustainable urban centers (Zhu et al., 2024).

To help the readers better understand how digital twins can enhance smart infrastructure and route optimization, we reflect on the potential cost benefits of digital twin systems compared to traditional traffic management systems as follows. Traditional traffic management systems rely on fixed infrastructure, such as traffic signals, surveillance cameras, inductive loop sensors, and traffic control centers. These systems are primarily reactive, responding to incidents or traffic patterns as they occur. In contrast, digital twin systems utilize virtual replicas of the traffic ecosystem.

They integrate real-time data from IoT devices, sensors, historical data, and AI or machine learning-based models to simulate, predict, and optimize traffic flows. Traditional traffic management systems often incur high costs for infrastructure installation, such as signals, cameras, and sensors, along with high labor expenses for setup and maintenance. In contrast, digital twin systems require a substantial initial investment in technologies, which includes IoT infrastructure, cloud computing, and digital twin software, along with the complexities of advanced modelling and simulation. Nevertheless, digital twin systems generally have reduced ongoing maintenance costs due to their reliance on software for performance optimization. Additionally, digital twin systems perform better in responding to incidents and adapting to climate change. They can quickly simulate different scenarios and provide effective response strategies, which reduce response times. Traditional systems, on the other hand, depend on manual methods and fixed rules for handling incidents, which can slow down their response efforts.

2.1.3 Smart and Sustainable Transport Planning

Sustainable transport planning is required to effectively manage transportation networks and services and balance economic, environmental, and social priorities to meet the needs of individuals, businesses, and society. Digital twin technology offers great opportunities for efficient, thoughtful mobility planning in smart cities. Optimizing route planning is essential for sustainable passenger and freight transport. Among other benefits, optimized routes reduce the distance travelled, directly lowering fuel consumption and GHG emissions and reducing congestion.

In the pursuit of advancing smart city transportation systems, Xiao et al. (2024) highlighted that embracing the digital twin model within smart cities offers a remarkable opportunity to reshape and revolutionize urban planning and management. To minimize operational costs and reduce emissions for smart cities, Xiao et al. (2024) proposed a new route planning approach for public transport that employs a digital twin model and a multi-objective analysis. The digital twin, as presented by Xiao et al. (2024), offers a virtual representation of the smart city's public transportation system, enabling comprehensive simulations and evaluations. The proposed route planning, with its unique feature of real-time adjustments to routes, schedules, and operational resources, ensures that the public transport service and infrastructure remain responsive to the dynamic changes of urban environments. This results in a more robust route planning scheme capable of handling unexpected events and fluctuations in commuter patterns (Xiao et al., 2024). While the authors argue that integrating a digital twin model significantly benefits transport planning by accurately representing the existing transportation system, there are challenges to consider. These include difficulties in integrating real-time data into the digital twin model, uncertainties in predicting urban transit dynamics, and the need to effectively adapt route schemes to different cities' unique infrastructures and conditions.

Another study by Hassan et al. (2024) also proposed a digital twin tool to optimize public transport routes in smart cities. This study created a digital twin using real-time data and reconstructed 3D CAD models that digitally represent physical vehicles in a virtual environment. This setup allows for the simulation and improvement of public transport vehicle routes to minimize travel times and maximize coverage. The proposed model utilized various data types, including weather and road conditions, which are critical for developing more resilient transport planning to address the challenges imposed by climate change. The simulation model by Hassan et al. (2024) suggested that the average vehicle speed decreases under adverse weather and road conditions, demonstrating the model's capability to capture traffic dynamics effectively.

Additionally, this study developed a virtual reality (VR) representation of vehicles in an interconnected digital environment (cyberspace), providing city stakeholders with an immersive experience. The route optimization digital twin algorithm evaluates the efficiency of potential routes. It selects the one with the highest efficiency while allowing stakeholders to prioritize their criteria, such as selecting the shortest total route length or maximizing population coverage. Despite this promising tool for transport planning and optimizing transit routes, it is important to recognize that its functionality is compromised by reconstructing vehicle models. Moreover, the algorithm relies on pre-trained models, which raises questions about the extent of its novel contributions beyond existing solutions.

The rise in vehicle traffic results in various externalities, such as increased congestion, lack of parking spaces, and higher pollutant emissions. This situation directly compromises human quality of life, primarily through increased air and noise pollution, particularly in urban areas. According to the World Health Organization, over 4 million premature deaths are attributed to air pollution. Urban transport planning can designate areas within urban environments where vehicle speeds are deliberately restricted. These areas, known as Limited Traffic Zones (LTZ) or Low-Speed Zones (LSZ), typically have speed limits of 20–30 km/h (12–20 mph). The goal is to enhance safety, reduce emissions and noise, and promote non-motorized transport, supporting cities in achieving their climate goals.

According to the study by Torrisi et al. (2024), traffic restrictions can be effectively implemented using various strategies. The authors developed a digital twin using microsimulation models to analyze various scenarios and assess quantitative factors such as service levels, queue lengths, and delay times across different Limited Traffic Zone (LTZ) configurations in Acireale, Italy. The configurations considered for the central area of the historic center included: (a) a permanent LTZ; (b) a temporarily restricted traffic zone; (c) a permanent LTZ covering the entire historic center (assuming a modal shift in travel demand); and (d) a 30 km zone that includes the entire historic center. This study highlighted the effectiveness of utilizing a digital twin model to simulate various scenarios and evaluate their real-world impacts. Torrisi et al. (2024) provided a helpful approach for calculating and comparing the key reference parameters, such as Level of Service, travel delays, emissions, and fuel consumption, across different scenarios.

Another study conducted by Shulajkovska et al. (2024) explored the potential of an open-source AI framework designed for smart cities to enhance urban sustainability. The researchers utilized microscopic traffic simulation tools to develop digital twin models for four cities, each grappling with unique mobility issues: reducing private vehicle access in Bilbao (Spain), assessing increased bicycle traffic in Amsterdam (Netherlands), planning a tunnel in Helsinki (Finland), and improving public transport in Messina (Italy). Focusing on Bilbao, the framework simulated the implications of closing Moyua Square to evaluate the effects of a Low Traffic Zone (LTZ). The findings indicated that an LTZ implemented from 16:00 to 20:00 significantly reduced emissions of pollutants like NO_x and CO_2, contributing to stronger mobility policies. However, it was found that accessibility to the city center decreased, adversely affecting the "Entry capacity" attribute, while public transport usage remained unchanged. Additionally, the study showcased how multi-label machine learning techniques can offer customized decision support for various urban settings, tackling issues such as pollution and public transport safety. Shulajkovska et al. (2024) demonstrated that the proposed framework significantly accelerates the decision-making process by three orders of magnitude, thereby enhancing the efficiency of urban planning.

Smart cities generate a wide range of data through various sensors, devices, and systems to improve urban living, optimize resource use, and enhance sustainability performance. Analyzing sustainability indicators, such as carbon footprints, climate resilience, circular economies, and biodiversity, provides valuable information on sustainability performance, risks, and opportunities within the built environment. The study published by Dovolil and Svítek (2024) devised an innovative approach by integrating sustainability indicators into digital twins to promote sustainability and resilience. This initiative has led to developing a versatile data platform that effectively incorporates Environmental, Social, and Governance (ESG) factors, enhancing the smart city concept.

Dovolil and Svítek (2024) employed mathematical models from various fields, including transportation, energy, and environmental science, along with the city simulation software, to assess "what-if" impact scenarios involving unpredictable events. The authors implemented a digital twin transport model to create a virtual reality space and develop a detailed 3D model of a specific urban unit, along with its services for different types of users. This model enabled the simulation and testing of various interventions by relevant stakeholders, helping to enhance urban resilience, improve sustainability, and provide tailored emergency services in critical situations. The authors suggested that the proposed digital twin could be a metaverse-based public consultation platform focused on ESG factors. This platform would allow residents to provide virtual feedback on planned transportation projects, ongoing transportation issues, and proposed policies. While it was indicated that such a tool could enhance transparency and public acceptance of mobility interventions, the study also emphasized the importance of continuous data collection. Ongoing ESG feedback from this metaverse would contribute to the ongoing enhancement and refinement of the transport digital twin, ensuring it remains responsive to the community's needs and concerns.

Despite the promising potential of digital twin technology application presented in this section, it is essential to recognize that the mobility sector is undergoing a significant transformation due to digitalization. The transformation in mobility, along with the increasing availability of new data sources, places additional pressure on improving the visualization of mobility environments and forecasting scenarios. Virtanen et al. (2024) made a significant observation, pointing out that to better represent mobility environments in urban digital twins, we must acknowledge that these digital twins are still in their early stages. According to Virtanen et al. (2024), at this point, they primarily consist of 3D city models that serve as foundational components.

2.2 *Future Mobility and Transportation Network Management Strategies for Intelligent Supply Chain Management Solutions*

The global supply chain plays a vital role in the world economy. However, its operations also contribute to climate change by releasing GHG emissions into the atmosphere, consuming resources, and causing environmental degradation. The negative impacts of climate change are particularly noticeable at various stages of the supply chain, including manufacturing processes and freight transportation via shipping, air, rail, and road freight. One promising solution to mitigate these impacts is the application of digital twins, which can enhance intelligent supply chain management. By utilizing digital twins, we can improve urban planning and operational efficiency while minimizing the adverse impacts of climate change.

2.2.1 Sustainable Urban Freight Transport

The decarbonization of urban freight primarily involves transitioning to electric and hybrid vehicles as well as utilizing alternative transportation methods, such as cargo bikes, tricycles, and drones. Last-mile delivery (LMD) represents the final stage in the supply chain, where products are delivered from a distribution hub to the end consumer. This aspect is crucial in logistics and contributes to the growing demand for urban transport. Therefore, it presents an opportunity to improve sustainable urban freight systems for more energy-efficient cities. A significant component of this transformation is the implementation of digital twins, a technology that will be discussed next for its potential to reshape urban logistics.

The increasing urban population, the rapid growth of e-commerce in recent years, and deficiencies in urban logistics planning, a lack of loading and unloading spaces, have created significant challenges for transportation. The high delivery cost per parcel arises from the inefficient use of various transportation modes and traffic restrictions in city centres. Additionally, there is intense pressure on service operators and logistics providers to reduce energy consumption and GHG emissions. Golinska-Dawson and Sethanan (2023) focused on developing solutions for data consolidation within smart cities, utilizing Artificial Intelligence (AI), IoT, and digital twins to enhance energy efficiency. By improving operational services, the goal was to lower the fuel demand for sustainable urban freight logistics.

The need to reduce both fuel and energy demand in smart cities is urgent. According to the study by Golinska-Dawson and Sethanan (2023), this can be achieved through the combination of parcels in micro-depots, the use of parcel lockers, and the implementation of mobile depots, which have the potential to leverage crowd-sourcing. The proposed consolidation can reduce the number of journeys and, consequently, energy demand (Golinska-Dawson & Sethanan, 2023). The authors introduced the Interconnectivity and Energy-Efficiency Matrix to evaluate potential solutions for sustainable urban freight in last-mile delivery. This innovative matrix tool could be helpful to assess and implement sustainable urban freight solutions. It was argued that embracing the most energy-efficient transportation modes can significantly reduce our carbon footprint. These modes include innovative solutions such as drones, autonomous delivery robots, and autonomous vehicles. Additionally, cargo bikes, particularly e-cargo bikes and e-tricycles, are emerging as sustainable alternatives. Electric vehicles, primarily vans, also play a crucial role alongside combined passenger-cargo transport and shared rapid transit systems, all promising a greener future for urban mobility. Despite these promising developments, Golinska-Dawson and Sethanan (2023) also emphasized that the key challenges lie in ensuring the economic performance of small consolidation centers, presenting an opportunity for innovation and progress.

Another study conducted by Royo et al. (2023) applied digital twins for city logistics, focusing on last-mile solutions while considering Madrid stakeholders' needs and expectations. For Madrid, a digital twin was developed using various data sources and models to represent the characteristics of city logistics and complex urban systems to achieve this purpose. This digital representation simulates potential decision scenarios for businesses and stakeholders involved in last-mile delivery, significantly improving the quality of decision-making. The Madrid digital twin has demonstrated its effectiveness in enhancing the decision-making process for city logistics stakeholders. Additionally, it has introduced the necessary data governance models to create these collaborative virtual replicas (Royo et al., 2023).

2.2.2 Improving the Global Trade Supply Chain

International trade and commerce primarily rely on maritime transportation, making ports essential for economic connectivity. Ports facilitate the movement of goods across international borders, acting as crucial entry and exit points for imports and exports. However, the physical infrastructure of ports, combined with rising global demand for goods and a limited supply of shipping containers, often leads to delays. Moreover, studies published by Partners (2016) and Klar et al. (2023) indicate that port operations are essential to global trade, handling about 90% of cargo shipped by sea. Notably, Partners (2016) found that the Port of Hamburg accounts for approximately 40% of the city's total energy consumption. Iris and Lam (2019) claimed that port operations have substantial environmental impacts and influence the Earth's climate due to their essential logistical roles, which include transportation, terminal operations, and warehousing. They also serve as hubs for industrial innovation, driving goods and energy production, assembly, and recycling (Iris & Lam, 2019). While the implementation of digital twins in port systems is still early in its development, a study by Klar et al. (2023) found that they can enhance efficiency. The authors suggested that digital twins can leverage optimization and machine learning to improve efficiency of port operations and facilities. By implementing this technology, ports can drastically reduce turnaround times, minimize congestion by reducing queues, and ensure energy-efficient ship movements prior to entering the port.

In March 2024, a container ship crashed into a bridge in the US city of Baltimore, causing immediate concerns in the global supply chain. Two years earlier, the COVID-19 pandemic caused large-scale social impacts and profoundly impacted the global supply chain, disrupting nearly every aspect of how goods are produced, transported, and consumed. The supply chain needs recovery strategies to overcome unexpected events and disruptions due to adverse weather (e.g., flooding), energy crises or geopolitical crises. Ad-hoc recovery plans are often adopted when disruption hits the supply chain. These disruptions can lead to the costly use of air transport, significantly raising CO_2 emissions, or they might result in shipping delays and higher freight costs due to necessary changes in navigation routes.

Ivanov (2023) developed a digital twin model using the anyLogistix supply chain software to represent standard supply chain designs, such as a hub system, and recovery strategies, including ad-hoc recovery through emergency air transport and preparedness-based recovery using 3D printing for product substitution, supported by low-emission electric trucks. The digital twin model was used to evaluate the effects of two sustainable recovery alternatives, additive manufacturing processes and the deployment of electric trucks, on the environmental characteristics and resilience potential of supply chains. The results indicated that these recovery strategies significantly reduced the environmental impacts associated with resilience strategies compared to ad-hoc methods, which resulted in higher CO_2 emissions (Ivanov, 2023). This study suggested that the proposed digital twin can assist decision-makers in selecting recovery policies based on responsiveness (expressed by on-time delivery metrics) and environmental impact (assessed by CO_2 emissions estimation).

Lastly, according to the study by Metcalfe et al. (2024), it is important to point out that despite the growing number of publications on this technology application in recent years, many case studies referred to as "digital twins" are actually examples of either a Digital Shadow or Digital Model. Therefore, they do not qualify as true digital twins.

3 Digital Twin Examples: Practical Applications in the Real-World Context

3.1 Thameslink Railway Program

The Thameslink Railway Program in London, one of the UK's largest railway infrastructure projects, has integrated digital twin technology to enhance its operations and reduce its carbon footprint. Thameslink, which operates through a busy corridor in London, sought to improve the efficiency and reliability of its services while aligning with the UK's sustainability goals (Barrow, 2024). The Thameslink digital twin key applications and outcomes can be summarized as follows.

- ***Real-time monitoring and predictive maintenance***: The Thameslink digital twin captures data from sensors installed on tracks, trains, and other infrastructure, allowing operators to monitor performance in a real-time manner. This capability facilitates predictive maintenance, identifying issues before they lead to breakdowns. Predictive maintenance not only improves service reliability but also reduces energy waste and emissions associated with emergency repairs and inefficient operations.
- ***Optimized train scheduling***: The digital twin models train schedules and track conditions to optimize energy use and reduce unnecessary idling times. By aligning schedules with real-time data, Thameslink reduces fuel consumption and emissions. This scheduling efficiency is especially beneficial in mitigating the environmental impact of the rail system, as trains idle less, consume less energy, and minimize delays.
- ***Energy consumption tracking***: Thameslink's digital twin technology also tracks the energy usage of each train, allowing for energy-optimized routing and speed adjustments. This reduces operational emissions while maintaining a high level of service. Over time, this model will support even greater reductions in carbon emissions, aligning with the UK's target of achieving net-zero greenhouse gas emissions by 2050.

While addressing the gains of implementing digital twin technology in complex systems like the Thameslink Railway, it is vital to underline that several challenges were faced throughout the deployment process:

1. ***Data integration and system compatibility***: Integrating digital twins with existing legacy systems was difficult due to compatibility issues. Ensuring seamless communication between new digital frameworks and traditional infrastructure required significant effort.
2. ***Data quality and management***: Maintaining high-quality, accurate, and timely data is essential for effective digital twin operation. Challenges included managing data complexity and ensuring consistency across various sources.

3. ***Technical expertise and resource allocation***: Developing and maintaining digital twins demand specialized technical skills and substantial resources. The Thameslink railway project often faced difficulties in recruiting or training personnel with the necessary expertise.
4. ***Cybersecurity risks***: The interconnected nature of digital twins increases vulnerability to cyber threats. Therefore, implementing robust cybersecurity measures was crucial to protect sensitive data and ensure system integrity.
5. ***Scalability and long-term maintenance***: Scaling digital twin solutions across large, complex systems like railway networks poses significant challenges. Ensuring consistent performance and managing ongoing maintenance require careful planning and resource allocation.

To conclude, this Thameslink digital twin case has contributed to lowering the environmental footprint of one of London's busiest rail lines. By providing a detailed model of operations, the system has allowed for more sustainable practices, helping to ease congestion and reduce transport emissions. Thameslink's digital twin approach provides a valuable framework for other cities looking to transition to sustainable public transport systems. The identified practical inhibitors are also valuable lessons to integrate advanced technological systems into a complex, busy urban railway corridor in the future while achieving ambitious operational and sustainability goals (Rail, 2021; Siemens, 2021).

3.2 Port of Rotterdam

The Port of Rotterdam, one of Europe's largest ports, has implemented a digital twin model to enhance port logistics, reduce emissions, and improve overall operational efficiency. This digital twin replicates real-time port activities, capturing data from ships, cargo handling, and transportation within and around the port (Port of Rotterdam, 2021). The digital twin applications at the Port of Rotterdam can be summarized as follows.

- ***Emission reduction through optimized vessel traffic***: By using real-time data from the digital twin, port operators optimize ship docking and cargo unloading schedules, minimizing waiting times and reducing fuel consumption. The twin also models the best entry and exit paths for vessels based on weather, tide, and port conditions, which reduces emissions by enabling ships to travel at more fuel-efficient speeds.
- ***Energy-efficient infrastructure***: The digital twin provides insights into energy consumption across port facilities, allowing for adjustments that cut energy waste. For instance, port lighting, heating, and cooling systems are optimized based on real-time operational needs, reducing the overall carbon footprint of port activities.
- ***Sustainable supply chain logistics***: The Port of Rotterdam's digital twin integrates with the logistics systems of rail and truck companies, enabling efficient loading and unloading schedules that reduce congestion and idling times. These

efficiencies lower emissions from transport vehicles and improve the port's overall carbon efficiency.

Since implementing its digital twin, the Port of Rotterdam has observed reductions in waiting times and emissions, moving closer to its target of becoming a carbon-neutral port by 2050. It's a leading example for sustainable, tech-enabled ports globally, supporting greener logistics through digital technology (Axians, 2024; IBM, 2022). Scaling Rotterdam's digital twin model to smaller or less technologically advanced ports offers significant opportunities for improved efficiency and sustainability. By adopting simplified and cost-effective versions of the technology, smaller ports can focus on essential applications like vessel traffic optimization and energy management. Challenges such as limited budgets, outdated infrastructure, and insufficient technical expertise can be addressed through modular solutions, workforce training, and partnerships with larger ports or technology providers. Scaled implementations can integrate basic logistics systems, reduce emissions by minimizing vessel idling and energy waste, and support sustainable supply chains. With careful adaptation and leveraging public–private collaborations, even smaller ports can achieve meaningful progress toward greener operations and contribute to the maritime industry's decarbonization goals.

4 Adaptive and Mitigative Strategies for Sustainability and Resilience in Transport Solutions

4.1 Adaptation and Mitigation Concepts

Adaption sustainability is closely related to the concept of resilience, defined as the capacity of a system, community, or nature to absorb shocks and maintain function, as well as renew, reorganize, and develop (Larsen & Gunnarsson-Östling, 2009). In the context of climate change, adaptive sustainability involves strategies and actions that enable systems to cope with the impacts of a changing climate and promote sustainable development (Dang, 2003). Literature highlights several key aspects of adaptive sustainability, starting with its focus on local contexts. Implementing adaptive sustainability requires understanding local vulnerabilities, capacities, and governance structures. Additionally, adaptive sustainability has strong synergies with sustainable development. It aligns with the broader goals of sustainable development, seeking to improve human well-being while preserving the environment for future generations (Field et al., 2014). Therefore, addressing climate change and achieving adaptive sustainability may necessitate transformative changes in systems, structures, and behaviours. This can involve shifting paradigms, adopting new technologies, and promoting innovative solutions (Field et al., 2014).

Mitigation sustainability, on the other hand, centres on reducing greenhouse gas emissions or enhancing natural sinks to limit the magnitude and rate of climate

Table 1 Comparison of adaptive and mitigative sustainability strategies

Aspect	Adaptive strategies	Mitigative strategies
Focus area	Local contexts, tailored to specific community needs and priorities	Global context, requiring international cooperation for collective benefits
Key objectives	Building resilience, adapting to climate impacts, and sustaining functionality	Reducing emissions, slowing global warming, and addressing the root causes of climate change
Key components	– Understanding local vulnerabilities and governance – Promoting transformative changes in systems and behaviours	– Financial and technological capacity – Robust institutions and community involvement
Temporal impact	Focuses on immediate and long-term adaptability to changing conditions	Emphasizes long-term solutions to reduce future climate risks
Challenges	Requires significant local data, resources, and governance	Faces issues of equity, free-riding, and resource allocation across nations

change (Bizikova et al., 2010; Swart & Raes, 2007). This approach is crucial for sustainable development because it eases climate-related pressures, such as extreme weather events (Field et al., 2014). Different groups and countries have varying abilities to implement mitigation, depending on their "mitigative capacity" (Field et al., 2014). Several factors shape the mitigative capacity, many of which are tied to a nation's level of development. First, countries with higher emissions may have more opportunities for reduction. Secondly, financial capacity is critical for implementing mitigation measures and investing in cleaner technologies. Next, access to and ability to adopt new mitigation technologies are essential (Field et al., 2014). Additionally, institutional strength of the countries matters; in other words, robust institutions and effective governance are needed to implement and enforce mitigation policies. Lastly, social capital and community involvement are also critical for success in mitigation efforts. Table 1 provides a summarized comparison of adaptive and mitigative sustainability strategies.

4.2 Adaptation and Mitigation Strategies in Transportation

Both adaptation and mitigation strategies found application areas in varying industries like urban planning, energy and agriculture sectors; mitigation strategies aiming at reducing existing emissions and environmental harms and adaption strategies creating future-resilient designs. Transportation has also been one of the core areas of these sustainability strategies.

From the adaptation point of view, strategies focus on adjusting systems, practices, and infrastructure to manage the effects of climate change and environmental challenges. They help transportation systems withstand climate impacts, such as extreme weather, rising sea levels, and temperature changes, ensuring ongoing functionality and safety. Some examples would be:

- ***Flood-resilient infrastructure***: Elevating roads, bridges, and railways in flood-prone areas to prevent disruptions. For example, the Netherlands has designed "climate-resilient roads" in areas vulnerable to flooding (Doorn-Hoekveld et al., 2022).
- ***Heat-resistant railways***: Upgrading rail infrastructure with heat-resistant materials to prevent track warping during extreme temperatures, as seen in parts of the United Kingdom (Thomas, 2023) and Australia (BITRE, 2023).
- ***Flexible public transit systems***: Implementing real-time route adjustments and rapid-response measures during extreme weather. Cities like New York and Tokyo have developed rapid response systems to reroute transit during hurricanes or typhoons (Su, 2016; Wilbanks & Fernandez, 2014; World Bank, 2019).
- ***Flood-prepared metro systems***: Installing waterproof gates, elevated entrances, and stormwater pumps in subway systems. Singapore's Mass Rapid Transit system incorporates flood protection measures like these in its stations (Singapore Public Utilities Board, 2024).

Mitigation strategies, on the other hand, focus more on reducing emissions, energy use, and other environmental impacts from transport, aiming to address the root causes of climate change. Relevant application examples would be:

- ***Electric and hybrid public transit***: Switching to electric buses, trams, and trains. Cities like Shenzhen, China have transitioned their entire public bus fleet to an electric one, significantly reducing emissions (World Bank, 2021).
- ***Bike-sharing programs and bike lanes***: Promoting cycling as an alternative to cars. Cities such as Amsterdam and Copenhagen have extensive bike lane networks, reducing reliance on fossil fuels (Emanuel, 2019; Nikolaeva, 2024).
- ***Incentives for Electric Vehicles (EVs)***: Offering subsidies, tax breaks, and charging infrastructure to encourage EV use, as done in Norway, where over half of new cars sold are electric (Bjerkan et al., 2016; Cincotta & Thomassen, 2023).
- ***Green freight and logistics***: Using biofuels, electric trucks, and optimized logistics for freight transport. Amazon and UPS have invested in electric delivery vehicles and efficient routing algorithms to cut emissions (Corrales & Gross, 2020; Holland et al., 2017; Wu et al., 2022).

4.3 Adaptation and Mitigation Strategies in Digital-Twin-Powered Transportation

With the vast applications of adaptation and mitigation strategies in transportation, the influence of digitalization in enhancing these strategies and creating new sustainability opportunities is profound. This section, building on the insights from previous sections that discussed the potential of digital twin technology (Sect. 2) and provided real-world application examples (Sect. 3), is specifically dedicated to examining the role of digital twin technology in shaping adaptation and mitigation strategies, thereby contributing to a more sustainable future in transportation.

Digital twin technology adoption can benefit adaptation strategies as follows. First, it enhances the ability to respond dynamically to disruptions through improved adaptive systems. This is exemplified by adaptive traffic management solutions designed to handle varying traffic demands and railway predictive maintenance, such as the Thameslink example in the UK. Second, digital twins contribute to climate resilience in transportation systems. As highlighted in the work of Hassan et al. (2024), weather-responsive planning can optimize public transport routes. Third, digital twins offer great flexibility in transport planning. Digital twins enable real-time adjustments to routes, schedules, and resource allocation, as demonstrated by the route planning approach proposed in Xiao et al. (2024). This flexibility is also evident in practical applications like the Rotterdam port operations in the Netherlands, where digital twin technology enhances insights into energy consumption and improves energy efficiency across port facilities, thereby reducing GHG emissions.

Fourth, digital twins aid in adapting infrastructure for resilient design. According to Bubalo (2024), various factors, such as parking limitations and drone operational speeds, can hinder the effective utilization of vertiport capacities. Digital twins can facilitate demand forecasting for future urban air mobility systems. Additionally, they can support resilient designs, such as constructing parallel taxiways alongside existing runways and upgrading infrastructure for high-speed rail networks, as noted in Krmac and Djordjevic (2024). Lastly, digital twins promote adaptive sustainability in transportation systems through enhanced public engagement and policy support. They provide visual simulations communicating climate and sustainable mobility initiatives to stakeholders and the public. This capability fosters the development of metaverse public consultation platforms, as noted by Dovolil and Svítek (2024), encouraging collaboration among government, private sectors, and communities and facilitating transparency in policy assessment for future decision-making.

Strategies leveraging digital twin technology also present significant advantages to mitigate the environmental impacts within the transportation sector First, digital twins can facilitate emission reduction through resource optimization and predictive maintenance. These capabilities minimize waste and enhance operational efficiency within transport systems. Specific applications include enabling real-time monitoring of the next generation of smart electric vehicles (SEV), optimizing train scheduling and energy consumption in urban rail systems based on track conditions, and modelling optimal entry and exit routes for vessels. These decisions can be made considering variables such as weather, tide, and port conditions in different locations, including the Port of Rotterdam. Second, the integration of digital twins in transportation supports enhanced energy efficiency. For instance, adaptive traffic signals exemplify smart infrastructure by adjusting their real-time operation according to current traffic conditions. Moreover, digital twin technology can provide critical insights into energy consumption patterns across various transportation facilities in ports and railways.

Third, digital twins are crucial in promoting sustainable urban mobility solutions and transport decarbonization strategies. This includes facilitating the decarbonization of urban freight logistics by optimizing cost-per-parcel metrics in last-mile delivery scenarios. Furthermore, digital twins can simulate potential future decision-making pathways, assisting business leaders and policymakers in improving the quality of their last-mile logistics decisions, as Royo et al. (2023) discussed. Last, digital twins can be instrumental in conducting impact assessments of diverse transport schemes and mobility interventions through scenario-based analysis and simulation methodologies. For example, they can evaluate various configurations of traffic restriction zones, such as Low-Speed Zones (LSZ) operating at speeds of 20–30 km/h, as illustrated by the study conducted by Torrisi et al. (2024) in Italy. This capability supports evidence-based policymaking by quantifying the potential effects of transportation strategies on urban environments.

Most importantly, some use cases of digital twins in transportation can also be example for the integrated use of adaptation and mitigation strategies. As historically demonstrated, both strategies have now started to be used in conjunction, and therefore, it is optimal to design systems that contribute two-fold. We can observe three relevant points that digital twin technology contributes to. First, its power to enable data-driven decision-making would contribute to both strategies. Application of simulations, scenario modelling, or multimodal integrations can contribute to both purposes: mitigating today's harms and adapting in the future. Secondly, by enabling standardization and collaboration, digital twin technology can help aligning the sustainability strategies throughout the supply chains. Some examples would be the isolated developments in Rail 4.0 and easing the data governance in the supply chains and projects like in the Madrid case. Moreover, digital twin technology would also support innovations in sustainable technologies like hydrogen-based systems. Portapas et al. (2021) explored a dock-to-dock use case, evaluating the safety and viability of alternative hydrogen-based fuels. Development of such innovations would bring both mitigative benefits in sustainability and also prepare the industries for the future business environment.

5 Conclusions and Future Direction

This chapter showcases the growing potential of digital twin technology by presenting the latest research and studies in the field. Digital twins are closely aligned with sustainability goals and provide dual benefits for both adaptation and mitigation strategies regarding climate change, transportation resilience, and sustainable development. They support the transition to low-carbon systems by optimizing energy, transportation, and industrial processes. Additionally, digital twins contribute to the development of smarter, greener cities by monitoring energy use and emissions and reducing them in real time. One significant advantage of digital twins is their ability to leverage real-time data, advanced simulations, and AI-driven insights to optimize transport management and supply chain logistics. Digital twins provide significant benefits for both adaptation and mitigation, as discussed in this chapter. In terms of adaptation, they aid in identifying vulnerable infrastructure (e.g. road, ports), prioritizing investments for improvements, and optimizing evacuation routes and supply chains during extreme events. For mitigation, digital twins facilitate real-time monitoring of energy consumption in transport services, which helps to reduce overall energy usage.

While digital twin applications have many benefits, there are challenges that need to be addressed. A primary concern involves scalability mainly due to the significant discrepancies in size between various systems, such as railways and ports. The complexity of ensuring that digital twins can accurately represent expansive regional or national transport networks poses a substantial technical challenge. In complex environments like ports, the efficacy of digital twins can be diminished when diverse operational requirements are not sufficiently managed. The development and deployment of digital twins necessitate substantial capital investments in technology, including sophisticated sensor systems, which raises concerns about affordability. Additionally, cybersecurity and data privacy pose substantial barriers to the effective use of digital twins. Given that these systems rely on interconnected technologies—specifically, the Internet of Things (IoT) and cloud computing—they are vulnerable to cyberattacks.

Future research should focus on advancements in urban mobility and the growing influence of digitalization. This involves creating a more comprehensive representation of the actual mobility environment by integrating new data sources, such as traffic volume from road segments and crowdsourced information with existing geospatial data. Besides, the digital twin for railways should prioritize the development of models for the real-time estimation of movable train service operations. This should include testing various unique use cases that fulfil the requirements of the rail industry. Lastly, it is essential to establish standardization and a common taxonomy that spans through technology classification, data collection and extraction, as well as quality decision-making.

References

Axians. (2024). How Rotterdam becomes the smartest port in the world. Retrieved January 21, 2025, from https://www.axians.com/use-case/how-rotterdam-becomes-the-smartest-port-in-the-world/

Barrow, K. (2024). Thameslink: Strengthening London's railway backbone. Rail J. Retrieved January 21, 2025, from https://www.railjournal.com/in_depth/thameslink-strengthening-londons-railway-backbone/

Bhatti, G., Mohan, H., & Singh, R. R. (2021). Towards the future of smart electric vehicles: Digital twin technology. *Renewable and Sustainable Energy Reviews, 141*, Article 110801.

BITRE. (2023). Road and rail supply chain resilience review-phase 1. Bureau of infrastructure and transport research economics

Bizikova, L., O'Brien, K., St. Clair, A. L., & Kristoffersen, B. (2010). Linking sustainable development with climate change adaptation and mitigation. In: K. O'Brien, A.L. St. Clair and B. Kristoffersen, eds. Climate Change, Ethics and Human Security. Cambridge: Cambridge University Press, pp. 157–179

Bjerkan, K. Y., Nørbech, T. E., & Nordtømme, M. E. (2016). Incentives for promoting Battery Electric Vehicle (BEV) adoption in Norway. *Transportation Research Part D: Transport and Environment, 43*, 169–180. https://doi.org/10.1016/j.trd.2015.12.002

Bubalo, B. (2024). Capacity and demand forecasting in urban air mobility-a simulation of air transportation in Hamburg. *Transportation Research Procedia, 80*, 11–19.

Cincotta, C., & Thomassen, Ø. (2023). Evaluating Norway's electric vehicle incentives. *SSRN Electronic Journal* [Preprint]. https://doi.org/10.2139/ssrn.4626980

Corrales, C. G., & Gross, G. (2020). The impact of amazon delivery van electrification. Universidad Pontificia Comillas

Dang, H. (2003). Synergy of adaptation and mitigation strategies in the context of sustainable development: the case of Vietnam. *Climate Policy, 3*, S81–S96.

Dovolil, P., Svítek, M. (2024) Integrating ESG into the smart city concept with focus on transport. In *2024 Smart City Symposium Prague (SCSP)* (pp. 1–7). IEEE.

Emanuel, M. (2019). Making a bicycle city: infrastructure and cycling in Copenhagen since 1880. *Urban History, 46*(3), 493–517. https://doi.org/10.1017/S0963926818000573

European Commission (EC). (2024). Climate action: Transport. Retrieved January 21, 2025, from https://climate.ec.europa.eu/eu-action/transport/overview_en

Field, C. B., Barros, V. R., Mach, K. J., & Mastrandrea, M. D. (2014). Climate-resilient pathways: adaptation, mitigation, and sustainable development. In: (Eds.), Climate Change 2014 Impacts, Adaptation, and Vulnerability. Cambridge: Cambridge University Press, pp. 1101–1131.

Golinska-Dawson, P., & Sethanan, K. (2023). Sustainable urban freight for energy-efficient smart cities-systematic literature review. *Energies, 16*(6), Article 6.

Hassan, M. U., Alaliyat, S., & Hameed, I. A. (2024). Towards the creation of a digital twin authoring tool: a smart mobility perspective in smart cities. *IEEE Access, 12*, 111280–111292.

Holland, C., et al. (2017). UPS optimizes delivery routes. *Interfaces, 47*(1), 8–23. https://doi.org/10.1287/inte.2016.0875

IBM. (2022). Port of Rotterdam embarks on digital transformation programme. Retrieved January 21, 2025, from https://supplychaindigital.com/technology/port-rotterdam-embarks-digital-transformation-programme-ibm

IEA. (2024a). Transport. Retrieved January 21, 2025, from https://www.iea.org/energy-system/transport

IEA. (2024b). Electric vehicles. Retrieved January 21, 2025, from https://www.iea.org/energy-system/transport/electric-vehicles

Iris, Ç., & Lam, J. S. L. (2019). A review of energy efficiency in ports: operational strategies, technologies and energy management systems. *Renewable and Sustainable Energy Reviews, 112*, 170–182.

Ivanov, D. (2023). Design and deployment of sustainable recovery strategies in the supply chain. *Computers & Industrial Engineering, 183*, Article 109444.
Klar, R., Fredriksson, A., & Angelakis, V. (2023). Digital twins for ports: derived from smart city and supply chain twinning experience. *IEEE Access, 11*, 71777–71799.
Krmac, E., & Djordjevic, B. (2024). Digital twins for railway sector: current state and future directions. *IEEE Access, 12*, 108597–108615.
Larsen, K., & Gunnarsson-Östling, U. (2009). Climate change scenarios and citizen-participation: mitigation and adaptation perspectives in constructing sustainable futures. *Habitat International 33*(3), 260–266. https://doi.org/10.1016/j.habitatint.2008.10.007.
Li, Y., Xu, J., Li, T., Mocanu, E., Jensen, C. S., Gao, D. W., & Zhang, Y. (2024). Digital twin-empowered autonomous driving for E-mobility: concept, framework, and modeling. *IEEE Electrification Magazine, 12*(3), 68–77.
Metcalfe, J., Ellul, C., Morley, J., & Stoter, J. (2024). Characterizing the role of geospatial science in digital twins. *ISPRS International Journal of Geo-Information, 13*(9), 320.
Mobility Siemens. (2021). Digital solutions and software. Retrieved January 21, 2025, from https://www.mobility.siemens.com/global/en/portfolio/digital-solutions-software/digital-services.html
Moloney, K. (2020). All the news that's fit to push: The New York Times Company and transmedia daily news. *International Journal of Communication, 14*, 20.
Network Rail. (2021). Sustainability at the heart of Network Rail as Thameslink programme wins double gold. Retrieved January 21, 2025, from https://www.networkrailmediacentre.co.uk/news/sustainability-at-the-heart-of-network-rail-as-thameslink-programme-wins-double-gold
Nikolaeva, A. (2024). Smart cities and (smart) cycling: exploring the synergies in Copenhagen and Amsterdam. *Journal of Urban Technology, 31*(1), 29–49. https://doi.org/10.1080/10630732.2024.2322007
Port of Rotterdam. (2021). The digital port. Retrieved January 21, 2025, from https://www.portofrotterdam.com/en/to-do-port/futureland/the-digital-port
Portapas, V., Zaidi, Y., Bakunowicz, J., Paddeu, D., Valera-Medina, A., & Didey, A. (2021) Targeting global environmental challenges by the means of novel multimodal transport: concept of operations. In *2021 Fifth world conference on smart trends in systems security and sustainability (WorldS4)* (pp. 101–106). IEEE.
Royo, B., Politaki, D., Gonzalez, J. N., & Batalla, A. (2023). Digital twin opportunities and benefits in last-mile logistics for Madrid value case. *Transportation Research Procedia, 72*, 1693–1699.
Shulajkovska, M., Smerkol, M., Noveski, G., & Gams, M. (2024). Enhancing urban sustainability: developing an open-source AI framework for smart cities. *Smart Cities, 7*(5), 2670–2701.
Sia Partners. (2016). The internet of things in transportation—port of Hamburg case study
Singapore Public Utilities Board. (2024). Managing Urban Runoff—Drainage Handbook. https://www.pub.gov.sg/Professionals/Resources/Guides-and-Handbooks
Statista. (2024). Transportation emissions worldwide. Retrieved January 21, 2025, from https://www.statista.com/topics/7476/transportation-emissions-worldwide/#topicOverview
Su, Y.-S. (2016). Urban flood resilience in New York City, London, Randstad, Tokyo, Shanghai
Swart, R., & Raes, F. (2007). Making integration of adaptation and mitigation work: mainstreaming into sustainable development policies? *Climate Policy 7*(4), 288–303. https://doi.org/10.1080/14693062.2007.9685657.
Taipei. (n.d.). *Journal of Management and Sustainability 6*(1), 92. https://doi.org/10.5539/jms.v6n1p92
Thomas. T. M. (2023). Heatwaves pushing the sector off its rails. Global Synthesis Report on Climate Action by Sector.
Torrisi, V., Rossetti, S., Barbagallo, A., Leonardi, P., & Ignaccolo, M. (2024). A digital twin simulation framework to assess the impact of street experiments: Transforming urban mobility in Acireale (Italy). In *International Conference on Computational Science and Its Applications.* Cham: Springer Nature Switzerland, pp. 197–210.

United Nations (UN). (2015). Transforming our world: The 2030 Agenda for Sustainable Development. Retrieved January 21, 2025, from https://sdgs.un.org/2030agenda

United Nations (UN). (2024). Facts and Figures. Retrieved November 16, 2024, from https://www.un.org/en/actnow/facts-and-figures

van Doorn-Hoekveld, W. J., et al. (2022). Adaptation to climate change in dutch flood risk management: innovative approaches and related challenges. *Utrecht Law Review 18*(2), 51–69. https://doi.org/10.36633/ulr.860

Verma, S., Sharma, A., Tran, B., & Alahakoon, D. (2024). A systematic review of digital twins for electric vehicles. *Journal of Traffic and Transportation Engineering (English Edition), 11*, 815–834.

Virtanen, J. P., Alander, J., Ponto, H., Santala, V., Martijnse-Hartikka, R., Andra, A., & Sillander, T. (2024). Contemporary development directions for urban digital twins. *The International Archives of the Photogrammetry, Remote Sensing and Spatial Information Sciences, 48*, 177–182.

Wilbanks, T. J. & Fernandez, S. (eds) (2014). Climate change and infrastructure, urban systems, and vulnerabilities. Washington, DC: Island Press/Center for Resource Economics. https://doi.org/10.5822/978-1-61091-556-4

World Bank. (2019). *Information and communication technology for disaster risk management in Japan.* World Bank. https://doi.org/10.1596/32797

World Bank. (2021). Electrification of public transport: a case study of the Shenzhen Bus Group. Washington, DC: World Bank. License: Creative Commons Attribution CC BY 3.0 IGO.

Wu, C., et al. (2022). Learning from drivers to tackle the Amazon Last Mile Routing Research Challenge. arXiv. http://arxiv.org/abs/2205.04001. Retrieved November 28, 2024

Xiao, M., Chen, L., Feng, H., Peng, Z., & Long, Q. (2024). Sustainable and robust route planning scheme for smart city public transport based on multi-objective optimization: Digital twin model. *Sustainable Energy Technologies and Assessments, 65*, Article 103787.

Zhu, H., Sun, F., Tang, K., Wu, H., Feng, J., & Tang, Z. (2024). Digital twin-enhanced adaptive traffic signal framework under limited synchronization conditions. *Sustainability, 16*(13), 5502.

Guilhermina Torrao is an Assistant Professor of Transport at Aston University (UK). She is a member of the Committee for the University Studies Transport Group. She is also a Fellow of the Higher Education Academy, recognized with a distinction. Dr. Torrao has significant experience in interdisciplinary projects, including the EU funding, such as the L3Pilot Project with pilot testing of SAE Level 3 automated driving in Europe and the SolutionsPlus project integrating urban electric mobility for passengers and freight. Her research assesses the impacts of new transportation technologies on sustainable, resilient, and inclusive mobility.

Ahmet Onur Agca is a Lecturer in Operations and Supply Chain Management at Aston University (UK). His studies focus on the intersection of three major areas of supply chain management: behavioral operations—how human force impacts the efficiency of operations; sustainability and circular design—as means of post-growth philosophy; and supply chain digitalization. He was previously involved in relevant projects, such as Intelligent Traffic Environments, Digitalized Supply Chain Integration, Industry 4.0 Maturity Assessment, and Future of Digital Manufacturing Ecosystem, in various institutions, including Durham Business School, Loughborough Business School, and Warwick Manufacturing Group.

Reem Hadeed is a Lecturer in engineering projects and supply chain management at Aston University (UK). She is a fellow of advanced higher education in the UK (FAHE) and the chartered institute of logistics and transport in the UK (FCILT). Her expertise is in organization capabilities, managing operations and supply chains, innovation in technology management, and digital transformation in the context of transportation, logistics, and manufacturing. Her current work focuses on supply chain resilience and sustainability in the textile industry and the future of digital manufacturing ecosystems.

A Comparative Analysis of Transport Operations, Practices, and Environmental Sustainability in the Rhine-Alpine and the Delhi-Mumbai Industrial Corridors

Bharadwaj Kasula and Marin Marinov

Abstract This study presents a comparative analysis of transport operations and environmentally sustainable practices in the Rhine-Alpine Corridor (RAC) and the Delhi-Mumbai Industrial Corridor (DMIC). Using a mixed-method approach, the research evaluates emission reduction strategies, renewable energy integration, and climate resilience measures across these corridors, revealing distinct approaches shaped by their developmental contexts. The findings demonstrate the RAC's successful implementation of multimodal integration and green technologies, achieving a 25% reduction in CO_2 emissions through systematic electrification and modal shift strategies. The DMIC showcases innovative approaches to environmental sustainability through comprehensive planning, targeting a 40% reduction in carbon emissions via renewable energy integration and smart city development. Key recommendations include prioritising rail transport electrification, implementing corridor-wide environmental monitoring systems, and establishing robust climate adaptation frameworks. This study emphasises the importance of context-specific environmental strategies while highlighting opportunities for knowledge transfer between developed and developing transport systems. The present research contributes valuable insights for policymakers and practitioners in sustainable transport infrastructure development, particularly in addressing climate change challenges while maintaining operational efficiency.

Keywords Transport corridors · Environmental sustainability · Climate change mitigation · Sustainable development · Multimodal transport · Rhine-Alpine corridor · Delhi-Mumbai industrial corridor · Green infrastructure · Policy implementation

B. Kasula
Supply Chain Management Expert, Aston University, Birmingham, UK
e-mail: bharadwaj9619@gmail.com

M. Marinov (✉)
SFHEA, Senior Lecturer, EM Research Lead, Aston University, Birmingham, UK
e-mail: m.marinov@aston.ac.uk

M. A. Dulebenets et al. (eds.), *Transport and Climate Change*,
https://doi.org/10.1007/978-3-032-00563-2_15

1 Introduction

In an era defined by global interconnectedness and mounting environmental concerns, transport corridors have emerged as critical arteries of international commerce and sustainable development (Quium, 2019). These corridors have evolved dramatically from their initial conception as simple physical routes linking economic centres to become sophisticated, integrated systems that shape regional development, environmental sustainability, and social progress. The transformation of transport corridors reflects a fundamental shift in our understanding of infrastructure development, as modern corridors integrate multiple transportation modes, advanced digital technologies, and comprehensive regulatory frameworks to facilitate not just the movement of goods and people, but also the sustainable development of entire regions. This evolution aligns with global imperatives for reducing carbon emissions, enhancing economic efficiency, and promoting social inclusivity.

The primary objective of this research is to conduct a comparative analysis of sustainable practices between mature and emerging transport corridors, specifically examining the Rhine-Alpine Corridor (RAC) and the Delhi-Mumbai Industrial Corridor (DMIC). This comparison aims to identify transferable best practices and develop context-specific recommendations for sustainable corridor development. The research addresses a critical gap in understanding how sustainable practices can be effectively implemented across different developmental contexts, particularly in the transfer of knowledge between established and emerging transport systems.

The study is motivated by several pressing factors. First, the urgent need to reduce transport-related carbon emissions, which currently account for a significant portion of global CO_2 emissions, demands innovative solutions that can be rapidly deployed in both existing and new transport corridors. Second, the increasing development of transport corridors in emerging economies presents a unique opportunity to implement sustainable practices from inception, potentially leapfrogging traditional development stages. Finally, the lack of comprehensive comparative analyses between mature and emerging transport corridors has limited our understanding of how sustainable practices can be effectively adapted across different contexts.

1.1 The Rhine-Alpine Corridor: A Mature European Model

The Rhine-Alpine Corridor (RAC) stands as a testament to successful European integration and sustainable transport development. Stretching approximately 1,400 kms from the bustling North Sea ports to the Mediterranean, this corridor serves as a vital economic lifeline for Europe, connecting major centres, including Rotterdam's maritime gateway, Frankfurt's financial hub, Basel's industrial centre, Milan's commercial heart, and Genoa's Mediterranean port. With an annual freight capacity of 1 billion tons, the RAC operates through highly electrified rail networks, integrated highway systems, and efficient inland waterways, supported by advanced digital infrastructure and cross-border coordination mechanisms. The RAC's development has been shaped by decades of European integration, benefiting from comprehensive policy frameworks, including the Trans-European Transport Network (TEN-T) initiative and the European Green Deal. These frameworks have fostered a strong focus on sustainability, emphasising emission reduction, modal shift towards environmentally friendly transport options, and enhanced energy efficiency across the corridor's operations.

1.2 The Delhi-Mumbai Industrial Corridor: An Emerging Vision

In contrast, the Delhi-Mumbai Industrial Corridor (DMIC) represents an ambitious vision for integrated industrial and transport development in India. Spanning 1,500 kms across six states, the DMIC encompasses twenty-four planned industrial regions and eight smart cities, anchored by a Dedicated Freight Corridor. The project's expected handling capacity of 182 million tons of freight annually by 2031 demonstrates its potential impact on India's economic landscape. The DMIC exemplifies a new approach to corridor development that integrates industrial planning, urban development, and transport infrastructure with environmental sustainability and digital technology from its inception. This comprehensive approach reflects a growing recognition of the need to balance rapid economic development with environmental sustainability and social inclusivity. The DMIC's integration of sustainable urban development, renewable energy systems, and advanced digital infrastructure aligns with India's national sustainability goals and international commitments, offering a unique model for developing economies.

1.3 Contemporary Challenges in Sustainable Corridor Development

The comparative analysis of these corridors reveals several critical challenges that must be addressed in the development of sustainable transport infrastructure. Development stage disparities between the two corridors influence their approach to sustainability, with the RAC focusing on upgrading existing systems while the DMIC has the opportunity to implement cutting-edge technologies from the ground up. These differences extend to governance frameworks, where the RAC's multi-national European approach contrasts sharply with India's centralised corridor management strategy. Technological integration presents another significant challenge, particularly in terms of ensuring interoperability and efficient system upgrades. Environmental considerations vary between the corridors, with each facing unique challenges in reducing emissions, managing resources, and preserving local ecosystems. Social impacts also differ significantly, reflecting the distinct socio-economic contexts of Europe and India.

1.4 Comparative Analysis: Scope and Significance

This comparative analysis offers valuable insights for policymakers, infrastructure developers, and researchers working on sustainable transport solutions globally. The study employs a comprehensive mixed-method approach, combining quantitative analysis of operational metrics with qualitative assessment of policy frameworks and implementation strategies. This methodology enables a thorough examination of both technical and socio-economic aspects of sustainable corridor development. The research examines how operational practices and sustainability initiatives differ between mature and developing transport corridors, identifying best practices that can be effectively transferred between different economic and geographical contexts.

It also explores how policy frameworks and technological innovations shape sustainable development in transport corridors, providing practical insights for future infrastructure projects. The study progresses through several interconnected sections, beginning with a theoretical framework that examines the evolution of transport corridors and sustainability concepts in infrastructure development. This is followed by a detailed comparative analysis of operational practices, environmental initiatives, and policy frameworks for both corridors. The analysis culminates in a discussion of future outlook and recommendations for sustainable corridor development globally.

Through this structured analysis, the chapter contributes to the growing body of knowledge on sustainable infrastructure development, offering practical insights for stakeholders working on similar projects worldwide. The lessons drawn from these case studies are particularly relevant as nations worldwide grapple with the challenge of developing efficient, sustainable transport networks while addressing climate change concerns. The comprehensive examination of these two distinct corridors provides unique insights into how mature and emerging economies can learn from each other in pursuing sustainable transport solutions. As the global community continues to emphasise the importance of sustainable development, the experiences of the RAC and DMIC offer valuable guidance for future transport corridor development, highlighting both universal principles and the importance of context-specific solutions in creating more sustainable and efficient transport networks.

1.5 Significance of This Study for Transport Sustainability

The significance of this comparative analysis extends far beyond academic contribution, addressing urgent practical challenges in sustainable transport development. As global commerce increasingly relies on efficient transport networks, understanding how different regions implement sustainable practices becomes crucial for both environmental preservation and economic development. This research arrives at a critical juncture when the transport sector faces mounting pressure to reduce its environmental footprint while maintaining operational efficiency. The study's examination of climate change mitigation through transport corridor development offers vital insights for policymakers and practitioners worldwide. By analyzing how major transport corridors can minimize their environmental impact while maintaining economic efficiency, this research directly addresses the transport sector's contribution to global greenhouse gas emissions, which currently accounts for approximately 24% of global CO_2 emissions (IPCC, 2022). The findings provide practical guidance for developing nations embarking on new infrastructure projects, demonstrating how environmental considerations can be integrated from the planning stage rather than retrofitted later.

The comparison between a mature European corridor and an emerging Indian one offers unique insights into the transferability of sustainable practices across different economic and cultural contexts. This cross-contextual learning proves particularly valuable as developing economies invest heavily in new transport infrastructure. The research reveals how emerging corridors can potentially leapfrog traditional development patterns by adopting advanced sustainable technologies and practices from inception, while established corridors can enhance their existing operations through systematic sustainability improvements. Innovation in sustainable infrastructure development emerges as another key contribution of this study. The findings demonstrate how technological advancement and policy frameworks can work together to enhance sustainability in transport operations. The research provides evidence-based insights into successful implementation strategies, particularly relevant for developing countries undertaking large-scale infrastructure projects. These insights help bridge the gap between theoretical sustainability goals and practical implementation challenges.

Furthermore, the study's examination of stakeholder engagement and social impact offers valuable lessons for effective multi-stakeholder collaboration in large-scale infrastructure projects. This aspect proves particularly crucial as transport corridors increasingly need to balance environmental protection with community needs and economic development goals. The research demonstrates how different governance approaches can effectively manage diverse stakeholder interests while maintaining focus on sustainability objectives. The timing of this research coincides with growing global emphasis on sustainable development and climate action. As countries worldwide grapple with the challenge of decarbonising their transport sectors, the insights from successful sustainable corridor practices provide valuable guidance for policymakers and practitioners. The study's findings contribute to our understanding of how transport infrastructure can evolve to meet both current operational needs and future sustainability requirements, making it an essential resource for those involved in transport corridor development globally.

1.6 Manuscript Structure

This chapter is organised into five sections. Following this introduction, Sect. 2 provides a comprehensive literature review examining the theoretical foundations of sustainable transport development and current best practices globally. Section 3 discusses the research methodology adopted in this study. Section 4 presents the results of the empirical investigation, focusing on operational indicators and sustainability metrics across both corridors. Section 5 discusses the implications of these findings, developing recommendations for policy and practice. Finally, Sect. 6 concludes with a synthesis of key insights and directions for future research. Each chapter builds upon the previous ones to create a coherent analysis of sustainable transport corridor development in different contexts.

2 Transport Corridors and Sustainable Development: A Global Perspective

2.1 Theoretical Foundation and Evolution of Environmental Sustainability in Transport

The theoretical understanding of environmental sustainability in transport corridors has evolved significantly over recent decades. Early approaches focused primarily on localised environmental impacts, but contemporary frameworks recognise transport corridors as complex socio-ecological systems with far-reaching environmental implications. This evolution reflects growing awareness of climate change and the critical role of transport infrastructure in environmental outcomes. The concept of "ecological modernisation", as developed by Mol et al. (2009), provides a useful theoretical framework for understanding how transport corridors can integrate environmental sustainability with technological advancement. This theory suggests that environmental protection and economic development can be mutually reinforcing through technological innovation and policy reform. In the context of transport corridors, this framework helps explain how infrastructure development can simultaneously serve economic needs and environmental goals.

Transport corridors have evolved dramatically from their initial conception as simple infrastructure networks to become complex, integrated systems shaping regional development and environmental sustainability. This evolution reflects broader changes in our understanding of infrastructure's role in sustainable development (Kunaka & Carruthers, 2014). The transformation of transport corridors mirrors the changing priorities of global development, from purely economic considerations to a more balanced approach incorporating environmental and social dimensions.

The initial infrastructure phase of the 1950s-1970s focused primarily on physical connectivity and national economic development. During this period, transport corridors were viewed primarily as conduits for movement, with limited consideration of their broader impacts on society and the environment. This perspective was shaped by post-war reconstruction needs and the dominance of national development planning (Arnold, 2006). The emphasis was on building robust physical infrastructure to support growing economies, often at the expense of environmental considerations.

The 1980s and 1990s marked a shift towards economic integration, as globalisation began to reshape international trade patterns. Transport corridors became critical facilitators of regional trade and cross-border cooperation. This period saw the emergence of sophisticated logistics concepts and the beginning of environmental awareness in transport planning (Regmi & Hanaoka, 2012). The integration of different transport modes and the consideration of supply chain efficiency became increasingly important during this phase.

The sustainability phase of the 2000s-2010s represented a fundamental shift in transport corridor development. Environmental impact considerations and social sustainability became central to corridor planning and operation. This period coincided with growing global awareness of climate change and the need for sustainable development practices. Transport corridors began to be viewed as integral components of sustainable development strategies, rather than merely economic infrastructure (Banister, 2008). The historical development of transport corridors has evolved through distinct phases, as summarised in Table 1 demonstrating the progression from basic infrastructure to smart, integrated systems.

The current smart corridor phase, emerging in the 2020s, represents the latest evolution in transport corridor development. This phase is characterised by the integration of digital technologies, artificial intelligence, and Internet of Things (IoT) applications. Modern transport corridors are increasingly becoming intelligent systems capable of real-time monitoring, predictive maintenance, and adaptive management (Zhang et al., 2019). This technological integration enables more efficient resource use and better environmental performance monitoring. The theoretical understanding of transport corridors has also evolved to encompass multiple dimensions of sustainability. Contemporary frameworks recognise the complex interplay between economic efficiency, environmental protection, and social equity. This more nuanced understanding has led to the development of integrated approaches to corridor planning and management, which consider both immediate operational needs and long-term sustainability goals.

Table 1 Evolution of transport corridor concepts

Period	Key concept	Focus areas
1950–1970s	Transport corridors	Infrastructure development
1980–1990s	Economic corridors	Trade facilitation, economic growth
2000–2010s	Sustainable corridors	Environmental considerations, social impact
2020s–Present	Smart corridors	Digital integration, resilience

2.2 *Sustainable Transportation Framework*

2.2.1 Conceptual Foundation

The concept of sustainable transportation in corridor development has emerged as a critical paradigm that integrates economic viability, environmental responsibility, and social equity. This framework, as articulated by Black (2010), represents a fundamental shift from traditional transport planning approaches that primarily emphasized economic efficiency. Modern sustainable transportation theory recognises the need to balance current operational requirements with long-term environmental preservation and social development goals. Contemporary understanding of sustainable transportation in corridor development encompasses multiple interconnected dimensions. The economic dimension focuses on operational efficiency and cost-effectiveness, while the environmental dimension addresses emission reduction, resource conservation, and ecosystem protection. The social dimension, often overlooked in earlier approaches, considers community impacts, accessibility, and social justice implications (Goldman & Gorham, 2006).

2.2.2 Implementation Strategies

The implementation of sustainable transportation principles in corridor development requires a comprehensive approach that addresses multiple aspects simultaneously. Modal integration has emerged as a crucial strategy, particularly in freight transport. The shift towards more environmentally friendly transport modes, such as rail and inland waterways, represents a key approach to reducing environmental impact while maintaining economic efficiency. Research by Woodburn et al. (2019) demonstrates that successful modal shift strategies can significantly reduce corridor-wide emissions while improving overall system efficiency. Technological innovation plays an increasingly important role in sustainable corridor development. The adoption of clean technologies, including electric vehicles and alternative fuels, offers significant potential for emission reduction. Grigoratos et al. (2019) highlight how technological advancement in areas such as hydrogen fuel cells and battery storage are creating new opportunities for sustainable transport operations. These innovations are particularly relevant in the context of long-distance freight movement, where traditional fossil fuel dependency has been challenging to overcome.

2.2.3 Policy Integration and Governance

The effectiveness of sustainable transportation initiatives depends heavily on supportive policy frameworks and governance structures. Banister (2008) emphasises the importance of aligning transport policies with broader sustainability goals, including climate change mitigation and social development objectives. This alignment requires coordination across multiple policy domains, including transport, energy, urban planning, and environmental protection. Cross-border coordination presents particular challenges in implementing sustainable practices, especially in international corridors. The experience of established corridors, such as the Rhine-Alpine Corridor, demonstrates the importance of harmonized regulations and standards across jurisdictions. Policy integration must address not only environmental standards but also operational procedures, technology adoption, and social impact management.

2.3 *Environmental Impact Analysis*

Transport corridors significantly contribute to global greenhouse gas emissions, with the transport sector accounting for approximately 24% of global CO_2 emissions from fuel combustion (IEA, 2020). The distribution of emissions varies considerably depending on modal split and technological adoption levels. Rail transport generally produces lower emissions per tonne-kilometre compared to road transport, while maritime transport can be more efficient for long-distance freight movement (Sims et al., 2014). Recent research has highlighted the complex relationship between corridor development and climate change impacts. Both direct emissions from transport operations and indirect emissions from infrastructure development and maintenance must be considered. The life-cycle approach to emissions assessment, as advocated by Arvidsson and Browne (2013), provides a more comprehensive understanding of environmental impacts and helps identify key intervention points for emission reduction.

Beyond greenhouse gas emissions, transport corridors generate significant local environmental impacts that require careful management. These impacts include air pollution from nitrogen oxides (NO_x) and particulate matter (PM), noise pollution affecting both human communities and wildlife, and habitat fragmentation along corridor routes. European Environment Agency (2020) has documented how these impacts are particularly acute in urban areas and sensitive ecosystems along corridor routes.

Recent research has significantly advanced our understanding of transport corridors' environmental impacts. Sims et al. (2014) provide a comprehensive analysis of transport-related emissions, highlighting how different modal choices affect carbon footprints. Their work demonstrates that rail transport typically produces 3–7 times lower emissions per tonne-kilometer compared to road transport, emphasising the environmental importance of modal shift strategies. The environmental impact of transport corridors extends beyond emissions. Abhijith et al. (2017) examine how transport infrastructure affects local air quality and ecosystem health, presenting evidence that proper design and green infrastructure can significantly mitigate negative environmental impacts. Their research shows that strategic placement of vegetation barriers can reduce particulate matter concentrations by up to 60% along transport routes.

2.4 Case Studies

2.4.1 The Rhine-Alpine Corridor: A Model of Mature Development

The Rhine-Alpine Corridor (RAC) represents one of Europe's most sophisticated transport networks, demonstrating the successful integration of sustainable practices in a mature corridor system. Stretching from the North Sea ports to the Mediterranean, this corridor serves as a critical economic artery handling approximately 1 billion tons of freight annually (European Commission, 2021). The corridor's development has been shaped by decades of European integration and environmental policy evolution. Recent studies by Witte et al. (2019) demonstrate how the RAC has achieved significant progress in multimodal integration, with a balanced modal split of 40% rail, 40% road, and 20% inland waterways. This distribution reflects successful implementation of EU transport policies aimed at reducing the environmental impacts while maintaining economic efficiency. The corridor's advanced implementation of the European Rail Traffic Management System (ERTMS) exemplifies how technological innovation can enhance both operational efficiency and environmental performance.

Environmental performance monitoring in the RAC reveals substantial achievements in emission reduction and energy efficiency. UIC (2020) reports that electrification of rail lines combined with renewable energy integration has reduced corridor-wide CO_2 emissions by approximately 25% over the past decade. These improvements demonstrate the potential for mature transport corridors to achieve significant environmental gains through systematic implementation of sustainable practices.

2.4.2 The Delhi-Mumbai Industrial Corridor: Emerging Sustainability Initiatives

The Delhi-Mumbai Industrial Corridor (DMIC) presents a contrasting case of corridor development in an emerging economy context. This ambitious project, spanning 1,500 kms across six Indian states, integrates industrial development with transport infrastructure while incorporating sustainability principles from its inception. The corridor's design reflects a comprehensive approach to sustainable development, as outlined by Aijaz and Hoelscher (2015), including smart city development principles, renewable energy integration, and advanced digital infrastructure.

The DMIC's approach to sustainability demonstrates how developing economies can potentially leapfrog certain developmental stages through innovative planning

Table 2 Comparison of the key characteristics of RAC and DMIC

Characteristic	Rhine-Alpine corridor	Delhi-Mumbai industrial corridor
Length	1,300 km	1,500 km
Countries	Netherlands, Germany, Switzerland, Italy, France	India (6 states—Uttar Pradesh, Haryana, Rajasthan, Madhya Pradesh, Gujarat, Maharashtra)
Key nodes	Rotterdam, Frankfurt, Genoa	Delhi, Mumbai, 8 planned smart cities
Current modal split	40% rail, 40% road, 20% inland waterways	30% rail, 70% road
Primary focus	Freight transport and logistics	Industrial development and freight transport
Development stage	Mature	Developing
Population impacted	~70 million	~180 million (projected)
Key sustainability initiatives	ERTMS, port automation, alternative fuels	Smart cities, renewable energy integration, water conservation
Governance	Multi-national (EU framework)	Centralised (special purpose vehicle)
Major challenges	Cross-border coordination, legacy infrastructure	Land acquisition, water scarcity, skill development

and technology adoption. The corridor's implementation of digital technologies and smart systems, as documented by DMICDC (2023), shows promise in achieving both operational efficiency and environmental performance improvements. The planned integration of renewable energy sources and water conservation measures represents a forward-thinking approach to resource management in corridor development. The fundamental characteristics of the RAC and DMIC corridors show distinct developmental patterns, as illustrated in Table 2, reflecting their different stages of maturity and operational contexts.

2.4.3 Additional Global Examples: Emerging Trends and Innovations

Recent developments in other global transport corridors provide valuable insights into innovative approaches to sustainability. The China–Europe Railway Express, part of the Belt and Road Initiative, demonstrates how international cooperation can facilitate sustainable transport solutions across vast distances. The study by Zhang et al. (2019) highlights how this corridor has achieved significant reductions in carbon emissions compared to traditional maritime routes while maintaining competitive operational efficiency. The North–South Transport Corridor connecting Russia with India via Iran presents another interesting case study in sustainable corridor development. This corridor's implementation of multimodal transport solutions and advanced logistics systems demonstrates how emerging corridors can incorporate sustainability principles while addressing complex geographical and political challenges.

2.5 Implementation Challenges and Solutions

The implementation of sustainable practices in transport corridors faces numerous technical and operational challenges. Infrastructure adaptation represents a significant challenge, particularly in mature corridors where legacy systems must be upgraded to meet contemporary sustainability standards. Research by Christodoulou and Demirel (2018) identifies how climate change impacts create additional complexity in infrastructure planning and maintenance. Technological integration presents another significant challenge, particularly in coordinating systems across different jurisdictions and operational environments. The experience of the RAC in implementing ERTMS demonstrates both the potential benefits and complexities of large-scale technological deployment in transport corridors. Cross-border coordination and policy harmonization remain significant challenges in sustainable corridor development. The multi-jurisdictional nature of many transport corridors creates complexity in implementing consistent environmental standards and operational procedures. The contrast between the EU's comprehensive policy framework for the RAC and the centralized governance approach of the DMIC highlights different models for addressing these challenges.

2.6 Future Directions in Sustainable Corridor Development

The future of sustainable transport corridors will be significantly shaped by technological advancement and digital integration. Emerging technologies, including artificial intelligence, blockchain, and autonomous systems, present new opportunities for enhancing both operational efficiency and environmental performance. Recent research by European Commission (2021) suggests that digital technologies could reduce corridor-wide emissions by up to 30% through optimized routing, improved load factors, and enhanced energy management. The integration of smart infrastructure systems represents a particularly promising direction for future development. The implementation of IoT sensors, real-time monitoring systems, and predictive maintenance capabilities can significantly improve resource efficiency and reduce environmental impacts. As demonstrated by early implementations in the DMIC, these technologies can enable more responsive and adaptive corridor management systems (DMICDC, 2023).

Climate change adaptation has emerged as a critical consideration in future corridor development. The increasing frequency and severity of extreme weather events necessitate new approaches to infrastructure design and operational planning. The research conducted by Reggiani et al. (2015) emphasises the importance of building resilience into transport corridor systems, suggesting that future development must consider both mitigation and adaptation strategies. The concept of "climate-proof" infrastructure is gaining prominence in corridor planning. This approach involves designing and operating transport systems that can withstand

and adapt to changing climatic conditions while maintaining operational efficiency. The experience of the RAC in implementing climate adaptation measures provides valuable insights for future corridor development globally.

The integration of social sustainability considerations into corridor development represents an increasingly important direction for future development. The study by Nayak (2015) highlights the need for more comprehensive approaches to addressing community impacts, labour rights, and social equity in corridor development. Future corridors must balance economic efficiency and environmental protection with social development objectives.

The integration of renewable energy into transport corridors represents a growing focus in sustainable infrastructure development. McKinnon (2018) examines how corridors can serve as platforms for renewable energy deployment, potentially transforming from energy consumers to energy producers. This concept of "energy-positive infrastructure" suggests that transport corridors could generate more renewable energy than they consume through strategic placement of solar, wind, and other renewable technologies.

The relationship between modal split and environmental performance has emerged as a critical area of research in sustainable transport development. Woodburn et al. (2019) demonstrate that shifting freight from road to rail can reduce carbon emissions by up to 75% per tonne-kilometer. Their analysis of European corridors shows that achieving a balanced modal split requires both infrastructure investment and policy intervention. The environmental benefits extend beyond emission reduction to include decreased noise pollution, reduced habitat fragmentation, and improved air quality in urban areas. Recent studies by International Energy Agency (2021) reveal that modal shift strategies must be complemented by technological innovation to maximise environmental benefits. For instance, the electrification of rail systems, when powered by renewable energy, can reduce corridor-wide emissions by up to 90% compared to conventional road transport. This finding emphasises the importance of considering both modal choice and energy sources in environmental planning.

Transport corridors significantly impact surrounding ecosystems, making green infrastructure integration crucial for environmental sustainability. The research by Clausen et al. (2019) identifies effective strategies for protecting biodiversity along transport routes, including wildlife crossings, habitat corridors, and buffer zones. Their work shows that well-designed green infrastructure can reduce wildlife mortality by up to 90% while maintaining corridor functionality. The concept of "ecological connectivity" has gained prominence in transport corridor planning. Reggiani et al. (2015) demonstrate how corridors can be designed to maintain or enhance ecosystem services rather than fragmenting habitats. Their research presents evidence that incorporating green infrastructure from the planning stage costs significantly less than retrofitting existing corridors and provides superior environmental outcomes.

2.7 Policy Implications and Recommendations

The development of effective policy frameworks for sustainable transport corridors requires careful consideration of multiple factors. Based on the analysis of successful implementations, particularly in the RAC, several key policy recommendations emerge as follows. First, the integration of environmental considerations into all aspects of corridor planning and operation is essential. This includes establishing clear environmental performance targets and monitoring mechanisms. Second, policy frameworks must support technological innovation while ensuring standardisation and interoperability across jurisdictions. Third, social impact assessment and community engagement should be integral components of corridor development policies. Successful implementation of sustainable practices requires carefully structured approaches that consider local contexts and capabilities. The experience of both mature and developing corridors suggests that phased implementation strategies, coupled with regular assessment and adaptation, are most effective. The transfer of best practices between corridors must be accompanied by careful adaptation to local conditions and capabilities.

3 Research Methodology: Assessing Environmental Performance in Transport Corridors

3.1 Methodological Framework Development

This research employs a comprehensive mixed-method approach designed to evaluate environmental sustainability in transport corridors. The methodology combines quantitative analysis of environmental performance metrics with qualitative assessment of sustainability practices, enabling deep understanding of both measurable outcomes and contextual factors. This approach acknowledges the complex nature of environmental impacts in transport infrastructure and the need for nuanced analysis across different developmental contexts. The methodological framework employed builds on established environmental assessment practices while introducing innovative approaches to comparative analysis. The design draws from the methodology used in traditional transport studies and emerging frameworks for sustainability assessment, creating a robust structure for examining environmental performance across distinctly different corridor contexts.

3.2 Research Design and Analytical Framework

The study adopts a comparative case study design that enables systematic analysis of environmental practices in both mature and emerging transport corridors, chosen specifically to capture the complexity of environmental challenges and solutions across different developmental stages. The Rhine-Alpine Corridor and Delhi-Mumbai Industrial Corridor serve as representative cases of mature and emerging systems respectively, offering rich opportunities for comparing environmental strategies and outcomes. The analytical framework incorporates three primary dimensions of environmental assessment: direct environmental impacts (emissions, resource consumption, ecosystem effects), climate resilience measures (adaptation strategies, infrastructure design, risk management), and systemic sustainability practices (renewable energy integration, circular economy approaches).

3.3 Data Collection and Environmental Metrics

The research relies on comprehensive secondary data sources chosen for their reliability and relevance to environmental performance assessment. Key data sources encompass Environmental Performance Data, including emissions monitoring reports from corridor management authorities, environmental impact assessments and audits, climate adaptation planning documents, and energy consumption and efficiency reports. Infrastructure and Operations Data comprises modal split statistics, infrastructure electrification records, renewable energy integration reports, and green infrastructure implementation records. Policy and Planning Documents include environmental regulations and standards, sustainability strategies and targets, climate action plans, and environmental management frameworks.

3.4 Analytical Methods and Tools

Our analysis employs multiple methods to ensure robust evaluation of environmental performance. The study uses statistical analysis to examine environmental performance indicators, including emissions data (CO_2, NO_x, PM) normalised by freight volume, energy efficiency metrics (kWh/tonne-km), modal split ratios and their environmental implications, and renewable energy utilisation rates. Content analysis of policy documents and environmental strategies provides insights into environmental management approaches, climate adaptation strategies, ecosystem protection measures, and stakeholder engagement in environmental planning. The integration of these methods enables comprehensive understanding of environmental performance while acknowledging the complex interplay between quantitative metrics and qualitative factors.

3.5 Comparative Assessment Framework

Our comparative assessment framework enables systematic evaluation of environmental performance across the two corridors while accounting for their different developmental contexts. The framework examines environmental practices at three distinct levels:

- ***Strategic level***: At this level, we analyze long-term environmental planning and climate adaptation strategies. Our assessment considers how each corridor approaches major environmental challenges, including emission reduction targets, biodiversity protection, and climate resilience. For instance, when comparing the RAC's established environmental management systems with the DMIC's emerging frameworks, we carefully consider how different developmental stages influence strategic environmental choices.
- ***Tactical level***: This level focuses on medium-term implementation of environmental initiatives. We examine specific programs and projects designed to enhance environmental performance, such as electrification schemes, renewable energy integration, and green infrastructure development. The analysis considers both successful deployment and implementation challenges, providing insights into effective environmental management approaches.
- ***Operational level***: At the operational level, we analyze day-to-day environmental management practices and their effectiveness. This includes examination of real-time environmental monitoring systems, emission control measures, and resource efficiency programs. Our analysis particularly focuses on how different operational approaches impact environmental outcomes.

3.6 Environmental Performance Indicators

To ensure comprehensive assessment of environmental impacts, we developed a structured set of performance indicators that capture multiple aspects of environmental sustainability. Primary Environmental Metrics include carbon emissions per tonne-kilometer, energy efficiency in transport operations, modal split ratios and their environmental implications, air quality impacts along corridor routes, noise pollution levels in sensitive areas, and biodiversity impact measures. Climate Resilience Indicators encompass infrastructure adaptation measures, extreme weather event preparation, climate risk assessment protocols, and resilience planning frameworks. Resource Efficiency Measures consist of renewable energy utilisation, water consumption patterns, material recycling rates, and waste management effectiveness.

3.7 Data Analysis and Validation

The data analysis employs rigorous validation procedures to ensure reliability of findings. We implement data triangulation by cross-referencing multiple sources and using both quantitative and qualitative evidence to support conclusions. Environmental performance data is normalised to account for differences in corridor scale and operational intensity, enabling meaningful comparisons despite different developmental contexts. Statistical analysis of environmental metrics follows established protocols for transport infrastructure assessment while incorporating emerging standards for sustainability evaluation. When analyzing trends in environmental performance, we carefully consider both absolute improvements and relative changes, acknowledging the different baseline conditions of each corridor.

3.8 Methodological Limitations and Mitigation Strategies

Some limitations can be acknowledged in the employed methodology. Specific strategies that were used to address methodological limitations include the following:

- ***Data Availability Challenges***: While the RAC provides extensive historical environmental data, the DMIC's developing nature means some metrics are projections rather than actual measurements. This limitation is addressed by clearly distinguishing between measured and projected data and by using conservative estimates in our comparative analyses.
- ***Contextual differences***: The distinct developmental stages of the two corridors present challenges for direct comparison. This is mitigated by developing normalised metrics that account for different operational scales and implementing stage-appropriate assessment criteria.
- ***Temporal considerations***: The dynamic nature of environmental challenges means that some findings may have limited long-term applicability. This is addressed by focusing on underlying principles and adaptable strategies rather than specific technical solutions.

4 Results: A Comprehensive Analysis of Transport Corridor Development

4.1 Transport Operations and Infrastructure Development

The analysis of transport operations reveals distinctive patterns in both corridors' development trajectories. The Rhine-Alpine Corridor (RAC) exemplifies a mature transport system with sophisticated operational integration. The corridor's balanced

Table 3 Comparison of the key performance indicators

Indicator	Rhine-Alpine corridor	Delhi-Mumbai industrial corridor
Modal split (Rail: Road: Water)	40:40:20	30:70:0 (Current), 60:40:0 (Projected)
Annual freight volume	1 billion tons	182 million tons (Projected)
Electrification	> 90% of rail network	100% of DFC (Planned)
Average transit time	24 h (rail)	72 h (current), 24 h (projected)

modal split, comprising 40% rail, 40% road, and 20% inland waterways, demonstrates successful implementation of sustainable transport principles (European Commission, 2021). This equilibrium supports the efficient movement of approximately 1 billion tons of freight annually, facilitated by extensive electrification of the rail network, which now exceeds 90% coverage. The corridor's operational efficiency is evident in its transit times, with journeys from Rotterdam to Genoa requiring only 24 h by rail and 48 h via inland waterway (Rhine-Alpine EGTC, 2022).

The Delhi-Mumbai Industrial Corridor (DMIC) presents a contrasting picture of emerging potential and ambitious transformation. Currently characterised by road dominance (70% road, 30% rail), the corridor's development plans envision a significant modal shift to achieve 60% rail and 40% road distribution by 2035 (DMICDC, 2023). The Dedicated Freight Corridor, serving as the project's backbone, aims to handle 182 million tons of freight annually by 2031. This projected capacity, supported by comprehensive electrification plans for the 1,504-km network, represents a substantial leap in sustainable transport infrastructure development. A detailed comparison of operational performance metrics between the two corridors is presented in Table 3, highlighting significant differences in modal split and freight volume capacity.

4.2 Implementation of Sustainable Practices

Sustainable practice implementation reveals distinct approaches shaped by developmental contexts. The RAC's mature sustainability framework encompasses comprehensive electrification, advanced multimodal integration, and innovative port technologies. Notable achievements include the implementation of shore power facilities at major ports and significant automation progress, with Rotterdam achieving 80% automation in container handling operations (Port of Rotterdam, 2023). The corridor's commitment to alternative fuels is demonstrated by the adoption of LNG and hydrogen systems in 20% of heavy-duty vehicles, marking a significant progress in reducing transport-related emissions. The DMIC's approach to sustainability demonstrates strategic foresight in infrastructure planning. The corridor's ambitious sustainability framework includes the development of five large-scale solar parks with a combined capacity of 5000 MW, complemented by eight smart cities incorporating advanced sustainability principles. The integration of comprehensive IoT and AI systems in logistics management targets operational efficiency improvements of 30–40%, representing a significant leap in sustainable transport technology adoption.

4.3 Environmental Performance and Emissions Analysis

Detailed analysis of environmental performance metrics reveals nuanced differences between the corridors. The RAC currently achieves sophisticated emissions management, with average CO_2 emissions of 80 g/tonne-km across all transport modes. Rail freight operations demonstrate particular efficiency, operating at 60–70 g CO_2/tonne-km, significantly lower than road freight at 110–120 g CO_2/tonne-km. The corridor's systematic implementation of sustainable practices has yielded a 20% reduction in the overall carbon footprint over the past decade (European Environment Agency, 2023).

The DMIC's environmental performance projections indicate potential for significant improvement from current levels. While present emissions average 85 g CO_2/tonne-km due to heavy reliance on road transport, the implementation of the Dedicated Freight Corridor is expected to reduce this to 35 g CO_2/tonne-km. Comprehensive emission reduction extends to other pollutants, with projected decreases in NO_x emissions from 0.7 to 0.2 g/tonne-km and particulate matter from 0.04 to 0.005 g/tonne-km, representing substantial environmental improvements. The comparative analysis of emissions data between the two corridors, as shown in Table 4, reveals distinct patterns in environmental performance and highlights opportunities for improvement.

4.4 Technological Innovation and Digital Integration

The technological landscape across both corridors reveals sophisticated yet distinct approaches to innovation. The RAC demonstrates leadership in implemented technologies, with the European Rail Traffic Management System (ERTMS) achieving 90% network coverage, significantly enhancing operational efficiency. This comprehensive system integration has yielded quantifiable benefits, including a 20% increase in network capacity and substantial reductions in operational delays (European Rail Agency, 2023). Port automation technologies have proven particularly successful, with automated systems handling 80% of container operations and contributing to a 35% reduction in port-related emissions.

The DMIC's technological strategy emphasises comprehensive digital integration from inception, potentially bypassing intermediate technological stages. The corridor's smart logistics framework incorporates advanced IoT and AI systems,

Table 4 Comparisons of emissions between RAC and DMIC

Emission type	Rhine-Alpine corridor (current)	Delhi-Mumbai industrial corridor (current)	Delhi-Mumbai industrial corridor (projected)
CO_2 (g/tonne-km)	80	85	35
CO_2—Rail Freight (g/tonne-km)	60–70	N/A	30–40
CO_2—Road Freight (g/tonne-km)	110–120	80–90	N/A
CO_2—Inland Waterway (g/tonne-km)	30–35	N/A	N/A
Particulate Matter (PM) (g/tonne-km)	0.015	0.04	0.005
Nitrogen Oxides (NO_x) (g/tonne-km)	0.45	0.7	0.2
Sulphur Dioxide (SO_2) (g/tonne-km)	0.02	0.08	0.01
Total GHG (g CO_2e/tonne-km)	90	100	45

projecting 30–40% improvements in operational efficiency upon full implementation. Smart city initiatives within the corridor demonstrate particular innovation in green building technologies, with energy consumption projected to decrease by 50% compared to conventional urban developments (Indian Green Building Council, 2023). The integration of blockchain technology for supply chain transparency and 5G networks for enhanced connectivity represents cutting-edge approaches to corridor management. The adoption of various technological solutions across both corridors demonstrates distinct approaches to innovation, as detailed in Table 5.

4.5 *Climate Resilience Measures*

Both corridors demonstrate increasing attention to climate resilience, though with different implementation approaches. The RAC has systematically upgraded existing infrastructure to enhance climate resilience, implementing flood protection systems along waterways, heat-resistant rail infrastructure, storm water management systems, and enhanced monitoring networks for extreme weather events. The DMIC incorporates climate resilience from the design stage, with particular emphasis on drought-resistant infrastructure design, comprehensive water management systems, green infrastructure integration, and climate-adaptive urban planning in logistics hubs.

Table 5 Comparison of technology adoption in RAC and DMIC

Technology area	Rhine-Alpine corridor	Delh–Mumbai industrial corridor
Traffic management	European Rail Traffic Management System (ERTMS) implemented on 90% of rail network	Planned implementation of Advanced Traffic Management System (ATMS) for the Dedicated freight corridor
Port automation	80% of container handling automated in major ports	Planned automation for new ports like Dholera and upgrades for existing ports
Alternative fuels	20% of heavy-duty vehicles use LNG or hydrogen	Plans for electric and hydrogen fuel cell vehicles in logistics
Smart logistics	IoT and AI-driven logistics optimization widely adopted	Comprehensive IoT and AI integration planned for freight management
Renewable energy integration	Extensive use of solar and wind power in operations, 30% of energy needs	Plans for 5 large-scale solar parks, aiming to meet 40% of energy demand by 2030
Green building technologies	Widespread adoption of energy-efficient buildings in urban nodes	Smart city plans incorporate green building standards, aiming for 50% energy reduction
Blockchain in supply chain	Pilot projects for enhancing transparency and traceability	Planned implementation for supply chain transparency and trade facilitation
5G Network	5G coverage in major economic centres and along key routes	5G trials in progress, full implementation planned in smart cities
Autonomous vehicles	Testing of autonomous trucks and platooning on highways	Plans for autonomous vehicle trials in controlled environments of smart cities
Drone technology	Use of drones for infrastructure inspection and last-mile delivery trials	Planned use of drones for surveillance and potentially for logistics in less accessible areas

4.6 Ecosystem Protection and Biodiversity Conservation

Our analysis reveals significant differences in how each corridor approaches ecosystem protection. The RAC has developed sophisticated biodiversity conservation measures through years of environmental monitoring and adaptation. Through systematic implementation of green infrastructure, the corridor has achieved notable success in reducing habitat fragmentation. Environmental monitoring data shows that wildlife crossings along the RAC have reduced animal mortality by 85% in key ecological zones while maintaining operational efficiency. The DMIC takes a preventive approach to ecosystem protection, incorporating environmental considerations into initial design. The corridor's environmental impact assessments have led to the identification of 32 critical ecological zones requiring special protection measures. The planned green infrastructure network includes wildlife corridors spanning 120 km and incorporates advanced monitoring systems to track ecosystem health in real-time. This proactive approach represents a significant advancement in how emerging transport corridors address biodiversity conservation.

4.7 Resource Efficiency and Circular Economy Integration

Resource efficiency analysis demonstrates how both corridors are moving toward circular economy principles, albeit through different pathways. The RAC has transformed existing operations to optimize resource use, achieving significant improvements in material efficiency. Water consumption in maintenance operations has decreased by 40% through implementation of closed-loop systems, while material recycling rates have reached 75% across major logistics hubs. The DMIC's approach to resource efficiency is built into its foundational design. The corridor's planned industrial zones incorporate circular economy principles. Advanced water recycling systems aim to reduce freshwater consumption by 60% compared to conventional systems. Integrated waste management facilities are designed to achieve 70% material recovery rates. Smart logistics centres optimize resource use through real-time monitoring and adaptive management. Construction standards require 30% recycled material use in infrastructure development at the minimum.

4.8 Environmental Policy Implementation Effectiveness

Our analysis reveals distinct approaches to environmental policy implementation between the two corridors. The RAC demonstrates the benefits of long-term policy evolution within the European Union's comprehensive environmental framework. The corridor's multi-national governance structure has successfully implemented environmental policies through coordinated action across borders. Environmental compliance data shows 95% adherence to EU environmental directives, with particularly strong performance in emissions control and habitat protection measures. The effectiveness of the RAC's policy implementation is evidenced by consistent improvements in environmental metrics over the past decade. For instance, the corridor has achieved a 40% reduction in regulated pollutants through systematic enforcement of increasingly stringent environmental standards. The policy framework's success stems from its integrated approach, combining mandatory requirements with market-based incentives for environmental improvements.

The DMIC's centralized governance model presents both advantages and challenges in environmental policy implementation. The corridor's ability to implement uniform environmental standards across six states demonstrates the benefits of coordinated national planning. Environmental policy effectiveness is enhanced by the integration of sustainability requirements into initial development contracts, ensuring that environmental considerations are addressed from the project's inception.

4.9 Stakeholder Engagement in Environmental Initiatives

Environmental management success heavily depends on effective stakeholder engagement, and our analysis reveals contrasting approaches between the corridors. The RAC has developed sophisticated stakeholder engagement mechanisms through years of operational experience. The corridor maintains regular environmental consultations with over 200 stakeholder groups, including local communities, environmental organizations, and industry representatives. This inclusive approach has resulted in successful implementation of environmental initiatives, with 85% of stakeholder-proposed environmental improvements being incorporated into corridor operations. The DMIC's stakeholder engagement strategy focuses on preventive consultation and early involvement in environmental planning. The corridor has established Environmental Monitoring Committees in all major development nodes, incorporating local expertise into environmental decision-making. This proactive approach has helped identify and address potential environmental concerns before they become operational issues. The engagement process has led to significant modifications in original plans, including the redesign of 15 critical sections to better protect local ecosystems.

4.10 Environmental Monitoring and Reporting Systems

The sophistication of environmental monitoring systems varies significantly between the corridors, reflecting their different developmental stages. The RAC operates an integrated environmental monitoring network that provides continuous data on key environmental parameters. The corridor's comprehensive monitoring approach has enabled early detection and rapid response to environmental issues, with an average response time of 2.4 h for critical environmental alerts. The system's effectiveness is demonstrated by a 60% reduction in environmental incidents over the past five years through preventive intervention based on monitoring data. The DMIC is implementing next-generation environmental monitoring technologies from the outset. The planned system will integrate satellite monitoring, ground sensors, and AI-driven analytics to provide comprehensive environmental oversight. The design incorporates lessons learned from established corridors while leveraging new technologies for enhanced environmental protection. Early implementation in completed sections shows promising results, with 95% accuracy in environmental impact prediction.

4.11 Implementation Challenges and Mitigation Strategies

The analysis reveals distinct implementation challenges reflecting each corridor's developmental context. The RAC faces sophisticated coordination challenges across multiple national jurisdictions, particularly in harmonizing technical standards and operational procedures. Legacy infrastructure modernization presents significant challenges, with upgrading costs for older sections requiring substantial investment (European Investment Bank, 2023). Urban congestion in major nodes, particularly Frankfurt and Milan, creates additional complexity in last-mile delivery optimization, affecting 80% of major cities along the corridor. The DMIC confronts fundamental development challenges that require innovative solutions. Land acquisition issues affect 30% of planned projects, resulting in average delays of 18 months in project implementation. Water scarcity presents a critical challenge, with 90% of planned industrial zones located in water-stressed areas, necessitating advanced water management solutions. The 40% shortfall in technical expertise for smart city operations highlights the need for comprehensive skill development programs (NITI Aayog, 2023).

4.12 Future Outlook and Sustainability Targets

Both corridors have established ambitious yet distinctly structured sustainability targets reflecting their developmental stages. The RAC's future framework targets a 55% reduction in transport-related emissions by 2030 compared to 1990 levels, progressing toward complete carbon neutrality in rail operations by 2040. The corridor's renewable energy integration strategy aims to achieve 80% renewable energy usage in operations by 2035, complemented by circular economy initiatives targeting 90% recycling rates in major nodes by 2040. The DMIC's transformative vision encompasses comprehensive changes across multiple dimensions. The targeted 60% rail freight share by 2035 is expected to reduce road congestion by 40% while achieving 20% reduction in logistics costs. Smart city developments along the corridor are projected to accommodate 15 million residents by 2040, incorporating advanced technologies to reduce urban energy consumption by 30% compared to conventional cities. The corridor's green manufacturing initiatives target 50% "green" classified manufacturing by 2035, supported by circular economy programs aiming for 70% waste recycling in industrial zones.

The anticipated trajectory of emission reductions across both corridors is shown in Fig. 1. This projection demonstrates the ambitious environmental targets set by both corridors, while accounting for their different starting points and implementation capabilities. The comparative sustainability targets are further elaborated in Fig. 2, which illustrates the specific goals across diverse measures for sustainable development initiatives of each corridor.

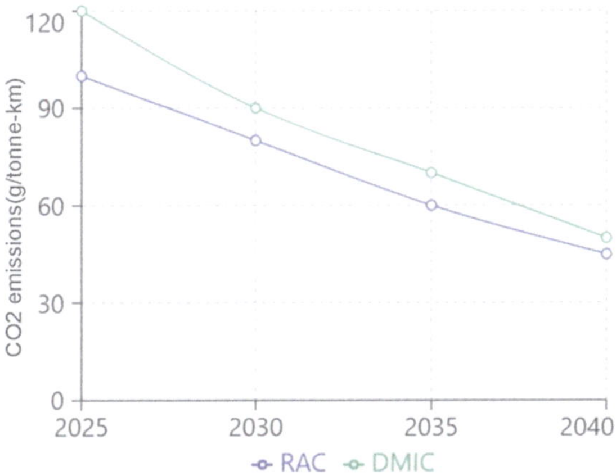

Fig. 1 Line graph showing projected emission reductions (2025–2040)

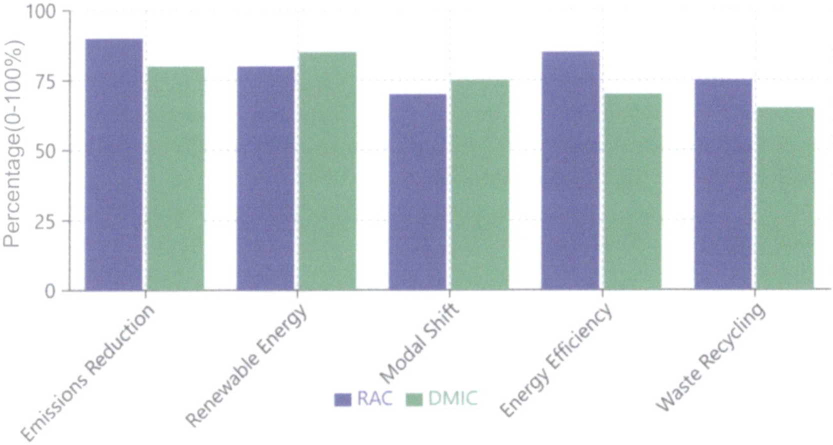

Fig. 2 Bar graph showing future sustainability targets for RAC and DMIC

5 Discussion and Recommendations

5.1 Introduction to Complex Transport Systems Development

The comparative analysis of the Rhine-Alpine Corridor (RAC) and Delhi-Mumbai Industrial Corridor (DMIC) reveals fundamental insights into sustainable transport corridor evolution across different developmental contexts. This discussion synthesises the key findings while situating them within broader theoretical frameworks and practical implications for future corridor development globally.

5.2 *Modal Split and Infrastructure Development: A Theoretical Analysis*

The striking contrast in modal split patterns between the RAC and DMIC exemplifies the evolutionary nature of sustainable transport development. The RAC's balanced distribution (40% rail, 40% road, 20% inland waterways) represents the culmination of decades of systematic infrastructure development and policy implementation. This aligns with the theoretical framework of Woodburn et al. (2019) on sustainable transport evolution, which emphasizes the importance of gradual system optimization through coordinated modal development. The DMIC's ambitious transformation from current road dominance (70% road, 30% rail) to targeted rail preference (60% rail, 40% road by 2035) represents what Perkins (2003) terms "developmental leapfrogging". This concept suggests that emerging systems can bypass intermediate developmental stages by implementing advanced solutions directly. However, this approach presents both opportunities and challenges that require careful consideration in implementation strategy.

The contrasting infrastructure development approaches reflect different temporal and resource contexts. The RAC's mature infrastructure network demonstrates the benefits of long-term, incremental development, supporting the findings of Kunaka and Carruthers (2014) on the importance of systematic infrastructure evolution. The corridor's high electrification rate (90%) and extensive intermodal connectivity (85% of major nodes) exemplify successful long-term infrastructure planning and implementation. The DMIC's greenfield development approach, while challenging, offers unique opportunities for implementing cutting-edge infrastructure solutions. This aligns with the research of Zhang et al. (2019) on technological leapfrogging in transport infrastructure, suggesting that emerging corridors can potentially achieve higher efficiency levels by adopting advanced technologies from inception.

5.3 *Environmental Performance and Sustainability*

The analysis of environmental performance reveals distinct approaches to sustainability and the environment. The RAC's achievements in emission reduction demonstrate the effectiveness of what Hollos et al. (2012) term "eco-efficiency" in mature systems. The corridor's comprehensive approach, combining rail electrification, stringent emission standards, and alternative fuel adoption, has yielded significant results, with CO_2 emissions averaging 80 g/tonne-km compared to conventional systems' 110–120 g/tonne-km. The DMIC's projected environmental performance aligns with the observations of Bai et al. (2018) on emerging economies' potential for environmental leapfrogging. The planned implementation of fully electric rail infrastructure and comprehensive renewable energy integration represents an ambitious approach to sustainability that could potentially surpass the current best practices.

Table 6 Best practices implemented in RAC and their transferability to DMIC

Best practice	Rhine-Alpine corridor	Transferability to DMIC
Multimodal integration	High level of integration	High potential for adoption
Green technologies	Widespread adoption	Opportunity for leapfrogging
Policy framework	EU-wide coordination	Need for interstate coordination
Stakeholder engagement	Well-established	Crucial for successful implementation

Both corridors demonstrate distinct approaches to technological innovation in pursuit of sustainability goals. The RAC's implementation of the European Rail Traffic Management System (ERTMS) and advanced port automation represents what Adams et al. (2016) describe as "sustainability-oriented innovation". This systematic approach has yielded measurable improvements in operational efficiency and environmental performance. The DMIC's comprehensive IoT and AI integration plans align with emerging concepts of smart corridor development. This approach supports the arguments of Zhang et al. (2019) regarding the transformative potential of digital technologies in transport efficiency and sustainability. The planned implementation of blockchain technology and advanced logistics management systems represents a more disruptive approach to innovation. The analysis of best practices and their potential for transfer between corridors, as presented in Table 6, suggests significant opportunities for knowledge sharing and adaptation.

5.4 *Critical Success Factors in Environmental Management*

Our analysis identifies several critical factors that determine environmental management success in transport corridors. First, the integration of environmental considerations into core planning and operational decisions proves more effective than treating environmental protection as an add-on requirement. The RAC's experience shows that environmental initiatives achieve greatest success when embedded within broader operational strategies. Second, technological innovation plays a crucial role in enabling environmental improvements. The DMIC's planned implementation of AI-driven environmental monitoring and smart grid systems represents a significant advance in how technology can support environmental management. However, our findings also indicate that technological solutions must be matched with appropriate governance structures and stakeholder engagement to be truly effective.

Third, the development of robust environmental monitoring systems proves essential for achieving sustained improvements in environmental performance. The RAC's sophisticated monitoring network has enabled rapid response to environmental issues and informed evidence-based decision-making. The DMIC's plans for next-generation monitoring systems build on these lessons while incorporating new capabilities for predictive environmental management.

5.5 Climate Resilience and Adaptation Strategies

Our analysis reveals evolving approaches to climate resilience in transport infrastructure. The RAC has developed sophisticated adaptation strategies through years of operational experience and climate impact assessment. These strategies encompass both physical infrastructure modifications and operational adjustments to address changing climate conditions. The corridor's success in reducing weather-related disruptions by 60% over five years demonstrates the effectiveness of this comprehensive approach to climate resilience. The DMIC's approach to climate resilience represents a significant advance in how emerging corridors address climate change. By incorporating climate projections into initial design specifications and developing adaptive management capabilities, the corridor establishes a new paradigm for climate-resilient infrastructure development. This proactive approach potentially offers greater long-term resilience than retrofitting existing infrastructure, though it requires significant upfront investment in climate-adaptive technologies and design.

5.6 Governance Models and Policy Integration

The contrasting governance models of the RAC and DMIC provide valuable insights into the relationship between institutional frameworks and sustainable development outcomes. The RAC's multinational framework, supported by EU-level coordination mechanisms, aligns with the research of Jordan and Lenschow (2010) on effective supranational environmental governance. This model facilitates standardization and cross-border cooperation but requires complex coordination mechanisms. The DMIC's centralized governance model offers advantages in the implementation of speed and resource mobilization but faces challenges in stakeholder engagement and local adaptation. This supports the arguments of Jordan and Lenschow (2010) regarding the importance of multi-level governance in environmental policy integration.

5.7 Resource Management and Operational Efficiency

The analysis of resource management strategies reveals fundamental differences in prioritization and implementation between the corridors. The RAC's mature resource management system demonstrates what McKinnon (2018) terms "integrated efficiency optimization", where multiple resource streams are managed through coordinated systems. The corridor's achievement of 0.06 kWh/tonne-km energy efficiency in rail operations exemplifies the benefits of long-term system optimization. The DMIC's approach to resource management reflects more immediate challenges, particularly in water-stressed regions. This aligns with the research of Nayak (2015) on environmental risks in rapid industrial development. The planned implementation of advanced water recycling systems and renewable energy integration (targeting 40% renewable energy by 2030) represents an ambitious attempt to address resource constraints through technological innovation.

The operational efficiency analysis reveals distinct approaches to system optimization. The RAC's highly integrated multimodal operations demonstrate the benefits of what Reggiani et al. (2015) describe as "network resilience through connectivity". The corridor's achievement of 85% intermodal facility coverage at major nodes has significantly enhanced operational flexibility and resource efficiency. The DMIC's planned operational systems, incorporating a comprehensive IoT and AI integration, represent what could be termed "digital-first efficiency optimization". This approach aligns with recent theoretical developments in smart logistics systems (Zhang et al., 2019) and offers potential for superior operational efficiency through predictive management and real-time optimization.

5.8 Skills Development and Capacity Building

The analysis of skills development needs and approaches reveals crucial differences between mature and emerging corridors. The RAC's focus on continuous skill upgrading and technological adaptation supports the emphasis of Bontekoning et al. (2004) on the evolutionary nature of transport workforce development. The corridor's successful implementation of advanced technologies has been supported by systematic professional development programs. The DMIC faces more fundamental challenges in skills development, with a reported 40% shortfall in required technical expertise for smart city operations. This gap highlights what can be termed as the "skills-innovation paradox" in emerging transport corridors, where technological leapfrogging creates immediate demand for advanced technical capabilities.

The potential for knowledge transfer between corridors presents significant opportunities for accelerated capability development. Following the framework developed by Arvis et al. (2018) for transport corridor capacity building, we identify three critical areas for knowledge transfer:

- Technical expertise in sustainable infrastructure development
- Operational management capabilities for complex multimodal systems
- Environmental performance monitoring and optimization skills.

5.9 Future Implications and Strategic Recommendations

The analysis suggests distinct but complementary pathways for sustainable corridor development. For mature corridors like the RAC, the focus should be on what we term "incremental innovation optimization", continuing to enhance efficiency while implementing cutting-edge technologies within established frameworks. This

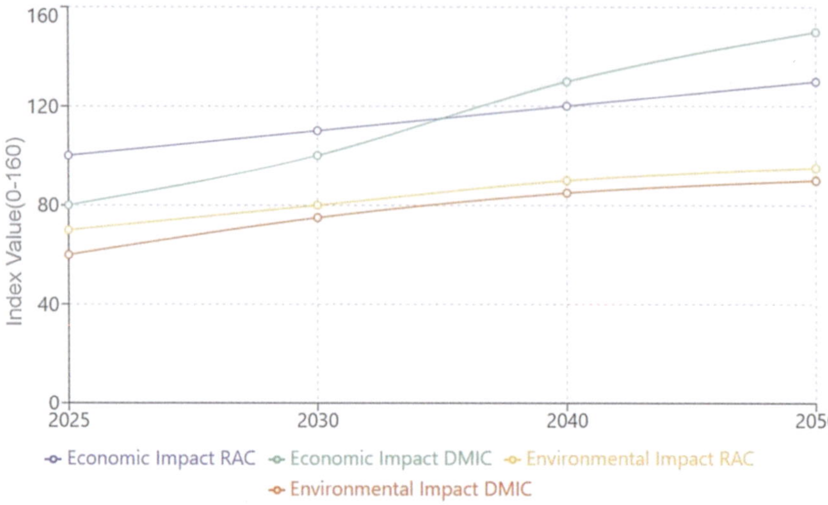

Fig. 3 Line graph showing projected economic and environmental impacts (2025–2050)

approach allows for systematic improvement while maintaining operational stability. For emerging corridors like the DMIC, we recommend a "strategic leapfrogging" approach, carefully selecting areas for technological advancement while ensuring foundational capabilities are established. This balanced approach can help avoid what Ivanov and Dolgui (2020) term "innovation-implementation gaps" in complex transport systems. The long-term implications of different development strategies are analyzed in Fig. 3. This comprehensive visualisation (based on the index values ranging from 0 to 160) demonstrates how economic and environmental impacts are projected to evolve over time, helping stakeholders understand the potential trade-offs and synergies between different development approaches.

Based on the comparative analysis, we propose a comprehensive implementation framework for sustainable corridor development. In particular, the RAC should focus on the following aspects: (1) accelerating the adoption of zero-emission technologies in remaining diesel operations; (2) enhancing cross-border digital integration for seamless operations; (3) developing advanced climate resilience strategies for existing infrastructure; and (4) implementing next-generation traffic management systems. In the meantime, the DMIC should prioritize the following areas: (1) building robust foundations for sustainable operations while implementing advanced technologies; (2) developing comprehensive water management systems; (3) creating innovative funding mechanisms for sustainable infrastructure; and (4) establishing effective stakeholder engagement platforms.

5.10 Synthesis and Future Research Directions

The comparative analysis of the RAC and DMIC provides valuable insights into sustainable transport corridor development across different contexts. The findings suggest that successful sustainable development requires careful balance between technological innovation, institutional capacity, and stakeholder engagement, adapted to specific developmental contexts. The future research should focus on the following areas:

- Long-term sustainability impacts of different development approaches
- Effectiveness of various governance models in promoting sustainable practices
- Role of emerging technologies in enhancing corridor sustainability
- Methods for measuring and optimizing social sustainability in transport corridors.

5.11 Global Implications

The conducted analysis demonstrates that while sustainable transport corridor development presents significant challenges, innovative approaches adapted to local contexts can yield substantial benefits. The experiences of both the RAC and DMIC offer complementary insights for global transport infrastructure development, particularly in balancing economic growth with environmental sustainability. As global trade continues to expand and climate change concerns intensify, the development of sustainable transport corridors becomes increasingly crucial. The lessons learned from these case studies provide valuable guidance for policymakers and practitioners worldwide, highlighting the importance of context-specific solutions while maintaining focus on universal sustainability principles. The success of future transport corridor development will depend on the ability to adapt these lessons to local conditions while maintaining the sight of global sustainability goals. This requires flexible yet robust frameworks that can accommodate both rapid technological change and varying developmental contexts.

6 Conclusions

6.1 Synthesis of Environmental Performance Analysis

This study provides a comprehensive assessment of environmental sustainability in transport corridors through detailed comparison of the Rhine-Alpine Corridor (RAC) and Delhi-Mumbai Industrial Corridor (DMIC). The analysis reveals that successful environmental management in transport infrastructure requires careful consideration of developmental context while maintaining focus on core sustainability principles. Our research demonstrates that both mature and emerging corridors can achieve significant environmental improvements through strategic approaches tailored to their specific circumstances. The findings indicate that environmental sustainability in transport corridors encompasses multiple interconnected dimensions. The RAC's experience shows how systematic implementation of environmental measures can transform existing infrastructure operations, achieving substantial reductions in environmental impact while maintaining operational efficiency. The corridor's success in reducing carbon emissions by 25% while increasing freight capacity demonstrates that environmental protection and economic performance can be mutually reinforcing.

The DMIC's innovative approach reveals how emerging corridors can potentially achieve superior environmental performance through comprehensive planning and advanced technology integration. By incorporating sustainability principles from inception and leveraging cutting-edge technologies, the DMIC establishes new paradigms for environmental management in transport infrastructure. This finding has profound implications for future transport corridor development globally.

6.2 *Key Contributions to Environmental Management Practice*

Our research makes several significant contributions to the understanding of environmental sustainability in transport corridors. First, it establishes a comprehensive framework for comparing environmental performance across corridors at different developmental stages. This framework enables meaningful assessment of environmental initiatives while accounting for varying contextual factors and operational requirements. Second, the study identifies critical success factors for environmental management in transport corridors. These include the integration of environmental considerations into core planning processes, development of robust monitoring systems, and effective stakeholder engagement mechanisms. The comparative analysis demonstrates how these factors manifest differently in mature and emerging corridors while maintaining their fundamental importance for environmental performance. Third, our research provides practical guidance for enhancing environmental sustainability in transport corridor development. The identification of transferable best practices and potential implementation pathways offers valuable insights for policymakers and practitioners working to develop environmentally sustainable transport infrastructure.

6.3 *Future Directions in Environmental Sustainability*

The environmental challenges facing transport corridors will continue to evolve, demanding innovative solutions and adaptive management approaches. Climate change emerges as a central concern that will reshape how we develop and operate transport infrastructure. Our analysis suggests three critical areas that will define the future of environmental sustainability in transport corridors. The first involves the transition to regenerative infrastructure design. Traditional approaches focusing solely on impact minimization are giving way to more ambitious goals of positive environmental contribution. The DMIC's integration of renewable energy generation and ecosystem enhancement demonstrates how transport corridors can become net contributors to environmental quality rather than simply minimizing harm. This evolution represents a fundamental shift in how we conceptualize the relationship between infrastructure and natural systems.

Second, the integration of advanced environmental technologies will increasingly shape corridor performance. The emergence of AI-driven environmental management systems, coupled with comprehensive sensor networks and predictive analytics, enables more sophisticated approaches to environmental protection. The RAC's success in implementing advanced monitoring systems, combined with the DMIC's ambitious plans for technological integration, indicates the transformative potential of these technologies when properly implemented. Third, climate resilience will become increasingly central to corridor development. Both corridors demonstrate evolving approaches to climate adaptation, from the RAC's systematic upgrading of existing infrastructure to the DMIC's integration of climate considerations in initial design. Future corridor development must embrace comprehensive climate resilience strategies that address both current impacts and projected future conditions.

6.4 Recommendations for Environmental Enhancement

Our research suggests several key recommendations for enhancing environmental sustainability in transport corridors. These recommendations acknowledge the different challenges faced by mature and emerging corridors while maintaining focus on core sustainability principles. For established corridors, we recommend prioritizing systematic infrastructure modernization with emphasis on environmental performance. This includes upgrading monitoring systems, enhancing energy efficiency, and strengthening climate resilience. The RAC's experience demonstrates the importance of maintaining continuous improvement in environmental performance even in mature systems. For emerging corridors, we emphasise the importance of comprehensive environmental planning from inception. The DMIC's approach shows how early integration of sustainability considerations can enable superior environmental performance. This includes developing robust environmental baseline data, implementing advanced monitoring systems, and creating effective stakeholder engagement mechanisms.

6.5 Implications for Global Infrastructure Development

The findings from this research have significant implications for global transport infrastructure development. As nations worldwide invest in new transport corridors, the experiences of the RAC and DMIC offer valuable lessons for achieving environmental sustainability. Our analysis demonstrates that while pathways to sustainability may vary, successful environmental management requires strong institutional commitment, adequate resource allocation, and effective stakeholder engagement. The transfer of environmental best practices between mature and emerging corridors requires careful consideration of local contexts. However, our research identifies several universal principles that support successful environmental outcomes, including comprehensive monitoring systems, integrated planning approaches, and adaptive management capabilities.

6.6 Final Reflections

The development of environmentally sustainable transport corridors represents one of the most significant challenges in modern infrastructure development. Our comparative analysis demonstrates that while this challenge is complex, it is not insurmountable. Through strategic planning, technological innovation, and committed implementation, transport corridors can achieve significant environmental improvements while maintaining operational efficiency. The experiences of the RAC and DMIC show that different pathways to environmental sustainability exist, each with its own advantages and challenges. The key to success lies in understanding these differences while maintaining focus on core sustainability principles. As global trade continues to grow and environmental pressures intensify, the lessons learned from these corridors will become increasingly valuable for developing transport infrastructure that serves both economic needs and environmental imperatives.

References

Abhijith, K. V., Kumar, P., Gallagher, J., McNabola, A., Baldauf, R., Pilla, F., Broderick, B., Di Sabatino, S., & Pulvirenti, B. (2017). Air pollution abatement performances of green infrastructure in open road and built-up street canyon environments–A review. *Atmospheric Environment, 162*, 71–86.

Adams, R., Jeanrenaud, S., Bessant, J., Denyer, D., & Overy, P. (2016). Sustainability-oriented innovation: A systematic review. *International Journal of Management Reviews, 18*(2), 180–205.

Aijaz, R., & Hoelscher, K. (2015). India's smart cities mission: An assessment. *ORF Issue Brief, 124*, 1–12.

Arnold, J. (2006). *Best practices in management of international trade corridors*. The World Bank.

Arvidsson, N., & Browne, M. (2013). A review of the success and failure of tram systems to carry urban freight: the implications for a low emission intermodal solution using electric vehicles on trams. *European Transport\Trasporti Europei, 54*, 1–18.

Arvis, J. F., Ojala, L., Wiederer, C., Shepherd, B., Raj, A., Dairabayeva, K., & Kiiski, T. (2018). *Connecting to compete 2018: Trade logistics in the global economy*. World Bank.

Bai, X., Dawson, R. J., Ürge-Vorsatz, D., Delgado, G. C., Barau, A. S., Dhakal, S., Dodman, D., Leonardsen, L., Masson-Delmotte, V., Roberts, D. C., & Schultz, S. (2018). Six research priorities for cities and climate change. *Nature, 555*(7694), 23–25.

Banister, D. (2008). The sustainable mobility paradigm. *Transport Policy, 15*(2), 73–80.

Black, W. R. (2010). *Sustainable transportation: Problems and solutions*. Guilford Press.

Bontekoning, Y. M., Macharis, C., & Trip, J. J. (2004). Is a new applied transportation research field emerging? A review of intermodal rail-truck freight transport literature. *Transportation Research Part a: Policy and Practice, 38*(1), 1–34.

Christodoulou, A., & Demirel, H. (2018). *Impacts of climate change on transport: A focus on airports, seaports and inland waterways*. Publications Office of the European Union.

Clausen, U., De Bock, J., & Lu, M. (2019). Logistics trends, challenges, and needs for further research and innovation. In: *Sustainable and efficient transport*. Emerald Publishing Limited.

Commission, E. (2021). *Trans-European Transport Network (TEN-T)*. European Commission.

DMICDC. (2023). *DMIC sustainability report*. Delhi Mumbai Industrial Corridor Development Corporation.

European Environment Agency. (2020). *Environmental noise in Europe—2020*. EEA Report No 22/2019. Publications Office of the European Union.

European Rail Agency. (2023). *ERTMS deployment progress report*. ERA.

European Investment Bank. (2023). *Investment Report 2023: Accelerating the twin transitions*. EIB.

European Environment Agency. (2023). *Transport and Environment Reporting Mechanism (TERM) report*. EEA.

Goldman, T., & Gorham, R. (2006). Sustainable urban transport: Four innovative directions. *Technology in Society, 28*(1–2), 261–273.

Grigoratos, T., Fontaras, G., Giechaskiel, B., & Zacharof, N. (2019). Real world emissions performance of heavy-duty Euro VI diesel vehicles. *Atmospheric Environment, 201*, 348–359.

Hollos, D., Blome, C., & Foerstl, K. (2012). Does sustainable supplier co-operation affect performance? Examining implications for the triple bottom line. *International Journal of Production Research, 50*(11), 2968–2986.

International Energy Agency (IEA). (2020). *CO2 Emissions from fuel combustion: Overview*. IEA.

International Energy Agency (IEA). (2021). *Net Zero by 2050: A roadmap for the global energy sector*. IEA.

IPCC. (2022). Climate change 2022: Impacts, adaptation and vulnerability. In: *Contribution of working group II to the sixth assessment report of the intergovernmental panel on climate change*. Cambridge University Press.

Indian Green Building Council. (2023). *Green Building standards for industrial corridors*. IGBC.

Ivanov, D., & Dolgui, A. (2020). Viability of intertwined supply networks: extending the supply chain resilience angles towards survivability. A position paper motivated by COVID-19 outbreak. *International Journal of Production Research, 58*(10), 2904–2915.

Jordan, A., & Lenschow, A. (2010). Environmental policy integration: A state-of-the-art review. *Environmental Policy and Governance, 20*(1), 147–158.

Kunaka, C., & Carruthers, R. (2014). *Trade and transport corridor management toolkit*. World Bank.

McKinnon, A. (2018). *Decarbonising logistics: Distributing goods in a low carbon world*. Kogan Page Publishers.

Mol, A. P. J., Sonnenfeld, D. A., & Spaargaren, G. (2009). *The ecological modernisation reader: Environmental reform in theory and practice*. Routledge.

Nayak, N. (2015). Displaced by development: A case study of the Delhi-Mumbai industrial corridor. *Economic and Political Weekly, 50*(49), 59–67.

NITI Aayog. (2023). *Transport report for India*. National Institution for Transforming India, Government of India.

Perkins, R. (2003). Environmental leapfrogging in developing countries: A critical assessment and reconstruction. *Natural Resources Forum, 27*(3), 177–188.

Port of Rotterdam. (2023). *Innovation report*. Port of Rotterdam Authority.

Quium, A. A. (2019). Transport corridors for wider socio–economic development. *Sustainability, 11*(19), 5248.

Reggiani, A., Nijkamp, P., & Lanzi, D. (2015). Transport resilience and vulnerability: The role of connectivity. *Transportation Research Part A: Policy and Practice, 81*, 4–15.

Regmi, M. B., & Hanaoka, S. (2012). Assessment of intermodal transport corridors: Cases from North-East and Central Asia. *Research in Transportation Business & Management, 5*, 27–37.

Rhine-Alpine EGTC. (2022). *Annual report 2021*. Rhine-Alpine EGTC.

Sims, R., Schaeffer, R., Creutzig, F., Cruz-Núñez, X., D'Agosto, M., Dimitriu, D., Figueroa Meza, M. J., Fulton, L., Kobayashi, S., Lah, O., McKinnon, A., Newman, P., Ouyang, M., Schauer, J. J., Sperling, D., & Tiwari, G. (2014). Transport. In: *Climate change 2014: Mitigation of climate change. Contribution of working group III to the fifth assessment report of the intergovernmental panel on climate change*. Cambridge University Press.

UIC. (2020). *Environmental report*. International Union of Railways.

Witte, P., Wiegmans, B., van Oort, F., & Spit, T. (2019). Governing inland ports: A multi-dimensional approach to addressing inland port-city challenges in European transport corridors. *Journal of Transport Geography, 78*, 149–161.

Woodburn, A., Allen, J., Browne, M., & Leonardi, J. (2019). *The impacts of globalisation on international road and rail freight transport activity: Past trends and future perspectives*. International Transport Forum Discussion Papers, 2019/07, OECD Publishing, Paris.

Zhang, Y., Ying, W., Hassan, G., Iftikhar, M., & Li, R. (2019). Industrial development, trade and knowledge transfer: Lessons from China's integrated transport network construction. *China Economic Journal, 12*(3), 303–320.

Bharadwaj Kasula is a passionate supply chain and operations professional who has recently finished his master's degree at Aston University (UK) in Supply Chain Management with distinction. His research focuses on sustainable transport infrastructure and environmental management in industrial corridors, with particular emphasis on comparing mature and emerging transport systems. Working closely with Dr. Marin Marinov, SFHEA, Senior Lecturer and EM Research Lead at Aston University, Kasula has conducted a groundbreaking comparative analysis of the Rhine-Alpine Corridor and Delhi-Mumbai Industrial Corridor, examining their approaches to environmental sustainability and climate change mitigation. His research provides valuable insights into how transport corridors can effectively balance operational efficiency with environmental protection across different developmental contexts. Kasula brings a unique perspective to transport sustainability research, combining academic excellence with practical industry experience. His professional background includes roles focusing on operational efficiency and environmental management at NxtWave Disruptive Technologies and Swadeshi Energy Solutions in India, where he implemented strategic sustainability initiatives achieving significant improvements in operational performance. As a member of both the Chartered Institute of Procurement and Supply Chain (CIPS) and the Chartered Management Institute (CMI), Kasula actively contributes to advancing sustainable practices in transport and logistics. His involvement with Enactus Aston's environmental initiatives, particularly the "A Drop in the Ocean" project, demonstrates his commitment to practical sustainability solutions. This combination of academic research and practical experience enables him to provide valuable insights into the development of sustainable transport infrastructure globally.

Marin Marinov is currently employed by Aston University (UK). He works within the Department of Engineering Management in the College of Engineering and Physical Sciences. He is a member of the department's management team, responsible for the research portfolio of the department. His educational background includes Technology, Organization, and Management of Train Movement. His research encompasses rail operations, freight transport, systems design, network performances, timetabling, planning, metro systems, urban freight, logistics and rail education. Recent research outputs on safety at level crossings, baggage handling, evaluation of metro systems performance, and innovating rail higher education have been published in SIMPAT (IF 5.35), Sustainability (IF 3.9), URRT (IF 1.8), PROMET (IF 1.10) Sustainable Rail Transport (book series), and Rail Engineering (IF 0.85). Current metrics of the research output include: 85 ORCiD validated works; Google Scholar: 162 registered works, 1776 citations to date; H-index: 22, i10-index: 39. Dr. Marinov serves as a reviewer for a number of subject-specific peer-reviewed journals, contributing to rail freight development and transport systems innovation. He is often acknowledged by Springer for his contributions as principal editor of Sustainable Rail Transport volumes. Dr. Marinov was a Senior Research Associate with NewRail, responsible for a Ph.D. opportunity scheme in Organization, Management, and Economics of Train Movement and Rail Systems Design. During his time at the NewRail's research center, Dr. Marinov had developed innovative concepts for rail freight and logistics-oriented services as well as new curriculum designs for modernizing rail higher education. Prior to his engagements at New Rail, Dr. Marinov had worked as a teaching assistant within the Transportation Systems branch of the MIT/Portugal program, where he lectured on Transportation Planning and Performance Assessment, Freight Transport and Logistics Processes as well as Rail Engineering.

Enhancing Maritime Supply Chain Resilience to Climate Change in the Greater Bay Area and Southeast Asia Region

Yui-Yip Lau and Qiong Chen

Abstract The maritime supply chain has faced unfolded challenges due to the COVID-19 pandemic, climate change, geopolitical conflicts, ever-changing trade routes and freight patterns, and keen competition. In particular, the Greater Bay Area and Southeast Asia region are identified as emerging areas that are located near the coastal area and associated with loose environmental regulations and the rapid growth of industrialization. As such, the maritime supply chain may face vulnerability that induces severe disturbance arising from maritime supply chain risks and adversely impacts the maritime supply chain's ability to deliver to the end customer market effectively. There is a need for a resilient maritime supply chain in the Greater Bay Area and Southeast Asia region. This chapter aims to conduct a systematic review on maritime supply chain resilience from 1997 to 2023. The systematic review identifies the most influential scholars, countries, and institutions working on this topic. Based on the systematic review findings, we can address the urgency of maritime supply chain resilience. In the meantime, there is an increasing trend of academic scholars who establish interdisciplinary research teams to generate impactful research outputs in response to the complex business environment. Moreover, a series of in-depth, semi-structured interviews are conducted with industrial practitioners, policymakers, logistics associations, researchers, and environmental representatives to investigate the key challenges of the maritime supply chain in the Greater Bay Area and Southeast Asia region, recognize port infrastructure disruptions because of climate change, and point out how to develop resilience-building capacities (i.e., absorptive, recovery, and adaptive).

Keywords Greater bay area · Southeast Asia region · Resilient maritime supply chain · Systematic review · Maritime supply chain risks

Y.-Y. Lau (✉)
Division of Business and Hospitality Management, School of Professional Education and Executive Development, The Hong Kong Polytechnic University, Yau Ma Tei, Kowloon, Hong Kong
e-mail: yuiyip.lau@cpce-polyu.edu.hk

Q. Chen
Navigation College, Jimei University, Xiamen 361001, China
e-mail: qchen@jmu.edu.cn

M. A. Dulebenets et al. (eds.), *Transport and Climate Change*,
https://doi.org/10.1007/978-3-032-00563-2_16

1 Introduction

Resilience was conceptually identified by Holling (1973) as "a measure of the ability of ecological systems to absorb changes of state variables, driving variables, and parameters, and persist". The notion of resilience has been widely adopted to explain the adaptive capacities of societies, human communities, and individuals. In the past decades, resilience has become a notable idea in the glossary of sustainable development and disaster risk management (Tiernan et al., 2019). Indeed, the rising number of high-impact epidemics has fundamentally changed the research direction from conventional risk-based methods to resilience-based approaches. Fekete et al. (2014) and Linkov et al. (2014) identified science trends, policy trends, and paradigm shifts as the reasons for investigating resilience and risk management. Resilience and risk management are urgently demanded because organizations and systems become more complicated, and the effects of incidents increase (Hollnagel, 2008; Johnsen, 2010). Moreover, Blockley et al. (2012) indicated that "resilience is logically regarded as necessary but not sufficient for sustainability, which implies a stricter requirement needed to achieve the sustainability of a system". This is crucial to the long-term development of economic and social systems under the associated and changeable international context. By defining resilience as planning and preparing for, absorbing, recovering from, and adapting to changes, this chapter aims to systematically explore the level of resilience of the ports and the maritime supply chain in the face of the pandemic while the memory is still fresh. To our knowledge, there is a lack of research on the resilience of ports and the maritime supply chain in the face of the pandemic, including COVID-19. Thus, this study would facilitate knowledge mobilization between geographies in response to the pandemic for future collaborations between them. Moreover, maritime resilience to other disruptions, such as climate change, will be investigated as well.

In general, the previous research on the integration of ports in the maritime supply chain has been understudied. Ports perform as nodes in the maritime supply chains across the globe. Ports provide different functions and are connected with intermodal transport networks, transformation processes, cargo handling, semi-industrial and industrial functions (e.g., recycling and assembly activities), storage, and distribution activities. Ports are regarded as one of the spatial clusters of firms, where various maritime supply chain operators participate in creating added value for the final consumers. Therefore, the associations between focal firms and ports have been identified (Notteboom et al., 2020; Song & Panayides, 2008). To a certain extent, the integration of ports in the maritime supply chain facilitates port development. Maritime supply chain actors can be at a higher level of cooperation and coordination, notably during various disruptive events, such as COVID-19. Also, ports are given a valuable opportunity to change themselves within an integrated and seamless maritime supply chain and improve their resilience to disruptions (Ng & Liu, 2014).

A significant portion of the existing supply chain literature on pandemics concentrates on the medical supply chain itself (Golan et al., 2020). Also, most works in infrastructure systems address 'system vulnerability' instead of 'system resilience', such as absorption, recovery, adaptation, and transformation (Gu & Liu, 2023; Hosseini & Kash, 2016; Lau et al., 2024). In addition, many do not illustrate the complex nature of resilience. In a more comprehensive definition, resilient infrastructure or organization's resilience management effectiveness depends on its ability to anticipate, absorb, adapt, and quickly recover from a potentially disruptive event (NIAC, 2009). There should be a particular focus on the resilience of the port and maritime supply chain as a whole. Leveraging the holistic outlook on the problem, interrelations among people, processes, systems, and external events can be fully captured, and vulnerable nodes can be identified based on their weight in the network. Consequently, resilience-building measures (risk prevention and reduction controls) can be evaluated and prioritized.

In this research, the Greater Bay Area and ASEAN playing a significant role in global maritime supply chains are taken into consideration. The Greater Bay Area was first addressed in the 13th Five-Year Plan (2016–2020) in 2016. Generally, the Greater Bay Area consists of nine cities in Guangdong Provinces from the Pearl River Delta and two special administrative regions (Hong Kong and Macao). It generates the largest city cluster globally (Ma et al., 2019; Zhou et al., 2018). The increase in connectivity in the Greater Bay Area fosters people, goods, and capital flows, to name a few. Greater Bay Area ports are strategic intermodal connectors for the global maritime supply chain with a GDP contribution of more than USD 1.6 billion and port container throughput of more than 67 million TEUs (HKTDC, 2020).

Additionally, several Southeast Asian countries were revitalized after obtaining independence from colonization starting in the 1950s (Keling et al., 2011). Southeast Asian countries have rapidly expanded since then. Finally, Southeast Asian countries were set up on 8 August 1967 in Bangkok, Thailand (ASEAN, 2021). In response to the significant global economic development and regional integration over the last few decades, the leaders of China and ASEAN are inclined towards creating an ASEAN-China Free Trade Area (Tongzon, 2005). Indeed, Foo et al. (2020) pointed out that President Xi Jinping developed the Belt and Road Initiative as a new global Chinese strategy in 2013. The establishment of the Belt and Road Initiative aims to build infrastructure, increase trading activities, and enhance connectivity between China and ASEAN countries. As such, ASEAN ports are vital players supporting the global maritime supply chain with a substantial GDP contribution (Statista, 2021) and port container throughput of 105.6 million TEUs (The World Bank, 2021). It is important to note that the ports taken into consideration in this research are among the world's busiest ports with direct connections to the major global and regional markets. Besides, the present analysis would be especially useful for a country in China and ASEAN, which has faced major challenges during the COVID-19 pandemic and experience climate change effects as well.

In general, this book chapter is split into four main sections. Section 1 provides the research context, settings, and the development of the Greater Bay Area and Southeast Asia region. Section 2 conducts a bibliometric analysis to gain strategic insights into active countries, authors, and academic institutions exploring the maritime resilience topic. Section 3 highlights the discussion points from the interviewee's perspective. Section 4 addresses the key findings and proposes the future research directions.

2 Bibliometric Analysis

Bibliometrics is described as an interdisciplinary area that combines mathematics, statistics, and linguistics. Its interpretation tracks research development and addresses related key trends in different disciplines. This approach is usually adopted to investigate the evolution of academic research areas, identify existing 'hot' issues of research, and forecast forthcoming research trends. To thoroughly explore the association between resilience and maritime transport, this study employed different data analysis tools, including Python and VOSviewer, to extensively filter the related Web of Science literature. As such, we set the search conditions in the following manner: the subject is 'Resilience' and 'Port' OR 'Resilience' and 'Maritime' OR 'Resilience' and 'Shipping' OR 'Resilience' and 'Ship' OR 'Resilience' and 'Water Transport' OR 'Resilience' and 'Container'. The analysis period was set from 1997 to 2023. The search outcomes were sorted and de-duplicated to identify the related literature. As of 20 March 2023, the total number of valid research articles for our study comprised 735.

The evolution of maritime transport resilience can be evaluated through the review of publications over the past 27 years. In general, we can divide it into two main stages. In the first stage (i.e., 1997–2009), a total of 27 research papers were published. In other words, 2 research papers were published on average yearly, representing 3.67% of the total number of research papers. In the second stage (i.e., after 2010), 708 research papers were published. Around 50 research papers were published on average yearly, representing 96.32% of the total number of research papers. The number of research papers has grown dramatically since 2010. The leading international maritime journals have published various research papers on maritime transport resilience, including *Ocean and Coastal Management*, *Maritime Policy and Management*, *Sustainability*, *Marine Policy*, and *Reliability Engineering and System Safety*. These journals have published almost 38% of the research papers related to maritime transport resilience. Interestingly, the number of research papers associated with maritime transport resilience has even more significantly risen since 2019. The driving force can be explained by the emergence of the COVID-19 pandemic and led to unexpected challenges in the maritime industry (e.g., lack of main personnel and truck drivers, freight rates, crew change services, shortage of containers, cargo handling delay, etc.). Also, a series of incidents like geopolitical implications of the Russia-Ukraine war, the Suez Canal blockage, China-US trade conflicts, and extreme weather changes have all induced rising freight rates, shipment cessation, port congestion, and interruption of maritime transport. Such incidents caused an increase in the maritime transport resilience research publications after 2010 (Yu et al., 2023).

2.1 Analysis of Countries and Authors

The co-occurrence network may give a constructive idea for assessing the impact of academics and exploring scholars' outstanding awareness. The size of the network nodes indicates the number of research articles published by the author with the links reflecting the internal association between them. The key author group and their collaborative associations in this discipline can be affected by examining authors' structural features and cooperative networks. This research study employed the VOSviewer software for the archive documents that were collected. By implying the limitations on node connections (a minimum of one collaboration with another author, and the minimum number of author occurrences is two), we identified 192 authors who fulfilled these criteria.

Figure 1 shows the academic cooperation between different authors in the form of a heat map. In this picture, the color represents the frequency of cooperation between the authors, from yellow to green to blue, and from strong to weak. The size of the highlights is the concentrated reflection of their research content. It can be observed that Becker A. and Yang Z. participated in a large number of research studies in this discipline. Also, sufficient collaborations among scholars can be noticed, with every scholar establishing their research team to explore particular research areas. Nevertheless, the connections between certain research teams are not apparent, and there are limited academic exchanges.

The countries that published research papers in this discipline were investigated as well (see Fig. 2). Figure 2 represents the academic cooperation network diagram. The nodes in the graph represent the country, the edges represent the cooperation relationship, and the color represents the research teams working together. Scholars

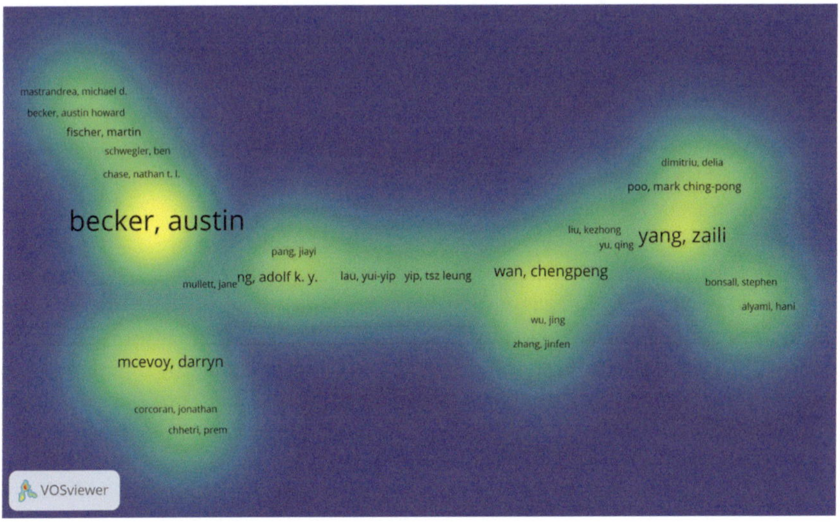

Fig. 1 Map of authors. *Source* Authors

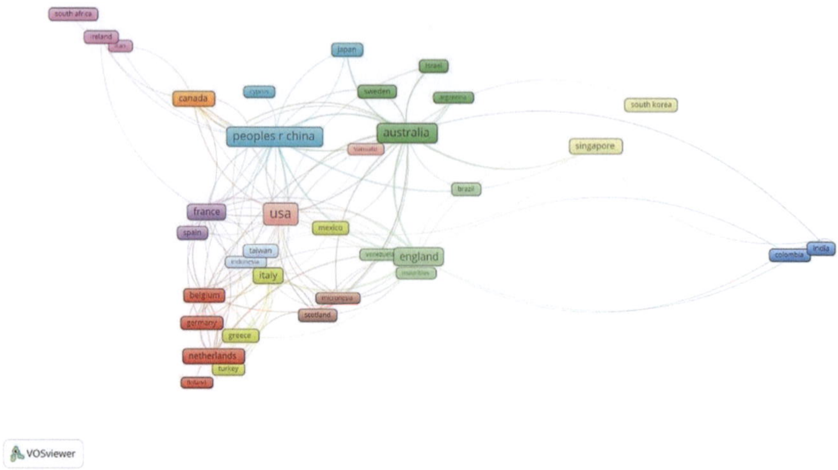

Fig. 2 Map of author countries. *Source* Authors

from the United States published most of the research articles, followed by China and the United Kingdom. According to Fig. 2, a frequent collaboration between countries belonging not only to the same continent but inter-continental collaborations can be observed as well.

2.2 *Analysis of Academic Institutions*

Python was used as a tool to analyze the relevant academic institutions that have been active in performing research related to maritime resilience. Figure 3 shows the world-leading academic institutions for maritime transport resilience research. In Fig. 3, the nodes correspond to particular academic institutions, whereas colors and links denote collaborative relationships between these institutions. Among these institutions, Dalian Maritime University is ranked the first to publish the most research articles (23 articles), followed by Griffith University, Nanyang Technology University, and Liverpool John Moores University. In general, the distinctive characteristics of the aforementioned academic institutions are associated with renowned scholars who are leading academic research on maritime resilience. They address the key problems that need interpretation, contribute considerable time and efforts to evolve research processes and focus on related policies and global topics. Additionally, these academic institutions are mainly maritime universities and colleges, remarkably connected with maritime affairs. Furthermore, the cities in which the schools are situated generally have major ports for export and import trading activities.

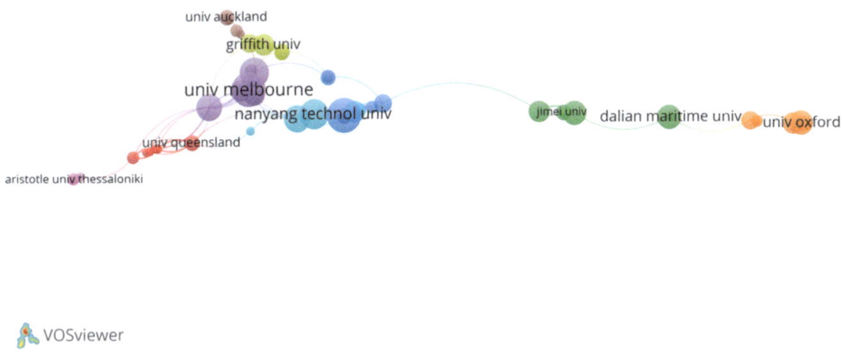

Fig. 3 Map of academic institutions. *Source* Authors

3 Interviews

3.1 Methodology

In the second part of this study, we conducted semi-structured, in-depth interviews with five relevant stakeholders, including logistics associations, environmentalists, and industrial practitioners (i.e., terminal operators, logistics firms, and transport operators), in October 2023. The participants were selected based on the following principles: (1) they have accumulated at least eight years of working experience in the maritime industry; (2) they have not only acquired the relevant bachelor's degree but also the professional qualifications from recognized bodies; and (3) they have a managerial position in a company. To ensure diverse perspectives, we selected two suitable participants from each relevant stakeholder group with a mixture of different age groups and departments. Based on the interviewees' availability, we arranged suitable timeslots for the participants to join the focus group interview. In addition, the interview contents focused on (1) the disruptions of the port infrastructure due to climate change; and (2) the resilience-building capacities (i.e., absorptive, recovery, and adaptive) to help ports to get through climate change. To align with the research objectives, the interview questions were based on previous research studies (Gu & Liu, 2023; Lau et al., 2024) to design the interview questions accordingly. The following interview questions were used:

Q1. How does climate change adversely impact port infrastructure?

Q2. What are the key elements of building a resilience assessment framework for port infrastructure systems in the face of climate change?

Q3. From your perspective, how can the technical resilience of port infrastructure be improved in response to climate change?

Q4. From your perspective, how can the organizational resilience of port infrastructure be improved in response to climate change?

Q5. From your perspective, how can the social resilience of port infrastructure be improved in response to climate change?

Q6. From your perspective, how can the economic resilience of port infrastructure be improved in response to climate change?

3.2 Interview Results

In this study, we mainly used the content analysis to summarize the key points from different stakeholders. As such, we point out the main observations in this section. The interviewees addressed the disruptions of port infrastructure due to climate change as follows.

- Environmentalists are concerned about damage to nearby ecosystems. Marine ecosystems and marine biodiversity will be adversely affected by climate change. Some endangered animals in the ocean will potentially disappear during our lifetime.
- Terminal operators are concerned about the increased risk of ship collisions due to severe weather events and other reasons in port areas, damage to port infrastructure, power outages, and increased maintenance costs. These unfavorable factors bring the financial burden for terminal operators to upgrade the facilities, increase the administrative work to improve the navigation safety in the terminal area, and motivate the shipping lines to call the port.
- Logistics associations, logistics firms, and transport operators expressed concerns about reduced navigational safety in port areas, increased downtime and delays, reduced cargo handling capacity, and disruption of maritime supply chains. Indeed, the overall maritime supply chain will suffer from different factors, for instance, increased supply uncertainty, rise in operating costs (e.g., insurance and freight rate), decreased customer service level because of shipment delays, redesign of the production workflow and schedules, and increased safety stock level to deal with unforeseeable or uncertain situations (Lau et al., 2023).

Disruptions caused by climate change threaten port infrastructure, trade networks, global trade, and coastal inundation. The firms will find it more difficult to design and implement the maritime supply chain networks in the forthcoming years. In response, the interviewees proposed some measures from the absorptive, recovery, and adaptive capacity perspectives to encounter the unfolded challenges. Regarding the absorptive capacity perspective, it was recommended to build close communication networks, establish an electronic exchange platform, use advanced equipment, recruit skilled labor with professional qualifications, and form a skillful emergency response team. Concerning the recovery capacity perspective, it was suggested to implement a crisis management policy, develop a business continuity plan, facilitate port innovation, establish port safety management, and sustain maintenance and reliability. Taking into consideration the adaptive capacity perspective, it was recommended to promote data interoperability/standardization in the future, construct a port community system in the following years, develop service restoration, explore various technology restoration alternatives, establish cyber port infrastructure, strengthen worldwide port network, and improve hinterland connections.

4 Closing Discussion and Concluding Remarks

This book chapter provides a systematic review of maritime supply chain resilience from 1997 to 2023. The systematic review identifies the most influential scholars, countries, and institutions working on the topic of maritime supply chain resilience. Based on the review results, we can observe the urgency of maritime supply chain resilience starting from 2010 and a further boost in the number of publications since 2019. This implies that the business environment becomes more complicated due to natural disasters, climate change, and human-made disasters. In response, academic scholars may form interdisciplinary research teams to produce impactful research outputs for the local community, industry stakeholders, environmentalists, and logistics associations and highlight the complex nature of maritime supply chain resilience. Furthermore, a comprehensive exploration, leading to a thorough understanding of the socioeconomic impacts of disruptions in the maritime supply chain and the integration of state-of-the-art technology to enhance resilience, is encouraged (Lau et al., 2024).

Through a series of semi-structured, in-depth interviews, we can identify the disruptions of the port infrastructure due to climate change and point out how to build up the resilience-building capacities (i.e., absorptive, recovery, and adaptive) to help ports to address climate change issues. The valuable insights obtained during the conducted interviews may help to develop a theoretical framework and propose possible ways to mitigate the negative impacts of climate change in emerging areas and coastal regions.

A synthesis of the study findings brings broader implications for the maritime supply chain. Through the development of a resilient maritime supply chain, policymakers and industrial practitioners (e.g., terminal operators, logistics firms, and transport operators) may come up with effective solutions for addressing disruptions and complicated maritime supply chain situations. The implementation of the resilient maritime supply chain is expected to minimize the associated risks promptly (Al Naimi et al., 2022; Tiernan et al., 2019).

The current study brings the implications of the motivation of interdisciplinary collaborations due to researchers from different disciplines being aware that they need to collaborate to identify sophisticated methodologies that will foster maritime transport resilience. In addition, the present study aligns with three Sustainable Development Goals (SDGs), especially SDGs 14, 13, and 9. As for SDG 14 ("*Life Below*

Water"), this study supports SDG 14 by identifying environmental challenges in maritime transport, for instance, damage to nearby ecosystems. Concerning SDG 13 ("*Climate Action*"), this study addresses the urgency for sustainable practices to minimize the effects of climate change on maritime supply chains. Regarding SDG 9 ("*Industry, Innovation, and Infrastructure*"), this study points out that the evolution of resilient maritime transport infrastructure via sustainable industrialization and innovation is vital in the forthcoming years. Strengthening resilience in maritime transport may improve maritime management, which can further ensure the preservation and sustainable deployment of marine and ocean resources (Gu & Liu, 2023; Panahi et al., 2022).

Along with important contributions, this study has several limitations that may provide opportunities for the future research directions. Firstly, this research study only focuses on the specific regions in the Greater Bay Area and Southeast Asia leading to a lack of generalization of the study outcomes. As such, we may consider other regions like North America and Europe to perform a more detailed comparative analysis with the next research effort. Secondly, the current study adopts the qualitative research methodology which may encounter a subjective bias. Due to the limited number of participants, we only carried out the content analysis. Thus, we may conduct additional interviews to do a thematic analysis. Also, the interview results are mainly determined by the interviewees' knowledge, experience, and willingness to answer questions. To overcome the weaknesses of research methodology, the research team may carry out quantitative research approaches like optimization, Bayesian network analysis, and fuzzy modelling. The mixed research methodology will offset the limitations of pure qualitative or quantitative research methodology. Thirdly, the present study only considers the environmental and climate change factors influencing the maritime supply chain. To construct a comprehensive maritime supply chain resilience model, we may identify other factors like maritime traffic, economics, and security, to name a few in the next research study.

Acknowledgements The study was supported by the Research Grants Council of the Hong Kong Special Administrative Region, China (UGC/FDS24/B07/22).

References

Al Naimi, M., Faisal, M. N., Sobh, R., & Sabir, L. B. (2022). A systematic mapping review exploring 10 years of research and supply chain resilience and reconfiguration. *International Journal of Logistics Research & Application, 25*(8), 1191–1218.

ASEAN. (2021). https://asean.org/about-asean/member-states/. Accessed on 27 Dec 2021.

Blockley, D., Godfrey, P., & Agarwal, J. (2012). Infrastructure resilience for high-impact low-chance risks. *Proceedings of the ICE—Civil Engineering, 165*(CE6), 13–19.

Fekete, A., Gabriele, H., & Sylvia, K. (2014). Benefits and challenges of resilience and vulnerability for disaster risk management. *International Journal of Disaster Risk Science, 5*(1), 3–20.

Foo, N., Lean, H. H., & Salim, R. (2020). The impact of China's one belt one road initiative on international trade in the ASEAN region. *The North American Journal of Economics and Finance, 54*, Article 101089.

Golan, M. S., Jernegan, L. H., & Linkov, I. (2020). Trends and applications of resilience analytics in supply chain modelling: Systematic literature review in the context of the COVID-19 pandemic. *Environmental Systems and Decisions, 40*, 222–243.

Gu, B., & Liu, J. (2023). A systematic review of resilience in the maritime transport. *International Journal of Logistics Research and Applications.* https://doi.org/10.1080/13675567.2023.2165051

HKTDC. (2020). https://research.hktdc.com/en/article/MzYzMDE5NzQ5. Access on 7 Dec 2020.

Holling, C. S. (1973). Resilience and stability of ecological systems. *Annual Review of Ecology and Systematics, 4*(1), 1–23.

Hollnagel, E., Nemeth, C. P., & Dekker, S. (2008). *Resilience engineering perspectives: Remaining sensitive to the possibility of failure.* Ashgate.

Hosseini, S., & Kash, B. (2016). Modeling infrastructure resilience using Bayesian Networks: A case study of inland waterway ports. *Computers and Industrial Engineering, 93*, 252–266.

Johnsen, S. (2010). Resilience in risk analysis and risk assessment. In T. Moore, & S. Shenoi, (Eds.), *Critical infrastructure protection IV, IFIP AICT 342* (pp. 2015–2027).

Keling, M. F., Md. Som, H., Saludin, M. N., Shuib, M. S., Ajis, M. N. (2011). The development of ASEAN from historical approach. *Asian Social Science, 7*(7), 169–189.

Lau, Y. Y., Chen, Q., Poo, M. C. P., Ng, A. K. Y., & Chung, C. Y. (2024). Maritime transport resilience: A systematic literature review on the current state of the art, research agenda, and future research directions. *Ocean & Coastal Management, 251*, Article 107086.

Lau, Y. Y., Ranjith, P. V., Chan, E. M. H., Kanrak, M., & Varma, A. J. (2023). A new shape of the supply chain during the COVID-19 pandemic. *Foresight, 25*(4), 578–596.

Linkov, I., Bridges, T., Creutzig, F., Decker, J., Fox-Lent, C., Kröger, W., Lambert, J. H., Levermann, A., Montreuil, B., & Nathwani, J. (2014). Changing the resilience paradigm. *Nature Climate Change, 4*(6), 407–409.

Ma, P., Wang, W., Zhang, B., Wang, J., Shi, G., Huang, G., Chen, F., Jiang, L., & Lin, H. (2019). Remotely sensing large-and small-scale ground subsidence: A case study of the Guangdong-Hong Kong-Macao Greater Bay Area of China. *Remote Sensing of Environment, 232*, Article 111282.

Ng, A. K. Y., & Liu, J. J. (2014). *Port-focal logistics and global supply chains.* Palgrave Macmillan.

NIAC. (2009). Critical Infrastructure resilience final report and recommendations. www.dhs.gov/xlibrary/assets/.../niac_critical_infrastructure_resilience.pdf

Notteboom, T., Van Der Lugt, L., Van Saase, N., Sel, S., & Neyens, K. (2020). The role of seaports in green supply chain management: Initiatives, attitudes, and perspectives in Rotterdoom. *Antwerp, North Sea Port, and Zeebrugge, Sustainability, 12*, 1688–1711.

Panahi, R., Gargari, N. S., Lau, Y. Y., & Ng, A. K. Y. (2022). Developing a resilience assessment model for critical infrastructures: The case of port in tackling the impacts posed by the COVID-19 pandemic. *Ocean & Coastal Management, 226*, Article 106240.

Song, D. W., & Panayides, P. M. (2008). Global supply chain and port/terminal: Integration and competitiveness. *Maritime Policy & Management, 35*(1), 73–87.

Statista. (2021). https://www.statista.com/statistics/804307/gross-domestic-product-gdp-per-capita-in-the-asean-countries/. Accessed on 27 Dec 2021.

The World Bank. (2021). https://data.worldbank.org/indicator/. Accessed on 27 Dec 2021

Tiernan, A., Drennan, L., Nalau, J., Onyango, E., Morrissey, L., & Mackey, B. (2019). A review of themes in disaster resilience literature and international practice since 2012. *Policy Design and Practice, 2*(1), 53–74.

Tongzon. (2005). ASEAN-China free trade area: A bane or boon for ASEAN countries. *The World Economy, 28*(2), 191–210.

Yu, H., Gong, Y., & Liu, J. (2023). A systematic literature review on port competitiveness. *International Journal of Logistics and Research Application*. https://doi.org/10.1080/13675567.2023.2187355

Zhou, Y., Shan, Y., Liu, G., & Guan, D. (2018). Emissions and low-carbon development in Guangdong-Hong Kong-Macao Greater Bay Area cities and their surroundings. *Applied Energy, 228*, 1683–1692.

Yui-Yip Lau is a Head of Student Affairs and Senior Lecturer from the School of Professional Education and Executive Development, The Hong Kong Polytechnic University. Until now, he has published more than 400 research papers in international journals and professional magazines, contributed over 25 book chapters, 8 books, and presented numerous papers at international conferences. He has collaborated with scholars from more than 20 countries and regions spreading over five continents on research projects. He has also secured over HK$ 10 million in research grants. Recently, he has been awarded a Certificate of Appreciation by the Institute of Seatransport in recognition of his outstanding performance on research and the Best Paper Award in leading international conferences. Moreover, he has participated in different consultancy projects with various (inter-)governmental organizations, academic institutions, and industrial associations. His research interests are cruise ships, ferries, impacts of climate change, shipping education and training, transport history, sustainability issues, resilient supply chain management, health logistics, human remains and regional development.

Qiong Chen is an Associate Professor at Jimei University and deeply engaged in research on ship pollution emissions, green shipping, and maritime policy. He has taken the lead in presiding over one provincial-level key social science fund and two municipal and department-level funds. In the academic field, he has published 30 scholarly papers in reputable journals, such as Transportation Research Part D, Maritime Policy and Management, and Ocean & Coastal Management. Among these, 26 are SCI/SSCI-indexed papers. Additionally, he has published 14 international conference papers. In terms of academic contributions, he has participated in the compilation of two English-language books. Moreover, he serves as a member of the editorial boards of several foreign journals, including Data in Brief and MethodsX. He is also highly recognized as a reviewer. He reviews for prestigious journals, such as Transportation Research Part D, Maritime Policy & Management, Cities, iScience, and Maritime Business Review. His outstanding review work was acknowledged with the Best Reviewer Award at the 2023 World Transportation Conference (WTC2023).

A Customized Hyperheuristic Algorithm for the Green Berth Allocation and Scheduling Problem with Carbon Dioxide Emission Considerations

Bokang Li, Fatemeh Shekoohi, Payam Afkhami, Razieh Khayamim, and Maxim A. Dulebenets

Abstract Berth allocation and scheduling at marine container terminals (MCTs) are crucial for optimizing port operations and maintaining the efficiency of maritime logistics under the pressures of increasing global trade flows and operational complexities. One of the most concerning challenges associated with berth allocation and scheduling is its environmental impact, particularly the carbon dioxide (CO_2) emissions generated during port operations. This study introduces a Customized Hyperheuristic Algorithm (CHA) to solve a mathematical optimization model formulated for the green berth allocation and scheduling problem, which integrates environmental considerations to minimize both the operational costs and the CO_2 emissions during vessel waiting and handling operations. CHA employs a dynamic selection mechanism that adaptively chooses heuristic operators based on their ongoing performance, enhancing the traditional framework of typical heuristic and metaheuristic algorithms. Detailed computational experiments compared CHA with established optimization techniques, such as CPLEX, ant colony optimization (ACO), simulated annealing (SA), tabu search (TS), and evolutionary algorithm (EA), which were all customized for the specific settings of the problem. It was found that CHA was able to match the objective values returned by CPLEX for small instances. Moreover, CHA outperformed ACO, SA, TS, and traditional EA by up to 8.22%, 6.96%, 9.32%, and 4.34% for large-scale instances in terms of the best values of the objective function

B. Li · F. Shekoohi · P. Afkhami · R. Khayamim · M. A. Dulebenets (✉)
Department of Civil and Environmental Engineering, Florida A&M University-Florida State University (FAMU-FSU) College of Engineering, Tallahassee, FL, USA
e-mail: mdulebenets@eng.famu.fsu.edu

B. Li
e-mail: bl22a@fsu.edu

F. Shekoohi
e-mail: fs24g@fsu.edu

P. Afkhami
e-mail: pa24@fsu.edu

R. Khayamim
e-mail: rk211@fsu.edu

M. A. Dulebenets et al. (eds.), *Transport and Climate Change*,
https://doi.org/10.1007/978-3-032-00563-2_17

recorded over five consecutive replications. These objective function improvements also resulted in lower emissions. Additional experiments presented some insights into port management. This study clearly demonstrated the applicability and effectiveness of CHA for sustainable and environment-friendly berth planning in real-world marine terminal operations.

Keyword Berth allocation and scheduling · Carbon dioxide emissions · Optimization · Hyperheuristic algorithms · Marine container terminals

1 Background

In today's interconnected global economy, maritime transportation is a fundamental pillar, handling the vast majority of international trade—around 90% by volume (Zhou et al., 2023). This mode of transport underpins global supply chains, ensuring the efficient movement of goods across continents. As a result, the smooth functioning of maritime transport is indispensable for the timely product delivery, making ports the critical nodes that facilitate the transfer between sea and land-based transport networks (Li et al., 2023; Verschuur et al., 2022). As global trade volumes continue to grow, ports are experiencing increased pressure to accommodate larger ships and manage the ever-increasing flow of goods efficiently. Port congestion, delays, and suboptimal management of resources can lead to severe disruptions in the global supply chain, resulting in elevated operational costs and delayed deliveries. Thus, port management, particularly in the context of berthing operations, has become a significant challenge that directly impacts both economic and environmental outcomes (Burns, 2018; Moon & Woo, 2014).

The Berth Allocation and Scheduling Problem (BASP) is at the core of efficient marine container terminal (MCT) operations. The primary objective of BASP is to assign incoming vessels to berths and schedule their handling in such a way that minimizes delays, waiting times, and operational costs. However, achieving these objectives simultaneously while considering dynamic practical constraints is computationally demanding, particularly in large-scale port operations where the sheer number of variables and constraints can overwhelm traditional optimization methods. Thus, BASP presents a highly complex combinatorial challenge with numerous dynamic elements (Bierwirth & Meisel, 2010, 2015; Dulebenets, 2023; Tang et al., 2022).

The stochastic nature of vessel arrivals introduces significant uncertainty in planning, as arrival times and cargo volumes can vary considerably, complicating scheduling accuracy. Furthermore, the finite availability of berths necessitates strategic management of these resources, often requiring prioritization among vessels based on urgency or compatibility with specific berths. Varying vessel sizes and service requirements mandate the alignment of physical and operational capabilities of berths and access channels to accommodate specific vessels, adding another layer of complexity to the problem (Song et al., 2019; Xu et al., 2012). These operational constraints require sophisticated strategies to ensure efficient utilization of port resources while optimizing for the key performance metrics, such as waiting times at sea, handling times at berths, and deviations from planned berth allocations (Imai et al., 2003).

One of the increasingly concerning challenges associated with BASP is its environmental impact, particularly the carbon dioxide (CO_2) emissions generated during port operations. Port operations are significant contributors to global CO_2 emissions (Sha et al., 2024; Wang et al., 2024). As such, managing CO_2 emissions in the context of berth allocation is a critical component of modern sustainable port management. Research has demonstrated that the waiting period for vessels is a major source of emissions, as vessels continue to burn fuel while idling, contributing substantially to their overall environmental footprint. The handling operations also add to the problem, with emissions arising from both the vessel's engines and the terminal's cargo-handling equipment. These emissions can vary depending on the intensity of operations, such as the number of quay cranes in use, the duration of container handling, and the specific fuel types involved (Kontovas & Psaraftis, 2011; Tran et al., 2017).

Moreover, deviations from optimal berthing positions can significantly exacerbate emissions, not only through additional vessel maneuvering but also by increasing the operational inefficiencies within the terminal. When vessels are assigned to suboptimal berths, cargo-handling equipment, such as internal transfer trucks, must travel longer distances to load or unload containers, leading to increased fuel consumption (Dulebenets et al., 2017). These vehicles, often diesel-powered, emit more CO_2 as they operate over extended distances. Additionally, terminal tractors and other ground transportation units may need to traverse greater lengths within the port, further contributing to the overall environmental impact.

Given the urgency of global climate goals, including the International Maritime Organization's (IMO's) targets for reducing the carbon intensity of shipping, the environmental impact of BASP cannot be overstated (IMO, 2020). The green berth allocation and scheduling problem is viewed as a complex optimization problem, which integrates CO_2 emission reduction considerations with traditional operational cost minimization, and is an important step toward more sustainable maritime logistics. By addressing both the operational complexity and environmental impact of berth scheduling, research in this topic provides a comprehensive solution to the challenges faced by ports in the modern era. As port operators strive to meet international sustainability standards, optimizing berth allocation and scheduling with CO_2 reduction considerations is a critical avenue for ensuring both economic efficiency and environmental responsibility (Wang et al., 2023; Zhang et al., 2023).

In recent years, hyperheuristic algorithms have been found to be capable of offering significant advantages in tackling complex optimization problems, particularly due to their adaptable, layered approach (Leng et al., 2020; Li et al., 2024; Mosadegh et al., 2020). Generally composed of a two-level heuristic structure, the first level represented by high-level heuristics (HLHs) oversees the selection and combination of low-level heuristics based on performance feedback, dynamically adjusting to improve solution quality. The second level represented by low-level heuristics (LLHs) consists of specific solution strategies that execute targeted actions to refine and enhance candidate solutions. This hierarchical structure enables hyperheuristics to balance exploration and exploitation effectively, providing a flexible and robust framework well-suited to dynamic and large-scale optimization challenges. Unlike hyperheuristics, typical heuristic and metaheuristic algorithms, which have been used in the BASP studies over the years, rely on the same operators during the entire search process, and this may cause some drawbacks in terms of the computational performance (Li et al., 2024).

This paper introduces a novel extension of hyperheuristic applications, aiming to present a feasible pathway and address the berth allocation and scheduling problem with emission considerations. The primary objective of this study is to develop a robust Customized Hyperheuristic Algorithm particularly designed to address the green berth allocation and scheduling problem, with an emphasis on minimizing CO_2 emissions during vessel waiting and handling operations. The adaptive nature of hyperheuristics offers flexible and efficient solutions to the computational complexities of dynamic discrete berth allocation, enabling the model to optimize both operational efficiency and environmental sustainability. The next sections of this chapter are structured as follows. Section 2 presents a thorough review of the most relevant literature on green berth allocation and scheduling problems with a focus on environmental considerations, as well as studies relevant with the application of hyperheuristic algorithms in other research domains. Section 3 defines the investigated decision problem and the proposed mathematical formulation. Section 4 provides a detailed presentation of the proposed hyperheuristic algorithm. Section 5 outlines the computational experiments conducted to evaluate the presented solution approach. Finally, this chapter concludes with a discussion of the results and directions for future research.

2 Literature Review

This section of the manuscript provides an in-depth review of relevant state-of-the-art studies. It specifically addresses two key topics: (1) green berth allocation and scheduling problems with a focus on environmental sustainability, particularly minimizing CO_2 emissions during port operations; and (2) the application of hyperheuristic algorithms in the state-of-the-art research. Furthermore, this section concludes by identifying the key research gaps and how they are addressed by the present study. Interested readers are recommended to refer to classic BASP survey studies by Bierwirth and Meisel (2010) and Bierwirth and Meisel (2015) along with a more recent survey study by Li et al. (2023) for a detailed overview of BASP studies, discussion of optimization models proposed, and important insights regarding the solution algorithms developed.

In 2010, Golias et al. (2010) presented a berth scheduling model to maximize productivity while reducing fuel consumption and emissions for vessels sailing to the following port of call. This model introduced a discrete berthing space approach to minimize vessel idle time, thus supporting environmental objectives by aligning berth assignments with slow steaming practices. Hu et al. (2014) formulated a nonlinear optimization model to address berth and quay crane allocation while considering fuel consumption and emissions. This study highlighted how quay crane allocation affects emissions and port costs, particularly emphasizing emissions produced during vessel wait times at berths. Later in 2017, Venturini et al. (2017) extended the classical BASP to a multi-port model with speed optimization. This model demonstrated that optimizing vessel speed along the route reduced fuel consumption along with the associated emissions, emphasizing the importance of cooperation between terminals and shipping liners for environmental sustainability.

In recent years, more advanced solution approaches have been employed to solve the problem studied herein. Dulebenets et al. (2017) proposed a variety of hybridized genetic algorithms to minimize emissions of carbon dioxide due to container loading and offloading at marine terminals. This study introduced a mixed-integer model for berth scheduling that integrated environmental impacts, demonstrating how emission costs could significantly influence the berth schedule design. Wang et al. (2019) focused on a multi-objective model for berth, quay crane, and yard truck assignments with an emphasis on minimizing carbon emissions. Using Non-Dominated Sorting Genetic Algorithm II (NSGA-II), the study explored trade-offs between emission reductions and service efficiency, showing the impact of operational decisions on port area emissions. Hu (2020) introduced a low-emission berth allocation model that optimized sailing speed and mooring time to reduce emissions. This study applied second-order cone programming to transform nonlinear functions and demonstrated how handling time adjustments could help minimizing emissions during both sailing and mooring.

Jiang et al. (2022) proposed a vessel scheduling and berth allocation model for ports with restricted channels, integrating carbon emissions into the scheduling framework. The model, solved using a double-population genetic algorithm, illustrated how comprehensive scheduling reduced vessel wait times and supported green shipping objectives. Lin et al. (2023) addressed the optimization of berth and yard space allocation in tidal ports under carbon tax policy constraints, focusing on the reduction of carbon emissions near ports. A mathematical model was proposed to minimize the overall costs due to carbon emission, ship waiting, and delayed departures of ships. Furthermore, a heuristic algorithm was introduced for efficient problem-solving. Jiang et al. (2023) explored quay crane and multiple terminal berth combined scheduling at container ports, considering carbon costs under uncertain environmental conditions. A nonlinear mixed-integer optimization formulation was developed that minimized expected total costs. An adaptive genetic algorithm enhanced by simulated annealing was presented as a solution approach to handle uncertainties effectively. This approach significantly reduced operational and carbon costs, optimizing resource utilization across multiple terminals.

Jauhar et al. (2023) introduced an innovative approach to optimize operations by minimizing the handling processes, costs due to demurrage, costs due to anchorage waiting time, costs imposed because of deviations from vessels' preferred positions, along with the costs due to carbon taxation. Two types of solutions were presented utilizing the ILOG CPLEX solver and several metaheuristics, which demonstrated effectiveness in reducing operational and carbon costs while maintaining high efficiency in dynamic and continuous allocation scenarios. Wen et al. (2023) developed a bi-objective fuzzy programming model that considered both economic and environmental sustainability aspects of maritime transport, affected by weather and emission control regulations. The model prioritized operational characteristics, such as two-speed strategies, fuel switching, and the use of LNG, to adapt to emission control areas while maintaining service reliability under uncertain weather conditions. Their findings indicated that slow steaming and strategic ship routing would be critical for reducing GHG emissions and ensuring service reliability.

Jiang et al. (2024) addressed the joint scheduling optimization of berths and cranes to minimize both carbon and operational costs. A nonlinear mixed-integer optimization formulation was proposed and solved by an enhanced adaptive spiral flying dung beetle algorithm, showing that a reasonable carbon tax policy effectively reduced carbon emissions at ports while maintaining operational efficiency. Wen et al. (2024) presented a multi-objective optimization model for ship scheduling that addresses port congestion and environmental impact through better berth allocation and service scheduling. Queuing theory was used to estimate uncertain waiting times at ports. The proposed grey wolf optimizer algorithm provided computational advantages, indicating the model's effectiveness in improving shipping schedules in the wake of port congestion.

It could be concluded that in the context of environmental and emission issues associated with berth allocation and scheduling at maritime container terminals, various solution methods have been implemented, as numerous studies have been conducted. However, hyperheuristic approaches have yet to be applied in the domain of green berth allocation and scheduling along with general berth allocation and scheduling at marine container terminals, despite their strong performance in other optimization problems.

As an example, Leng et al. (2020) developed a decomposition-based multi-objective hyperheuristic for a bi-objective cold chain problem, targeting the reduction of logistics costs and carbon emissions. The framework incorporated adaptive selection and acceptance strategies to coordinate multiple low-level heuristic operators. Mosadegh et al. (2020) introduced a Q-learning-based simulated annealing hyperheuristic for the sequencing of mixed-model assembly lines with uncertain processing times. The presented method effectively improved sequencing optimization under multi-station environments with stochastic settings. Cruz-Duarte et al. (2021) presented a customizable hyperheuristic framework for continuous optimization using simulated annealing to combine multiple low-level search operators. This approach demonstrated significant performance gains on high-dimensional continuous benchmark functions.

In 2023, Song et al. (2023) introduced a hyperheuristic-based memetic algorithm to address the distributed assembly permutation flow-shop scheduling problem, focusing on completion time reduction. The algorithm used a high-level estimation of distribution strategy to enhance local search efficiency, significantly outperforming other algorithms for large-scale scheduling. Zhu et al. (2023) presented a genetic programming hyperheuristic for the surgical cases assignment problem, improving scheduling accuracy in operating rooms. The method dynamically managed low-level heuristics for efficiency, surpassing previous approaches in operating cost minimization. Zhao et al. (2023) employed an estimation of distribution algorithm-based hyperheuristic for flowshop scheduling, incorporating critical resource allocation to enhance scheduling in distributed manufacturing. The method effectively minimized completion time, proving advantageous for high-scale industrial tasks.

Duan et al. (2024) proposed a genetic programming hyperheuristic for dynamic, energy-efficient scheduling in hybrid flow shops with machine breakdowns. By generating scheduling rules adaptable to real-time shop events, this approach showed flexibility and high performance across diverse scheduling scenarios. Rodríguez-Esparza et al. (2024) developed a hyperheuristic approach for traffic surveillance, applying a genetic algorithm with Thompson Sampling to optimize vehicle segmentation in complex urban environments. This method demonstrated robust performance in handling dynamic traffic data through effective hyperheuristic strategies.

In summary, while extensive research has investigated the emission reduction in berth allocation and scheduling problems through various algorithmic methodologies, there remains a notable gap in applying hyperheuristic algorithms in this research field (Golias et al., 2010; Hu et al., 2014; Jiang et al., 2022, 2024; Jiang et al., 2023; Lin et al., 2023; Venturini et al., 2017; Wang et al., 2019; Wen et al., 2024). Hyperheuristic algorithms, known for their inherent adaptability to different dynamic contexts and ability to coordinate multiple low-level heuristics dynamically, present significant potential for enhancing solutions for the problem studied herein.

Therefore, this study introduces a Customized Hyperheuristic Algorithm (CHA) to solve a mathematical optimization model developed for the green berth allocation and scheduling problem, which integrates environmental considerations through minimizing both operational costs and CO_2 emissions during vessel waiting and handling operations. CHA employs a dynamic selection mechanism that adaptively chooses heuristic operators based on their ongoing performance, enhancing the traditional framework of typical heuristic and metaheuristic algorithms. Detailed computational experiments are conducted to compare CHA with established optimization techniques, such as CPLEX, ant colony optimization (ACO), simulated annealing (SA), tabu search (TS), and an evolutionary algorithm (EA), and showcase the CHA's superior capability in reducing emissions while providing effective berth allocation and scheduling plans in a timely manner. Additional experiments are implemented to provide valuable insights into port management.

3 Problem Description and Optimization Model

This section presents an optimization model formulated for the Green Berth Allocation and Scheduling Problem (**GBASP**), which can be classified as a mixed-integer linear programming (MILP) model. This model optimizes berth allocation and scheduling while incorporating environmental considerations, particularly the minimization of CO_2 emissions. The model expands upon traditional berth allocation and scheduling frameworks presented by the state-of-the-art studies and introduces environmental considerations to minimize emissions during vessel waiting and handling periods alongside traditional cost elements.

Notations

Sets	
$V = \{1, \ldots, n\}$	Set of vessels to be serviced at the considered MCT
$B = \{1, \ldots, m\}$	Set of berths for handling operations at the considered MCT
$S = \{1, \ldots, k\}$	Set of considered service orders
Decision variables	
$\boldsymbol{x}_{vbs}, v \in V, b \in B, s \in S$	= 1 if vessel v is handled as the *sth* vessel at berth b (= 0 otherwise)
Auxiliary variables	
$\boldsymbol{T}_{vbs}, v \in V, b \in B, s \in S$	Time of idling for berth b between the service of vessel v served as the *sth* vessel and its immediate predecessor served as the $(s-1)$th vessel (h)
$\boldsymbol{T}_v^{st}, v \in V$	Service starting time for vessel v at the allocated berth (h)
$\boldsymbol{T}_v^{et}, v \in V$	Service ending time for vessel v at the allocated berth (h)
$\boldsymbol{T}_v^{wt}, v \in V$	Waiting time of vessel v at the allocated berth (h)
$\boldsymbol{T}_v^{del}, v \in V$	Hours of delays associated with the departure of vessel v (h)
$\boldsymbol{C}_v, v \in V$	Total amount of carbon dioxide emission generated during the waiting and handling stages of vessel v and port equipment during the handling stage of vessel v (tons)
$\boldsymbol{C}_v^{vessel}, v \in V$	Total amount of carbon dioxide emission generated by vessel v during its waiting and handling stages (tons)
$\boldsymbol{C}_v^{hand}, v \in V$	Total amount of carbon dioxide emission generated by port equipment during the handling stage of vessel v (tons)
$\boldsymbol{C}_v^{dev}, v \in V$	Total amount of generated carbon dioxide emission associated with deviation from the preferred berthing position during the handling stage of vessel v (tons)
$\boldsymbol{L}_v^{dev}, v \in V$	Deviation between the desired berth and the actual berth for vessel v (m)
Parameters	
$T_b^{av}, b \in B$	The time within the planning period considered when berth b becomes available for the first time (h)
$T_v^{at}, v \in V$	Projected time of arrival for vessel v (h)
$T_v^{sd}, v \in V$	Scheduled departure time of vessel v (h)
$L_b^{loc}, b \in B$	Location of berth b at the port considered (m)
$T_{vb}^{ht}, v \in V, b \in B$	Handling operation time of vessel v at berth b (h)
$L_v^{des}, v \in V$	Preferential berth of vessel v (m)
$d_v^{sc}, v \in V$	Unit cost of service for vessel v (USD/h)
$d_v^{wc}, v \in V$	Unit cost of waiting for vessel v (USD/h)

(continued)

(continued)

Sets	
$d_v^{cld}, v \in V$	Unit cost of delay in departure of vessel v (USD/h)
$d_v^{cd}, v \in V$	Unit cost associated with deviation between the desired and actual berths of vessel v (USD/m)
d^c	Unit cost due to carbon dioxide emission (USD/ton)
$N_v, v \in V$	Amount of TEUs to be handled on vessel v
$HP_{vb}, v \in V, b \in B$	Handling productivity for vessel v at berth b (TEU/h)
$EC_v^{AE}, v \in V$	Unit energy consumption of auxiliary engines during the waiting and handling stages of vessel v (kWh/h)
$EF_v^{AE}, v \in V$	Emission factor of auxiliary engines of vessel v (ton/kWh)
$L_v^{AB}, v \in V$	Load of auxiliary boilers during the waiting and handling stages of vessel v (kWh/h)
$EF_v^{AB}, v \in V$	Emission factor of auxiliary boilers of vessel v (ton/kWh)
$EF_{vb}^{hand}, v \in V, b \in B$	Unit emissions produced from the port equipment during the handling operations of one TEU for vessel v at berth b (ton/TEU)
$EF_{vb}^{dev}, v \in V, b \in B$	Unit emissions produced due to the deviation form the preferred berthing position during the handling operations of one TEU of vessel v at berth b (ton/TEU)
Ω	A large positive number

3.1 Decision Problem

The **GBASP** model investigates the allocation of berths and the scheduling of vessels at an MCT with a discrete berthing layout, where the quay is partitioned into specific berths, each capable of serving only one vessel at a time. The arrival times of vessels at the terminal are dynamic, but stochastic uncertainties (e.g., delays due to adverse weather conditions or disruptions at previous ports) are not modeled in this problem. The sequence of vessel servicing is determined prior to execution, based on vessel characteristics and terminal constraints. Upon arriving at the terminal, vessels are assigned to a specific berth. If a berth is occupied, the vessel must wait in a designated holding area until it can be serviced, incurring waiting costs. This waiting period is also a major source of CO_2 emissions, as vessels continue to run auxiliary engines and auxiliary boilers, thus burning fuel while waiting, contributing significantly to their environmental impact. Congestion due to waiting vessels can further hinder terminal operations by delaying both incoming and departing vessels, reducing the overall efficiency of the terminal. These inefficiencies not only increase operational costs but also raise CO_2 emissions as vessels spend additional time idling at sea or within the terminal's waiting/holding area.

Once the vessel is berthed, handling costs are incurred for the loading and unloading operations, which are conducted using the terminal equipment (e.g., quay cranes, yard cranes, internal trucks). During the handling phase, both the vessels (that continue running auxiliary engines and auxiliary boilers) and the terminal equipment generate additional CO_2 emissions. Emissions from handling operations depend on the handling productivity and duration of equipment use, particularly the heavy machinery involved in moving cargo, which consumes fuel and electricity. It is critical that vessels adhere to their scheduled departure times to avoid penalties for late departures, as delays can disrupt shipping schedules and lead to delay costs for shipping companies. Prolonged delays not only have economic repercussions but also result in increasing CO_2 emissions from both the vessels and the terminal equipment as these resources continue to be engaged beyond their scheduled service times.

Additionally, every arriving vessel has a desired berthing position that is located close to the reserved storage space in the marshaling yard, and deviations from this ideal berthing position result in deviation costs and increased fuel consumption due to the additional maneuvering required reaching a suboptimal berth. This maneuvering, which typically involves additional travel distance by the internal transport units to the designated storage area in the marshaling yard, generates significant CO_2 emissions, further contributing to the environmental cost of terminal operations. Similar to the present research, deviations from the desired berthing positions have been captured by numerous BASP studies (Bierwirth & Meisel, 2010, 2015; Li et al., 2023).

The **GBASP** model captures emissions from three primary sources: vessel waiting periods, handling operations, and deviations from the desired berth. First, during waiting periods, CO_2 emissions result from the continuous operation of auxiliary engines and boilers as vessels idle in the holding area. Second, emissions arise from handling operations, where port equipment causes emission. This is directly related to the number of TEUs handled and the emission factors of the equipment used. In the meantime, vessel's own systems continue operating and produce emissions (unless shore power is provided by the port). Third, berthing deviations lead to additional emissions as vessels maneuver to suboptimal positions, further increasing fuel consumption by the internal transport equipment units assigned to these vessels. By accounting for these sources, **GBASP** aims to minimize the environmental impact of port operations while balancing economic efficiency.

3.2 Optimization Model

The **GBASP** model is formulated as a mixed-integer linear program, with the objective of minimizing the total turnaround costs, including both operational and environmental costs. The model's objective function is subject to a series of constraints that ensure feasibility and compliance with practical considerations in vessel scheduling and berth assignment. **GBASP** is formulated as follows:

GBASP: Green Berth Allocation and Scheduling Problem

$$\min \boldsymbol{Z} = \left[\sum_{v \in V} \left(d_v^{wc} \boldsymbol{T}_v^{wt} \right) + \sum_{v \in V} \sum_{b \in B} \sum_{s \in S} \left(d_v^{sc} T_{vb}^{ht} \boldsymbol{x}_{vbs} \right) + \sum_{v \in V} \left(d_v^{cld} \boldsymbol{T}_v^{del} \right) + \sum_{v \in V} \left(d_v^{cd} \boldsymbol{L}_v^{dev} \right) + \sum_{v \in V} \left(d^c \boldsymbol{C}_v \right) \right] \quad (1)$$

Subject to:

$$\sum_{b \in B} \sum_{s \in S} \boldsymbol{x}_{vbs} = 1, \forall v \in V \quad (2)$$

$$\sum_{v \in V} \boldsymbol{x}_{vbs} \leq 1, \forall b \in B, s \in S \quad (3)$$

$$\sum_{\hat{v} \in V: \hat{v} \neq v} \sum_{\hat{s} \in S: \hat{s} < s} \left(T_{\hat{v}b}^{ht} \boldsymbol{x}_{\hat{v}b\hat{s}} + \boldsymbol{T}_{\hat{v}b\hat{s}} \right) + \boldsymbol{T}_{vbs} - \left(T_v^{at} - T_b^{av} \right) \quad (4)$$
$$\boldsymbol{x}_{vbs} \geq 0, \forall v \in V, b \in B, s \in S$$

$$\boldsymbol{T}_v^{st} \geq \sum_{\hat{v} \in V: \hat{v} \neq v} \sum_{\hat{s} \in S: \hat{s} < s} \left(T_{\hat{v}b}^{ht} \boldsymbol{x}_{\hat{v}b\hat{s}} + \boldsymbol{T}_{\hat{v}b\hat{s}} \right) + \boldsymbol{T}_{vbs} + T_b^{av} - \Omega(1 - \boldsymbol{x}_{vbs}), \forall v \in V, b \in B, s \in S \quad (5)$$

$$\boldsymbol{T}_{vbs} \leq \Omega \boldsymbol{x}_{vbs}, \forall v \in V, b \in B, s \in S \quad (6)$$

$$\boldsymbol{T}_v^{wt} \geq \boldsymbol{T}_v^{st} - T_v^{at}, \forall v \in V \quad (7)$$

$$\boldsymbol{T}_v^{et} \geq \boldsymbol{T}_v^{st} + \sum_{b \in B} \sum_{s \in S} \left(T_{vb}^{ht} \boldsymbol{x}_{vbs} \right), \forall v \in V \quad (8)$$

$$\boldsymbol{T}_v^{del} \geq \boldsymbol{T}_v^{et} - T_v^{sd}, \forall v \in V \quad (9)$$

$$\boldsymbol{L}_v^{dev} \geq \sum_{b \in B} \sum_{s \in S} \left(L_b^{loc} \boldsymbol{x}_{vbs} \right) - L_v^{des}, \forall v \in V \quad (10)$$

$$\boldsymbol{L}_v^{dev} \geq L_v^{des} - \sum_{b \in B} \sum_{s \in S} \left(L_b^{loc} \boldsymbol{x}_{vbs} \right), \forall v \in V \quad (11)$$

$$\boldsymbol{C}_v^{vessel} = \left(EC_v^{AE} EF_v^{AE} + L_v^{AB} EF_v^{AB} \right) \left[\boldsymbol{T}_v^{wt} + \sum_{b \in B} \sum_{s \in S} \left(T_{vb}^{ht} \boldsymbol{x}_{vbs} \right) \right], \forall v \in V \quad (12)$$

$$\boldsymbol{C}_v^{hand} = \sum_{b \in B} \sum_{s \in S} \left(EF_{vb}^{hand} N_v \boldsymbol{x}_{vbs} \right), \forall v \in V \quad (13)$$

$$\boldsymbol{C}_v^{dev} = \sum_{b \in B} \sum_{s \in S} \left(EF_{vb}^{dev} N_v \boldsymbol{x}_{vbs} \right), \forall v \in V \tag{14}$$

$$\boldsymbol{C}_v = \boldsymbol{C}_v^{vessel} + \boldsymbol{C}_v^{hand} + \boldsymbol{C}_v^{dev}, \forall v \in V \tag{15}$$

$$T_{vb}^{ht} = \frac{N_v}{HP_{vb}}, \forall v \in V, b \in B \tag{16}$$

$$\boldsymbol{x}_{vbs} \in \mathbb{B}, \forall v \in V, b \in B, s \in S \tag{17}$$

$$\boldsymbol{L}_v^{dev}, L_b^{loc}, L_v^{des} \in \mathbb{N}, \forall v \in V, b \in B \tag{18}$$

$$\begin{aligned} &\boldsymbol{T}_{vbs}, \boldsymbol{T}_v^{st}, \boldsymbol{T}_v^{et}, \boldsymbol{T}_v^{wt}, \boldsymbol{T}_v^{del}, \boldsymbol{C}_v, \boldsymbol{C}_v^{vessel}, \boldsymbol{C}_v^{hand}, \\ &\boldsymbol{C}_v^{dev}, T_b^{av}, T_v^{at}, T_v^{sd}, T_{vb}^{ht}, d_v^{sc}, d_v^{wc}, d_v^{cld}, d_v^{cd}, d^c, N_v, HP_{vb}, \\ &EC_v^{AE}, EF_v^{AE}, L_v^{AB}, EF_v^{AB}, EF_{vb}^{hand}, EF_{vb}^{dev}, \Omega \in \mathbb{R}^+, \forall v \in V, b \in B, s \in S \end{aligned} \tag{19}$$

The **GBASP** mathematical model is structured around an objective function and 18 sets of constraints. The objective function (1) is designed to minimize the overall operational and environmental costs of scheduling vessels and allocating berths. It incorporates several cost components:

(i) the waiting costs incurred when vessels must delay their service until the allocated berth is available;
(ii) the handling costs associated with the loading and unloading of vessels while they are serviced at the berth;
(iii) the late departure penalties, which account for the financial losses incurred if vessels depart later than their scheduled time;
(iv) the deviation costs, representing the penalties for berthing vessels in positions that differ from their desired or optimal berthing locations; and
(v) the carbon dioxide emission costs, which take into account the environmental impact of emissions produced during vessel waiting, handling, and deviations from the optimal berth.

The constraint sets ensure that the model operates within realistic limits and adheres to operational requirements. Constraints (2) ensure that each vessel is assigned to only one berth and is serviced once, while constraints (3) maintain berth exclusivity, ensuring that no berth handles more than one vessel at a time. Constraints (4) ensure that vessels are not serviced before their arrival at the terminal, aligning the scheduling with real-world time constraints. Constraints (5) define the start time of service for each vessel, ensuring that operations do not commence before the previously scheduled operations are complete. Constraints (6) ensure that the idle time of a berth between the operation of a vessel and its immediate predecessor applies only when the vessel is assigned to that berth in a given sequence. Constraints (7) ensure that waiting times are correctly computed as the difference between the vessel's start of service and arrival times. Constraints (8) calculate the service end time, while constraints (9) account for any delays in departure. Constraints (10) and (11) handle the deviations from the desired berthing positions.

Constraints (12) calculate the total CO_2 emissions generated by each vessel during its waiting and handling phases. The total emissions from the vessel's auxiliary engines and boilers are determined by multiplying the energy consumption of the auxiliary engines and load of the auxiliary boilers with their respective emission factors. These values are then multiplied by the time the vessel spends waiting and the total handling time which accounts for the vessel's active operation at the assigned berth. Constraints (13) calculate the CO_2 emissions from the port equipment used during the handling operations of the vessel. These emissions depend on the number of TEUs handled and the emission factor associated with the port equipment at the assigned berth. The emissions are computed by summing across all berths and service orders where the vessels will be active for the considered planning period.

Constraints (14) account for emissions resulting from deviations from the vessel's preferred berthing position. These deviation-related emissions are computed by multiplying the number of TEUs handled with the emission factor related to berth deviation. Similar to handling emissions, this component is estimated based on the summation across all berths and service orders where the vessels will be berthed for the considered planning period. The total emissions for each vessel are then computed in constraints (15) as a sum of emissions from all three sources: waiting and handling phases from the vessel itself; emissions from the port equipment; and emissions caused by berthing deviations. Constraints (16) calculate the handling time based on the number of TEUs and the handling productivity. Constraints (17) to (19) define the domain of decision variables and parameters, ensuring that all model elements are bound with respect to operational limits.

4 Algorithm Description

This section of the manuscript details the development of the Customized Hyperheuristic Algorithm (CHA), which is specifically designed to tackle the green berth allocation and scheduling problem with a focus on minimizing carbon emissions. The CHA framework is inherently inspired by evolutionary computation strategies but introduces a distinct approach that emphasizes dynamic selection and adaptation of heuristic operators throughout the optimization process. Unlike traditional evolutionary algorithms that rely on a fixed set of operations, CHA continuously adjusts its heuristic selection based on the record of their performance, allowing for more flexible and efficient exploration of the solution space.

The CHA algorithm employs a variety of evolutionary strategies, dynamically selecting from a pool of crossover and mutation operators. This selection is governed by an adaptive mechanism that evaluates the effectiveness of each operator after each *Epoch* of generations is completed. After each *Epoch*, the high-level heuristic within the CHA algorithm updates the selection of low-level heuristics based on their recent performance, allowing it to adapt dynamically to the evolving solution landscape. The low-level heuristics in CHA include three recombination operators (cycle crossover, order crossover, and partial-mapped crossover) along with three mutation operators (swap, insert, and invert mutations). These operators, which have been widely used in evolutionary computing and proved their effectiveness, yield a total of nine possible combinations, providing the algorithm with flexibility in selecting the most effective heuristics for different stages of the optimization process. By leveraging a diverse set of heuristic strategies, the algorithm ensures both thorough exploration of new regions and effective exploitation of promising solutions, ultimately leading to high-quality results (Li et al., 2024). This enables the algorithm to efficiently navigate through the large and complex solution space of **GBASP**, where traditional methods often struggle with computational demands.

4.1 Main Algorithmic Steps

The major steps of the Customized Hyperheuristic Algorithm (CHA) framework are presented in Fig. 1, which begins with Step 1, where the required input data for the **GBASP** decision model and CHA are defined. This includes data related to emissions, vessel scheduling costs and constraints, berth allocation parameters, and algorithmic settings, such as population size, mutation and recombination probabilities, as well as termination criteria. In Step 2, two generation counters, g and g^*, are initialized. Here, g tracks each successive generation, while g^* designates specific generations as "generations of update" where recombination and mutation operators will be reviewed and potentially updated depending on their performance. Instead of performing updates in every generation, which would increase computational time complexity, the algorithm applies updates after a designated interval, known as "*Epoch*", allowing operator performance to stabilize and be assessed over multiple generations.

With initialization complete, Step 3 generates an initial population of solutions using a combination of deterministic and stochastic methods (i.e., traditional first-come-first-served policy to initialize the first 50% of the population and random initialization for the rest of population). This ensures a diverse set of initial solutions, allowing the algorithm to explore a wide range of possibilities from the outset. In Step 4, the fitness of each individual solution in this population is evaluated based on the **GBASP** objective function, which includes minimizing CO_2 emissions and optimizing berth allocation and scheduling plans. Following the evaluation, Step 5 initiates a warm-up phase, during which each recombination and mutation operator combination is preliminarily assessed to establish baseline performance metrics. These baselines are set using objective function improvements over an *Epoch* for each operator combination. Operators that achieve substantial objective function improvements are marked as promising, and these baseline values will guide the algorithm in selecting operators in future generations.

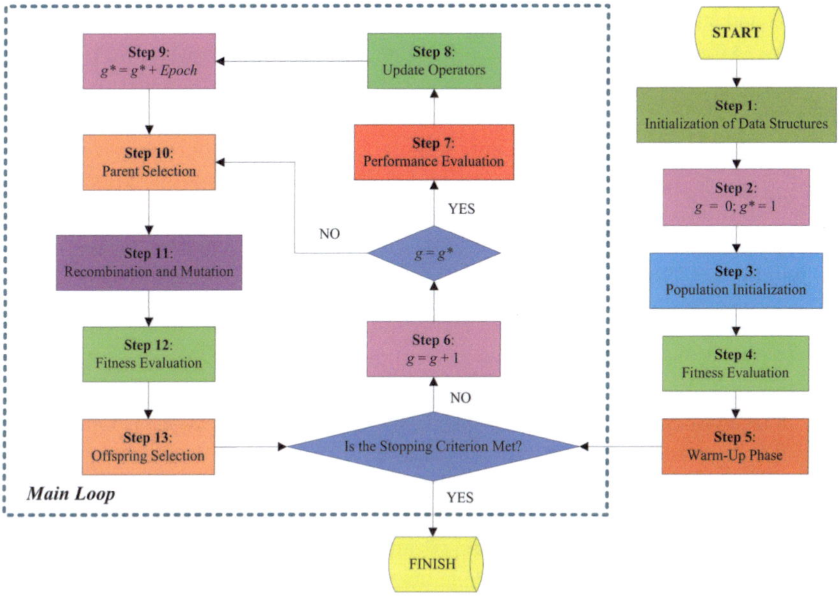

Fig. 1 Algorithm framework for the customized hyperheuristic algorithm

With steps 1 through 5 complete, the algorithm enters its main optimization loop, spanning Steps 6 through 13, which encompasses all procedures that are performed in each consecutive generation. At the beginning of the main loop, the algorithm checks if the termination criterion, such as a maximum number of generations, has been met. If this condition is not met, the generation counter g is incremented by one (i.e., $g = g + 1$) in Step 6, allowing the algorithm to progress to the next generation. Following Step 6, the algorithm checks whether g matches g^*, suggesting a "generation of update". This leads to two potential paths:

1. If $g = g^*$ (g is a "generation of update"), the algorithm proceeds to Step 7 instead of moving to parent selection. In Step 7, it updates the performance of the combination of recombination and mutation operators selected for deployment during the past *Epoch* based on the cumulative data collected over the last *Epoch* (or from the warm-up phase if $g = 1$). This step provides a comparison of each operator combination performance, evaluating their contributions to objective improvements. In Step 8, a roulette wheel selection mechanism uses these updated metrics to select the operator combination for the next *Epoch*, ensuring that the most effective operators are prioritized. As an example, if there are two operator combinations considered, and the first operator combination yields 20% in the objective function improvements after *Epoch* generations, whereas the second operator combination yields 15% in the objective function improvements, then the first operator combination will have a higher probability of being

selected by the roulette wheel selection mechanism for the next *Epoch* generations. After completing the operator update, the algorithm moves to Step 9, which adjusts g^* by setting $g^* = g^* + Epoch$, scheduling the next operator update after another *Epoch*. This adjustment ensures that the next operator review occurs only after the specified interval, maintaining the periodic update of the algorithm.
2. If $g \neq g^*$ (g is not a "generation of update"), the algorithm skips Steps 7, 8, and 9 and proceeds directly to Step 10. In this case, the algorithm uses the recombination and mutation operators that were selected during the last update without any change.

In Step 10, regardless of whether an operator update has occurred, the algorithm selects parent chromosomes for the current generation using a binary tournament selection method. This method promotes population diversity while giving fitter solutions a higher chance to contribute to the next generation. In Step 11, the selected recombination and mutation operators are applied to these parent chromosomes, producing a new set of offspring chromosomes. Step 12 then evaluates the fitness of these offspring chromosomes, determining their quality according to the **GBASP** objective function. Following the fitness evaluation, the algorithm prepares the population chromosomes for the next generation. In Step 13, the algorithm employs the Boltzmann selection strategy to determine which offspring chromosomes will proceed to the next generation, balancing exploration with exploitation by favoring solutions with more promising fitness value while preserving some diversity within the population. Then, the termination criterion is checked, and CHA concludes once the termination criterion has been satisfied.

4.2 Low-Level Heuristics for Recombination

Three recombination operators were developed within the proposed CHA: the order recombination operator, the cycle recombination operator, and the partially mapped recombination operator. An example of the cycle recombination operator functionality is illustrated in Fig. 2. The process initiates by identifying the gene at the first locus of Parent P1, where berth "1" corresponds to ship "1". Ship "1" is added to the cycle. The corresponding locus in Parent P2 is then examined, revealing ship "4" at berth "1", which is also included in the cycle. Subsequently, ship "4" is located in Parent P1 at berth "2" at the fourth locus. Referring to the fourth locus in Parent P2, ship "8" is identified and added to the cycle. Ship "8" is positioned at the eighth locus in Parent P1, corresponding to ship "1" in Parent P2 at berth "3". Since ship "1" is already part of the cycle, the cycle is completed. For Offspring O1, the genes involved in this cycle (the ones that have ships "1", "4", and "8") are transferred directly from Parent P1. The remaining ships (ships "7", "6", "2", "3", and "5") are sequentially filled from Parent P2 to ensure completeness of Offspring O1. Offspring O2 is generated using an analogous cycle detection and gene transfer method, with roles of the parents reversed.

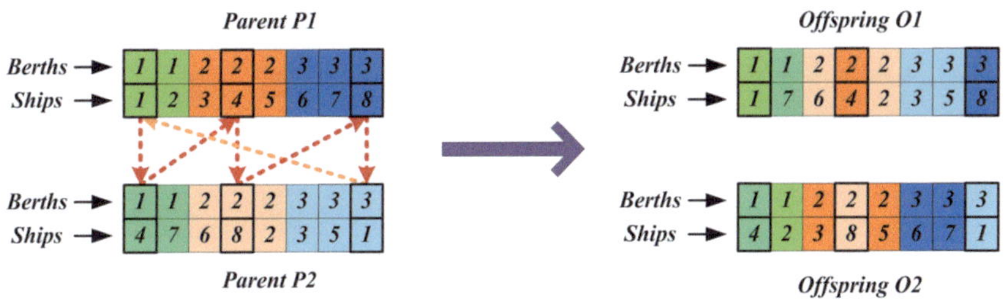

Fig. 2 Cycle recombination

The order recombination operator employs a two-step procedure. As depicted in Fig. 3, initially, a random segment of genes from Parent P1, such as the genes containing ships “3”, “4”, “5”, and “6”, is selected and placed into Offspring O1, preserving the original order. Subsequently, the remaining ships (ships “7”, “8”, “2”, and “1”) from Parent P2 are inserted into the vacant positions of Offspring O1, maintaining the sequence from Parent P2. Offspring O2 follows the same approach but begins with a different segment from Parent P2, such as the segment of genes with ships “8”, “2”, and “3”. These genes are directly inserted into Offspring O2, with the remaining genes from Parent P1 filling the gaps in their original order. This ensures that each offspring contains a complete and unique set of genes without duplicates.

An example of the partially mapped crossover is illustrated in Fig. 4, which starts by selecting and transferring a subset of genes from Parent P1 to Offspring O1. During the example crossover operation, the segment from the third to the sixth locus in Parent P1 is directly copied to the offspring (i.e., the genes with ships “3”, “4”, “5”, and “6”), preserving positional integrity. The next step is to analyze the same segment in Parent P2 to ensure the offspring maintains its feasibility. Ships “8” and “2” within the selected loci of Parent P2 are not present in the equivalent loci of Parent P1, necessitating a mapping procedure. This mapping identifies that ship

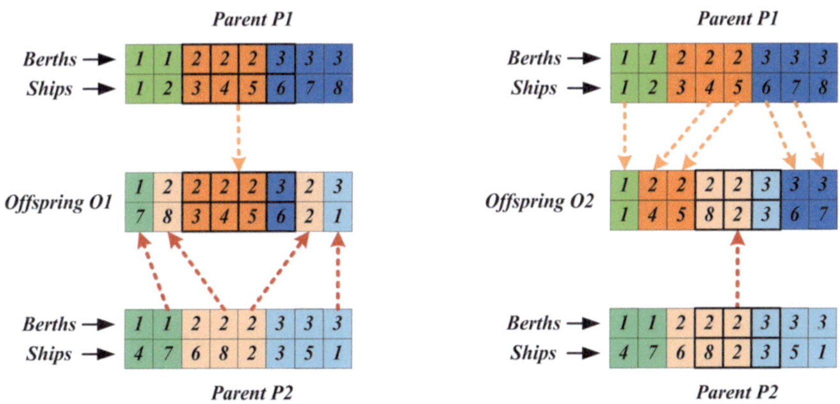

Fig. 3 Order recombination

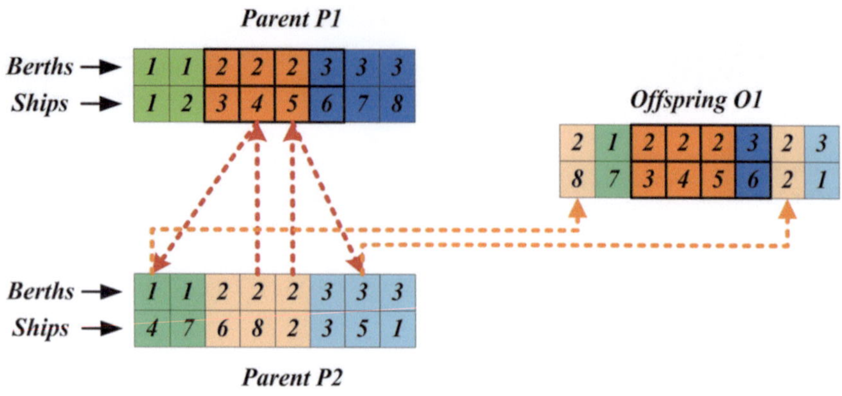

Fig. 4 Partially-mapped recombination

"8" in Parent P2 corresponds to the locus of ship "4" in Parent P1. The algorithm then locates the position of ship "4" in Parent P2, found at the first locus. Thus, the corresponding berth and ship pair (i.e., berth "2" and ship "8" from the fourth locus in Parent P2) is mapped and placed into the first locus of the offspring. Similarly, the other berth and ship pair (i.e., berth "2" and ship "2") follows the same mapping logic and is placed into the seventh locus of the offspring. The genes at loci in Parent P2 that do not require mapping (such as the genes placed at the second and eight loci) are directly copied to their corresponding positions in the offspring.

4.3 Low-Level Heuristics for Mutation

In the proposed CHA framework, three mutation operators are applied: insert mutation, invert mutation, and swap mutation. Figure 5 demonstrates an instance of the insert mutation operator. This operation selects a specific gene, representing ship "6" in the considered instance and relocates it within the sequence. Initially, ship "6" was scheduled first for service at berth "3". Through the insertion mutation, ship "6" is repositioned to follow ship "3" at berth "2". In the meantime, ship "4" and ship "5" are moved to new loci accordingly, effectively modifying their service sequence with appropriate changes in the ship-to-berth assignment structure.

Figure 6 depicts the invert mutation operation. This mutation operator inverts the order of a selected sequence of genes, effectively reversing the service order of a subset of ships. In this example, the segment consisting of ships "3", "4", "5", "6", and "7" is selected. Initially, these ships were arranged in the service order "3" → "4" → "5" → "6" → "7". After the inversion mutation, the sequence is reversed, resulting in the new service order "7" → "6" → "5" → "4" → "3", altering their relative order with appropriate changes in the ship-to-berth assignment structure.

Fig. 5 Insert mutation

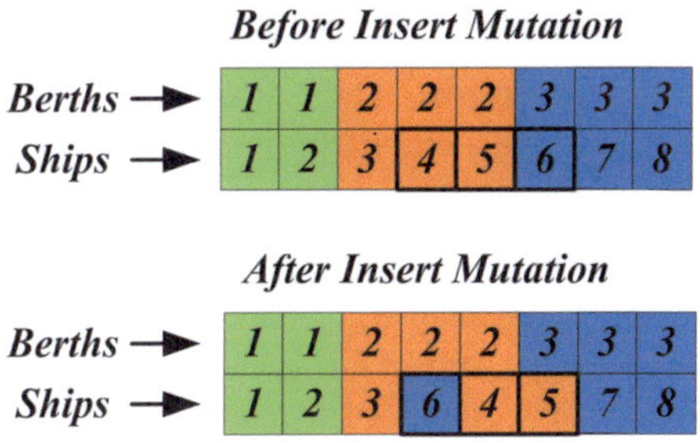

Fig. 6 Invert mutation

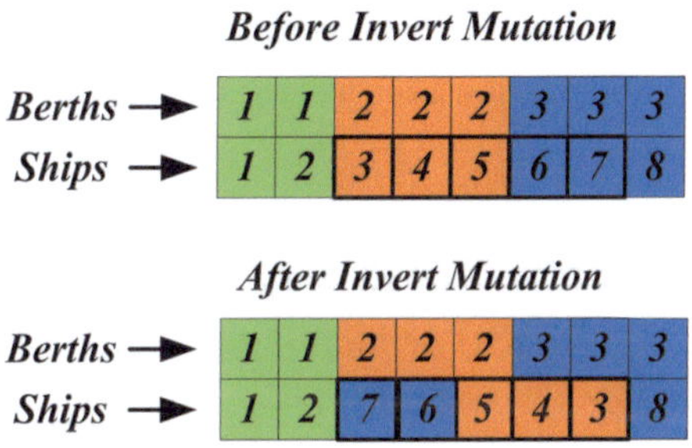

Figure 7 provides an example of the swap mutation operator. This mutation swaps the positions of two selected genes within the chromosome. In this case, ships "2" and "7" are chosen for the swap mutation. Ship "2", originally positioned to follow ship "1" at berth "1", is swapped with ship "7", initially succeeding ship "6" at berth "3". This exchange results in ship "2" moving to the position after ship "6" at berth "3", while ship "7" takes the initial position of ship "2", following ship "1" at berth "1".

Fig. 7 Swap mutation

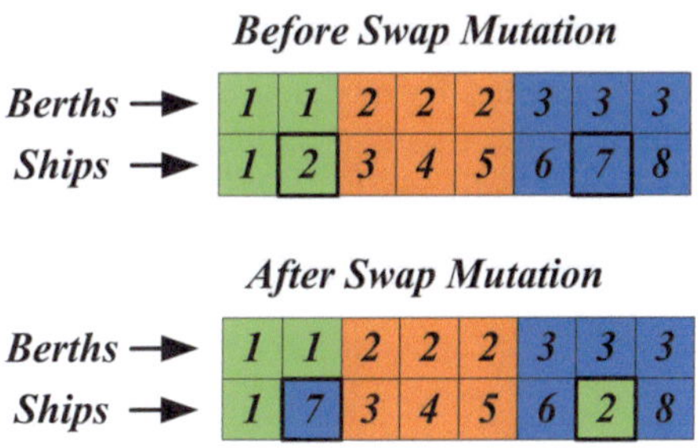

4.4 High-Level Heuristics for Solution and Operator Selection

High-level heuristics (HLHs) are integral to the CHA algorithm, particularly in optimizing operator selection as the algorithm evolves. In this study, the roulette wheel selection mechanism, a staple in evolutionary computation, serves as the primary HLH for its effective distribution of selection probabilities across various fitness levels. This operator assigns selection probabilities proportionally to fitness, with fitter individuals occupying larger sections of the wheel. Within the operator selection context, search operators that yield superior objective values receive an increased representation on the wheel (Qian et al., 2018). Given the similar time complexity of the recombination and mutation operators under consideration, the roulette wheel selection operator prioritizes solution quality improvements across each epoch as the principal criterion, reinforcing the progressive enhancement of solution fitness.

In the parent selection stage, the proposed CHA algorithm implements a binary tournament selection strategy, a technique well-established in evolutionary computation for reliably identifying superior candidates for reproduction (Soong et al., 2022). This method operates by pairing individuals for competition, with the fitter individual advancing, thus promoting a higher overall quality within the chromosome pool. To further enhance population diversity, each tournament begins with a random sampling of two solutions. The selection process continues iteratively until the algorithm reaches the required amount of parental chromosomes. Such a methodology not only sustains a balanced level of competitive pressure inside the population but also fosters genetic diversity, which further leads to enhanced optimization outcomes and reducing the likelihood of premature convergence on suboptimal solutions.

In determining which offspring chromosomes move forward to the following generation, the CHA algorithm employs the Boltzmann selection strategy, a method widely recognized for its dynamic adaptability and effectiveness in evolutionary frameworks (Chen et al., 2021). This selection process carefully balances individual fitness levels, ensuring that both high-performance solutions and genetic variation are maintained. By modulating the selection pressure through iterative adjustments to the temperature parameter, the algorithm adapts over successive generations to retain diversity—particularly crucial during the early stages. Sustained genetic variety is crucial for averting premature convergence and supporting the discovery of globally optimal solutions towards the end of the algorithmic run.

5 Computational Experiments

5.1 Key Experimental Settings

In this study, a comprehensive set of computational experiments was conducted to evaluate the performance of the proposed CHA in solving **GBASP**. The CHA was compared against several well-known metaheuristic techniques, including ant colony optimization (ACO), simulated annealing (SA), and tabu search (TS), as well as evolutionary algorithm (EA), which were all customized for problem-specific settings of **GBASP**. The EA was implemented as a controlled benchmark to illustrate the effect of adaptive operator selection methodology of CHA. Unlike CHA, which periodically evaluates and adjusts operator combinations based on real-time performance data, EA applies a static combination of one recombination and one mutation operator (i.e., cycle recombination and swap mutation operators which were found to be the most promising based on initial test runs) chosen from the operator pool, serving as a non-adaptive comparison benchmark. The exact optimization solver CPLEX was also included in the experiments to provide a baseline, given its wide application in solving mixed-integer linear programming optimization models.

All algorithms deployed in the experiments were implemented using the MATLAB 2023a suite. The mathematical model for **GBASP** was formulated within the General Algebraic Modeling System (GAMS) and solved using the CPLEX solver. These experiments were conducted on a Dell computer with an Intel(R) Core™ i7 processor, running Windows 10 with 32 GB of RAM, ensuring consistent computational resources across all solution methods. The parameters for **GBASP** were derived from the current literature related to berth scheduling and allocation optimization (Imai et al., 2003; Dulebenets, 2017; Tran et al., 2017; Sha et al., 2024; Wang et al., 2024). Table 1 presents the comprehensive list of these parameters. The reliance on data generation methodologies detailed in the past berth scheduling studies stems from the significant difficulty in obtaining direct operational data from MCT operators due to stringent confidentiality policies and concerns over operational security. Consequently, researchers have established a range of robust methods to create synthetic datasets that closely simulate real-life operational conditions. Adopting these data generation strategies ensures that this research aligns with prior studies, providing consistency and credibility while using representative and realistic data in the analysis.

To assess the effectiveness of CHA and alternative solution approaches across instances with various complexities, three categories of problem instances were developed for testing: (1) 10 small-scale instances, featuring 12 to 20 vessels with 2 to 3 berthing positions; (2) 10 medium-scale instances, involving 52 to 60 vessels with 2 to 4 berthing positions; and (3) 10 large-scale instances, consisting of 92 to 100 vessels also with 2 to 4 berthing positions. This structured approach enabled a layered examination of algorithmic effectiveness, solution quality, and computational efficiency across varying levels of problem complexity.

Table 1 Values adopted for the parameters of the **GBASP** mathematical model

Parameter	Adopted values
Initial berth availability: $T_b^{av}, b \in B$ (h)	$T_b^{av} = 0, \forall b \in B$
Scheduled arrival time of vessel v at the port: $T_v^{at}, v \in V$ (h)	$T_v^{at} = T_{v-1}^{at} + EXP(2)$*
Planned departure time of vessel v : $T_v^{sd}, v \in V$ (h)	$T_v^{sd} = T_v^{at} + 5 + EXP(10)$*
Possible berth locations at a given port: $L_b^{loc}, b \in B$ (m)	[100; 200; 300; 400; 500; 600; 700 800; 900; 1,000; 1,100; 1,200; 1,300; 1,400; 1,500]
Preferential berth of ship v : $L_v^{des}, v \in V$ (m)	Randomly chosen from the berthing positions available
Unit cost of handling for vessel v: $d_v^{sc}, v \in V$ (USD/h)	U[45,000; 55,000]**
Unit cost of waiting for vessel v: $d_v^{wc}, v \in V$ (USD/h)	U[800; 1,200]**
Unit cost of delays in departure for vessel v: $d_v^{cld}, v \in V$ (USD/h)	U[4,000; 6,000]**
Unit cost for deviation between the desired and actual berths for vessel v: $d_v^{cd}, v \in V$ (USD/m)	U[40; 55]**
Unit cost due to carbon dioxide emission: d^c (USD/ton)	32
Amount of TEUs to be handled on vessel v: $N_v, v \in V$	U[1,000; 2,000]**
Handling productivity for vessel v at berth b: $HP_{vb}, v \in V, b \in B$ (TEU/h)	U[100; 200]**
Unit energy consumption of auxiliary engines during the waiting and handling stages of vessel v: $EC_v^{AE}, v \in V$ (kWh/h)	U[2,000; 2,500]**
Emission factor of auxiliary engines of vessel v: $EF_v^{AE}, v \in V$ (ton/kWh)	U[671; 721]**
Load of auxiliary boilers during the waiting and handling stages of vessel v: $L_v^{AB}, v \in V$(kWh/h)	U[900; 1,100]**
Emission factor of auxiliary boilers of vessel v: $EF_v^{AB}, v \in V$ (ton/kWh)	U[912; 1,012] $**$
Unit emissions produced from the port equipment during the handling operations of one TEU for vessel v at berth b: $EF_{vb}^{hand}, v \in V, b \in B$ (ton/TEU)	U[0.01729; 0.02129]**

(continued)

Table 1 (continued)

Parameter	Adopted values
Unit emissions produced due to the deviation from the preferred berthing position during the handling operations of one TEU for vessel v at berth b: $EF_{vb}^{dev}, v \in V, b \in B$ (ton/TEU)	$U[0; 0.004258]$**
A sufficiently large positive number: Ω	10,000

Notes * $EXP(2)$ is used to denote ship arrivals that are exponentially distributed with a mean of 2 h
** $U[a; b]$ is used to denote the random numbers that are uniformly distributed and range between values a and b

The parameter tuning phase of the considered solution algorithms adopted Taguchi's method. Instead of evaluating all possible parameter sets exhaustively, Taguchi's approach narrows down the focus to essential combinations that are anticipated to yield desirable outcomes (Dulebenets, 2023). The tuning process involved five randomly selected problem instances, each tested across five independent runs to compute average performance metrics (focusing mainly on solution quality and average computational time). The results of this comprehensive parameter tuning are presented in Table 2 outlining: (1) the parameters tested for each algorithm; (2) the values considered during the tuning; and (3) the best parameter values determined.

5.2 *Evaluation Against Other Common Solution Methods*

Each one of the developed instances was solved five times using CHA, the benchmark algorithms CPLEX and EA, as well as ACO, SA, and TS. For each algorithm, the best and average objective function values were recorded, alongside the mean computational time required to reach the final solution. The outcomes of the conducted experiments are illustrated with Table 3, whereas the convergence plots of CHA, ACO, SA, TS and EA are presented in Fig. 8. The convergence criterion preset for CPLEX was defined as reaching the solution with an objective value within 0.1% of the projected optimal value. As shown in Table 3, in all of the ten small-scale experimental cases, CPLEX and CHA achieved the best solutions after completing the test executions. However, CPLEX's computation time was often close to acceptable limits, as it frequently reached the imposed maximum runtime of 7,200 s without meeting the predefined solution optimality gap. Only in case S6, with 12 vessels and 3 berths, CPLEX achieved the target solution optimality gap at 6,142.34 s within the time limit (see Table 3). Due to decline in computational performance with increasing problem size, CPLEX was not implemented to solve the medium and large-scale instances.

The comparative analysis of all five algorithms, CHA, ACO, SA, TS, and EA, reveals apparent differences in their performance and computational efficiency in solving **GBASP** (see Table 3). ACO and SA demonstrated the shortest average computational times at 18.14 and 17.92 s, respectively, showcasing their ability to

Table 2 Results of the parameter tuning analysis

Algorithm	Parameter	Tested values	Selected value	Algorithm	Parameter	Tested Values	Selected Value
CHA	Size of population	[30; 40; 50]	40	ACO	Initial pheromone amount	[0.10; 0.15; 0.20]	0.10
CHA	Initial temperature	[2,000; 3,000; 4000]	3,000	ACO	Pheromone trail parameter	[1.00; 1.50; 2.00]	1.50
CHA	Temperature interval	[0.50; 0.75; 1.00]	1.00	ACO	Heuristic parameter	[1.00; 1.50; 2.00]	1.50
CHA	Rate of recombination	[0.50; 0.60; 0.70]	0.60	ACO	Maximum number of iterations	[3,000; 4,000; 5,000]	4,000
CHA	Rate of mutation	[0.04; 0.05; 0.06]	0.04	SA	Size of population	[30; 40; 50]	40
CHA	Number of generations between operator combination updates (*Epoch*)	[100; 150; 200]	150	SA	Initial temperature	[2,000; 3,000; 4,000]	2,000
CHA	Maximum number of generations	[3,000; 4,000; 5,000]	4,000	SA	Temperature interval	[0.50; 0.75; 1.00]	0.50
EA	Size of population	[30; 40; 50]	40	SA	Rate of recombination for solution search	[0.50; 0.60; 0.70]	0.70
EA	Initial temperature	[2,000; 3,000; 4000]	3,000	SA	Rate of mutation for solution search	[0.04; 0.05; 0.06]	0.04
EA	Temperature interval	[0.50; 0.75; 1.00]	0.75	SA	Maximum number of iterations	[3,000; 4,000; 5,000]	4,000
EA	Rate of recombination	[0.50; 0.60; 0.70]	0.60	TS	Number of solutions evaluated during the TS solution search	[30; 40; 50]	40

(continued)

Table 2 (continued)

Algorithm	Parameter	Tested values	Selected value	Algorithm	Parameter	Tested Values	Selected Value
EA	Rate of mutation	[0.04; 0.05; 0.06]	0.04	TS	Rate of recombination for solution search	[0.50; 0.60; 0.70]	0.50
EA	Maximum number of generations	[3,000; 4,000; 5,000]	4,000	TS	Rate of mutation for solution search	[0.04; 0.05; 0.06]	0.06
ACO	Size of population	[30; 40; 50]	40	TS	Size of the tabu list	[10; 15; 20]	15
ACO	Rate of recombination for solution search	[0.50; 0.60; 0.70]	0.60	TS	Maximum iterations a solution stays in the tabu list	[10; 20; 30]	30
ACO	Rate of mutation for solution search	[0.04; 0.05; 0.06]	0.06	TS	Total number of iterations a solution is not used after being added to the tabu list	[10; 15; 20]	15
ACO	Evaporation rate	[0.10; 0.15; 0.20]	0.10	TS	Maximum number of iterations	[3,000; 4,000; 5,000]	4,000

Notes * The temperature in generation *gen* was set based on the following mathematical equation: $T_{gen} = max(1, T^0 - \Delta T \cdot gen)$ (units of temperature)

quickly converge to effective solutions. The efficiency of ACO is largely attributed to its constructive probabilistic search technique, which simulates the pheromone trail-laying and following behavior of ants, effectively guiding the search towards promising regions of the solution space. SA method of allowing occasional acceptance of worse solutions prevents the algorithm from becoming trapped in local optima, thereby facilitating a broader exploration of the solution space. TS showed a slightly longer average runtime of 20.69 s. This increased time is due to its strategy of using a tabu list to avoid cycling back to previously explored solutions. While this strategy provides thorough search capabilities and robustness against local optima entrapment, it is associated with increasing computational time.

EA, with an average runtime of 18.65 s, has very similar performance compared to the CHA algorithm in terms of the overall time efficiency. On average, CHA converged in 19.27 s, slightly longer than EA, ACO, and SA, but still within competitive time ranges. The additional time consumed by CHA compared with EA is justified by its hyperheuristic structure, which adaptively selects and updates the deployed heuristic operators based on real-time performance. This capability not

Table 3 Evaluation of the considered solution approaches for all the problem instances

Instance	|V|	|B|	CHA			CPLEX			ACO			SA			TS			EA		
			Z, 10^3 USD	$\overline{Z}$, 10^3 USD	CPU, sec	Z, 10^3 USD	$\overline{Z}$, 10^3 USD	CPU, sec	Z, 10^3 USD	$\overline{Z}$, 10^3 USD	CPU, sec	Z, 10^3 USD	$\overline{Z}$, 10^3 USD	CPU, sec	Z, 10^3 USD	$\overline{Z}$, 10^3 USD	CPU, sec	Z, 10^3 USD	$\overline{Z}$, 10^3 USD	CPU, sec
S1	12	2	**8,389**	8,389	6.14	**8,389**	8,389	7200	8,397	8,412	4.54	**8,389**	8,518	4.88	**8,389**	8,439	8.8	**8,389**	8,434	5.33
S2	14	2	**9,627**	9,627	6.55	**9,627**	9,627	7200	9,680	9,680	5.2	9,680	9,680	5.52	9,680	9,680	8.91	9,680	9,680	5.82
S3	16	2	**11,412**	11,431	7.24	**11,412**	11,412	7200	11,564	11,564	5.72	11,564	11,564	6.11	11,564	11,564	9.33	11,564	11,564	6.43
S4	18	2	**12,959**	12,959	7.62	**12,959**	12,959	7200	13,167	13,199	7.19	12,984	13,142	7.03	13,167	13,167	9.06	13,167	13,167	7.74
S5	20	2	**14,731**	14,731	9.84	**14,731**	14,731	7200	14,748	14,790	8.87	14,738	14,738	7.63	14,741	14,751	10.09	14,738	14,738	8.16
S6	12	3	**7,954**	7,954	6.1	**7,954**	7,954	6142.34	**7,954**	7,954	4.57	**7,954**	7,954	4.85	**7,954**	7,954	8.78	**7,954**	7,954	5.2
S7	14	3	**9,181**	9,181	6.46	**9,181**	9,181	7200	9,257	9,317	5.13	**9,181**	9,237	5.56	9,217	9,234	8.73	9,254	9,255	5.86
S8	16	3	**10,844**	10,850	7.75	**10,844**	10,844	7200	11,015	11,067	7.05	**10,844**	10,924	6.99	11,099	11,138	8.3	10,945	11,027	6.4
S9	18	3	**11,351**	11,351	7.82	**11,351**	11,351	7200	**11,351**	11,432	6.33	11,417	11,417	6.66	**11,351**	11,410	10.61	11,417	11,417	7.04
S10	20	3	**12,432**	12,432	10.22	**12,432**	12,432	7200	12,448	12,508	9.92	12,448	12,512	10.25	12,448	12,544	12.31	12,464	12,465	7.7
M1	52	2	**45,171**	45,216	18.41				45,784	45,945	16.63	45,847	45,943	16.69	45,732	45,827	21.36	45,633	45,694	17.41
M2	54	2	**51,286**	51,306	19.07				52,320	52,550	17.29	52,248	52,455	17.68	51,888	52,036	19.9	52,013	52,087	17.78
M3	56	2	**56,080**	56,140	20.46				57,024	57,246	19.45	56,874	57,107	19.62	56,482	56,630	20.52	56,595	56,631	18.37
M4	58	2	**58,375**	58,441	19.56				59,244	59,571	18.53	59,223	59,456	18	59,168	59,357	20.98	58,854	58,982	19.02
M5	60	2	**68,645**	68,746	20.39				69,963	70,349	19.03	70,375	70,480	18.84	69,764	70,013	21.79	69,770	69,941	19.59
M6	52	4	**32,024**	32,178	18.02				32,985	33,163	16.71	32,919	33,138	16.33	33,046	33,199	19.47	32,841	32,890	17.18
M7	54	4	**34,643**	34,735	18.76				35,681	35,898	17.14	35,484	35,795	17.15	35,160	35,379	20.02	35,415	35,520	17.75
M8	56	4	**36,107**	36,305	19.25				37,261	37,925	17.88	37,119	37,648	17.67	37,415	37,704	20.28	37,211	37,282	18.52
M9	58	4	**36,393**	36,524	19.77				37,146	37,414	18.61	37,186	37,461	18.33	36,991	37,366	20.71	37,034	37,180	19.12
M10	60	4	**39,798**	39,938	20.24				40,845	41,184	19.08	40,860	41,246	18.81	40,674	40,878	21.52	40,504	40,657	19.94

(continued)

Table 3 (continued)

Instance	$\|V\|$	$\|B\|$	CHA			CPLEX			ACO			SA			TS			EA		
			Z, 10^3 USD	$\overline{\mathbf{Z}}$, 10^3 USD	CPU, sec	**Z**, 10^3 USD	$\overline{\mathbf{Z}}$, 10^3 USD	CPU, sec	**Z**, 10^3 USD	$\overline{\mathbf{Z}}$, 10^3 USD	CPU, sec	**Z**, 10^3 USD	$\overline{\mathbf{Z}}$, 10^3 USD	CPU, sec	**Z**, 10^3 USD	$\overline{\mathbf{Z}}$, 10^3 USD	CPU, sec	**Z**, 10^3 USD	$\overline{\mathbf{Z}}$, 10^3 USD	CPU, sec
L1	92	2	**107,842**	108,193	29.66				114,436	115,160	28.25	115,124	115,757	27.65	111,857	112,977	30.2	112,525	112,950	29.35
L2	94	2	**115,924**	116,030	30.2				122,499	123,139	28.76	121,765	122,557	28.44	119,310	119,531	30.73	119,856	120,326	30.09
L3	96	2	**123,710**	124,245	31.23				128,742	129,242	29.18	129,039	129,318	28.87	126,225	126,912	31.57	126,760	127,282	30.22
L4	98	2	**134,100**	134,417	30.93				139,261	139,892	29.78	139,630	140,603	29.31	136,876	137,901	31.9	137,373	137,912	31.15
L5	100	2	**144,663**	144,749	31.67				149,273	149,554	30.2	149,670	150,697	29.74	147,376	147,844	32.93	148,369	148,466	31.67
L6	92	4	**61,773**	62,131	30.11				63,901	64,196	28.41	64,514	65,027	27.7	63,291	63,923	30.73	63,377	63,685	31.28
L7	94	4	**64,019**	64,506	29.95				65,782	66,179	28.78	66,779	67,223	28.12	65,625	66,008	31.29	65,790	66,075	29.96
L8	96	4	**69,386**	70,182	30.89				71,265	71,564	29.45	72,486	73,122	28.92	75,853	76,709	32.56	72,062	72,147	30.58
L9	98	4	**71,399**	72,019	31.45				75,407	76,465	33.42	75,139	75,683	31.58	76,300	77,179	33.13	73,052	73,304	32.23
L10	100	4	**72,910**	73,329	32.34				78,903	79,809	33.08	77,988	78,542	32.63	77,912	78,737	34.04	75,747	75,853	32.56
Average			**51,104**	51,275	19.27	10,888	10,888	7094.23	52,910	53,212	18.14	52,982	53,298	17.92	52,552	52,866	20.69	52,345	52,486	18.65

* *Note* "**Z**" is adopted to show the best value of the objective function recorded from five consecutive replications; "$\overline{\mathbf{Z}}$ is adopted to show the average value of the objective function recorded from five consecutive replications; "CPU" is adopted to show the average value of the computational time recorded from five consecutive replications; bold font is adopted to show the best value of the objective function among all the values of objective function recorded for all the algorithms considered for every instance of the problem

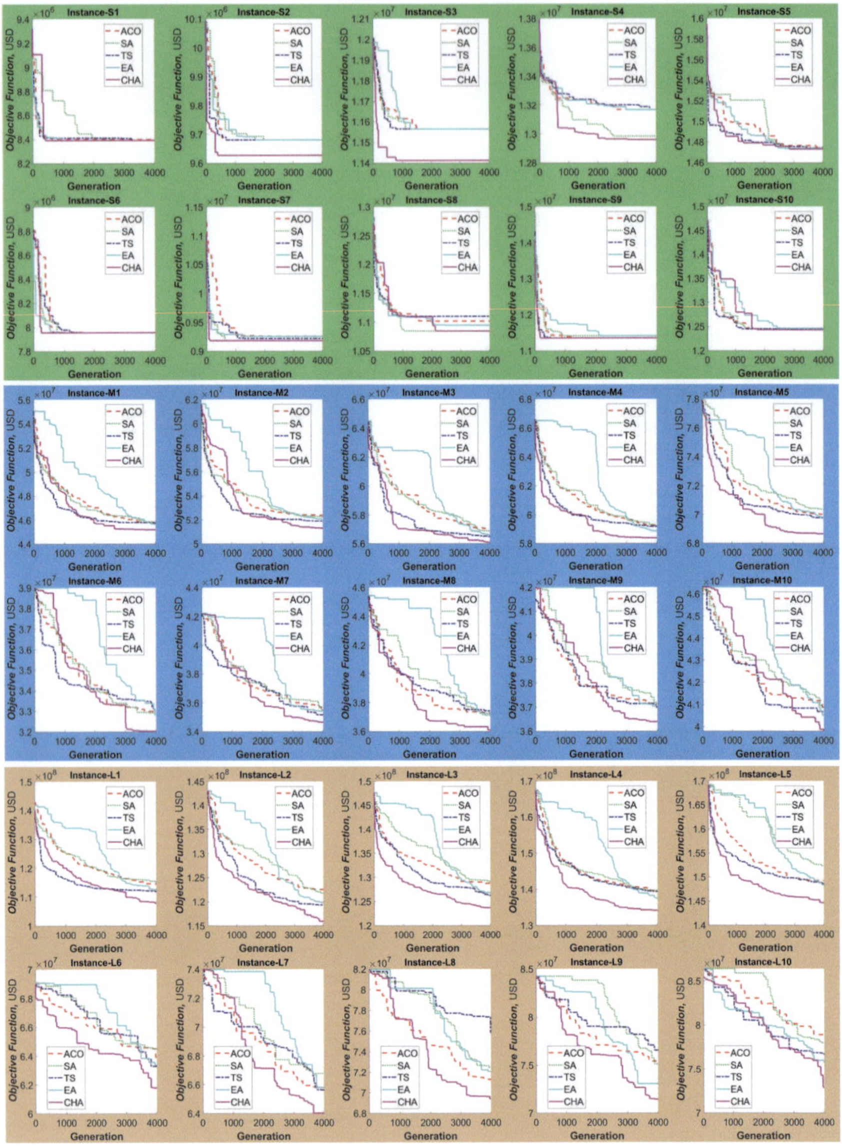

Fig. 8 Convergence plots of CHA, ACO, SA, TS, and EA

only enables it to effectively respond to changes in problem search space but also enhances its ability to find robust solutions across a wide range of scenarios.

It can be observed from the convergence plots (see Fig. 8) that without the ability to adjust operators dynamically, EA, ACO, SA, and TS showed slower convergence and reduced solution quality compared with CHA, particularly in medium and large-scale instances. The convergence curves reveal distinct differences in search behavior and convergence speed. More specifically, CHA demonstrated a steep initial improvement in objective values, followed by a gradual and stable convergence toward high-fitness solutions in most of the instances. This pattern, characterized by the rapid early progress and smooth final convergence, reflects CHA's balanced approach to exploration and exploitation, allowing it to adaptively navigate complex solution spaces. By contrast, ACO, SA, TS, and traditional EA exhibited slower and less stable convergence rates, indicating that their fixed search strategies lack the flexibility required to handle the problem's search space complexity as effectively as CHA (see Fig. 8).

CHA significantly outperformed alternative metaheuristic algorithms and EA in optimizing **GBASP**, showcasing its robust capabilities across a diverse set of operational scales of the problem. While ACO, SA, TS, and EA can achieve promising solutions in some of the small-scale instances, their effectiveness diminished with increasing size and complexity of the problem increases. This distinction becomes particularly apparent in medium and large-scale instances, where the dynamic adaptability and strategic heuristic adjustments of CHA lead to consistently lower costs and enhanced performance (see Table 3 and Fig. 8). CHA achieved the most promising results in all the 30 tested instances.

In detailed comparisons, CHA maintained an impressive record, producing the lowest average best objective values of 51,104 USD and the most promising average objective values of 51,275 USD across all 30 test cases. CHA outperformed ACO, SA, TS, and traditional EA by up to 8.22%, 6.96%, 9.32%, and 4.34% for large-scale instances in terms of the best values of the objective function recorded over five consecutive replications. The superior performance of CHA in these complex environments underscores its ability to effectively minimize costs while navigating the intricacies of berth scheduling with a precision that other algorithms are not able to match. Moreover, during the computational experiments, the stability of algorithms in returning reliable solutions was assessed as well by comparing the best and the average objective function values. A more reliable algorithm will generally have smaller differences between the best and the average objective function values. The average differences between the best and the average objective function values comprised 0.26%, 0.53%, 0.58%, 0.50%, and 0.22% for CHA, ACO, SA, TS, and EA over all the considered problem instances. This finding underscores high stability of all the algorithms, as they were able return solutions with similar objective function values from one replication to another for each problem instance. It is important to note that CHA and EA showcased smaller differences between the best and the average objective function values compared to ACO, SA, and TS, which is another indicator of their superiority.

Furthermore, the flexibility of CHA in managing the complexities of small to large-scale green berth allocation and scheduling instances underscores its significant utility for port operations aiming to reduce total turnaround cost while reducing the environmental impact. The CHA capability to dynamically adapt heuristic strategies in response to evolving port conditions makes it a promising tool for optimizing the operational flow and minimizing CO_2 emissions during vessel waiting and handling phases at marine terminals. This proficiency in balancing operational demands with stringent environmental goals distinguishes CHA from traditional methods, which often struggle to address both aspects effectively in complex scheduling environments.

5.3 Managerial Insights

For the second part of the computational experiments, CHA was applied to evaluate the operational impacts of variations in the unit penalty on CO_2 emissions and the unit penalty on delayed time in departures. The objective was to gain strategic understanding of how these parameters affect the key performance indicators, such as total vessel turnaround cost, total waiting time, handling time, total delayed time in departures, total deviations from desired berthing positions, as well as the total CO_2 emissions. To assess the effects of the unit penalty of CO_2 emissions, 25 scenarios were generated by increasing the unit penalty of CO_2 emissions from 500 to 2,500 with 500 USD/ton increments and the number of vessels from 52 to 60 with 2 vessel increments. The CHA algorithm was employed for each of these scenarios, with five independent runs conducted per scenario to compute the average values for the main variables in the **GBASP** optimization model, including: (1) overall vessel turnaround cost: $\boldsymbol{TTC} = \sum_{v\in V}\left(d_v^{wc}\boldsymbol{T}_v^{wt}\right) + \sum_{v\in V}\sum_{b\in B}\sum_{s\in S}\left(d_v^{sc}T_{vb}^{ht}\boldsymbol{x}_{vbs}\right) + \sum_{v\in V}\left(d_v^{cld}\boldsymbol{T}_v^{del}\right) + \sum_{v\in V}\left(d_v^{cd}\boldsymbol{L}_v^{dev}\right) + \sum_{v\in V}(d_c\boldsymbol{C}_v)$ (USD); (2) overall waiting time $\boldsymbol{TWT} = \sum_{v\in V}\boldsymbol{T}_v^{wt}$ (h); (3) overall handling time: $\boldsymbol{THT} = \sum_{v\in V}\sum_{b\in B}\sum_{s\in S}\left(T_{vb}^{ht}\boldsymbol{x}_{vbs}\right)$ (h); (4) overall delayed time in departures: $\boldsymbol{TDT} = \sum_{v\in V}\boldsymbol{T}_v^{del}$ (h); (5) overall deviation from the desired berthing positions $\boldsymbol{TDEV} = \sum_{v\in V}\boldsymbol{L}_v^{dev}$ (m); and (6) overall carbon dioxide emission $\boldsymbol{TCO}_2 = \sum_{v\in V}\boldsymbol{C}_v$ (tons).

The results shown in Fig. 9 reveal that increasing unit penalty on CO_2 emissions notably increased the total turnaround cost and led to a clear reduction in the total amount of CO_2 emissions ($\boldsymbol{TCO}_2$) across all vessel scenarios. The total waiting time ($\boldsymbol{TWT}$) and the total delayed time in departure ($\boldsymbol{TDT}$) were apparently decreased across all vessel scenarios accordingly. The total turnaround time decreased as it is directly correlated with the total amount of emissions (due to the fact that the vessels emit consistently during the waiting and handling operations at the MCT). Hence, increasing the unit penalty on CO_2 emissions resulted in reduced waiting times and delays. It can be concluded that the total amount of CO_2 emissions could be decreasing by focusing on altering the unit penalty on CO_2 emissions; however, the total turnaround cost must stay within an acceptable range to keep the competitiveness of the MCT while ensuring meaningful efforts on environmental protection.

It is also important to notice that both $\boldsymbol{THT}$ and $\boldsymbol{TDEV}$ did not display consistent patterns with changes in the unit penalty on CO_2 emissions, mostly stabilized with a small range, indicating that while the unit penalty on CO_2 emissions has a significant effect on waiting time and delayed departures, it does not necessarily affect the total

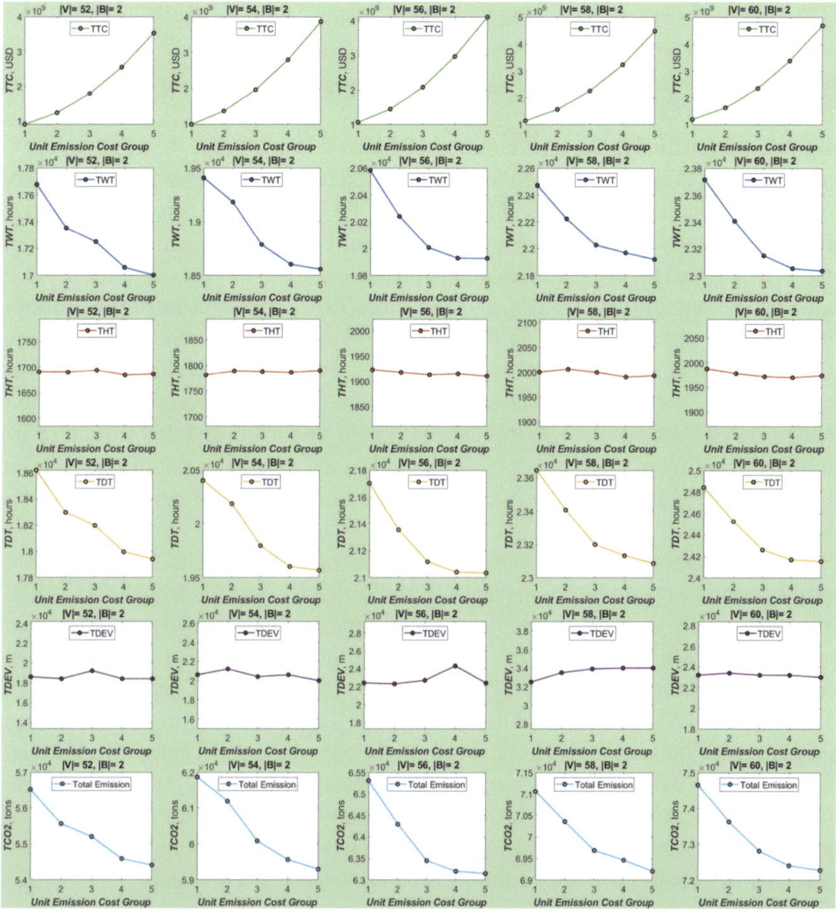

Fig. 9 The impact of increasing unit emission cost on the overall cost, overall waiting time, overall handling time, overall delayed departures, deviation from preferential berths, and overall emissions

handling time and accuracy of berthing assignments significantly. ***THT*** did not vary significantly primarily because the total service cost constitutes a major component of the objective function, leading to an already optimized berthing position allocation to minimize handling expenses in each scenario. Shorter handling times also result in lower emissions from vessels. However, comparing with the service cost, the cost of emissions seems to be less influential. Additionally, emissions from port equipment are only associated with the number of TEUs handled, not the duration of handling. Since the total number of TEUs is constant and emission factors across different berths vary within a narrow range, emissions from port equipment remain nearly fixed in the considered scenarios, leaving a minimal room for improvement as well. Additionally, both ***TDEV*** and emissions due to berthing position deviations had less substantial effects on the total turnaround cost compared to other cost components.

The effects of varying unit penalty on delayed time in departures was also studied, with 25 scenarios generated by decreasing the average unit penalty on delayed time from 6,250 to 1,250 with 1,250 USD/h increments and increasing the number of vessels from 152 to 160 with 2 vessel increments. The CHA algorithm was used in each scenario, with five independent runs conducted to determine the averages of the main variables of the **GBASP** model. As indicated in Fig. 10, the outcomes indicated a consistent increase in $\boldsymbol{TWT}$, $\boldsymbol{TDT}$, and $\boldsymbol{TCO_2}$ as the unit penalty on delayed time in departures decreased. This suggests that higher unit penalties on delayed time in departures facilitate smoother operations and alleviate port congestion, while reducing the amount of emissions. In contrast, $\boldsymbol{THT}$ remained mostly unaffected by changes in the unit penalty on delayed time in departures. Such a pattern can be justified by the fact that vessels were allocated to the same or similar cost-effective berths in each scenario, as it was explained in the previous experiment for the unit penalty on CO_2 emissions. $\boldsymbol{TDEV}$ also exhibited minor fluctuations similar to the previous experiment, which can be attributed to the relatively minor contribution of $\boldsymbol{TDEV}$ and associated emissions to the total turnaround cost compared to other factors.

These numerical experiments provide critical practical insights into optimizing marine terminal operations using the **GBASP** model supported by the CHA algorithm. The analysis demonstrated that increasing unit penalties on emissions and late departures were influential in enhancing port service allocation and scheduling. Increasing the unit penalty of CO_2 emissions and the unit cost of late departures led to substantial reductions in the total waiting time, total delayed time in departures, and total emissions, emphasizing the potential operational gains achievable through the alternation of these penalties, as long as the total turnaround cost remains within an acceptable range. Emission reduction is particularly important considering the existing initiatives for mitigating climate change effects and ambitious emission reduction targets in different countries around the world. This study demonstrates the applicability and effectiveness of CHA for sustainable and environment-friendly berth planning in real-world marine terminal operations.

6 Conclusion

Berth allocation and scheduling at maritime container terminals are crucial for optimizing port operations and maintaining the efficiency of maritime logistics under the pressures of increasing global trade flows and operational complexities. One of the most concerning challenges associated with berth allocation and scheduling is its environmental impact, particularly the carbon dioxide emissions generated during port operations. This paper introduced a Customized Hyperheuristic Algorithm (CHA) for the green berth allocation and scheduling problem, which integrates environmental considerations to minimize both the operational costs and the CO_2 emissions during vessel waiting and handling operations. The CHA algorithm incorporated various evolutionary strategies and dynamically applied different genetic

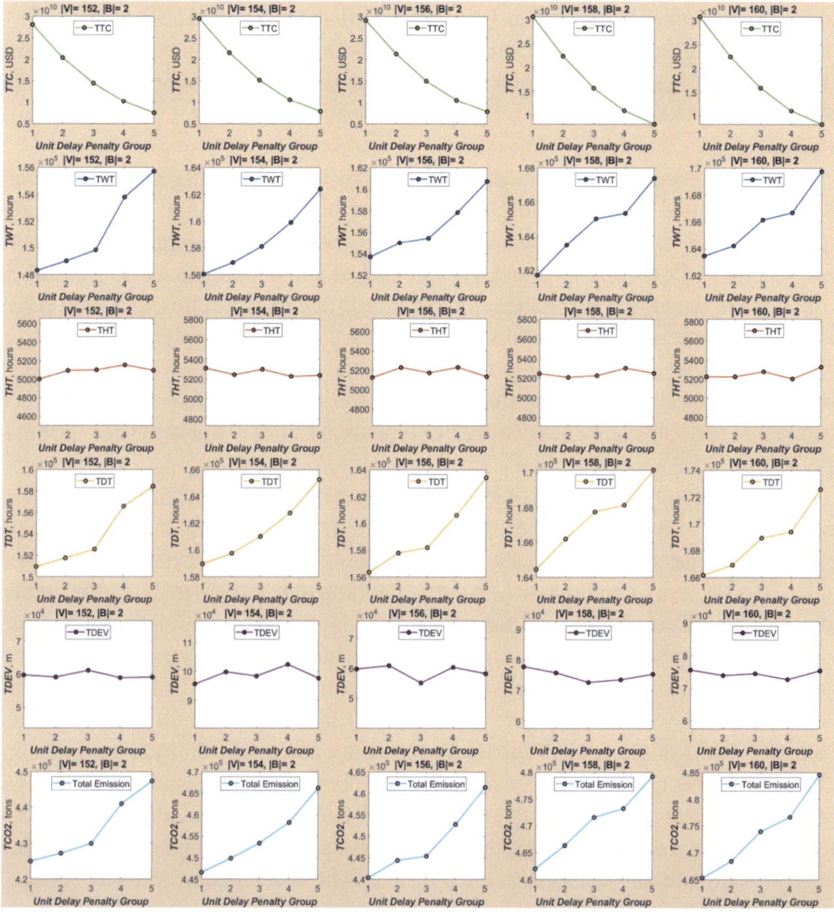

Fig. 10 The impact of increasing unit penalty on delayed time in departures on the overall cost, overall waiting time, overall handling time, overall delayed departures, deviation from preferential berths, and overall emissions

operators to solve the optimization problem. CHA employed a dynamic selection mechanism that adaptively chose heuristic operators based on their ongoing performance, enhancing the traditional framework of typical heuristic and metaheuristic algorithms.

Detailed computational experiments compared CHA with established optimization techniques, such as CPLEX, ant colony optimization, simulated annealing, tabu search, and evolutionary algorithm that were all customized for the decision problem of interest. The findings revealed that the CHA's adaptive approach in selecting and employing recombination and mutation operators delivered substantial improvements in the search process. For small-scale problem instances, CHA consistently matched exact optimization results (i.e., CPLEX), and for all tested scenarios, it surpassed the competing metaheuristics along with the traditional evolutionary algorithm while maintaining very promising computational time. The results underscored CHA's superior capability in reducing emissions while improving operational efficiency via providing effective berth allocation and scheduling plans in a timely manner.

Additional experiments provided some insights into port management. The analysis demonstrated that increasing unit penalties on emissions and late departures were influential in enhancing port service allocation and scheduling. Increasing the unit penalty of CO_2 emissions and the unit cost of late departures led to significant reductions in the total waiting time, total delayed time in departures, and total emissions, emphasizing the potential operational gains achievable through the alternation of these penalties, as long as the total turnaround cost remains within an acceptable range. This study demonstrated the applicability and effectiveness of CHA for sustainable and environment-friendly berth planning in real-world marine terminal operations.

Future research avenues could explore several promising directions. First, enhancing the adaptability of the CHA algorithm by incorporating machine learning techniques for more adaptive decision-making could further improve its performance under varying port conditions. Second, extending the model to integrate multi-objective optimization that simultaneously considers other operational aspects (e.g., noise pollution and energy consumption) alongside CO_2 emissions would offer a more comprehensive approach to sustainable port management. Third, applying the CHA framework to other port operations, such as yard crane scheduling and container stacking optimization, could validate its versatility and applicability. Fourth, incorporating unexpected changes, such as weather disruptions or vessel arrival delays, for dynamic berth allocation with environmental considerations could enhance practical significance of this work. Fifth, collaborative scheduling approaches involving multiple terminals and interconnected ports could be developed to optimize regional or global maritime logistics networks, thereby broadening the algorithm's impact on the sustainability of larger supply chains. Lastly, although this study performed quite extensive experiments and compared CHA with established optimization techniques (i.e., CPLEX, ant colony optimization, simulated annealing, tabu search, and evolutionary algorithm that were all customized for the decision problem of interest), alternative solution approaches could be further used for a more thorough evaluation, including more recent metaheuristics, Benders decomposition-based methods, branch-and-price algorithm, and column generation-based approaches.

References

Bierwirth, C., & Meisel, F. (2010). A survey of berth allocation and quay crane scheduling problems in container terminals. *European Journal of Operational Research, 202*(3), 615–627.

Bierwirth, C., & Meisel, F. (2015). A follow-up survey of berth allocation and quay crane scheduling problems in container terminals. *European Journal of Operational Research, 244*(3), 675–689.

Burns, M. G. (2018). *Port management and operations. CRC Press* (1st ed.). Boca Raton.
Chen, M. R., Huang, Y. Y., Zeng, G. Q., Lu, K. D., & Yang, L. Q. (2021). An improved bat algorithm hybridized with extremal optimization and Boltzmann selection. *Expert Systems with Applications, 175*, Article 114812.
Cruz-Duarte, J. M., Amaya, I., Ortiz-Bayliss, J. C., Conant-Pablos, S. E., Terashima-Marín, H., & Shi, Y. (2021). Hyper-heuristics to customise metaheuristics for continuous optimisation. *Swarm and Evolutionary Computation, 66*, Article 100935.
Duan, J., Liu, F., Zhang, Q., Qin, J., & Zhou, Y. (2024). Genetic programming hyper-heuristic-based solution for dynamic energy-efficient scheduling of hybrid flow shop scheduling with machine breakdowns and random job arrivals. *Expert Systems with Applications*, 124375.
Dulebenets, M. A. (2023). A Diffused Memetic Optimizer for reactive berth allocation and scheduling at marine container terminals in response to disruptions. *Swarm and Evolutionary Computation, 80*, Article 101334.
Dulebenets, M. A., Moses, R., Ozguven, E. E., & Vanli, A. (2017). Minimizing carbon dioxide emissions due to container handling at marine container terminals via hybrid evolutionary algorithms. *IEEE Access, 5*, 8131–8147.
Golias, M., Boile, M., Theofanis, S., & Efstathiou, C. (2010). The berth scheduling problem: Maximizing berth productivity and minimizing fuel consumption and emissions production. *Transportation Research Record, 2166*(1), 20–27.
Hu, Q. M., Hu, Z. H., & Du, Y. (2014). Berth and quay-crane allocation problem considering fuel consumption and emissions from vessels. *Computers & Industrial Engineering, 70*, 1–10.
Hu, Z. H. (2020). Low-emission berth allocation by optimizing sailing speed and mooring time. *Transport, 35*(5), 486–499.
Imai, A., Nishimura, E., & Papadimitriou, S. (2003). Berth allocation with service priority. *Transportation Research Part B: Methodological, 37*(5), 437–457.
IMO. (2020). *Initial IMO strategy on reduction of GHG emissions from ships*. IMO.
Jauhar, S. K., Pratap, S., Kamble, S., Gupta, S., & Belhadi, A. (2023). A prescriptive analytics approach to solve the continuous berth allocation and yard assignment problem using integrated carbon emissions policies. *Annals of Operations Research*, 1–32.
Jiang, M., Feng, J., Zhou, J., Zhou, L., Ma, F., Wu, G., & Zhang, Y. (2023). Multi-terminal berth and quay crane joint scheduling in container Ports considering carbon cost. *Sustainability, 15*(6), 5018.
Jiang, M., Ma, F., Zhang, Y., Lv, S., Pei, Z., & Wu, G. (2024). Collaborative scheduling optimization of container Port Berths and cranes under low-carbon environment. *Sustainability, 16*(7), 2985.
Jiang, X., Zhong, M., Shi, J., Li, W., Sui, Y., & Dou, Y. (2022). Overall scheduling model for vessels scheduling and berth allocation for ports with restricted channels that considers carbon emissions. *Journal of Marine Science and Engineering, 10*(11), 1757.
Kontovas, C., & Psaraftis, H. N. (2011). Reduction of emissions along the maritime intermodal container chain: Operational models and policies. *Maritime Policy & Management, 38*(4), 451–469.
Leng, L., Zhang, J., Zhang, C., Zhao, Y., Wang, W., & Li, G. (2020). Decomposition-based hyperheuristic approaches for the bi-objective cold chain considering environmental effects. *Computers & Operations Research, 123*, Article 105043.
Li, B., Elmi, Z., Borowska-Stefańska, M., Wiśniewski, S., Lau, Y.Y., Chen, Q., & Dulebenets, M.A. (2024). A comprehensive framework for hyperheuristic algorithms for Berth allocation and scheduling at Marine Container Terminals. In: Principles and applications of adaptive artificial intelligence (pp. 1–19). IGI Global.
Li, B., Elmi, Z., Manske, A., Jacobs, E., Lau, Y. Y., Chen, Q., & Dulebenets, M. A. (2023). Berth allocation and scheduling at marine container terminals: A state-of-the-art review of solution approaches and relevant scheduling attributes. *Journal of Computational Design and Engineering, 10*(4), 1707–1735.
Lin, S., Zhen, L., Wang, W., & Tan, Z. (2023). Green berth and yard space allocation under carbon tax policy in tidal ports. *Maritime Policy & Management, 50*(8), 1080–1101.

Moon, D. S. H., & Woo, J. K. (2014). The impact of port operations on efficient ship operation from both economic and environmental perspectives. *Maritime Policy & Management, 41*(5), 444–461.

Mosadegh, H., Ghomi, S. F., & Süer, G. A. (2020). Stochastic mixed-model assembly line sequencing problem: Mathematical modeling and Q-learning based simulated annealing hyper-heuristics. *European Journal of Operational Research, 282*(2), 530–544.

Qian, W., Chai, J., Xu, Z., & Zhang, Z. (2018). Differential evolution algorithm with multiple mutation strategies based on roulette wheel selection. *Applied Intelligence, 48*, 3612–3629.

Rodríguez-Esparza, E., Ramos-Soto, O., Masegosa, A. D., Onieva, E., Oliva, D., Arriandiaga, A., & Ghosh, A. (2024). Optimizing road traffic surveillance: A robust hyper-heuristic approach for vehicle segmentation. *IEEE Access, 12*, 29503–29524.

Sha, M., Zhai, J., Wan, Z., Wu, Z., Jin, Y., & Zhu, E. (2024). Insights into the global characteristics of shipping exhaust emissions at Berth. *Journal of Marine Science and Engineering, 12*(9), 1527.

Song, H. B., Yang, Y. H., Lin, J., & Ye, J. X. (2023). An effective hyper heuristic-based memetic algorithm for the distributed assembly permutation flow-shop scheduling problem. *Applied Soft Computing, 135*, Article 110022.

Song, Y., Zhang, J., Liu, M., & Chu, C. (2019). The berth allocation optimisation with the consideration of time-varying water depths. *International Journal of Production Research, 57*(2), 488–516.

Soong, C. J., Rahman, R. A., Ramli, R., Manaf, M. S. A., & Ting, C. C. (2022). An evolutionary algorithm: An enhancement of binary tournament selection for fish feed formulation. *Complexity*, 2022.

Tang, S., Jin, J. G., & Lu, C. (2022). Investigation of berth allocation problem in container ports considering the variety of disruption. *Computers & Industrial Engineering, 172*, Article 108564.

Tran, N. K., Haasis, H. D., & Buer, T. (2017). Container shipping route design incorporating the costs of shipping, inland/feeder transport, inventory and CO2 emission. *Maritime Economics & Logistics, 19*(4), 667–694.

Venturini, G., Iris, Ç., Kontovas, C. A., & Larsen, A. (2017). The multi-port berth allocation problem with speed optimization and emission considerations. *Transportation Research Part D: Transport and Environment, 54*, 142–159.

Verschuur, J., Koks, E. E., & Hall, J. W. (2022). Ports' criticality in international trade and global supply-chains. *Nature Communications, 13*(1), 4351.

Wang, J., Li, H., Yang, Z., & Ge, Y. E. (2024). A novel scheme for shore power data to enhance containership-at-berth emission estimation. *Transportation Research Part D: Transport and Environment, 134*, Article 104353.

Wang, T., Cheng, P., & Zhen, L. (2023). Green development of the maritime industry: Overview, perspectives, and future research opportunities. *Transportation Research Part E: Logistics and Transportation Review, 179*, Article 103322.

Wang, T., Li, M., & Hu, H. (2019). Berth allocation and quay crane-yard truck assignment considering carbon emissions in port area. *International Journal of Shipping and Transport Logistics, 11*(2–3), 216–242.

Wen, X., Chen, Q., Yin, Y. Q., & Lau, Y. Y. (2023). Green Vessel Scheduling with Weather Impact and Emission Control Area Consideration. *Mathematics, 11*(24), 4874.

Wen, X., Chen, Q., Yin, Y. Q., Lau, Y. Y., & Dulebenets, M. A. (2024). Multi-Objective Optimization for Ship Scheduling with Port Congestion and Environmental Considerations. *Journal of Marine Science and Engineering, 12*(1), 114.

Xu, D., Li, C. L., & Leung, J. Y. T. (2012). Berth allocation with time-dependent physical limitations on vessels. *European Journal of Operational Research, 216*(1), 47–56.

Zhang, Z., Song, C., Zhang, J., Chen, Z., Liu, M., Aziz, F., Kurniawan, T.A., & Yap, P.S. (2023). Digitalization and innovation in green ports: A review of current issues, contributions and the way forward in promoting sustainable ports and maritime logistics. *Science of the Total Environment*, 169075.

Zhao, F., Zhu, B., & Wang, L. (2023). An estimation of distribution algorithm-based hyper-heuristic for the distributed assembly mixed no-idle permutation flowshop scheduling problem. *IEEE Transactions on Systems, Man, and Cybernetics: Systems, 53*(9), 5626–5637.

Zhou, X., Ruan, X., Wang, H., & Zhou, G. (2023). Exploring spatial patterns and environmental risk factors for global maritime accidents: A 20-year analysis. *Ocean Engineering, 286*, Article 115628.

Zhu, L., Zhou, Y., Sun, S., & Su, Q. (2023). Surgical cases assignment problem using an efficient genetic programming hyper-heuristic. *Computers & Industrial Engineering, 178*, Article 109102.

Bokang Li is a Doctoral Candidate in the Department of Civil and Environmental Engineering at Florida A&M University-Florida State University (FAMU-FSU) College of Engineering. He holds a B.S. degree in Engineering with concentration in Harbor, Waterway and Coastal Engineering from Wuhan University (Wuhan, China). His research interests include, but are not limited to, operations research, optimization, metaheuristics, hybrid algorithms, hyper-heuristics, supply chain management, transportation engineering, freight terminals, and marine container terminals. He has been actively involved in multiple research projects funded by federal and state agencies, including the U.S. Department of Defense and the Florida Department of Transportation. His research findings have been published in the leading international journals, such as Reliability Engineering and System Safety, Transportation Research Part D, Journal of Industrial Information Integration, and Computers and Industrial Engineering.

Fatemeh Shekoohi is a Ph.D. student in the Department of Civil and Environmental Engineering at Florida State University (FSU). She holds bachelor's and master's degrees in industrial engineering from Shiraz University of Technology, Iran. Her research focuses on supply chain management, sustainability, logistics, and optimization under uncertainty, with applications in transportation and resource allocation. Her expertise includes operations research, mathematical modeling, and decision-making under uncertainty, particularly in supply chain networks, logistics efficiency, and transportation systems. She has contributed to research on sustainable pharmaceutical supply chains, emphasizing robust optimization, perishability considerations, and environmental impact reduction. Her work involves the application of optimization techniques to improve decision-making in large-scale, complex systems. Fatemeh has published research in sustainable supply chain design, focusing on resource allocation, and environmental considerations. Her current research explores the intersection of supply chain sustainability and transportation optimization, integrating mathematical modeling and operational strategies to enhance efficiency and reduce environmental impacts. She aims to contribute to the development of efficient, sustainable, and resilient logistics systems, aligning her work with emerging challenges in green supply chain management and transportation planning.

Payam Afkhami is a Ph.D. candidate in the Department of Civil and Environmental Engineering at Florida State University (FSU). He holds a Bachelor of Science in Industrial Engineering from the University of Tehran, Iran and a Master of Science in industrial engineering from Shiraz University of Technology, Iran. His research focuses on operations research, supply chain management, transportation, logistics, and multi-criteria decision-making. His expertise includes optimization techniques, scheduling, and mathematical modeling for complex decision-making processes in supply chain management and transportation. He has contributed to multiple research projects, including waste management, biofuel supply chain optimization, and resource allocation for safety enhancement at level crossings. His work integrates heuristic and meta-heuristic algorithms with mathematical programming to develop efficient solutions for real-world problems. Payam has published research in biofuel supply chain optimization and transportation network efficiency, addressing key challenges in sustainable logistics and infrastructure planning. His current work explores the applications of operations research in transportation systems, focusing on resource allocation, scheduling, and optimization models for improving sustainability and efficiency. His long-term goal is to develop advanced decision-support systems that enhance transportation planning, infrastructure resilience, and sustainable supply chain operations.

Razieh Khayamim is a Ph.D. candidate in Transportation Engineering at Florida State University, where she serves as a Graduate Research Assistant at the Resilient Infrastructure and Disaster Response (RIDER) Lab. She holds a B.Sc. and M.Sc. in Civil Engineering from Isfahan University of Technology, Iran. Her research lies at the intersection of operations research and transportation systems, with a strong focus on optimizing emergency evacuation planning, sustainable mobility, and disaster resilience. With a passion for solving real-world challenges, Razieh has authored and contributed as a co-author to multiple peer-reviewed journal articles and conference proceedings. Her work explores diverse critical areas, such as optimizing hurricane evacuation strategies integrating engineering and social performance indicators to enhance decision-making, mobile electric vehicle (EV) charging stations deployment, and optimizing transportation network expansion to promote sustainable, low-carbon mobility. She leverages advanced optimization techniques, artificial intelligence, and data-driven decision-making to develop practical solutions that improve transportation system efficiency and resilience. Beyond her research, Razieh is actively engaged in professional organizations. She serves as the Treasurer of the FAMU-FSU Institute of Transportation Engineers (ITE) chapter and is a member of the TRB Young Members Council for Sustainability and Resilience. Her research interests include, but are not limited to, transportation network analysis, operation research, sustainable and resilient transportation systems, advanced computational optimization methodologies, and artificial intelligence applications in transportation engineering.

Maxim A. Dulebenets is an Associate Professor and Graduate Program Director in the Department of Civil and Environmental Engineering at Florida A&M University-Florida State University (FAMU-FSU) College of Engineering. Dr. Dulebenets holds B.Sc. and M.Sc. degrees in Railway Construction from the Moscow State University of Railroad Engineering (Moscow, Russia) and M.Sc. and Ph.D. degrees from the University of Memphis (Memphis, TN – USA) in Civil Engineering with concentration in Transportation. His research interests include, but are not limited to, operations research, optimization, simulation modeling, NP-hard problems, mathematical programming, metaheuristics, hybrid algorithms, evolutionary computation, transportation engineering, freight transportation, intermodal freight facilities, railroads, liner shipping scheduling, and GPS data processing. Dr. Dulebenets has been involved in a variety of research projects with the overall value ~$19.5 million (~$2.9 million as a PI), sponsored by the United States Department of Defense, United States Department of Transportation, National Science Foundation, Center for Accessibility and Safety for an Aging Population, Florida Department of Transportation, Tennessee Department of Transportation, Center for Freight and Infrastructure Research and Education, Federal Highway Administration, Canadian National Railway, Federal Express, Intermodal Freight Transportation Institute, North Caucasus Railroad Authority, and Moscow Railroad Authority.

He serves as a referee for 250+ international journals. According to the 2024 Scopus citation report, he is ranked as #14 in the world out of 28,891 scholars in Logistics and Transportation based on the composite c-score c(ns) and last year performance and #148 based on the lifetime performance. His work has been published in the leading international journals, including the International Journal of Production Economics, Information Sciences, Reliability Engineering and System Safety, and IEEE Transactions on Intelligent Transportation Systems. Dr. Dulebenets is actively involved in activities of more than 10 Standing Committees and Subcommittees of the Transportation Research Board of the National Academies of Sciences, Engineering, and Medicine. He is an Affiliate Member of the INFORMS Optimization Society and a Senior Member of the IEEE. Dr. Dulebenets is a licensed Professional Engineer registered in the State of Michigan (License № 6201063266).

Decarbonization Transition Strategy of Container Vessels with Vessel Size, Route, and Policy Considerations

Anne Xaviera Go, Tomoya Kawasaki, Rei Miratsu, and Ryuji Miyake

Abstract Stricter decarbonization targets force shipping companies to reassess their transition strategy that needs to be economical and compliant with the regulatory targets of the future. This study aims to illustrate the steps that companies may take to achieve these goals for various sizes of container vessels and routes. The decarbonization strategy to be implemented must consider retrofit costs, new shipbuilding prices, efficiencies in new ship design, maintenance, and downtime costs. Through scenario analysis, the study aims to highlight how the decision changes based on the implementation of Carbon Intensity Indicator (CII) and penalties related to the violation of different potential emissions reduction policies. The study reveals that the transition strategies of shipping companies vary based on ship sizes, highlighting the need for tailored approaches in the decarbonization process. The strategies for the same ship size operating on different routes are nearly equivalent. Future fuel price forecasts are crucial in influencing decisions related to ship purchases and retrofits, as economic viability remains a primary concern for the industry. For penalties to be effective in driving change, they must be substantial enough to compel shipping companies to adopt cleaner technologies and fuels. Moreover, clarity on potential future greenhouse gas (GHG) policies and regulations is essential for companies to navigate their decarbonization strategies effectively. The study underscores the complex interplay of environmental regulations, financial considerations, and fuel choices in shaping the decisions of shipping companies for various container vessel sizes.

A. X. Go · T. Kawasaki (✉)
Department of Systems Innovation, Graduate School of Engineering, The University of Tokyo, Bunkyo-Ku, Hongo, Tokyo, Japan
e-mail: kawasaki@sys.t.u-tokyo.ac.jp

A. X. Go
e-mail: annego@g.ecc.u-tokyo.ac.jp

R. Miratsu · R. Miyake
Nippon Kaiji Kyokai Administration Center Annex, Research Institute, Nipon Kaiji Kyokai (ClassNK), Tokyo, Japan
e-mail: r-miratsu@classnk.or.jp

R. Miyake
e-mail: miyake@classnk.or.jp

M. A. Dulebenets et al. (eds.), *Transport and Climate Change*,
https://doi.org/10.1007/978-3-032-00563-2_18

Keywords Decarbonization · Container vessels · Alternative fuels · Optimization · Maritime emission policies · Net-zero transition

1 Introduction

The shipping industry regulators have become increasingly concerned about its environmental impact and are more persistent in their plans to meet the long-term net zero-emissions commitment. Ships are significant sources of greenhouse gas emissions, accounting for 2.9% of the total GHG emissions (OECD, 2022). Since its establishment, the International Maritime Organization (IMO) has created policy frameworks to consolidate and guide efforts to reduce emissions. Maritime shipping policies have evolved significantly, with IMO leading efforts since the 1990s. In 2003, the IMO adopted Annex VI of MARPOL, introducing regulations to control air pollution from ships, including sulfur oxide (SOx) and nitrogen oxide (NOx) emissions. In 2011, the Energy Efficiency Design Index (EEDI) was introduced, setting mandatory energy efficiency standards for new ships to reduce carbon emissions. In 2018, the IMO adopted an initial strategy to reduce greenhouse gas emissions by at least 50% by 2050 compared to 2008 levels, with a long-term goal of reaching net-zero emissions. This was further strengthened in 2021 with the introduction of the Carbon Intensity Indicator (CII) and the Energy Efficiency Existing Ship Index (EEXI), both aimed at improving the energy efficiency of ships.

The IMO's commitment to decarbonization was reinforced in 2023 with a revised strategy targeting the reduction in GHG emissions by 2050 and emphasizing the adoption of low- and zero-carbon fuels. The updated strategy aims for a reduction of GHG emissions by at least 70% by 2050, with a strong emphasis on achieving net-zero emissions by or around 2050, depending on developments in alternative fuels and energy sources (IMO, 2023). This strategy also underscores the importance of mid-term measures, such as the adoption of low- and zero-carbon fuels and the integration of new technologies, to ensure the shipping industry stays on track to meet these targets. The European Union (EU) has also made significant strides in decarbonizing maritime transport. In 2023, the EU extended its Emissions Trading System (ETS) to include shipping, requiring companies to monitor, report, and pay for their CO_2 emissions. Additionally, the FuelEU Maritime initiative, part of the European Green Deal, aims to reduce the carbon intensity of maritime transport by setting GHG intensity limits on energy used by ships starting in 2025. These measures create financial incentives for adopting cleaner fuels and technologies, supporting the goal of climate neutrality by 2050.

In light of recent changes in regulatory policies, the shipping industry is exploring alternative fuel types to reduce its environmental impact and move towards more efficient and environmentally friendly solutions. Among the existing options, liquefied natural gas (LNG) is the most feasible alternative fuel for ships. Having been used by other industries, LNG has achieved competitive pricing and increased supply due to growing infrastructure and production capacities. Methanol is being evaluated as a viable alternative fuel source, especially for smaller vessels and short-sea shipping. Despite its potential to replace current fuel oil, the energy consumption to produce methanol from feedstock requires significantly more than traditional heavy fuel oil (HFO) production (Law et al., 2021). Apart from LNG and methanol, biofuels have also been extensively developed as an alternative fuel for the shipping industry. Despite their market presence, the availability and pricing of biofuels can vary depending on feedstock availability, production volumes, and government policies. Ammonia is a potential alternative fuel type for companies looking to propel their ships with zero carbon emissions. The future price and supply projections of ammonia remain uncertain due to ongoing research and development, but its potential to be produced from renewable energy sources at scale may help stabilize prices in the future. Lastly, hydrogen, another emerging fuel being evaluated for the maritime sector, is in its early stages of development and commercialization. Hydrogen can also be stored in ammonia through the Haber–Bosch process (Serpell et al., 2023).

While the overall direction of maritime GHG policy is clear, with frameworks like FuelEU Maritime and the inclusion of shipping in the EU ETS, there remains uncertainty regarding the medium-term goals and specific annual targets. Although reduction milestones have been set for the coming decades, detailed annual targets are not yet fully defined, and there is variability in how emissions are accounted for (Tank-to-Wake vs. Well-to-Wake) as well as in the implementation of penalties and compliance mechanisms. As the industry adapts to these evolving regulations, a degree of uncertainty persists, prompting a cautious wait-and-see approach among stakeholders. These policies can significantly influence the strategic decisions of shipping companies as they balance economic viability and regulatory compliance. Options include investing in new vessels that are designed to meet GHG reduction targets, retrofitting existing ships with cleaner technologies, or accepting the financial burden of paying penalties for non-compliance. The choice will depend on various factors, including the availability of new technologies, the cost of retrofitting, and the projected penalties under the greenhouse gas reduction policies and regulations. Furthermore, these decisions are not made in isolation; they have direct implications for customers and shippers. If the cumulative costs of compliance—ranging from capital investments in new ships or retrofitting to operational changes and potential penalties—become too high, companies may pass these costs on to their customers. This could result in increased shipping rates, which would ripple through the supply chain and affect the cost of goods for consumers. In this evolving regulatory environment, shipping companies will need to be agile and forward-thinking to maintain competitiveness and avoid disrupting their service offerings.

The decision-making process for each ship may vary based on its characteristics, such as engine type, fuel consumption, service routes, and twenty-foot equivalent unit (TEU) capacity. Consequently, each decision—whether to retrofit, replace, or potentially pay penalties—must be carefully tailored to the individual ship's profile and the broader context of its operations. This approach ensures that companies can optimize their fleet management strategies in line with both regulatory demands and operational efficiency. A thorough assessment of ship management options, such as retrofitting or complete replacement, is essential for companies. The first decision factor is cost, as companies need to weigh the financial implications of each option, considering maintenance expenses, potential operational inefficiencies, and upfront investments. These are coupled with the challenges and uncertainties associated with alternative fuel supply, infrastructure, and pricing. Regulatory compliance is another important consideration, with companies mandated to evaluate the compliance of their existing fleet to evolving targets. Technological advancements also play a role, as newer ships often incorporate advanced technologies that enhance operational efficiency and reduce the environmental impact. Evaluating these factors allows companies to determine whether retrofitting existing ships with new technologies is feasible or if complete replacement is needed to benefit from the latest innovations. Moreover, aligning ship management decisions with long-term strategic goals, such as market demand and environmental targets, ensures sound fleet composition and competitiveness in the maritime industry. This study builds on the existing literature on how companies in liner shipping can meet their net-zero objectives and minimize costs by providing a model for engine and fuel choice while considering vessel sizes, routes, and future regulatory policies.

Section 2 presents a literature review covering the relevant studies. Section 3 outlines the mathematical model used in the analysis. In Sect. 4, the case study, data, and scenarios are introduced. Section 5 details the results of the model optimization, while Sect. 6 offers a discussion of the key insights derived from the model outcomes and the various scenarios. Finally, Sect. 7 concludes the study and suggests directions for future research.

2 Literature Review

Given the importance of achieving the net-zero target by 2050, several efforts have aimed to understand what pathways must be taken by shipping companies (see Table 1). In most of these studies, cost minimization was the main driver in the decision (Lagemann et al., 2022, 2023; Loennechen et al., 2024; Zhao et al., 2023, 2024; Zwaginga et al., 2024). In the research by Lagemann et al. (2022) and Lagemann et al. (2023), the optimization model was formulated as a bi-objective integer optimization problem using both cost and emissions minimization. In the other studies, the emissions targets in certain years were considered as constraints in the model. Among the alternative fuels analyzed, biofuels, LNG, LPG, methanol, ammonia, and hydrogen were considered in most of the related studies. As the starting fuel, heavy fuel oil (HFO), very low sulfur fuel oil (VLSFO), and marine diesel oil (MDO) were often considered. Most of the studies used dry bulkers as the vessel type in their analysis (Lagemann et al., 2022, 2023; Loennechen et al., 2024; Zhao et al., 2024; Zwaginga et al., 2024). In the study conducted by Zhao et al. (2023), a fleet of container vessels were used in analyzing the decarbonization pathways. They used a fleet of 1,600-TEU and 8,500-TEU vessels, covering both short-sea and deep-sea routes.

Table 1 Vessel type, scope, and methods used in recent studies

References	Vessel	Scope	Method
Lagemann et al. (2022)	Dry bulker	Single vessel	Bi-objective integer optimization for cost and emissions
Lagemann et al. (2023)	Dry bulker	Single vessel	Two-stage stochastic optimization model considering fuel and carbon price
Zhao et al. (2023)	Container	Fleet	Cost minimization integrating ship purchase and sale, retrofit, deployment, and speed optimization under various policies for fleet composition
Loennechen et al. (2024)	Dry bulker	Fleet	Two-stage stochastic optimization model considering fuel and carbon price for fleet composition
Zhao et al. (2024)	Dry bulker	Single vessel	Interval quadratic programming model integrating power systems, alternative fuels, and air emission control measures
This paper	Container	Single vessel	Modelling ship technology and alternative fuel choice to assess optimal decision strategy while accounting for various vessel sizes, routes, and policy implementations

It has been noted that fuel prices play a significant role in the decarbonization transition strategy of shipping companies. In the study by Zhao et al. (2023), several price scenarios were considered alongside varied GHG target implementations, including 2018 IMO GHG strategic goals, "Fit for 55" package, demand-based GHG targets, and 100% reduction by 2050.To incorporate the uncertainty inherent in future fuel price forecasts, Lagemann et al. (2023), Loennechen et al. (2024), and Zwaginga et al. (2024) used stochastic parameters to model fuel price as uncertain factors. It is interesting to note that the probability distributions used for the fuel price followed a triangular distribution between the upper and lower bound prices for each fuel type. According to Lagemann et al. (2023), the triangular distribution was chosen as it is not possible to accurately forecast the probability distribution of prices for future fuels presently. Moreover, the carbon price was implicitly considered in the models by incorporating it with the fuel price (Lagemann et al., 2022, 2023; Zwaginga et al., 2024). In the study by Lagemann et al. (2023), carbon emission prices of 0, 100, and 1,000 USD per ton of CO_2eq were considered, alongside a stochastic beta-variate carbon price.

In the study by Chen et al. (2023), a multinomial logit model was used to find the main determinants that affect the choices of shipowners with respect to alternative fuels when they decide on new vessel orders. Alternative fuels (e.g., liquefied natural gas) were found to be the most attractive choice for gas ships and ro-ro carriers, whereas conventional fuels would be preferential for other vessel types. Although vessel type is deemed significant in shipowners' selection, ship size may have limited influence on the decision. Moreover, the study emphasizes the role of policy mechanisms in shaping shipping behaviors. That is, recent IMO policies have increased the construction demand for new and energy-efficient ships. Another study investigated the economic performance of alternative fuels and energy efficiency measures using the marginal abatement cost curve (MACC) (Tu et al., 2024). MACCs were designed for every ship type and diverse GHG abatement alternatives, such as technical measures, operational measures, alternative fuels, and renewable energy sources. The cost-effectiveness rankings of different alternative fuel types considered for the ships are as follows: LNG, bio-methanol, bio-LNG, e-ammonia, e-methanol, and hydrotreated vegetable oil (HVO). Carbon pricing mechanisms can effectively improve the cost-effectiveness of alternative fuels with high fuel costs such as bio-LNG, bio-methanol, e-methanol, e-ammonia, and HVO. A carbon price ranging from 300 to 800 USD per ton of CO_2eq can be imposed to improve competitiveness in terms of cost.

It should be noted that in these studies, the transition strategy of container vessels, given their size and routes, has not been examined in full. In the single vessel analysis in related studies, only one specified ship servicing a single route was analyzed (Lagemann et al., 2023: Lagemann et al., 2022). In the fleet models by Zhao et al. (2023) and Loennechen et al. (2024), a mix of various vessel sizes were considered. However, the discussions focused on the fleet composition as a whole, rather than the change in engine and fuel choice due to variations in vessel size and route. Zhao et al. (2024) describe the emission reduction pathways for various sizes of dry bulker in the analysis. Although vessel size was considered in the analysis, the study does not specify how the decision changes for various emission reduction policy implementations and different routes.

This study contributes to the existing literature by examining how ship technology and fuel choices vary across different decarbonization policy implementations. Only a few studies have considered engine and fuel choice among different ship sizes and routes. Furthermore, the present research provides new insights by focusing specifically on container vessels, in contrast to previous studies that have mostly concentrated on dry bulkers. With the diverse capacities and routes being served by container vessels, it is crucial to understand how these differences might influence the shipping companies in setting their decarbonization transition strategy. Moreover, as future net-zero policies, especially medium-term, are still developing, it is important to conduct the necessary analyses under strict GHG target compliance and various emissions penalty scenarios and determine how these will affect the strategy for various container vessels from now until 2050. The main objective of this study is to develop a transition strategy for container vessels with the use of maritime alternative fuels accounting for both economic and emissions costs while considering various vessel sizes, routes, and policy implementations. The objective function is to minimize costs until 2050, considering the IMO GHG reduction measure, CII annually. Uncertainties in future market, regulatory, and technology conditions are assessed using sensitivity analysis to understand how the behavior of companies may change.

3 Model Formulation

The model minimizes the cost for running a ship from 2024 to 2050, which includes the capital cost, comprised of new shipbuilding price and retrofit costs, fuel cost, maintenance cost, and downtime cost in Eq. 1 (Lagemann et al., 2022; Zhao et al., 2023). The other cost parameters, such as crew cost, insurance cost, and lubrication costs, were not considered in the objective function as these figures make a smaller contribution to the total cost compared to the aforementioned factors. Table 2 provides a list of the main symbols used in mathematical modeling, including sets, parameters, and variables.

$$\min V = \sum_{t\in T}\sum_{k\in K} C^B_{kt}x^B_{kt} + \sum_{t\in T}\sum_{k\in K}\sum_{k'\in K} C^R_{k'kt}x^R_{k'kt} + \sum_{t\in T}\sum_{f\in F} P_{ft}F_M\eta_f\left(T_{sail}\cdot y_{ft}\right) + \sum_{t\in T}\sum_{f\in F} P_{ft}F_A\eta_f\left(T_{sail}\cdot y_{ft}\right) + \sum_{t\in T}\sum_{k\in K} C^{main}_t z_{kt} + \sum_{t\in T}\sum_{k\in K}\sum_{k'\in K} R_t x^R_{k'kt} - \sum_{t\in T}\sum_{k\in K} E_t C^S_t x^B_{kt} + U\sum_{t\in T}(H_t + G_t) \quad (1)$$

Subject to :

$$\sum_{f\in F} y_{ft} = 1\ \forall t \in T \quad (2)$$

$$\sum_{k\in K} z_{kt} = 1\ \forall t \in T \quad (3)$$

$$y_{ft} + z_{kt} \le 1 + W_{fk} \forall t \epsilon T, f \epsilon F, k \epsilon K\# \quad (4)$$

$$x^B_{kt} + x^R_{k'kt} \le 1 \forall t \epsilon T, k\prime, k \epsilon K\# \quad (5)$$

$$-Z_{k',\,t-1} + \sum_{k\in K} x^R_{k'kt} \le 0\ \forall k' \in K,\ t \in T \quad (6)$$

$$C^S_t = \sum_{f\in F} P_{ft}F_M\eta_f\left(T_{sail}\cdot y_{ft}\right) + \sum_{f\in F} P_{ft}F_A\eta_f\left(T_{sail}\cdot y_{ft}\right)\ \forall t \in T \quad (7)$$

Table 2 Model sets, parameters, and variables

Class	Symbol	Description
Sets	T	Set of discrete time periods t (annual)
	K	Set of k (current) and k′ (previous) engine technologies used
	F	Set of fuels f
Parameters	C^B_{kt}	New build cost at time t for technology k
	$C^R_{k'kt}$	Retrofit cost at time t for technology k from k′
	P_{ft}	Price of fuel f at time t
	F_M, F_A	Fuel consumption rate for main and auxiliary engines
	η_f	Mass conversion coefficients of fuels in reference to VLSFO
	Q_f	Carbon dioxide-ton per ton of fuel
	W_{fk}	Fuel compatibility of engine k to fuel f
	C^{main}_t	Cost of maintenance at time t
	R_t	Charter rate per year
	E_t	Efficiency or savings from buying a new ship versus base fuel efficiency performance
	C^S_t	Fuel cost at time t
	DWT	Deadweight of ship
	l	Capacity factor of ship based on DWT, set at 1 and 0.7
	T_{sail}	Sailing time of a vessel in a specified route
	D_{sail}	Sailing distance in a specified route
	CII^{red}_t	Percentage reduction for CII target annually vs 2019 reference
	CII^{target}_t	CII target of the ship at time t
	CII_t	Estimated attained CII from sailing ship at time t
	U	Penalty cost per ton-emission
	J_t	Difference in ton-emissions between CII_t and CII^{target}_t annually
	H_t	Estimated excess ton-emissions vs CII target
	GWP_{100}	Global warming potential over a 100-year time frame for CH_4
	$\%CH_4$	Percentage of CH_4 content for LNG and green-methane fuels
	$\%Slip$	Percent methane slip for dual-fuel slow speed engines
	G_t	Methane slip in tons annually
Variables	x^B_{kt}	$\epsilon\{0, 1\}, = 1$ if technology k is bought at time t
	$x^R_{k'kt}$	$\epsilon\{0, 1\}, = 1$ if technology k is retrofit from k′ at time t
	y_{ft}	$\epsilon\{0, 1\}, = 1$ if fuel f is used for sailing at time t
	z_{kt}	$\epsilon\{0, 1\}, = 1$ if engine k is used for sailing at time t derived from x^B_{kt} and $x^R_{k'kt}$

$$CII_t = \frac{\sum_{f\epsilon F} Q_f F_M \eta_f \left(T_{sail} \cdot y_{ft}\right) + \sum_{f\epsilon F} Q_f F_A \eta_f \left(T_{sail} \cdot y_{ft}\right)}{l * DWT * D_{sail}} \forall t \epsilon T \tag{8}$$

$$CII_t^{target} = 1984 * DWT^{-0.489} * \left(1 - CII_t^{red}\right) \forall t \epsilon T \tag{9}$$

$$CII_t \leq CII_t^{target} \forall t \epsilon T \tag{10}$$

$$J_t = 10^{-6}\left(CII_t - CII_t^{target}\right) * (l * DWT * D_{sail}) \forall t \epsilon T \tag{11}$$

$$H_t = \begin{cases} 0, J_t < 0 \\ J_t, J_t \geq 0 \end{cases} \forall t \epsilon T \tag{12}$$

$$G_t = GWP_{100} * \%CH_4 * \%Slip \left(\sum_{f \in F} F_M \eta_f \left(T_{sail} \cdot y_{ft}\right) + \sum_{f \in F} F_A \eta_f \left(T_{sail} \cdot y_{ft}\right) \right) \forall t \in T \tag{13}$$

An efficiency factor is also incorporated in Eq. 1, which reduces the fuel cost vs the base scenario, if a new ship is purchased. This factor is incorporated to determine at which level of fuel efficiency from new ships buying new ships will be more favorable than retrofitting existing ships. This assumption is in line with recent technological developments in ship design, in which wind-assisted propulsion technologies are being considered to cut down on fuel consumption while maintaining the same operational capacity (DNV, 2022). It should be noted that structural integrity as a factor of ship age and lifecycle costs of building new ships and retrofitting ships is not considered in the analysis but is pertinent to having a holistic perspective in investment decisions.

The constraints of the model ensure that only one fuel and engine type is selected for each year (Eqs. 2 and 3). The fuel selection is based on the compatibility of the fuel to the engine technology decision (Eq. 4). Another constraint is declared to ensure that the decision for doing nothing, retrofitting, and buying a new ship is only one for each year (Eq. 5). To ensure that the retrofit decision is only based on the available engine technology from the previous year, Eq. 6 is used. Looking at the objective function, the model is nonlinear due to the efficiency factor incorporated into the fuel cost when a new ship is bought (Eq. 7). Linearization is used to run the model as a linear programming problem. The Carbon Intensity Indicator (CII) is calculated using the estimated fuel consumption of container vessels when serving its assigned route for a year (Eq. 8). The estimated CII for each year in this model does not include allowances for voyage exclusion and emissions at the port. The CII targets (Eq. 9) are based on the ship deadweight and the required CII C rating to ensure compliance based on the 2021 CII Rating Guidelines (IMO, 2021). The various d values are the boundaries of the rating bands: d1 and d2 are the boundaries to the B rating, d2 and d3 are the boundaries to the C rating, with the required CII in the middle, and d3 and d4 are the boundaries to the D rating. Ratings below d1 belong in the A rating, while ratings above d4 belong in the E rating. The estimated CII is then compared to the CII target at the d3 boundary, the minimum to achieve the C rating. The study sets the CII targets stricter as 2050 approaches, with 20% reduction in 2030, 70% in 2040, and 100% in 2050, in order to achieve 2023 IMO GHG strategy. Figure 1 shows the assumed reduction trend of the required CII and its scale from 2023 to 2050 in this study.

In model version 1, Eq. 10 is enforced as a hard constraint, in which the attained CII must always be less than the target CII for all years. The penalty cost per ton-emission U in Eq. 1 is also set to zero in this version (see Table 3). For the penalty model version (i.e., version 2), Eq. 10 is relaxed, and any excess from the target will be imposed with a penalty U (see Eqs. 11 and 12). CII is defined on a Tank-to-Wake basis for CO_2 emissions only. For the sake of evaluating other possible implementations of greenhouse gas (GHG) reduction policies, the indicator was expanded to account for Well-to-Tank emissions and other GHG emissions in CO_2-equivalents (CO_2eq). To calculate methane slip emissions, Eq. 13 is used only when fuel choice for a certain year is LNG or green-methane, where GWP_{100} is at 29.8 ton of CO_2eq per ton of CH_4; percentage of CH_4 is 0.9 for LNG and 1 for green-methane; and percent methane slip for dual-fuel slow speed engines is at 0.2% mass of fuel used (Mærsk Mc-Kinney Møller Center for Zero Carbon Shipping, 2024a).

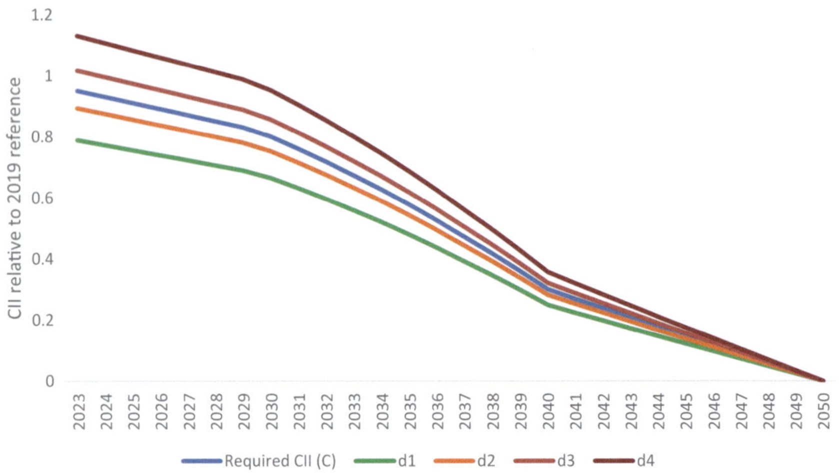

Fig. 1 Reduction trend of the required CII rating and scale from 2023 to 2050

Table 3 Model versions and their corresponding details

Model	Model adjustment	Emissions calculation
Strict CII	Equation 10 is enforced, U set to 0	CO_2 emissions
Penalty	Equation 10 is relaxed, $U \epsilon \{100, 250,1000\}$	CO_2, GHG, and methane slip

4 Case Study: Data and Scenarios

The characteristics of container vessels and the routes used in the model are detailed in Table 4. The study aims to understand how the differences in the size and the route of the ship may impact the timing of decisions related to meeting the decarbonization targets. Feeder, panamax, and neopanamax container vessels are evaluated using different trade routes. All the vessels in this study use ME-C engines that can be retrofitted to dual-fuel use. Design speed and average speed obtained from Automatic Identification System (AIS) are used in this study. The required power is proportional to the cubic of the speed.

Table 5 shows more details regarding the routes used in the analysis. The route information was obtained from Ocean Network Express (ONE, 2023). Feeder vessels were assigned to inter-Asia routes 1 and 2, panamax vessels were assigned to intra-Asia route 2 and inter-continental routes 3 and 4, and neopanamax vessels were assigned to inter-continental routes 3 and 4. The route assignment was based on AIS data from ClassNK registered ships. Route 1 has 19% of its route in emission control areas (ECAs), and route 2 has 9% mainly due to the ECA implemented in China, while route 3 only has 7% compared to route 4, which has 20% due to the ECA in the EU.

The new shipbuilding price and charter rates were obtained from Clarksons (2023). The maintenance cost was assumed at 2% of new shipbuilding price annually (see Table 6). Table 7 shows the retrofit costs obtained from available data provided by MAN Energy (2023b) and estimated for each container vessel size using the cost assumptions from Lagemann et al. (2022). A discount rate of 2% was used in calculating the net present value of the total cost. The fuels considered in this study are VLSFO, biofuel, LNG, methane, methanol, blue-ammonia, and green-ammonia.

Table 4 Ship characteristics and routes

Characteristics	Feeder	Panamax	Neopanamax
TEU	1,700	5,100	15,500
DWT (ton)	23,400	63,500	15,5000
Design speed (kn)	18.5	22	23
AIS speed (kn)	14.8	16.1	16.1
SMCR (kW)	11,200	27,690	56,800
Power—main (kW)—90%	10,080	24,921	51,120
T3_SFOC (g/kWh)	159.1	157.1	154.4
T2_SFOC (g/kWh)	158.1	156.1	153.4
Power—Auxiliary (kW)	530	942	1,670
Routes	JP-CN-VN JP-CN-VN-SG-MY	JP-CN-VN-SG-MY JP-SG-US JP-SG-FR-NL-DE	JP-SG-US JP-SG-FR-NL-DE

Source MANEnergy (2023a)

Table 5 Routes and total distance per trip

Route	Total distance (nm)	Tier 3/ECA distance (nm)
(1) JP-CN-VN	5,240	1,036
(2) JP-CN-VN-SG-MY	6,791	642
(3) JP-SG-US	16,721	1,246
(4) JP-SG-FR-NL-DE	24,128	4,856

Sources ONE (2023), Maritime Optima (2023)

Table 6 Cost parameters

Ship	New shipbuilding price (M USD)	Charter rate (M USD/day)	Maintenance cost (M USD/yr)
Feeder	30.9	0.010	0.618
Panamax	81.2	0.020	1.624
Neopanamax	176.0	0.058	3.520

Source Clarksons (2023)

Table 7 Retrofit costs for container vessels

M USD	Feeder				Panamax				Neopanamax			
Engine	(1)	(2)	(3)	(4)	(1)	(2)	(3)	(4)	(1)	(2)	(3)	(4)
(1) VLSFO	0.0	12.0	8.6	19.4	0.0	25.2	18.0	40.6	0.0	35.0	25.0	56.5
(2) LNG	N/A	0.0	9.6	14.6	N/A	0.0	20.1	30.6	N/A	0.0	27.9	42.5
(3) Methanol	N/A	12.0	0.0	19.4	N/A	25.2	0.0	40.6	N/A	35.0	0.0	56.5
(4) Ammonia	N/A	N/A	N/A	0.0	N/A	N/A	N/A	0.0	N/A	N/A	N/A	0.0

Dual-fuel engine retrofit: (1) VLSFO; (2) LNG; (3) Methanol; (4) Ammonia
Sources MAN Energy (2023b) and Lagemann et al. (2022)

Hydrogen is not included in the study due to its storage and transportation requirements, low energy coefficient per unit mass, and their corresponding implications to cost (Serpell et al., 2023).

As future fuel price forecasting is not included in the scope of the study, the fuel price forecasts were obtained mainly from the Planning and Design Center for Greener Ships (GSC, 2022) and Fuel Cost Calculator of Mærsk Mc-Kinney Møller Center for Zero Carbon Shipping (2024b)—see Table 8. It can be observed that the price for LNG is cheaper than VLSFO for price versions 1 and 2, which was the case towards the end of 2020. Price version 3 is then devised to have a scenario, in which methanol prices are competitive vs LNG. The price of green-methanol is higher than green-ammonia in this version to assess its potential impact on the choice of net-zero fuel as the calculation approaches 2050. It can also be observed that the price of VLSFO in this version is cheaper than LNG. The different price assumptions are used to explore the impact of fuel prices on the transition strategy of container

Table 8 Mass conversion and fuel price forecasts in USD/GJ

		Version 1 (USD/GJ)			Version 2 (USD/GJ)			Version 3 (USD/GJ)		
Fuel type	Mass conversion coefficient	2030	2040	2050	2030	2040	2050	2030	2040	2050
VLSFO	1.00[a]	6.8[b]	5.6[b]	5.0[b]	13.5[e]	13.5[e]	13.5[e]	9.9	9.9	9.9
LNG	0.84[a]	4.7[b]	4.5[b]	4.3[b]	9.0[e]	9.4[e]	9.2[e]	16.4	17.2	16.8
Green-Methane	0.84[a]	49.2[b]	45.8[b]	42.4[b]	41.3[e]	33.2[e]	23.5[e]	48.8	39.2	27.8
Methanol	2.06[a]	8.6[c]	7.9	7.3	14.7[c]	12.4	11.2	17.3	14.7	13.2
Green-Methanol	2.06[a]	44.7[b]	41.2[b]	37.7[b]	48.8[e]	38.2[e]	25.9[e]	38.4	30.1	20.4
Biofuel	1.10[a]	20.4[d]	16.9	14.9	55.8[e]	45.3[e]	29.6[e]	26.5	24.4	22.5
Green-Ammonia	2.20[a]	40.8[b]	36.8[b]	32.8[b]	31.3[e]	25.0[e]	17.5[e]	31.3[e]	25.0[e]	17.5[e]
Blue-Ammonia	2.20[a]	21.8[b]	21.2[b]	20.4[b]	22.8[e]	22.7[e]	21.9[e]	22.8[e]	22.7[e]	21.9[e]

Sources European Commission (2021)[a], GSC (2022)[b], IEA (2024)[c], IEA (2020)[d], Mærsk Kinney Møller Center for Zero Carbon Shipping (2024b)[e]

Table 9 Emission factors for CO_2 and CO_2eq

	CO_2 Only		Greenhouse gases in CO_2eq	
Fuel type	WtT	TtW	WtT	TtW
VLSFO	0.55[a]	3.19[a]	0.69[a]	3.41[a]
LNG	0.53[a]	2.75[a]	1.08[a]	2.86[a]
Green-Methane	0.00[b]	0.00[b]	0.00[c]	0.08[c]
Methanol	0.40[e]	1.38[b]	0.62[c]	1.42[c]
Green-Methanol	0.00[b]	0.00[b]	0.00[c]	0.05[c]
Biofuel	0.00[b]	2.83[b]	0.00[d]	2.88[d]
Green-Ammonia	0.00[b]	0.00[b]	0.00[d]	0.10[d]
Blue-Ammonia	2.25[b]	0.00[b]	2.25[d]	0.10[d]

Sources Comer and Osipova (2021)[a], European Commission (2021)[b], Nelissen et al. (2022)[c], Lindstad et al. (2021)[d], and van Lieshout (2020)[e]

vessels. Table 9 shows the emission factor (E.F.) assumptions for GHG policies used in the scenario analysis. For the scenario analysis, four (4) versions of emission factor assumptions were used: 1—carbon dioxide only; 2—carbon dioxide only with biofuel as net-zero fuel; 3—greenhouse gases in CO_2eq; and 4—greenhouse gases in CO_2eq with biofuel as net-zero fuel.

In the case of CII, two capacity factors were used (1 and 0.7). CII was calculated based on supply-based transport work (IMO, 2021) due to the absence of data on actual transport work. For container vessels, bulk carriers, and tankers, deadweight tonnage (DWT) was used as capacity. In the case of design indices (EEDI and EEXI), 70 per cent of the deadweight was used as capacity for calculating the design index

(IMO, 2013, Nippon Kaiji Kyokai, 2021). To make the analysis comprehensive, the study incorporated capacity factors 1 and 0.7 to understand how the decision strategy for container vessels changes with the transport work assumption.

The scenario analysis accounted for the broad range of uncertainties that shipping companies may face in the coming years, particularly in relation to future regulatory conditions (see Table 10). As the shipping industry faces increasing pressure to reduce emissions, evolving environmental regulations and market shifts play a significant role in shaping vessel operations, fuel choices, and investment decisions. The list of scenarios covers variations in policy implementation aspects, which include emission factor versions (1, 2, 3, 4) and accounting scenarios (Well-to-Wake [WtW] and Tank-to-Wake [TtW]). This study also evaluated the strict implementation of the Carbon Intensity Indicator (CII) and penalty scenarios per ton of CO_2eq and methane slip. For the emissions, the penalty scenarios considered were 0 USD, 100 USD, 250 USD, and 1,000 USD per ton-emission.

To evaluate the decision strategy for various vessel sizes and routes and future fuel prices, the plan was constructed as follows:

- **Vessel Size and Route Combinations**: The scenario analysis was done for each vessel size (feeder, panamax, and neopanamax) and their respective route combinations (see Table 4). These different vessel types and routes represent various operational realities within the global shipping industry. By doing this, the analysis allows for a tailored understanding of how specific combinations of vessel characteristics and routes are impacted by different regulatory conditions.
- **Transport Work Assumptions**: The attained CII was calculated based on two transport work assumptions relative to vessel DWT (at capacity factors 1 and 0.7).
- **Fuel Price Versions**: To account for the uncertainty in fuel prices, three different price scenarios were considered as discussed in Table 8. Each of these price scenarios was applied in the scenario analysis for every vessel size and route combination.

5 Results

The model results show that the strict implementation of the CII target will force shipping companies to push through with retrofits to the engines and ships to ensure compliance. The following figures show the engine decisions until 2050 for feeder, panamax, and neopanamax vessels for routes 1, 3, and 4 respectively (see Figs. 2, 3 and 4). Figure 2 shows the engine and fuel choice for feeder vessels at route 1. From the engine choice, feeder vessels tend not to choose to retrofit if the GHG policy is less stringent and when biofuel is considered a net-zero fuel. In these scenarios, the fuel choice is either retained as is (VLSFO) or shifted to biofuels. In strict CII implementation and 1,000 USD per ton-CO_2 and CO_2eq penalty, retrofit is not recommended only when biofuels are considered as net-zero fuel. In penalty scenarios 0–100 USD per ton-CO_2 penalty, feeder vessels tend to do nothing and incur costs until 2050.

Table 10 Details of scenarios for analysis

Scenario	Model	Penalty (U)	E. F. version	Emission	WtW-TtW	Speed
1	Strict CII	0 USD	1	CO_2 only	TtW	Design
2	Strict CII	0 USD	1	CO_2 only	TtW	Reduced
3	Strict CII	0 USD	1	CO_2 only	WtW	Reduced
4	Strict CII	0 USD	2	CO_2 only	TtW	Reduced
5	Strict CII	0 USD	2	CO_2 only	WtW	Reduced
6	Penalty	0 USD	1	CO_2 only	WtW	Reduced
7	Penalty	100 USD	1	CO_2 only	WtW	Reduced
8	Penalty	250 USD	1	CO_2 only	WtW	Reduced
9	Penalty	1000 USD	1	CO_2 only	WtW	Reduced
10	Penalty	100 USD	2	CO_2 only	WtW	Reduced
11	Penalty	250 USD	2	CO_2 only	WtW	Reduced
12	Penalty	1000 USD	2	CO_2 only	WtW	Reduced
13	Penalty	100 USD	3	GHG	WtW	Reduced
14	Penalty	250 USD	3	GHG	WtW	Reduced
15	Penalty	1000 USD	3	GHG	WtW	Reduced
16	Penalty	100 USD	4	GHG	WtW	Reduced
17	Penalty	250 USD	4	GHG	WtW	Reduced
18	Penalty	1000 USD	4	GHG	WtW	Reduced
19	Penalty	100 USD	3	GHG w/CH_4 slip	WtW	Reduced
20	Penalty	250 USD	3	GHG w/CH_4 slip	WtW	Reduced
21	Penalty	1000 USD	3	GHG w/CH_4 slip	WtW	Reduced
22	Penalty	100 USD	4	GHG w/CH_4 slip	WtW	Reduced
23	Penalty	250 USD	4	GHG w/CH_4 slip	WtW	Reduced
24	Penalty	1000 USD	4	GHG w/CH_4 slip	WtW	Reduced

In strict CII implementation and higher penalty scenarios, retrofit to LNG is considered. Depending on the penalty imposed, the choice of fuel can be more flexible. At the 1,000 USD penalty scenarios, the vessels retrofit to LNG while incurring penalties after the CII actual estimate exceeds the CII target projected. In other penalty scenarios that do not consider biofuel as a net-zero fuel, the vessels could ultimately choose alternative fuels, such as green-methane, blue-ammonia, and green-methanol depending on the emissions policy scenario. It can be also observed that vessels tend to incur the highest costs when the policy implementations are the most stringent (strict CII implementation, 1,000 USD penalty, E.F. 3, WtW accounting). This pattern can also be observed in the other calculations for panamax and neopanamax vessels (see Table 9).

Figure 3 shows the engine and fuel choices for panamax vessels at route 3. Like feeder vessels, panamax vessels will consider biofuels when they are treated as net-zero fuel. Due to the lower price assumption of blue-ammonia to green-ammonia,

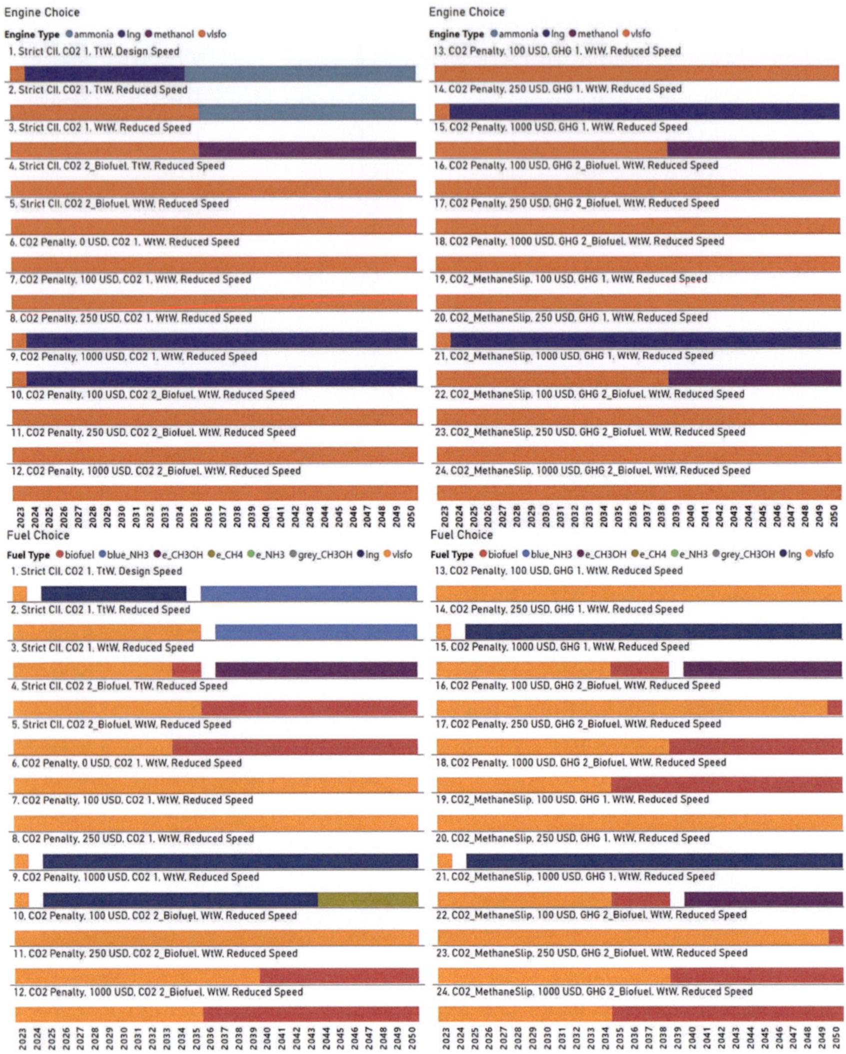

Fig. 2 Engine and fuel choices for feeder vessels at route 1 using 1 as the capacity factor and price version 1. *In the Fuel Choice figures, the space indicates retrofitting operations, which means no fuel is used during the time period

ships prefer blue-ammonia when TtW accounting is used in emissions policy implementation because of its significantly lower price. It should also be noted that green-methanol was chosen as net-zero fuel from VLSFO due to lower retrofit cost assumptions to methanol vs ammonia, despite green-methanol being more expensive than green-ammonia. Although most of the strategy is similar for feeder and panamax vessels, LNG was chosen as an intermediary fuel for panamax vessels at route 3 before changing to blue-ammonia.

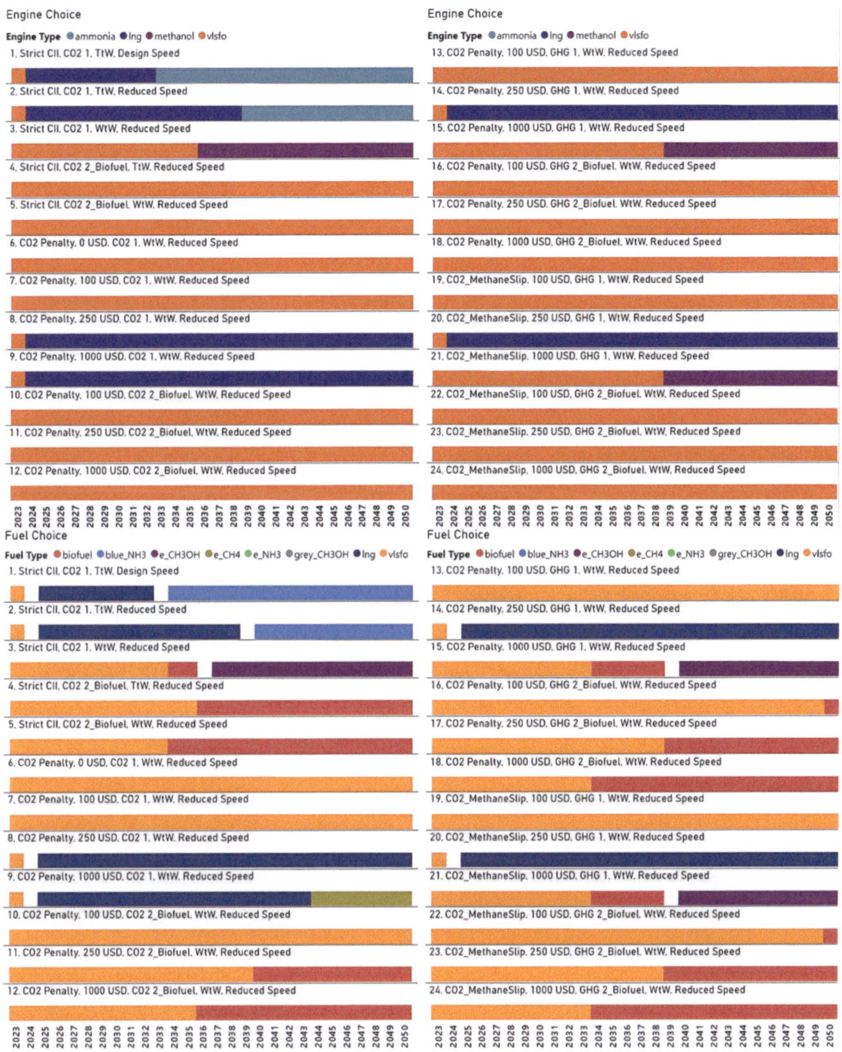

Fig. 3 Engine and fuel choices for panamax vessels at route 3 using 1 as the capacity factor and price version 1

Figure 4 shows the engine and fuel choices for neopanamax vessels at route 4. The findings for neopanamax vessels are different from those of feeder and panamax vessels in several scenarios. One difference is the earlier transition to net-zero fuels due to the stricter CII targets for larger vessels. Another difference is the choice of net-zero fuels in scenarios 3 and 15, showing the preferences of green-ammonia and green-methane over green-methanol. However, in scenario 21, where methane slip is accounted for, green-methanol is chosen over green-methane. Across all vessel sizes, the role of biofuels greatly affects the engine and fuel choices. Retrofits to

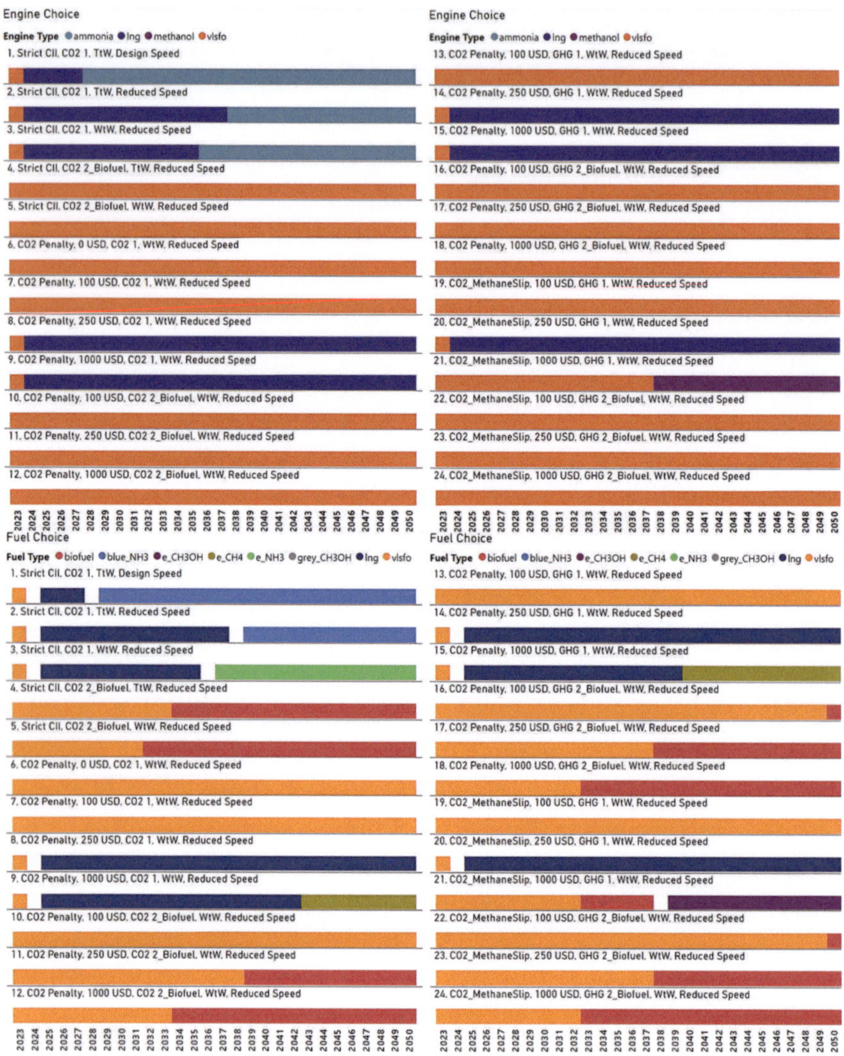

Fig. 4 Engine and fuel choices for neopanamax vessels at route 4 using 1 as the capacity factor and price version 1

technologies that cater to other alternative fuels are not considered, and vessels ultimately switch to biofuel when it is considered as a net-zero fuel option. It should be noted that biofuel in price version 1 is cheaper than its net-zero counterparts. The impact of price assumptions on the engine and fuel choices are discussed further in the succeeding section.

The decision for each vessel varies significantly based on the type of policy implemented. Depending on the stringency of policies, the total costs for each vessel-route change (see Table 11). The more stringent the emissions reduction policies are, the

Table 11 Total cost for selected routes using price version 1 in M USD

Scenario	Description	Neopanamax route 4	Panamax route 3	Feeder route 1
1	Strict CII, E.F. 1, TtW, Design sp	710	312	124
2	Strict CII, E.F. 1, TtW, Reduced sp	419	212	96
3	Strict CII, E.F. 1, WtW, Reduced sp	521	259	117
4	Strict CII, E.F. 2, TtW, Reduced sp	319	154	67
5	Strict CII, E.F. 2, WtW, Reduced sp	334	162	71
6	CO_2 penalty, 0 USD, E.F. 1, WtW, Reduced sp	228	112	47
7	CO_2 penalty, 100 USD, E.F. 1, WtW, Reduced sp	274	135	58
8	CO_2 penalty, 250 USD, E.F. 1, WtW, Reduced sp	324	162	71
9	CO_2 penalty, 1000 USD, E.F. 1, WtW, Reduced sp	471	235	105
10	CO_2 penalty, 100 USD, E.F. 2, WtW, Reduced sp	274	135	58
11	CO_2 penalty, 250 USD, E.F. 2, WtW, Reduced sp	306	151	65
12	CO_2 penalty, 1000 USD, E.F. 2, WtW, Reduced sp	326	159	69
13	CO_2 penalty, 100 USD, E.F. 3, WtW, Reduced sp	283	139	60
14	CO_2 penalty, 250 USD, E.F. 3, WtW, Reduced sp	350	175	77
15	CO_2 penalty, 1000 USD, E.F. 3, WtW, Reduced sp	517	256	115
16	CO_2 penalty, 100 USD, E.F. 4, WtW, Reduced sp	283	139	60

(continued)

Table 11 (continued)

Scenario	Description	Neopanamax route 4	Panamax route 3	Feeder route 1
17	CO_2 penalty, 250 USD, E.F. 4, WtW, Reduced sp	316	155	67
18	CO_2 penalty, 1000 USD, E.F. 4, WtW, Reduced sp	339	165	71
19	CO_2_CH_4 slip, 100 USD, E.F. 3, WtW, Reduced sp	283	139	60
20	CO_2_CH_4 slip, 250 USD, E.F. 3, WtW, Reduced sp	354	177	78
21	CO_2_CH_4 slip, 1000 USD, E.F. 3, WtW, Reduced sp	527	256	115
22	CO_2_CH_4 slip, 100 USD, E.F. 4, WtW, Reduced sp	283	139	60
23	CO_2_CH_4 slip, 250 USD, E.F. 4, WtW, Reduced sp	316	155	67
24	CO_2_CH_4 slip, 1000 USD, E.F. 4, WtW, Reduced sp	339	165	71

higher the costs for the shipping companies will be. It can also be observed that the 1,000 USD per ton CO_2eq penalty scenarios mimic the strict CII implementation scenarios.

6 Discussion

6.1 Penalties

The results show the role of the implementation of CII targets, CO_2 factor assumptions, and associated penalties for non-compliance in changing the transition strategy. When the model was run with the CII constraint relaxed, the best solution was to do no retrofit to the vessels and to keep using VLSFO as the main fuel. If penalties are to be put in place for violations in meeting the requirements, they must be substantial enough for the shipping companies to take action. When the penalty per ton of CO_2 increases to 250 USD, the decision changes for all vessels. At this level, vessels begin retrofitting to LNG to mitigate their CO_2 emissions and reduce penalty costs, when biofuel is not considered as net-zero fuel. However, even with these retrofits, the analysis shows that these vessels will still incur penalties until 2050, indicating that LNG alone may not be sufficient to meet long-term emission targets. A significant shift occurs when the penalty per ton of CO_2 reaches 1,000 USD. At this high penalty level, all ships are compelled to transition to net-zero fuels by 2050. When the methane slip is included in the penalty calculation, LNG is still chosen when the penalty per ton-CO_2eq is at 250 USD. If the penalty is increased to 1,000 USD, biofuel can be selected as well. The steep financial penalties make it economically unviable to continue using conventional fuels, pushing companies to adopt cleaner alternatives. This analysis illustrates the powerful influence of high penalties in accelerating the adoption of green technologies and fuels, driving the shipping industry towards substantial emission reductions.

6.2 Transport Work Assumption

Figure 5 shows the fuel choices for feeder vessels using two different assumptions for transport work in the CII calculation. Because of the higher attained CII for $l = 0.7$, the switch to less-carbon intensive fuels from VLSFO was made earlier, explaining the choice of LNG as intermediary fuel before shifting to net-zero fuels, such as green-methane, green-ammonia, and biofuels. In the $l = 1$ assumption, vessels could still utilize VLSFO for longer periods before shifting to net-zero fuels. This is due to the lower attained CII when compared to the CII target for each year. Green-methanol is preferred over green-ammonia in several of the scenarios when $l = 1$ because only one retrofit is required (diesel to methanol dual-fuel) as opposed to having two retrofit operations when $l = 0.7$ (diesel to LNG to ammonia dual-fuel). This is despite green-methanol being more expensive than green-ammonia and LNG in price version 1.

6.3 Routes

The decarbonization strategy can be influenced by the characteristics the route served. For the panamax vessels in scenario 2, three routes were explored, in which routes 3 and 4 adopted LNG before retrofitting to use blue-ammonia, as opposed to route 2 which shifted to blue-ammonia without retrofitting to LNG first (see Fig. 6). Although the routes had varying ECA profiles, the key difference in the approach lies in the distance sailed, where panamax vessels on route 2 had a shorter total sailing distance as opposed to routes 3 and 4. The other scenarios generally had the same fuel choices for the three routes.

The same pattern was observed for feeder vessels. It was only in scenario 2 where the fuel choices at routes 1 and 2 were different. Figure 7 shows the fuel choices at route 1, with feeder vessels retrofit immediately to ammonia as opposed to feeder

Fig. 5 Fuel choices for feeder vessels in two transport work assumptions with the capacity factors of 1 and 0.7 using price version 1

vessels taking route 2 retrofitting to LNG first and then to ammonia. Because route 2 operates for longer distances than route 1, it is more cost-effective to invest in another retrofit operation than to use the more expensive conventional fuel.

All scenarios had the same fuel choices and transition timing for neopanamax with routes 3 and 4. Unlike feeder and neopanamax vessels, there were several instances in which the timing of retrofit and change in fuel choices was different for panamax when servicing different routes. Although the timing was different, the decision strategy for engine and fuel choices was consistent for most scenarios

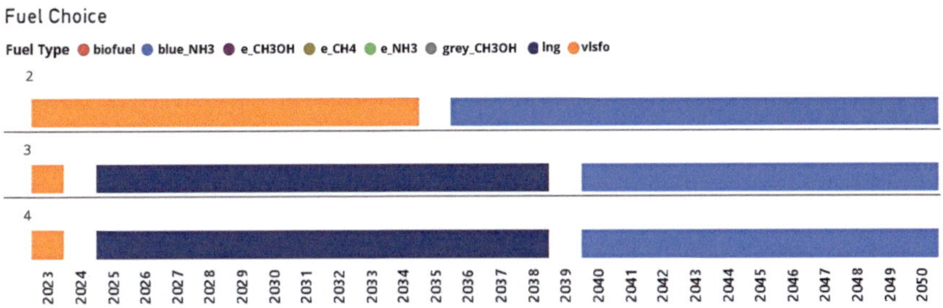

Fig. 6 Difference in fuel choices for panamax vessels based on routes in scenario 2

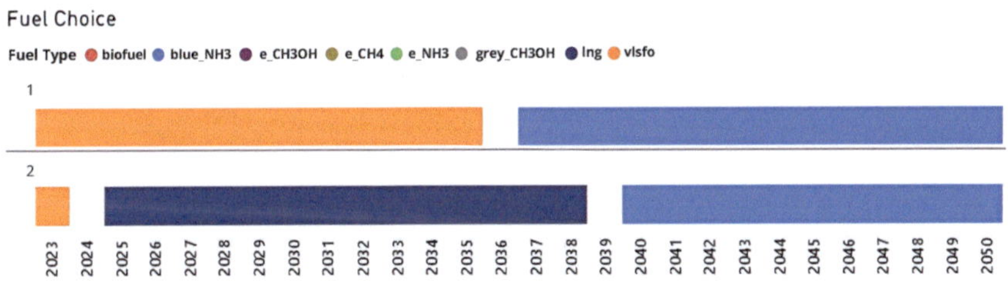

Fig. 7 Difference in fuel choices for feeder vessels based on routes in scenario 2

(see Fig. 8). Although there are a few scenarios where the decarbonization approach was different, most of the scenarios showed the same strategy despite changing the routes taken. This indicates that the engine and fuel choices for the same vessel size at different routes are relatively robust.

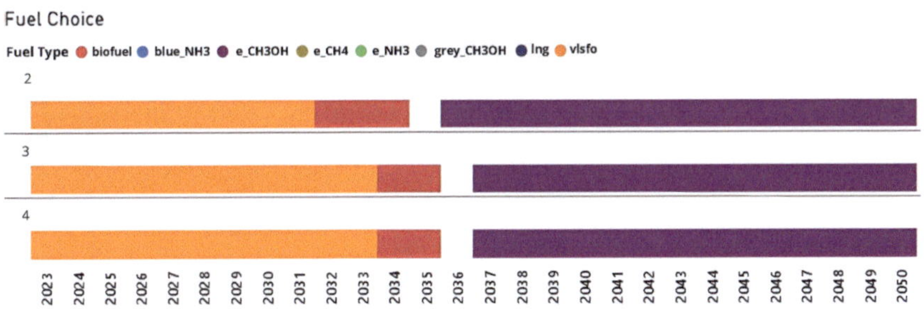

Fig. 8 Difference in transition timing for panamax vessels based on routes in scenario 3

6.4 Sizes

For the same route, vessel sizes have differences in the approach for decarbonization. In both transport work assumptions, the difference in size translates to variations in transition timing. Using the capacity factor of 1, panamax and neopanamax vessels have distinct engine and fuel choices under certain policy scenarios. Moreover, using the capacity factor of 0.7 reflects an earlier adoption of less-carbon intensive fuels, such as LNG or biofuels, due to a higher estimated CII attained. These differences in approach are highlighted in the cases below. At $l = 1$, panamax vessels have an earlier transition to net-zero fuels compared to feeder vessels as shown in Fig. 9 due to the stricter CII target for large vessels in strict CII implementation scenarios and in penalty scenarios. Earlier transition is also observed for scenarios where biofuel is considered as a net-zero fuel.

The difference in approach for vessel sizes is more noticeable when comparing between panamax and neopanamax vessels (see Fig. 10). In strict CII implementation, panamax vessels retrofit to use green-methanol, while neopanamax vessels retrofit to LNG first before shifting to green-ammonia as fuel. For the 1,000 USD penalty per ton-CO_2 emission scenario, neopanamax vessels shift to green-methane from LNG without conducting another retrofit operation. The panamax vessels in this scenario have the same strategy as the strict CII scenario with a later transition to green-methanol.

At $l = 0.7$, panamax vessels also have an earlier transition to net-zero fuels as opposed to feeder vessels (see Fig. 11). The same pattern can be observed when comparing the transition timings of panamax and neopanamax vessels at the same route. However, in scenarios where biofuel is considered as a net-zero fuel, the approach differs for feeder and panamax vessels (see Fig. 12). Depending on the policy scenario, feeder vessels switch to biofuels but may or may not choose to retrofit to LNG. The same can be said for panamax vessels. The difference in the approach lies in the implementation of policy where a steep penalty (at 1,000 USD per ton-emission) can force early adoption of net-zero fuels, as opposed to strict compliance to the CII target, which allows for a gradual shift to biofuels in the succeeding years.

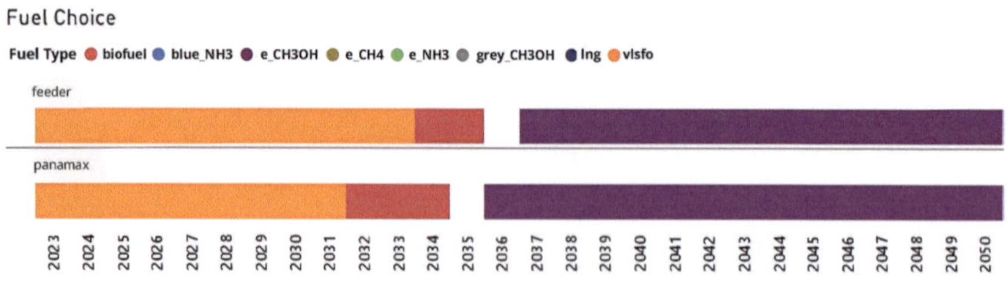

Fig. 9 Feeder and panamax transition with route 2 in strict CII scenario using 1 as the capacity factor

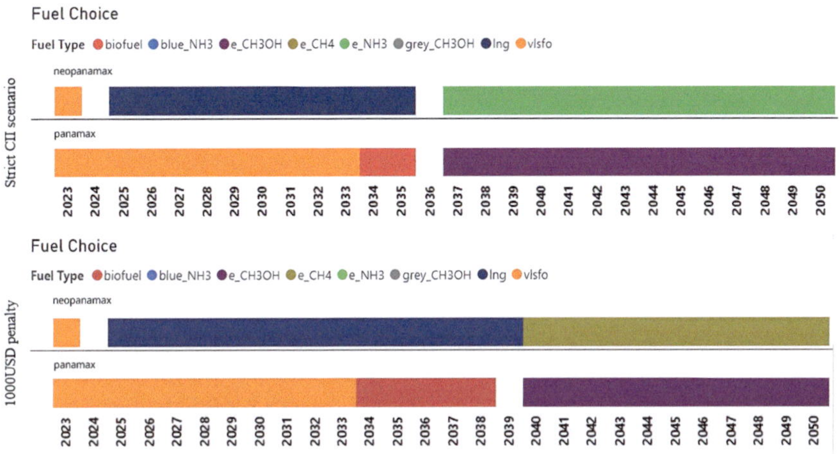

Fig. 10 Neopanamax and panamax transition with route 4 in strict CII and 1,000 USD penalty scenario using 1 as the capacity factor

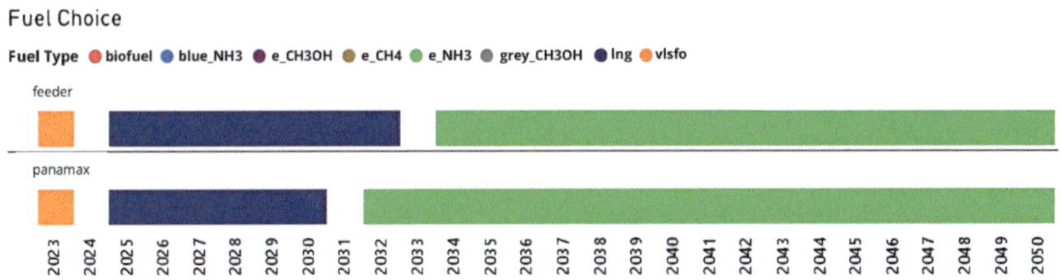

Fig. 11 Feeder and panamax transition with route 2 in strict CII scenario using 0.7 as the capacity factor

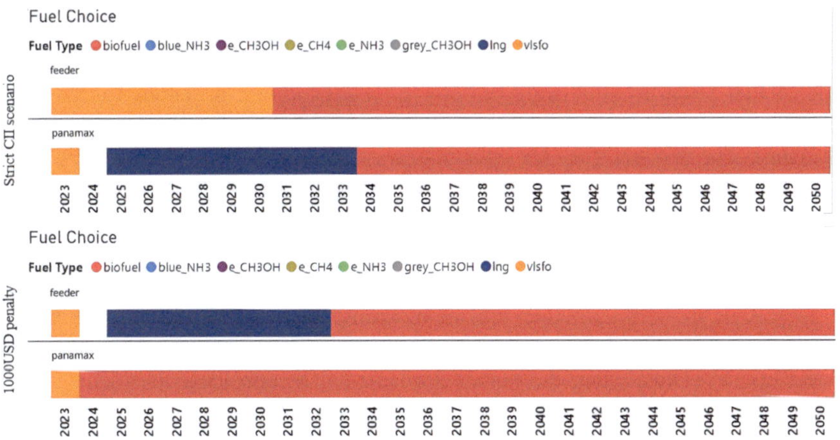

Fig. 12 Feeder and panamax vessels at route 2 in strict CII and 1,000 USD penalty scenario with biofuel as net-zero fuel using 0.7 as the capacity factor

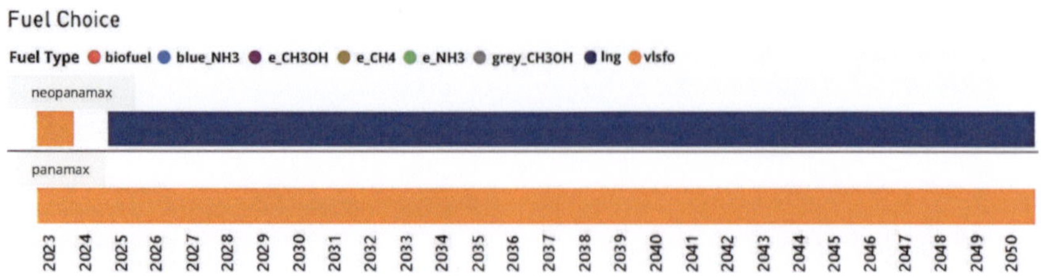

Fig. 13 Neopanamax and panamax vessels with route 4 in 100 USD penalty scenario using 0.7 as the capacity factor

Aside from an earlier transition timing for larger vessels, the penalty imposed on excess emissions influences the engine and fuel choices for different vessel sizes. At the 100 USD penalty, panamax vessels opt to incur penalties until 2050 (see Fig. 13). This contrasts with neopamanax vessels, which opt to retrofit to LNG and incur penalties when LNG as fuel no longer meets the stricter net-zero target. At the 250 USD penalty, both vessel sizes retrofit to LNG while incurring penalties as 2050 approaches.

6.5 *Fuel Prices*

With the fuel cost at 40–60% of the total calculated cost, fuel cost assumptions can significantly change the decision for container vessels. The earlier discussions focused on price version 1. Although there are similar patterns that emerge with the other two price versions, there are distinctions in the engine and fuel choices in the other two price scenarios. Price version 2 assumes LNG to be significantly cheaper than VLSFO. Even in no penalty scenarios for emissions, a retrofit to LNG is chosen by all vessels regardless of ship type and route (see Fig. 14). This highlights the role of fuel price in influencing vessel capital expenditure decisions for shipping companies. With the earlier decision to retrofit to LNG, green-methane is ultimately chosen as the net-zero fuel for most scenarios at $l = 1$. Moreover, in this price version, biofuel is the most expensive among the fuel alternatives. It is not chosen despite being considered as a net-zero fuel.

At $l = 1$, green-methane is chosen as net-zero fuel despite the inclusion of methane slip in the emissions calculation for all penalty structures. This is different at $l = 0.7$, in which another retrofit to ammonia is considered to use green-ammonia from LNG when the 1,000 USD penalty per ton-CO_2eq was implemented (see Fig. 15). Thus, when the policy implementation considers the methane slip in the calculation, green-methane may not be used, while LNG still serves as an intermediary fuel choice because of its low fuel price in this price scenario.

Price version 3 assumes the price difference of grey methanol and LNG to be smaller, and green-methanol to be cheaper than green-methane per GJ of energy. The price of VLSFO in this scenario is cheaper than LNG. In this price version,

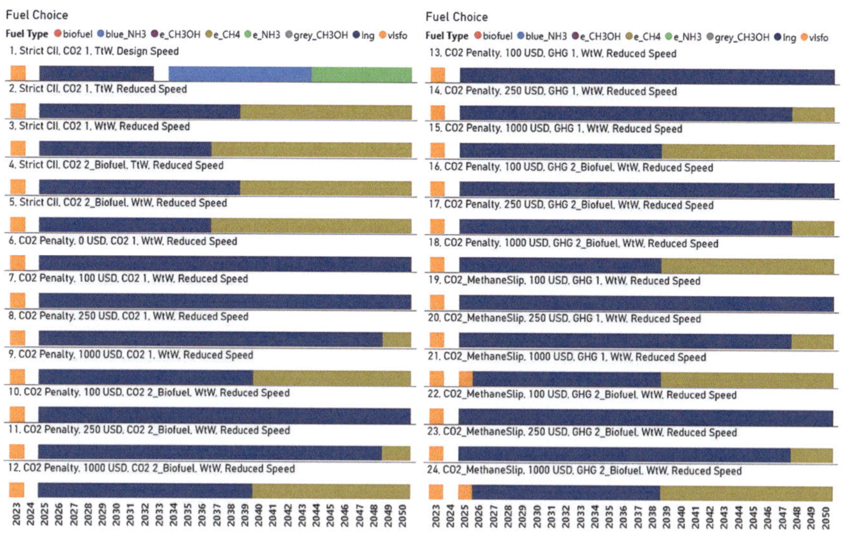

Fig. 14 Fuel choices for panamax vessels at route 4 using price version 2 and 1 as the capacity factor

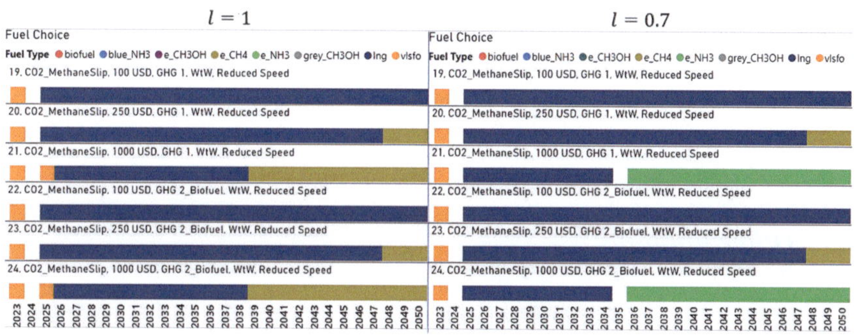

Fig. 15 Fuel choices for panamax vessels using price version 2 and capacity factors 1 and 0.7

the switch to biofuel is done for scenarios at the 1,000 USD penalty with biofuel considered as net-zero fuel (see Fig. 16). However, when biofuels were not considered as net-zero fuel, methanol and green-ammonia were ultimately chosen in strict CII implementation and high penalty scenarios. Compared with the other two price versions, LNG was not considered as an intermediary fuel choice in most scenarios. It is more economical to incur emissions penalty when the price of VLSFO is low before eventually shifting to a net-zero fuel, namely biofuel, green-ammonia, and green-methanol. The difference between $l = 1$ and $l = 0.7$ is the choice of green-ammonia in some cases. This is influenced by the attained CII calculation vs CII target as explained in the earlier section.

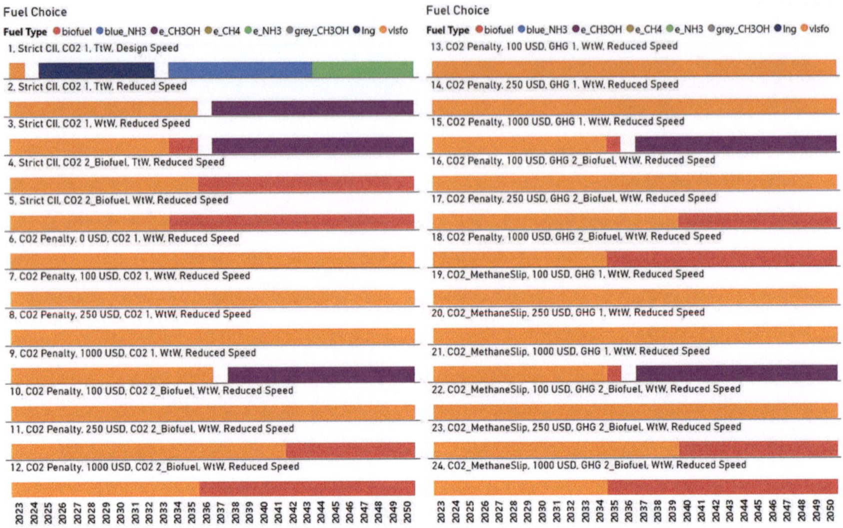

Fig. 16 Fuel choices for panamax vessels at route 4 using price version 3 and 1 as the capacity factor

The uncertainty in future fuel prices may significantly affect the operating cost of vessels, once retrofit or ship purchase decisions are made. Depending on the scenario, the total cost significantly varies. This may have an impact on the cost-effectiveness and competitiveness of containerized shipping.

6.6 *Other Findings*

With the high capital cost of buying new ships, retrofit is always considered versus purchasing new ships. Moreover, consideration of a fuel efficiency factor for new ships to make the comparison between retrofitting existing ships and buying new ships is also important. For feeder vessels, a fuel efficiency factor of at least 28% is required when considering a new ship purchase. The required efficiency factor is higher for larger ships, which is at least 37% and 46% for panamax and neopanamax, respectively. This requirement is significantly higher than the current fuel savings attained from wind-assisted propulsion systems at 20% (DNV, 2022). Thus, it is necessary to incorporate analysis on the maintenance cost over age, structural integrity, service life extension (Liu et al., 2021), and lifecycle cost to provide a more holistic view when deciding between retrofit and new ship purchase.

One of the primary findings from the scenario analysis conducted in this research is that speed reduction significantly reduces fuel costs and extends the transition timings and strategies for shipping companies. By operating vessels at lower speeds, fuel consumption decreases substantially, which directly translates to lower fuel expenses. This cost-saving measure, however, comes with trade-offs. Lower speeds result in longer transit times, which can impact the overall efficiency of the shipping operation. As a result, annual throughput and revenue are affected, as ships can make fewer trips within the same period. Shipping companies must balance the economic benefits of reduced fuel costs with the potential loss in revenue due to decreased operational efficiency.

In Tank-to-Wake (TtW) scenarios, the analysis reveals a preference for blue-ammonia over green-ammonia. Blue-ammonia, produced from natural gas with carbon capture and storage, is considered more economically viable in the current market conditions compared to green-ammonia, which is produced using renewable energy sources. The preference is driven by the lower production costs and higher availability of blue-ammonia. This choice also highlights the complexities in transitioning to greener fuels, as it depends not only on environmental benefits but also on economic feasibility and market readiness. The firmness of GHG emission factor implementation for various alternative fuels plays a critical role in shaping the transition strategies of shipping companies. Different implementation policies can lead to significant variations in how companies approach reducing their carbon footprints. For instance, policies that have more stringent emission factors might force companies to adopt cleaner technologies and fuels more rapidly, while more lenient policies could allow for a gradual transition. The analysis underscores the importance of regulatory environments in influencing corporate strategies and investments in sustainable practices.

The operating costs of vessels can vary significantly depending on the policy environment. Policies that impose stricter emission standards or higher carbon pricing can increase operational costs, whereas supportive policies and subsidies for green technologies can reduce these costs. The analysis highlights that the cost-competitiveness of containerized shipping is closely tied to the type of GHG policy implemented. Shipping companies must incorporate the developing GHG regulations in their decision-making to maintain their market position while striving to meet environmental goals.

Given its low price, LNG is considered the intermediary fuel choice before shifting to net-zero fuels to meet GHG targets. This is because it provides immediate carbon reduction vs conventional marine fuels, and it is a relatively mature technology. However, this is not environmentally sustainable as LNG is still a fossil fuel, and methane slips from using LNG can have even more severe effects than CO_2 emissions. Potential leaks from LNG production, transport, and combustion can significantly undermine the benefits of using LNG instead of conventional fuels. In the meantime, LNG may provide short-term environmental benefits, its widespread adoption can potentially postpone the transition to net-zero fuels, such as green-methanol, green-ammonia, and e-hydrogen, as significant investments in LNG infrastructure at ship and port levels may facilitate the use of this fuel for several years. Current policies and regulations have not fully considered the methane emissions from LNG in shipping, which has contributed to the consideration of LNG as a low-carbon fuel (Carr et al., 2023). This issue can be mitigated through investing in technologies that minimize or eliminate methane slips or through investing in methane capture technologies.

Green-ammonia offers a net-zero alternative to conventional fuels; however, it faces challenges in infrastructure investment, energy-intensive production, and safety concerns related to its toxicity. Unlike LNG, methane, and methanol, the use of ammonia as fuel is not yet technologically mature and the bunkering infrastructure for ammonia remains to be limited (Scarbrough & Longva, 2023). Although hydrogen is not included in the scope of this analysis, it is an important net-zero alternative fuel. Hydrogen can be used as liquid hydrogen, which requires extremely low temperatures for storage, or as ammonia, produced via the Haber–Bosch process. One challenge with liquid hydrogen is its low volumetric energy density (Fan et al., 2024), meaning larger storage tanks are needed for long voyages. Similar to green-ammonia, hydrogen faces several challenges, including high infrastructure investment, energy-intensive production, safety concerns, maturity, and supply and distribution infrastructure. According to Scarbrough and Longva (2023), there is no existing distribution and bunkering infrastructure for hydrogen yet in the shipping industry. While both green-ammonia and hydrogen offer promising net-zero alternatives to conventional fuels, their widespread adoption in the shipping industry may be hindered by the lack of established infrastructure and technological maturity. These complicate their transition to mainstream use in maritime shipping.

Finally, the analysis emphasizes the importance of future policies and regulations, along with future fuel price assumptions, in capital expenditure (CAPEX) decisions. The uncertainty surrounding future policy developments, fuel prices, and technological maturity add complexity to investment decisions in the shipping industry. Companies must consider these in making strategic investments, such as retrofitting existing vessels or acquiring new ones designed for alternative fuels. The interplay among regulatory frameworks, market conditions, and maturity of net-zero technologies will significantly influence the economic viability and timing of these investments.

7 Conclusion and Future Work

The study highlights the role of emissions reduction policies and fuel price assumptions in shaping the decarbonization transition strategies of various container vessel sizes. The strict implementation of the Carbon Intensity Indicator (CII) and high penalty costs for non-compliance drive significant changes in retrofitting and fuel choice decisions. In scenarios with the low penalty cost, companies often opt to maintain the same fuel and engine choice, while incurring penalties rather than investing in retrofits or transitioning to cleaner fuels. As penalties reach 250 USD per ton-CO_2, retrofitting to LNG becomes viable. At the 1,000 USD penalty, the transition to net-zero fuels, such as green-methanol, green-methane, green-ammonia, and biofuels, becomes mandatory. This emphasizes the role of policy implementation in accelerating decarbonization efforts.

Large vessels, including panamax and neopanamax, generally transition to net-zero fuels earlier than feeder vessels. This is driven by higher fuel consumption and stricter CII targets for larger vessels. Differences in retrofitting and fuel choice strategies between vessel sizes highlight the importance of tailoring decarbonization approaches to vessel-specific operational profiles. In addition, the decision strategy is not greatly influenced by the route served for the same vessel size. While there are a few scenarios in which the engine and fuel choice differed, most of the scenarios showed the same strategy despite changing the routes.

As demonstrated in the analysis, fuel cost assumptions significantly affect technology selection for container vessels. Lower LNG prices may encourage early retrofitting and its use as an intermediary fuel before transitioning to net-zero fuels. Conversely, high prices of biofuels and green alternative fuels may result in reliance on penalty payments until stricter policies foster clean technology adoption. The results from different price scenarios demonstrate the potential variability in decarbonization pathways due to market and technological uncertainties at this moment.

Lastly, the treatment of biofuels in policy implementation dramatically changes the transition strategy of container vessels. When considered as net-zero fuel, biofuels become an attractive option, thereby reducing the need for retrofits in certain cases. However, when it is not recognized as a net-zero option, alternative fuels such as green-ammonia and green-methanol are preferred.

This research can be enhanced by incorporating the uncertainty of future supply and maturity of alternative fuel technologies. Moreover, the forecast of fuel prices until 2050 can be incorporated using a probability distribution function and used in the problem optimization to make the results more robust given the volatilities in price forecast (Lagemann et al., 2023; Loennechen et al., 2024). By incorporating all these factors, shipping companies can make informed, strategic decisions that balance economic considerations with environmental goals, ultimately contributing to a greener future for global shipping.

References

Carr, E., Mccabe, S., Elling, M., Winebrake, J., & Comer, B. (2023). *Options for reducing methane emissions from new and existing LNG-fueled ships*. https://theicct.org/wp-content/uploads/2023/10/Options-for-Reducing-Methane-Emissions-from-New-and-Existing-LNG-Fueled-Ships-FINAL-926.pdf

Chen, S., Wang, X., Zheng, S., & Chen, Y. (2023). Exploring drivers shaping the choice of alternative-fueled new vessels. *Journal of Marine Science and Engineering, 11*(10), 1896. https://doi.org/10.3390/jmse11101896

Clarksons. (2023). *World fleet register*. https://www.clarksons.net/wfr/

Comer, B., & Osipova, L. (2021). *Update: Accounting for well-to-wake carbon dioxide equivalent emissions in maritime transportation climate policies*. https://theicct.org/wp-content/uploads/2021/08/update-well-to-wake-co2-aug21-1.pdf

DNV. (2022). *DNV releases update to WAPS technical standard*. https://www.dnv.com/news/dnv-releases-update-to-waps-technical-standard-235023/

European Commission. (2021). *Use of renewable and low-carbon fuels in maritime transport and amending Directive 2009/16/EC*. https://eur-lex.europa.eu/legal-content/EN/TXT/HTML/?uri=CELEX:52021PC0562

Fan, H., Xu, X., Abdussamie, N., Chen, P.S.-L., & Harris, A. (2024). Comparative study of LNG, liquid hydrogen, and liquid ammonia post-release evaporation and dispersion during bunkering. *International Journal of Hydrogen Energy, 65*, 526–539. https://doi.org/10.1016/j.ijhydene.2024.04.039

GSC. (2022). *Efforts of the planning and design center for greener ships (GSC)*. https://www.classnk.or.jp/hp/pdf/research/rd/2022/05_e02.pdf

IEA. (2020). *Indicative shipping fuel cost ranges—charts—data & statistics*. https://www.iea.org/data-and-statistics/charts/indicative-shipping-fuel-cost-ranges

IEA. (2024). *Simplified levelised cost of competing low-carbon technologies in chemicals production–charts—data & statistics*. https://www.iea.org/data-and-statistics/charts/simplified-levelised-cost-of-competing-low-carbon-technologies-in-chemicals-production

IMO. (2013). Annex 14 resolution. *MEPC, 231*(65). https://wwwcdn.imo.org/localresources/en/KnowledgeCentre/IndexofIMOResolutions/MEPCDocuments/MEPC.231(65).pdf

IMO. (2021). Annex 13 resolution. *MEPC, 339*(76). https://wwwcdn.imo.org/localresources/en/OurWork/Environment/Documents/Air%20pollution/MEPC.339(76).pdf

IMO. (2023). *Revised GHG reduction strategy for global shipping adopted*. https://www.imo.org/en/MediaCentre/PressBriefings/pages/Revised-GHG-reduction-strategy-for-global-shipping-adopted-.aspx

Lagemann, B., Lagouvardou, S., Lindstad, E., Fagerholt, K., Psaraftis, H. N., & Erikstad, S. O. (2023). Optimal ship lifetime fuel and power system selection under uncertainty. *Transportation Research Part D: Transport and Environment, 119*, Article 103748. https://doi.org/10.1016/j.trd.2023.103748

Lagemann, B., Lindstad, E., Fagerholt, K., Rialland, A., & Ove Erikstad, S. (2022). Optimal ship lifetime fuel and power system selection. *Transportation Research Part d: Transport and Environment, 102*, Article 103145. https://doi.org/10.1016/j.trd.2021.103145

Law, L., Foscoli, B., Mastorakos, E., & Evans, S. (2021). A comparison of alternative fuels for shipping in terms of lifecycle energy and cost. *Energies, 14*(24), 8502. https://doi.org/10.3390/en14248502

Lieshout, P.S., de Jonge, V., Verbeek, R., & Vredeveldt, A. W. (2020). *Green maritime methanol: WP3 factsheet and comparison with diesel and LNG*. https://publications.tno.nl/publication/34637491/xDLPzK/TNO-2020-R11822.pdf

Lindstad, E., Lagemann, B., Rialland, A., Gamlem, G. M., & Valland, A. (2021). Reduction of maritime GHG emissions and the potential role of E-fuels. *Transportation Research Part d: Transport and Environment, 101*, Article 103075. https://doi.org/10.1016/j.trd.2021.103075

Liu, L., Yang, D., & Frangopol, D. (2021). Ship service life extension considering ship condition and remaining design life. *Marine Structures, 78*(0951–8339). https://doi.org/10.1016/j.marstruc.2021.102940.

Loennechen, O., Fagerholt, K., Lagemann, B., & Stålhane, M. (2024). Maritime fleet composition under future greenhouse gas emission restrictions and uncertain fuel prices. *Maritime Transport Research, 6*, Article 100103. https://doi.org/10.1016/j.martra.2024.100103

MAN Energy. (2023a). *CEAS engine calculations*. https://www.man-es.com/marine/products/planning-tools-and-downloads/ceas-engine-calculations

MAN Energy. (2023b). *Dual-fuel conversion | two-stroke engines|MAN PrimeServ*. https://www.man-es.com/services/new-service-solutions/dual-fuel-conversion

Mærsk Mc-Kinney Møller Center for Zero Carbon Shipping. (2024a). *Tackling methane slip in shipping insights into options for regulation and quantification*. https://cms.zerocarbonshipping.com/media/uploads/documents/Tackling-Methane-Slip-Insights-into-options-for-regulation-and-quantification.pdf

Mærsk Mc-Kinney Møller Center for Zero Carbon Shipping. (2024b). *Fuel cost calculator*. https://www.zerocarbonshipping.com/cost-calculator/?s=0

Maritime Optima. (2023). *Sea route calculator*. https://maritimeoptima.com/shipatlas/feature/sea-route-calculator
Nelissen, D., Kleijn, A., & Faber, J. (2022). *FuelEU Maritime and EU ETS Sound incentives for the fuel choice*. https://cedelft.eu/wp-content/uploads/sites/2/2022/02/CE_Delft_210396_FuelEU_Maritime_and_EU_ETS_Def_February-2022.pdf
Nippon Kaiji Kyokai (ClassNK). (2021). *Outlines of EEXI regulation*. https://www.classnk.or.jp/hp/pdf/activities/statutory/eexi/eexi_rev5e.pdf
OECD. (2022). *Ocean shipping and shipbuilding*. https://www.oecd.org/ocean/topics/ocean-shipping/
ONE. (2023). *Service maps*. https://www.one-line.com/en/routes/current-services
Scarbrough, T., & Longva, T. (2023). Study on the readiness and availability of low- and zero-carbon technology and marine fuels. https://wwwcdn.imo.org/localresources/en/OurWork/Environment/Documents/FFT%20Project/Preliminary%20outputs%20for%20the%20study.pdf
Serpell, O., Hsain, Z., Chu, W., & Johnsen, W. (2023). Ammonia's role in a net-zero hydrogen economy. https://kleinmanenergy.upenn.edu/research/publications/ammonias-role-in-a-net-zero-hydrogen-economy/
Tu, H., Liu, Z., & Zhang, Y. (2024). Study on cost-effective performance of alternative fuels and energy efficiency measures for shipping decarbonization. *Journal of Marine Science and Engineering, 12*(5), 743. https://doi.org/10.3390/jmse12050743
Zhao, Y., Chen, Y., Fagerholt, K., Lindstad, E., & Zhou, J. (2023). Pathways towards carbon reduction through technology transition in liner shipping. *Maritime Policy & Management*, 1–23. https://doi.org/10.1080/03088839.2023.2224813
Zhao, Y., Ge, R., Zhou, J., & Notteboom, T. (2024). Decarbonization pathways for bulk vessels: Integrating power systems, fuels, and control measures. *Ocean Engineering, 300*(0). https://doi.org/10.1016/j.oceaneng.2024.117488
Zwaginga, J., Lagemann, B., Erikstad, S. O., & Pruyn, J. (2024). Optimal ship fuel selection under life cycle uncertainty. *Sustainability, 16*(5), 1947. https://doi.org/10.3390/su16051947

Anne Xaviera Go is a master's student in the Department of Systems Innovation, Graduate School of Engineering at the University of Tokyo, Japan. Her research focuses on supporting the maritime shipping industry in achieving its decarbonization goals through optimization and resource management. Prior to pursuing her graduate studies, she held various roles with increasing scope and leadership in supply chain management of multinational consumer goods and local aquaculture from 2016 to 2023, specializing in demand and supply planning, logistics operations, and process improvements.

Tomoya Kawasaki is an Associate Professor in the Department of Systems Innovation, Graduate School of Engineering at the University of Tokyo, Japan. He earned his Doctor of Engineering degree from Tokyo Institute of Technology, Japan in 2012. His career includes being a researcher at the Japan Maritime Center, an Assistant Professor at Nihon University, and an Assistant Professor at the Tokyo Institute of Technology before assuming his current position in August 2020. Dr. Kawasaki's research focuses on logistics systems engineering, supply chain management, and the application of advanced data analytics in logistics. He is particularly interested in optimizing container management, improving logistics efficiency, and integrating innovative technologies into supply chain operations. His work bridges academia and industry, contributing to practical advancements in logistics and transportation. He has been actively involved in academic societies and collaborative research projects, receiving several prestigious awards for his contributions. These include best paper awards from the Japan Logistics Society (2014, 2020), the Japan Society of Logistics and Shipping Economics (2014, 2023), and the T-LOG conference (2018, 2020). His research has been recognized for its practical implications and innovative approaches to addressing complex logistics challenges.

Rei Miratsu is a researcher at the Research Institute of Nippon Kaiji Kyokai (ClassNK). He earned his master's degree in environmental studies from the Department of Ocean Technology, Policy, and Environment at the University of Tokyo in 2016 and commenced his career at the Research Institute of ClassNK in the same year. His research is dedicated to advancing safety and environmental protection in the maritime industry. Specifically, his work encompasses the analysis of ship operations through the application of big data, involvement in multiple full-scale ship measurement projects, and the development of fleet prediction simulations for alternative fuel vessels. The outcomes of his research have significantly contributed to the development of classification rules and guidelines.

Ryuji Miyake is a principal researcher at Research Institute of Nippon Kaiji Kyokai (ClassNK). He has a Doctor of Engineering degree in Naval Architecture and Ocean Engineering from the University of Tokyo. Dr. Miyake started his carrier at Research Institute of ClassNK in 2001. From 2011 to 2022, he was responsible for verification of EEDI, EEXI, SEEMP, IMO-DCS, and CII as a manager of plan approval division in ClassNK. Since 2023, he has studied Life Cycle GHG Intensity and the verification scheme on marine fuel at Research Institute of ClassNK. He has been attending the IMO MEPC as one of Japan delegation representatives since 2014, where he contributed to the development of Guidelines on EEDI, EEXI, SEEMP, IMO-DCS, CII, and LCA. As a verifier, expert, or researcher, he has participated in many international conferences and committees by IMO, ISO, IACS, ITTC, and ISSC.

Green Port Transformation: A Case Study for the Port of Xiamen

Qiong Chen, Jiaqi Gao, Wentao Liao, Yui-Yip Lau, Guoqiang Chen, Maneerat Kanrak, Pengfei Zhang, and Lijuan Deng

Abstract With the surge of maritime trade and the development of port infrastructure, the pollution of the environment by shipping activities is becoming more and more obvious. Therefore, this chapter discusses the research progress of ship carbon emissions and makes an in-depth analysis of the green port initiatives at Xiamen Port. At present, the measures for the green initiatives for ships at ports are mainly divided based on three aspects: alternative fuels, onshore power supply, and policy improvement. All three aspects have different perspectives. Most scholars put forward emission reduction measures for onshore power supply and alternative fuels, whereas policy improvement is viewed as a comprehensive embodiment of these two groups of measures and also the basis for the development of green ports. This study evaluates the effectiveness of these three aspects and analyzes the existing green initiatives of Xiamen Port according to these three aspects. The importance of Xiamen

Q. Chen · J. Gao (✉) · W. Liao · G. Chen · P. Zhang · L. Deng (✉)
Navigation College, Jimei University, Xiamen, China
e-mail: 86353940@qq.com

L. Deng
e-mail: lindadeng2002@sina.com

Q. Chen
e-mail: qchen@jmu.edu.cn

W. Liao
e-mail: Liaospy@foxmail.com

G. Chen
e-mail: chengq@jmu.edu.cn

P. Zhang
e-mail: shippinglaw@163.com

Y.-Y. Lau
Division of Business and Hospitality Management, College of Professional and Continuing Education, The Hong Kong Polytechnic University, Yau Ma Tei, Kowloon, Hong Kong
e-mail: yuiyilau@cpce-polyu.edu.hk

M. Kanrak
Faculty of Interdisciplinary Studies, Khon Kaen University, Nong Khai, Thailand
e-mail: maneerat@kku.ac.th

M. A. Dulebenets et al. (eds.), *Transport and Climate Change*,
https://doi.org/10.1007/978-3-032-00563-2_19

Port in the cruise shipping network is assessed as well. In addition, the economic and social benefits brought by the green initiatives at Xiamen Port are evaluated in detail. Finally, a set of constructive recommendations are proposed to facilitate the green transformation of Xiamen Port. These recommendations focus on monitoring of ship emissions, green shipping infrastructure development, improvement of green shipping policies, financial incentives, and collaborative initiatives.

Keywords Complex networks · Carbon emissions · Environmental policy · Xiamen port

1 Introduction

The boom in maritime trade is a significant force driving the economic development of port cities (Lane and Peters, 2020). However, with the advancement of waterway transportation and port construction, the environmental pollution caused by shipping activities has become increasingly apparent. Amidst the growing severity of global climate change and environmental issues, the increase in extreme weather events and rising sea levels pose threats to port infrastructure and operational safety (Verschuur et al., 2023). Concurrently, the international community's concern for greenhouse gas emissions has been escalating, prompting governments and international organizations to introduce policies that demand ports reduce their carbon footprint and undergo a green transformation. Ports, as pivotal nodes in international shipping trade, are increasingly recognized for their role in environmental sustainability (Verschuur et al., 2022). Xiamen Port, a crucial gateway on the southeast coast of China and a key hub in the global supply chain, is facing an urgent need to upgrade and transform into a green port. The establishment of a green port is not only an inevitable response to the global trend that supports emission reduction but also a strategic move to enhance the port's competitiveness and achieve sustainable development.

Current research on ship pollution emissions primarily employs two methods: (1) carbon emission models; and (2) shipping network analysis. Carbon emission models serve as essential tools for assessing and predicting ship pollution emissions. These models provide a scientific basis for formulating emission reduction strategies by quantifying the carbon emissions of ships under various operating conditions. With the enhancement of computational power and the ease of data acquisition, the precision and scope of carbon emission models have seen significant improvements. Miola and Ciuffo (2011) divide these models into two main categories: a top-down approach, which uses ship fuel sales data and fuel emission factors to estimate emission inventories, and a bottom-up approach, aimed at achieving greater accuracy in calculating emission inventories taking into account ship characteristics. However, the bottom-up method requires a range of data related to ship traffic. As an example, Chen et al. (2021) used Automatic Identification System (AIS) data along with ship static data (e.g., maximum designed speed, gross tonnage, rated engine power) to estimate the overall amount of emissions produced by Arctic passenger ships. Gan et al. (2022) estimated the ship emissions in the west area of the Port of Shenzhen (China) based on the AIS data and ship characteristics. A variety of other studies also utilized AIS data, combined with the bottom-up approach, to assess ship emissions. The use of AIS data, integrated with the bottom-up method, to calculate pollution emissions for ports or certain areas has become fairly popular but is still quite challenging. These methodologies and data provide a foundation and theoretical support for the assessment of ship pollution emissions along China's coastal regions.

Another approach to studying ship pollution emissions involves the analysis of shipping networks. This method examines the complex and global nature of shipping activities, considering not only the emissions of individual ships but also the efficiency and environmental impact of the entire shipping system. By identifying key shipping nodes and routes, shipping network analysis offers guidance for optimizing route layouts, enhancing port efficiency, and reducing carbon emissions. Since ports are viewed as important hubs connecting sea and land transportation, the optimization of the network's structure plays an important role in the development of the port and the entire regional economy. In order to further analyze the role of the shipping network, this paper will take the Xiamen's shipping network as a case study and analyze the complex network topology to visually demonstrate its contribution to the study of ship pollution emissions.

According to the above methods, from the estimation of ship carbon emissions to the development of green ports, the current research of scholars is mainly divided based on three aspects: alternative fuels, onshore power supply, and policy improvement. The analysis of carbon emissions caused by sailing or berthing ships can be used to put forward corresponding solutions and improvement methods. This paper will discuss the effectiveness of each one of the aforementioned aspects, analyze the achievements and shortcomings of the current green port development practices in Xiamen Port, and then propose several constructive solutions.

The structure of this chapter is as follows. The background and importance of this study are presented in Sect. 1. Section 2 provides a literature review on the network modeling, including the relevant research on shipping networks, and the analysis of current ship pollution modeling methods. Section 3 presents the key emission reduction measures considered for the analysis in this study. Section 4 performs the cruise shipping network analysis with an emphasis on the role of Xiamen Port. Section 5 outlines some of the major challenges of Xiamen Port in terms of its green transformation. Section 6 provides the major recommendations developed based on this study that would be of importance to Xiamen Port and relevant stakeholders. Section 7 presents the key concluding remarks and the future outlook.

2 Literature Review

2.1 Complex Networks

Graph theory began its formal journey with a problem known as the "Seven Bridges of Königsberg", studied by Leonhard Euler in 1736 (see Fig. 1). Konig's work on this problem laid the foundational mathematical principles of graph theory, introducing

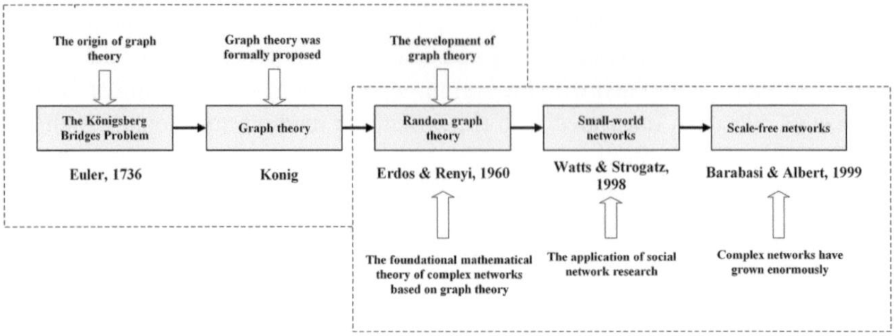

Fig. 1 Evolution of complex networks

the concept of a graph as a structure consisting of vertices and edges. The solution to the Königsberg Bridges Problem demonstrated that it was impossible to traverse all seven bridges exactly once and return to the starting point, marking the birth of the field. Moving forward, Erdos and Renyi (1960), Hungarian mathematicians known for their prolific work in number theory and combinatorics, made significant contributions to graph theory, particularly in the area of random graph theory, which was originally developed in 1959.

In the late twentieth century, graph theory leaped forward with the introduction of concepts such as "small-world networks" by Watts and Strogatz (1998). This concept describes highly clustered networks that have short average path lengths between nodes, reflecting the surprizing interconnectedness of many real-world networks. Another pivotal development was the introduction of "scale-free networks" by Barabási and Albert (1999). This model suggests that some networks, unlike random networks, have a few nodes that are highly connected, while the majority of nodes have only a few connections. This model has been applied to various complex systems, including the Internet, social networks, and biological networks.

In shipping networks, complex network analysis is often used to evaluate the underlying characteristics of networks. Kaluza et al. (2010) proved that global and regional container shipping networks have small-world characteristics along with broad degree and weight distributions. Wan et al. (2021a) formulated a complex network model for China's Maritime Silk Road, examined its characteristics, and determined that the model possesses attributes similar to the ones of a small-world network. LaRock et al. (2022) presented a path-based approach for the analysis of the global liner shipping network. The navigational trajectories were computed through the network. The objective was to minimize the overall number of cargo transfers along the route, which can be viewed a desirable goal in practical applications. Li et al. (2024) quantitatively analyzed the changes in the spatial structure of the container transport network on international routes affected by the pandemic. The study indicated that hub ports of the main shipping routes along with long-distance routes were more susceptible to the COVID-19 effects.

Generally, the current studies on China's transportation networks are predominantly focused on the analysis of container ships, with a lack of research on the transportation networks of other types of vessels. Moreover, only a fraction of the literature addresses the impact of the pandemic and other disruptions (e.g., climate change) on shipping networks, and most do not delve into the changes and reasons behind the transportation network alterations under disruptive conditions. In light of this, the present study constructs a Xiamen cruise network model based on complex network theory, which consists of multiple cruise ports and the routes between them. The chapter analyzes the evolution of the cruise network under the influence of disruptions from the perspective of network centrality characteristics and proposes relevant policy recommendations to improve the Xiamen Port cruise network's resilience to the impact of disruptions with the main focus on climate change.

2.2 Emission Modeling

Chen et al. (2017) indicated that the Automatic Identification System (AIS) is a vessel monitoring system capable of transmitting precise dynamic data, such as a ship's real-time location and speed. The study developed the emission inventory for the Port of Qingdao in North China. With the continuous advancement of big data technology, the AIS data applications have become more and more popular. For instance, Longépé et al. (2018) improved the efficiency and accuracy of the regulatory oversight in the shipping industry by utilizing a vessel detection system combined with AIS data. Wang et al. (2020) used the AIS data collected for Xiamen Port in order to estimate emissions produced by ships in 2018. It was found that more than 80% of emissions were generated by ships that were leaving or entering the port area, and around 20% were produced by ships sailing outside the port.

Ju and Hargreaves (2021) performed a detailed analysis of carbon dioxide emissions in the Western Singapore Straits using the AIS data along with the static information from the Singapore Maritime Data Hub. The findings can be beneficial to the relevant Singapore agencies to manage emissions. In another effort, Chen et al. (2023) used the AIS data to quantify emissions produced by passenger ships in Danish waters. It was found that the COVID-19 pandemic had major effects on the amount of emissions produced. The number of ships and their sailing speed reduced during the COVID-19 time period, which caused a decrease in the amount of emissions produced.

Integrating AIS data with bottom-up approaches has improved the accuracy of pollution emission calculations for specific ports and regions. These technologies and datasets provide a foundation of data and theoretical support for assessing ship pollution emissions along China's coastline. The conducted studies demonstrate the broad applications of AIS technology, offering robust technical support and decision-making references for the sustainable development of the shipping industry and the management of marine resources in the future.

Another group of studies focused on the environmental aspects of maritime transportation and ship emissions without specifically using the AIS data in their analyses. As an example, Yi et al. (2011) studied carbon dioxide intensity reduction targets in

China. A comprehensive index was developed that can be potentially used to achieve ambitious emission reduction goals. The index was based on responsibility, potential, and emission reduction capacity. Chen and He (2016) proposed a sector-based approach that can be adopted for carbon dioxide emission control. It was suggested that the burden sharing should be performed based on capacity, responsibility, and efficiency. Tang et al. (2017) presented a multi-agent model that can be used to establish an emissions trading scheme in China. The main economic agents of the model were the government and various companies from diverse sectors. The analysis confirmed the validity of the proposed methodology and its potential for emission trading systems.

Wu et al. (2022) performed a detailed review of the literature on emission trading strategies in shipping. The key drivers of emission trading schemes were pointed out, including limitations of the available solutions, potential of market-based options, and stakeholder attitudes. However, along with drivers, various barriers were identified as well (e.g., sectorial and geographical coverage, free emission quote percentage, conflict between responsibilities). Fan et al. (2023) conducted a micro–macro analysis using a carbon footprint model for inland shipping. It was found that the LNG alternative was quite efficient in terms of emission reduction compared to other alternatives.

Zhou et al. (2023) investigated carbon emission hotspots for oceangoing ships in the vicinity of the Port of Wuhan. Spatial and temporal dynamics of ship emissions were evaluated, and important recommendations were proposed. Zhu et al. (2023) presented a bi-level multi-objective optimization model that can be used for carbon emission allowance in maritime shipping. The proposed model can be effectively applied to analyze interactions of shipping companies with the government. Jin et al. (2024) explored the evolution of carbon emissions for over 100 cities belonging to the Yangtze River Economic Belt. The results of the study confirmed that economic growth, governmental interventions, and technological developments influenced the distribution of emissions.

The results from the conducted literature review show that quite a few models have been developed over the years to study ship emissions. Some of these models can be considered as shipping network models, while the others can be viewed as alternative ship emission models. Previous studies have focused mainly on container shipping networks, leaving cruise shipping networks underexplored. Moreover, there is a lack of studies specifically focusing on Xiamen Port and its green initiatives despite the fact that this port plays a critical role for China's foreign trade. In order to address this limitation in the state of the art, the present study aims to analyze Xiamen's cruise network and associated environmental aspects. Potential economic benefits and social benefits that can be achieved from green initiatives at Xiamen Port are discussed as well.

3 Emission Reduction Measures

A substantial amount of port carbon emissions now comes from operations related to ship berthing (Deniz et al., 2010). Therefore, reducing carbon emissions from ships has become an important focus to minimize port carbon emissions. At the same time, the ship carbon emission reduction strategy can be applied to ocean transportation and contribute to the emission reduction plan of the whole life cycle of ships. The existing emission reduction measures for the daily activities of ships in port areas can be divided into three main types: onshore power supply (OPS), policy improvement, and alternative fuels. This section will analyze the effectiveness of these three current measures for emission reduction to further evaluate the existing measures and shortcomings of Xiamen Port in terms of these three aspects. Table 1 summarizes the key features of representative studies comparing port ship activities and research focus along with the emission reduction strategies adopted.

Based on the representative research on ship emission modeling and reduction listed in Table 1, various activities have been investigated over the years, including ship berthing, maneuvering, and handling process (e.g., loading and unloading of containers using quay cranes). Different emission reduction strategies have been evaluated by the reviewed studies. Onshore power supply and alternative fuels seem to be quite popular strategies, and their efficiency has been proven in various investigations. Policy implementation has been considered as well. Based on this analysis, the present study will evaluate the effectiveness of onshore power supply, policy improvement, and alternative fuels for Xiamen Port.

Table 1 Representative studies on berthing ships' carbon emission reduction strategies

References	Emission reduction strategies	Activity process	Research topic
Wang et al. (2024)	OPS	Berthing	Utilization of OPS by different ship types
Tang et al. (2023)	OPS	Berthing	Assessment of emission reduction effect
Vichos et al. (2022)	OPS, hydrogen storage system	Berthing	Assessment of emission reduction effect
Wan et al. (2021b)	OPS, low-sulfur fuels	Maneuvering, Berthing, Handling	Assessment of emission reduction effect
Jiao and Wang (2021)	OPS, low-sulfur fuels	Berthing	Cost optimization
Yang (2017)	Policy implementation, equipment replacement	Berthing, yard, gate	Assessment of emission reduction effect
López-Aparicio et al. (2017)	OPS, low-sulfur fuels	Maneuvering, Berthing, at sea	Assessment of emission reduction effect

3.1 Alternative Fuels

When it comes to maritime transportation, energy options have significant potential for decreasing ship carbon emissions, including electrification technologies, alternative fuels, and hybrid options. Clean or renewable fuels may substantially decrease carbon emissions at ports. The actual transition to biofuels from fossil fuels is a milestone in the pursuit of energy-efficient approaches. The key alternative energy methods considered in the current research efforts are liquefied natural gas (LNG), mixed biodiesel, low-sulfur fuel oil, and hydrogen energy. The use of LNG as fuel to decrease carbon emissions is becoming more and more popular in practice. LNG may be adopted as a dual-fuel engine to power ships. When comparing with conventional diesel fuel, oceangoing ships can reduce carbon emissions after transitioning to LNG (Balcombe et al., 2022).

Furthermore, conversion of fossil fuels into biofuels is another promising alternative to reduce emissions at ports. Biomass fuels may be mixed with other fossil fuels, and then biodiesel, which can be viewed as a mixture of fuels, can efficiently decrease carbon emissions. Different mixing ratios can lead to different emission reduction effects. For example, a 30% biofuel and diesel mixture can reduce total emissions by over 20% (Geerlings & Van Duin, 2011). In addition, a mix of diesel and electricity can also reduce emissions. As an example, replacing conventional diesel-powered ferries with diesel-electric hybrid ferries may enhance energy efficiency by approximately 12% and achieve a 10% reduction in carbon emissions (Chou et al., 2021). Methanol, hydrogen, and low-sulfur fuel oil can be considered as some of the promising energy alternatives for carbon emission reduction.

3.2 Onshore Power Supply

Ships anchored at berthing positions mainly rely on auxiliary engines and produce carbon emissions. The use of onshore power supply (OPS) is an effective option to decrease noise and enhance air quality at ports, as it allows berthing ships to shut off auxiliary engines, thereby reducing carbon emissions. Nevertheless, the efficiency of OPS in reducing emissions is also affected by the electricity source. If coal generates electricity, the port will generate more emissions. On the contrary, if clean energy such as wind or hydropower is used, port carbon emissions will be significantly reduced. Based on its environmental and economic benefits, the use of OPS is effective in decreasing carbon emissions at ports.

However, the promotion and implementation of OPS faces multiple challenges. First, infrastructure construction and technical compatibility are the key issues to be overcome. Ships and ports need corresponding equipment and technical support for the connection and use of OPS. Second, high initial investment costs are another obstacle, which includes the construction of onshore power facilities and the cost of upgrading ships. To overcome these challenges, policy support and incentives play a crucial role. In addition, before the implementation of the OPS, it is necessary to consider the cleanliness of the power source. If electricity comes from high-carbon sources such as coal, OPS will be less effective. Therefore, promoting the use of clean energy, such as wind and solar energy, is crucial to the realization of OPS emission reduction effect. Through policy guidance and incentives, port and ship operators can be encouraged to adopt cleaner sources of electricity, thus improving the emission reduction effect of OPS.

3.3 Policy Improvement

Policy improvement is a key factor to promote carbon emission reduction in ports. By developing and implementing effective policies, ports, ship operators, and relevant stakeholders can be guided and encouraged to take emission reduction actions. This includes the establishment of a comprehensive port carbon emission reduction policy framework using diverse policy tools, such as tax incentives, subsidies, emissions trading systems, and green port certification. Strengthening international cooperation and coordinating port emission reduction policies of different countries and regions can assist with the achievement of the global carbon emission reduction targets.

The effectiveness of existing policies is regularly assessed and adapted to inform new scientific discoveries, technological advances, and market changes. Ensuring the participation of stakeholders from port operators, shipping companies, environmental organizations, and local communities in the policy development process can improve policy acceptance and implementation effectiveness. Through these measures, we can more comprehensively analyze the potential and challenges of onshore power supply and policy improvement in reducing port carbon emissions and provide scientific basis along with practical guidance for the development of effective emission reduction strategies.

In practical applications, it is often difficult to achieve the best emission reduction effect using a single strategy. Therefore, an integrated strategy combining alternative fuels, OPS, and policy improvements may be more effective. For example, some ports in China consider berthing fee discounts to foreign trade ships using new energy fuels and using port power. Such policy incentives help promote the use of OPS while reducing carbon emissions at ports. In addition, policies can support the research and development and promotion of new alternative fuels to reduce their costs and improve their market competitiveness. Through this comprehensive strategy, the problem of port carbon emissions can be solved more comprehensively, and the sustainable development of the port can be realized.

4 Shipping Network Analysis and Evaluation of Benefits

4.1 *Cruise Shipping Network*

Figure 2 presents the graph of the cruise shipping network consisting of 84 ports and 308 links. Bigger dots present ports with larger degree values, and the thickness of the lines connecting between ports is the frequency of cruise shipping between neighboring cruise ports. A greater degree of a port indicates that this port is directly connected to a larger number of other ports. So, this port is more important and is a key port in the network.

The average connectivity degree of the network is 3.609, which is relatively low, reflecting the closeness of the connection between ports. That is each port can be connected to other three ports on average. This indicates that there is a certain degree

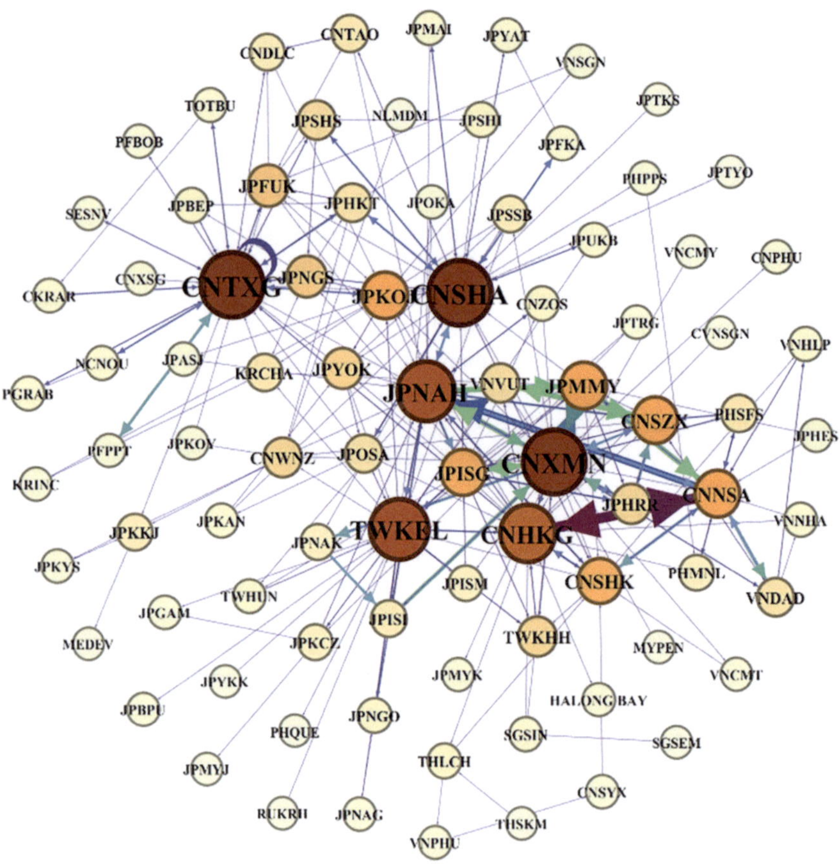

Fig. 2 The cruise shipping network in 2019

of connectivity, but there is still room for further improvement. The low average degree also indicates that the ports do not have much connection to other ports, suggesting that more can be done to expand the cruise routes to connect to Xiamen. This can increase the connectivity degree of Xiamen, which will optimize the structure and connectivity of the network. Such an alternative can also enhance network stability and efficiency for information dissemination.

Centrality degree measures the number of direct connections a port has with other ports. A higher degree of centrality indicates a greater number of direct links to other ports, signifying a more pivotal role and status as a critical node within the cruise network. Based on Fig. 3, Naha (Japan) had the largest proportion of cruise ship calls in 2019 accounting for 16% of the total, followed by Shanghai (China) and Xiamen (China) with the same proportion of 11%. Ishigaki Island (Japan) and Keelung (Taiwan) rank third with the same proportion of 10%. The top five ports, including Xiamen, handle nearly 60% of all cruise calls. This suggests that Xiamen is not only an important part of the cruise network, but to some extent, Xiamen has a greater centrality degree in the network because it is directly connected to a large number of ports. Thus, Xiamen is the center and an essential port in the network.

The analysis results show the important position of Xiamen Port in the cruise shipping network, with the second largest percentage of cruise ship calls after Naha. Similar to Shanghai, Xiamen Port accounts for 11% of all cruise ship calls, which can be viewed as a significant share. However, Shanghai and Xiamen are quite different in terms of economic development, ship monitoring, and environmental initiatives. In

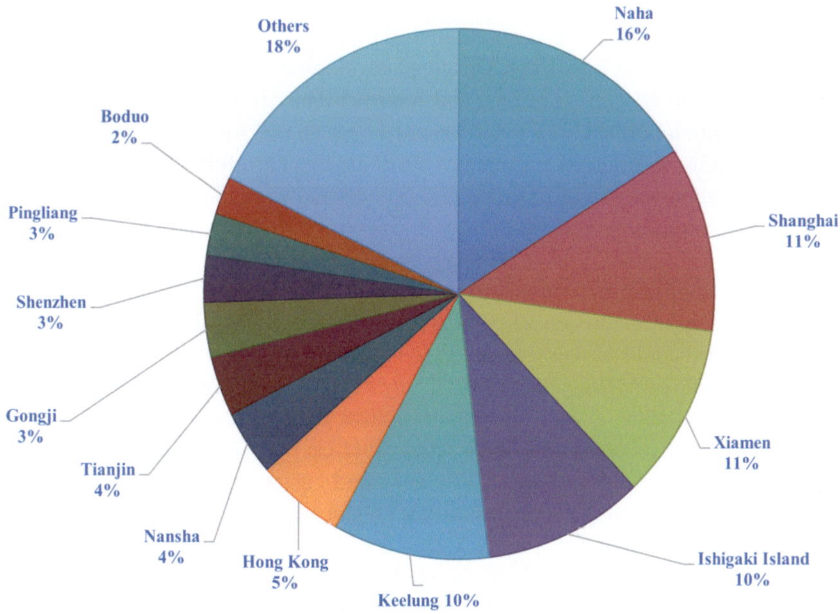

Fig. 3 The proportion of cruise ship calls in the cruise shipping network

addition, as can be seen from the complex network, Xiamen Port is connected to many ports in other countries, which requires Xiamen Port to continuously improve its own operational efficiency and service quality to meet the needs of different countries. For example, Japan, the UK, and other major countries have been paying attention to new energy ships in recent years. Therefore, Xiamen Port is also expected to actively introduce and apply green technologies, design innovative terminal facilities suitable for the new generation energy ships, such as intelligent energy management systems, improve the management of port energy, and optimize the energy allocation.

Through the conducted shipping network analysis, we can further examine the carbon emissions. The economic benefits of implementing carbon reduction strategies are reflected on multiple levels, and their outcomes can bring a profound economic value to Xiamen Port. By adopting carbon reduction measures, Xiamen Port can significantly reduce energy consumption, decrease reliance on fossil fuels, thereby saving operational costs and enhancing the port's economic efficiency. Furthermore, the implementation of carbon reduction measures can further promote the transformation and upgrading of port operations, pushing the port towards a cleaner and more efficient direction. This not only improves the operational efficiency of the port but also strengthens its competitiveness on the shipping market.

4.2 Social Benefit Analysis

The social benefits of Xiamen Port's transformation into a green port are reflected in many aspects. The first point is about environmental protection through the implementation of green port initiatives. The green development of Xiamen Port can effectively reduce port carbon emissions and improve the port air quality. In 2018, for example, Xiamen Port ship pollutants reached over 16,000 tons, including a significant proportion of sulfur and nitrogen oxide emissions harmful to living organisms. A further implementation of new green initiatives and strengthening of the existing ones can effectively reduce the amount of emissions produced by port activities every year.

Furthermore, in terms of energy efficiency improvement, Xiamen Port has achieved a comprehensive reduction of energy consumption by 20.5%, saving 5,664 tons of standard coal and reducing 9,725 tons of carbon dioxide emissions. In terms of technological innovation and industrial upgrading, Xiamen Port relies on information technology means, combined with the construction of smart ports, and realizes the application of the green port ecosystem, such as the i-TOS terminal operating system, by improving the logistics efficiency and customer convenience. In addition, the transformation of Xiamen Port has enhanced the corporate sense of social responsibility. For example, Xiamen Container Terminal Group organizes themed activities, such as afforestation and garbage classification, every year to improve the construction and management level of the green port and benefit the surrounding communities.

4.3 Economic Benefit Analysis

The transformation of Xiamen Port into a green port brings a set of economic benefits that are mainly reflected in three aspects. The first point is associated with the improvement of energy efficiency. For example, through the implementation of the "container terminal electricity replacement action" and the "full power terminal", Xiamen Port has realized the 100% electricity operation of gantry cranes, and the total proportion of electric equipment has exceeded 55%. This not only reduces energy consumption, but also reduces carbon emissions and improves energy efficiency. In addition, 88% of gantry cranes have implemented energy utilization transformation to further improve the energy consumption reduction effect. The energy consumption of the gantry cranes has decreased by 59% compared with the traditional operation mode. The transformation of the port power facilities not only makes the full use of energy but also greatly decreases the carbon emissions produced.

The second point is the cost savings. Through measures such as "oil to electricity", Xiamen Port has realized energy cost savings. For example, the electric reactor machine is leased for trials and is expected to be fully promoted after the feasibility analysis. This project will not only reduce the overall operating cost but will also realize a win–win situation of social and economic benefits. The power cost of electric reactor is only half of the original fuel, which can be recovered in 6 years, greatly reducing the operating cost of the port. The third point is the promotion of port competitiveness. The green port transformation will help to enhance the competitiveness of Xiamen Port, attract more green shipping enterprises and related industries to work together, and promote the high-quality development of regional economy. Through the development of smart ports, we can improve the social logistics efficiency and realize the green port ecosystem, so as to improve the overall service level and market competitiveness of the ports.

5 Analysis of the Inefficiencies in the Green Development of Xiamen Port

5.1 Ship Pollution Issues

In response to the national "dual carbon" policy, Xiamen Port has set clear emission reduction targets based on the "Xiamen Port Low-Carbon Development Action Plan": by 2025, the CO_2 emissions per unit of port throughput and per unit of operating ship turnover volume should decrease by 10% and 5% compared to 2020, respectively. Xiamen Port has made progress in reducing ship pollution, including the implementation of shore power projects, to lower emissions during berthing. Hainate Electric (Xiamen) New Energy Technology Co., Ltd. has promoted the electrification of port handling equipment, effectively reducing carbon emissions during operations.

Additionally, Xiamen Port has significantly improved energy efficiency and reduced reliance on fossil fuels through intelligent transformations, including the use of fully electric gantry cranes. Despite these notable achievements, there is still considerable room for improvement compared to the goals set forth in the "Xiamen Port Low-Carbon Development Action Plan". For instance, although the application rate of electrified equipment in Xiamen Port has exceeded 55%, the energy consumption per unit of gantry crane operation has only decreased by 59% compared to traditional operation modes, indicating significant potential for further improvement in reducing ship carbon emissions and promoting low-carbon technologies.

5.2 *Insufficient Integration of Scientific Research and Innovation Resources*

Xiamen Port has made significant progress in promoting the integration of scientific research and innovation resources, including the construction of the country's first automated terminal and the completion of a fully intelligent transformation project for traditional container terminals. These milestone advancements have not only enhanced the port's operational efficiency but also laid the foundation for its green development. However, there is still a gap between Xiamen Port and international advanced levels in terms of scientific research resource integration and innovation platform design.

For example, some internationally advanced ports, such as the Port of Hamburg and the Port of Copenhagen, have established comprehensive carbon emission monitoring and reporting systems capable of real-time tracking and reporting the environmental impact of port activities. While Xiamen Port has begun monitoring carbon emissions, the coverage and accuracy of the monitoring system still need improvement. Furthermore, ports like the Port of Antwerp and the Port of Dubai have established transnational port cooperation networks to promote green innovation in the port industry through knowledge sharing and joint research and development projects, an area where Xiamen Port has significant room for enhancement in international cooperation.

5.3 *The Need for Further Refinement of Green Shipping Policies*

The "Xiamen Port Low-Carbon Development Action Plan" has proposed a series of long-term goals, the achievement of which requires support from a range of innovative solutions, involving all aspects of port operations, such as energy use, cargo handling, ship operations, and the upgrading of port infrastructure. Therefore, Xiamen City has actively implemented a series of policies to ensure a successful transition of Xiamen Port to a green port. For example, Xiamen City has issued the "Xiamen City Port and Shipping Field 'Green Dynamics of Xiamen Bay' Three-Year Action Plan (2023–2025)", aiming to promote the use of new and clean energy sources, reduce port ship pollution, and lower carbon emissions, facilitating a green and low-carbon transformation of Xiamen Port's shipping industry.

The plan includes 20 key measures, such as supporting government and corporate innovation, strengthening the foundation for development, improving ecological governance, and establishing institutional mechanisms. Additionally, Xiamen City has enacted local legislation for the prevention and control of ship pollution, establishing a "ship-port-city" collaborative model to advance the construction of an emergency response system for hazardous emissions. These policies and measures reflect Xiamen City's high priority on green and low-carbon development in the port and shipping sector. However, compared to other provinces, Xiamen City's green shipping policies still have room for further refinement.

5.4 Coordination Issues Between Policies and the Market

Xiamen City has proposed a series of shipping policies to promote the development of green shipping and has balanced the coordination between green shipping and the shipping economy during policy implementation. This balance allows Xiamen Port to achieve green shipping without affecting the economic development of its shipping industry. For example, Xiamen City has encouraged financial institutions to establish green finance divisions and specialized agencies to support green investment and financing activities in the shipping industry. Financial institutions are encouraged to launch green ports, green ships, and green shipping financial products and to support shipping companies in issuing green bonds and blue bonds, exploring the development of green asset-backed securities.

Additionally, Xiamen City has provided policy and financial support to encourage port and ship operators to adopt green technologies and equipment, enhancing the energy efficiency of port operations. Although these policies demonstrate Xiamen City's commitment to the coordinated development of policy and the shipping economy, there is still a need for more detailed implementation at the operational level of shipping companies, including specific application standards for green technologies, subsidy policies for new energy equipment, and methods for assessing and rewarding emission reduction achievements.

6 Recommendations

6.1 Promoting the Control of Ship Pollution Issues

Firstly, it is necessary to improve the usage rate and coverage of shore power facilities. Xiamen Port should expand the availability of shore power facilities and establish incentives to ensure their consistent usage. For example, Shanghai has set a usage rate target in its "Carbon Peaking Implementation Plan", requiring a 30% usage rate for shore power facilities at container terminals and striving for 100% usage rates at cruise terminals (Shanghai Municipal Government, 2022). What distinguishes Xiamen Port from Shanghai Port is that Shanghai Port is a critical point in global trade, handling the goods from around the world, while Xiamen Port mainly connects the trade routes with Southeast Asia and East Asia. This makes Shanghai Port more demanding than other ports in both scale and supervision.

However, Xiamen Port still has places to learn from. The document "Xiamen Special Economic Zone Ship Pollution Prevention and Control Regulations" mentions that Xiamen's current shore power facilities utilization rate reached 90%. On this basis, Xiamen can increase its shore power facilities to 95% or more. Xiamen Port can better respond to the existing national policies with their environmental targets by setting some specific thresholds for carbon emission reduction on a yearly basis. This basis is expected to facilitate a more effective transition to the green port, achieving a low-carbon operational environment.

Secondly, Xiamen Port can explore and promote the use of green fuels, such as methanol and LNG, offering port service fee reduction for ships using green fuels. Such incentives will further encourage more vessels to adopt green energy. Thirdly, promoting the phasing-out of old ships would be a good alternative to further decrease ship pollution. Fujian Province can implement a classified management of old ships based on age, condition, and pollution risk levels, establishing different phasing-out schedules and incentive measures. The Xiamen Port Ship Carbon Emission Green Research Platform can be utilized to effectively categorize pollution levels of new and old ships at Xiamen Port, facilitating the resolution of ship pollution issues.

6.2 Strengthen Green Shipping Infrastructure

Firstly, it is essential to establish a carbon emission green research platform. Xiamen Port can learn from the advanced experiences of the Port of Hamburg (Hamburg Port Authority, 2023) and the Port of Copenhagen (Automds Green Compliance Specialist, 2022) to create a more comprehensive carbon emission monitoring and reporting system, including the use of high-precision sensors and automated monitoring equipment. Secondly, promoting transnational port cooperation networks would be another important step. Drawing on the transnational cooperation networks of the Port of Antwerp and the Port of Dubai, Xiamen Port can seek partnerships with international ports to share best practices and innovative technologies in green shipping. For instance, through cooperation with the Port of Singapore, Xiamen Port can learn from its advanced experiences in port automation and intelligence. Thirdly, it is necessary to enhance the level of intelligent transformation. Xiamen Port can further promote intelligent transformations, such as the fully intelligent transformation project of Hairun Terminal, to improve loading and unloading efficiency and safety. At the same time, more advanced technologies, such as 5G, artificial intelligence, and the Internet of Things, can be explored to enhance the intelligence level of port operations.

6.3 Refine Green Shipping Policies

Firstly, relevant stakeholders may consider strengthening inter-departmental coordination and integration. Xiamen City can establish a policy coordination mechanism set up by the maritime sector, transportation, environmental protection, and other departments. During the policy formulation stage, ensuring policy compatibility and consistency through inter-departmental working groups or meetings would be critical. Secondly, acceleration of the improvement of green shipping laws and regulations should be considered as well. At present, Shandong Province has issued policies for new energy ship routes and crew members with specific subsidy measures. For a single route with an amount of between 300,000 and 1.5 million, the minimum base salary for crew members has rose by at least 5% (China Shipping Network, 2024). Xiamen, where Xiamen Port is located, currently has insufficient subsidies for new energy ships. The existing rules and policies require improvements. Xiamen Port has perennial LNG ships and low-sulfur fuel ships docking at the port, but the subsidies for these ships do not exceed 175 yuan per ton, which is comparatively lower than the ones offered by other large ports in China.

Xiamen Port can consider increasing subsidies to 215 yuan per ton without affecting the port development and GDP. The green transformation of Xiamen Port is essential, as this port serves as an important node leading to Southeast Asia. Xiamen Port representatives are encouraged to refer to the existing policies adopted by other larger ports in China, such as the Port of Shandong and the Port of Shanghai, to promote green transformation.

Thirdly, port ship pollution response and management policies should be developed and implemented. Xiamen City can continue to carry out law enforcement and supervision for ship pollution prevention, enhance the port's ship pollution response and management activities, and promote the landing of innovative technology along with research projects related to ship ballast water treatment and monitoring in Fujian Province.

6.4 Collaborative Innovation for Carbon Emission Reduction and Ship Trade Flows

Firstly, increasing the subsidy standards for green technology applications would be an important step. Xiamen City can refer to Shanghai's subsidy policies and increase the subsidy efforts for electric ships. For example, Shanghai offers subsidies of up to 5 million yuan for electric ship projects (Shanghai Municipal Transportation Commission, 2024). As indicated earlier, Xiamen City can develop similar subsidy standards to encourage more shipping companies to adopt green technologies. Secondly, another essential action item would be assessing and rewarding emission reduction achievements. Xiamen City can establish an assessment system to evaluate the emission reduction effectiveness of shipping companies and provide rewards. This can refer to practices from other ports, such as tax reductions and financial rewards for companies that meet certain emission reduction standards. Thirdly, it is necessary to implement green shipping equipment subsidy policies. Xiamen City can introduce specific subsidy policies for green shipping equipment, offering financial subsidies to companies that adopt green shipping equipment. For instance, subsidies for ships equipped with shore power systems can encourage the use of shore power during port stays, which will further reduce carbon emissions.

6.5 Strategic Implementation of Green Initiatives

As the discourse on the green transformation of ports unfolds, it often paints an optimistic picture of the numerous benefits that such transitions can bring. Yet, it is essential to recognize that the actual implementation of these transformations is a complex process that entails various trade-offs. The reality is that certain measures may not produce immediate returns. While acknowledging that some green measures may not offer immediate financial rewards, Xiamen Port has adeptly balanced its strategy. The port has prioritized initiatives that deliver both environmental and operational benefits in the short term, such as state-of-the-art waste management systems that enhance both sustainability and operational efficiency. Concurrently, investments in research and development are paving the way for future innovations that will secure the port's long-term sustainability. This balanced approach ensures that Xiamen Port's green transformation is not only environmentally sound but also economically viable and socially acceptable.

In essence, Xiamen Port's green transformation serves as a testament to the feasibility and desirability of sustainable port operations. Its strategic and holistic approach to green initiatives underscores the potential for ports to lead the way towards a more sustainable and efficient future.

7 Conclusion

In this study, we have delved deeply into the issue of ship emissions at Xiamen Port and proposed diverse countermeasures that are expected to alleviate emissions produced in the following years. Through a comprehensive analysis of the current state of ship carbon emissions at Xiamen Port, we have found that despite a certain progress in promoting the development of a green port, the problem of carbon emissions remains significant. To achieve carbon reduction targets, we have proposed a series of innovative management mechanisms and technical solutions aimed at optimizing ship operations through technological means, enhancing the regulatory efficiency of the port, and ensuring the fair and reasonable allocation of carbon emissions. Based on the findings and conclusions of this research, we put forward the following policy recommendations. Firstly, port management departments should strengthen cooperation with scientific research institutions to jointly develop and promote low-carbon technologies and improve the energy efficiency of port operations. Secondly, the government and port enterprises should work together to establish a fair and reasonable carbon emission quota allocation mechanism to encourage ship operators to take emission reduction measures.

Technological innovation is key to achieving carbon reduction. By introducing advanced monitoring technologies, such as the Automatic Identification System (AIS) and Internet of Things (IoT) devices, it would be possible to collect real-time carbon emission data from ships, providing a basis for the accurate calculation of carbon emissions. At the same time, by utilizing big data analytics and artificial intelligence algorithms, it would be possible to delve into the patterns and trends behind the data to provide scientific guidance for the formulation of emission reduction strategies. Furthermore, carbon allocation plans can be designed based on a multi-criteria decision analysis approach, ensuring that the allocation of carbon emission quotas is both fair and efficient.

In conclusion, although this study is mainly focused on Xiamen Port, its methods and findings can be extrapolated to provide valuable insights to other ports around the world. As demonstrated in this study, complex network analysis provides an integrated framework that can adapt to a variety of port environments. By recognizing the unique operational characteristics and environmental context of different ports, alternative fuels, onshore power supply, and policy improvement strategies can be tailored to meet their specific needs. This study can serve as a blueprint, highlighting the potential of the global initiative to achieve green port operations. The broader impact of this study encourages cooperative approaches between ports, facilitating the exchange of knowledge and the development of standardized practices, contributing to the collective goal of reducing carbon emissions within the maritime industry. As the maritime sector continues to evolve, the principles outlined here can guide future research and decision-making, ensuring that ports play a key role in the transition to sustainable global trade and transport.

References

Automds Green Compliance Specialist. (2022). Copenhagen Port in Carbon Capture and Storage Project, [EB/OL]. https://www.sohu.com/a/541843156_120588350

Balcombe, P., Heggo, D. A., & Harrison, M. (2022). Total methane and CO2 emissions from liquefied natural gas carrier ships: The first primary measurements. *Environmental Science & Technology., 56*(13), 9632–9640.

Barabási, A. L., & Albert, R. (1999). Emergence of scaling in random networks. *Science, 286*(5439), 509–512.

Chen, D. S., Wang, X. T., Nelson, P., et al. (2017). Ship emission inventory and its impact on the PM2.5 air pollution in Qingdao Port, North China. *Atmospheric Environment, 166*, 351–361.

Chen, Q., Ge, Y. E., Lau, Y. Y., Dulebenets, M. A., Sun, X., Kawasaki, T., Mellalou, A., & Tao, X. (2023). Effects of COVID-19 on passenger shipping activities and emissions: Empirical analysis of passenger ships in Danish waters. *Maritime Policy & Management, 50*(6), 776–796.

Chen, Q., Lau, Y. Y., Ge, Y. E., Dulebenets, M. A., Kawasaki, T., & Ng, A. K. (2021). Interactions between Arctic passenger ship activities and emissions. *Transportation Research Part D: Transport and Environment, 97*, Article 102925.

Chen, W., & He, Q. (2016). Intersectoral burden sharing of CO2 mitigation in China in 2020. *Mitigation and Adaptation Strategies for Global Change, 21*, 1–14.

China Shipping network. (2024). Shandong official declaration: new energy ship award standard will be 20% higher than conventional power, [EB/OL]. https://info.chineseshipping.com.cn/cninfo/News/202404/t20240402_1388645.shtml

Chou, C., et al. (2021). Analysis of energy efficiencies of in-port ferries and island passenger-ships and improvement policies to reduce CO2 emissions. *Marine Pollution Bulletin, 172*, Article 112826.

Deniz, C., Kilic, A., & Cıvkaroglu, G. (2010). Estimation of shipping emissions in Candarli Gulf. *Turkey. Environ Monit Assess, 171*, 219–228.

Erdos, P., & Renyi, A. (1960). On the evolution of random graphs. *Publ Math Inst Hung Acad Sci, 5*(1), 17–60.

Fan, A., et al. (2023). Carbon footprint model and low–carbon pathway of inland shipping based on micro–macro analysis. *Energy, 263*, Article 126150.

Gan, L., Che, W., Zhou, M., et al. (2022). Ship exhaust emission estimation and analysis using Automatic Identification System data: The west area of Shenzhen port, China, as a case study. *Ocean & Coastal Management, 226*, Article 106245.

Geerlings, H., & van Duin, R. (2011). A new method for assessing CO2-emissions from container terminals: A promising approach applied in Rotterdam. *Journal of Cleaner Production, 19*(6), 657–666.

Hamburg Port Authority. (2023). For a greener port. [EB/OL]. https://www.hafen-hamburg.de/cn/port-of-hamburg-magazine/energy-hub/for-a-greener-port

Jiao, Y., & Wang, C. (2021). Shore power vs low sulfur fuel oil: Pricing strategies of carriers and port in a transport chain. *International Journal of Low-Carbon Technologies, 16*(3), 715–724.

Jin, Y., Zhang, K., Li, D., Wang, S., & Liu, W. (2024). Analysis of the spatial–temporal evolution and driving factors of carbon emission efficiency in the Yangtze River economic Belt. *Ecological Indicators., 165*, Article 112092.

Ju, Y., & Hargreaves, C. A. (2021). The impact of shipping CO2 emissions from marine traffic in Western Singapore Straits during COVID-19. *Science of the Total Environment, 789*, Article 148063.

Kaluza, P., Kölzsch, A., Gastner, M. T., et al. (2010). The complex network of global cargoship movements. *Journal of the Royal Society Interface, 7*(48), 1093–1103.

Lane, J. M., & Pretes, M. (2020). Maritime dependency and economic prosperity: Why access to oceanic trade matters. *Marine Policy, 121*, Article 104180.

LaRock, T., Xu, M., & Eliassi-Rad, T. (2022). A path-based approach to analyzing the global liner shipping network. *EPJ Data Sci, 18*, 1045–1063.

Li, Z., et al. (2024). Data-driven research on the impact of COVID-19 on the global container shipping network. *Ocean & Coastal Management, 248*, Article 106969.
Longépé, N., Hajduch, G., Ardianto, R., et al. (2018). Completing fishing monitoring with spaceborne Vessel Detection System (VDS) and Automatic Identification System (AIS) to assess illegal fishing in Indonesia. *Marine Pollution Bulletin, 131*, 33–39.
López-Aparicio, S., et al. (2017). Shipping emissions in a Nordic port: Assessment of mitigation strategies. *Transportation Research Part D: Transport and Environment, 53*, 205–216.
Miola, A., & Ciuffo, B. (2011). Estimating air emissions from ships: Meta-analysis of modelling approaches and available data sources. *Atmospheric Environment, 45*(13), 2242–2251.
Shanghai Municipal Government. (2022). Notice of the Shanghai Municipal People's government on issuing the implementation plan for carbon peaking in Shanghai, [EB/OL]. https://www.shanghai.gov.cn/nw12344/20220728/75468067a4a848139d2a2eed16ce9e11.html
Shanghai Municipal Transportation Commission. (2024). Shanghai Port shore power project to be subsidy program publicly announced, [EB/OL]. https://jtw.sh.gov.cn/gsgg/20241029/e3c33cf044184d5db57c2d6808c79451.html
Tang, D., Jiang, T., Xu, C., Chen, Z., Yuan, Y., Zhao, W., & Guerrero, J. M. (2023). Assessing the potential for energy efficiency improvement through cold ironing: A monte carlo analysis with real port data. *J Mar Sci Eng, 11*, 1780.
Tang, L., Wu, J. Q., Yu, L. A., et al (2017) Carbon allowance auction design of China's emissions trading scheme: A multi-agent-based approach. *Energy Policy, 102*, 30–40
Verschuur, J., Koks, E. E., & Hall, J. W. (2022). Ports' criticality in international trade and global supply-chains. *Nature Communications, 13*, 4351.
Verschuur, J., Koks, E. E., & Hall, J. W. (2023). Systemic risks from climate-related disruptions at ports. *Nature Clinical Practice Endocrinology & Metabolism, 13*, 804–806.
Vichos, E., Sifakis, N., & Tsoutsos, T. (2022). Challenges of integrating hydrogen energy storage systems into nearly zero-energy ports. *Energy, 241*, Article 122878.
Wan, C., et al. (2021a). Identifying important ports in maritime container shipping networks along the Maritime Silk Road. *Ocean & Coastal Management, 211*, Article 105738.
Wan, Z., et al. (2021b). Evaluation of emission reduction strategies for berthing containerships: A case study of the Shekou Container Terminal. *Journal of Cleaner Production, 299*, Article 126820.
Wang, J., Huang, Z., Liu, Y. Y., Chen, S. Y., Wu, Y. C., He, Y. Y., Yang, X. Y. (2020). Vessels' air pollutant emissions inventory and emission characteristics in the Xiamen emission control area. *Huan Jing ke Xue= Huanjing Kexue, 41*(8), 3572–3280
Wang, W., et al. (2024). Investigating relationships between shore power utilization and characteristics of ships and berths. *Transportation Research Part D: Transport and Environment, 136*, Article 104446.
Watts, D. J., & Strogatz, S. H. (1998). Collective dynamics of 'Small-World' networks. *Nature, 393*(6684), 440–442.
Wu, M., Li, K. X., Xiao, Y., & Yuen, K. F. (2022). Carbon Emission Trading Scheme in the shipping sector: Drivers, challenges, and impacts. *Marine Policy, 138*, Article 104989.
Yang, Y. (2017). Operating strategies of CO2 reduction for a container terminal based on carbon footprint perspective. *Journal of Cleaner Production, 141*, 472–480.
Yi, W. J., Zou, L. L., Guo, J., et al. (2011). How can China reach its CO2 intensity reduction targets by 2020? A regional allocation based on equity and development. *Energy Policy, 39*(5), 2407–2415.
Zhou, C., et al. (2023). Identification and analysis of ship carbon emission hotspots based on data field theory: A case study in Wuhan Port. *Ocean & Coastal Management, 235*, Article 106479.
Zhu, M., Shen, S., & Shi, W. (2023). Carbon emission allowance allocation based on a bi-level multi-objective model in maritime shipping. *Ocean & Coastal Management, 241*, Article 106665.

Qiong Chen is an Associate Professor at Jimei University and deeply engaged in research on ship pollution emissions, green shipping, and maritime policy. He has taken the lead in presiding over one provincial-level key social science fund and two municipal and department-level funds. In the academic field, he has published 30 scholarly papers in reputable journals, such as Transportation Research Part D, Maritime Policy and Management, and Ocean and Coastal Management. Among these, 26 are SCI/SSCI-indexed papers. Additionally, he has published 14 international conference papers. In terms of academic contributions, he has participated in the compilation of two English-language books. Moreover, he serves as a member of the editorial boards of several foreign journals, including Data in Brief and MethodsX. He is also highly recognized as a reviewer. He reviews for prestigious journals, such as Transportation Research Part D, Maritime Policy and Management, Cities, iScience, and Maritime Business Review. His outstanding review work was acknowledged with the Best Reviewer Award at the 2023 World Transportation Conference (WTC2023).

Jiaqi Gao is currently a Ph.D. candidate at the Navigation College of Jimei University, where he is deeply engaged in the field of traffic engineering. His expertise lies in the intricate study of shipping networks, which involves optimizing maritime routes and analyzing the efficiency of global shipping systems. Gao is also exploring innovative technologies that can reduce the environmental impact of maritime transportation. He develops sophisticated frameworks to quantify and mitigate greenhouse gas emissions from shipping activities.

Wentao Liao is an undergraduate student at the College of Navigation of Jimei University. He does a lot of work in data analysis and processing, and his main research direction is shipping networks. His research focuses on analyzing the structure and operational efficiency of maritime transportation systems. He explores the structure of cruise ship networks, the degree of network nodes, and the average path length. He is committed to combining theoretical knowledge with industry needs, aiming to provide practical solutions to modern shipping challenges.

Yui-Yip Lau is a Head of Student Affairs and Senior Lecturer from the School of Professional Education and Executive Development, The Hong Kong Polytechnic University. Until now, he has published more than 400 research papers in international journals and professional magazines, contributed over 25 book chapters, 8 books, and presented numerous papers at international conferences. He has collaborated with scholars from more than 20 countries and regions spreading over five continents on research projects. He has also secured over HK$ 10 million in research grants. Recently, he has been awarded a Certificate of Appreciation by the Institute of Sea transport in recognition of his outstanding performance on research and the Best Paper Award in leading international conferences. Moreover, he has participated in different consultancy projects with various (inter-)governmental organizations, academic institutions, and industrial associations. His research interests are cruise ships, ferries, impacts of climate change, shipping education and training, transport history, sustainability issues, resilient supply chain management, health logistics, human remains and regional development.

Guoqiang Chen received the M.S. degree in Traffic Information Engineering and Control from Beijing Jiaotong University, Beijing, China in 2018 and the Ph.D. degree in cartography and geographical information systems from Aerospace Information Research Institute, Chinese Academy of Sciences, and from the University of Chinese Academy of Science, Beijing. He is now an Associate Professor at Navigation College, Jimei University. His current research interests include the Moon-based Earth observation, SAR image processing, SDG evaluation, remote sensing of earth resources and environment.

Maneerat Kanrak is a lecturer of international transport and logistics management at Khon Kaen University, Thailand. She obtained a Ph.D. in maritime and logistics on cruise shipping network analysis from the Australian Maritime College, University of Tasmania, Australia. She also got a Bachelor of Business Administration in International Business with First-Class Honours and Gold Medal and a Master of Science in Logistics Management. Maneerat worked in purchasing and inventory management before becoming a lecturer. Her research interests include supply chain and transportation systems, logistics management, transport and logistics networks and operations research. She has published papers in international journals and attended international conferences.

Pengfei Zhang has been working in the maritime industry for over 20 years as a master mariner, maritime lawyer dually qualified in England and Wales and China, and a senior executive in a large shipping group. He has gained hands-on experience covering almost the full spectrum of maritime business, from shipbuilding to ship operations and to ship recycling. He has practiced as a maritime lawyer at Shanghai All Brights Law offices since 2008, specializing in shipbuilding and financing, ship collision and SNP disputes. With a Ph.D. in international maritime studies, Pengfei has a distinguished track record in publications and conference speeches. He has authored/co-authored five books on maritime management and over 100 articles carried by prestigious journals, including (inter alia) Economic and Labour Relations Review, Industrial Relations Journal, Marine Policy, Journal of Navigation, and International Economic Law. With professional experience in the industry at a senior management level, Pengfei is well informed about the latest developments of the Logistics and International Trade (LIT) industry and well connected with a wide range of international organizations, non-government organizations, and professional associations, such as the ILO, IMO, ITF, BIMCO, ISMA, ISF, NI, and so on. He is a senior member of China-Europe Commercial Collaboration Association (CECCA), a supporting member of the London Maritime Arbitrators Association (LMAA), and a member of the Chartered Management Institute (CMI).

Lijuan Deng is an Associate Professor at Jimei University and is deeply engaged in the research of shipping policy, admiralty, and maritime law. She has co-edited 12 textbooks, three of which were national programming teaching materials, and one served as a provincial programming teaching material. In the academic field, she has published a monograph and around 50 academic papers. In the area of social service, she has taken part in more than 50 institution-funded research programs.

MIX
Papier aus verantwortungsvollen Quellen
Paper from responsible sources
FSC® C105338

If you have any concerns about our products,
you can contact us on
ProductSafety@springernature.com

In case Publisher is established outside the EU,
the EU authorized representative is:
Springer Nature Customer Service Center GmbH
Europaplatz 3, 69115 Heidelberg, Germany

Printed by Libri Plureos GmbH
in Hamburg, Germany